Materials, Technology and Reliability for Advanced Interconnects and Low-k Dielectrics—2004

MATERIALS RESEARCH SOCIETY
SYMPOSIUM PROCEEDINGS VOLUME 812

Materials, Technology and Reliability for Advanced Interconnects and Low-k Dielectrics—2004

Symposium held April 13–15, 2004, San Francisco, California, U.S.A.

EDITORS:

R.J. Carter
LSI Logic Corporation
Gresham, Oregon, U.S.A.

C.S. Hau-Riege
Advanced Micro Devices Inc.
Sunnyvale, California, U.S.A.

G.M. Kloster
Intel Corporation
Hillsboro, Oregon, U.S.A.

T.-M. Lu
Rensselaer Polytechnic Institute
Troy, New York, U.S.A.

S.E. Schulz
TU Chemnitz
Chemnitz, Germany

Materials Research Society
Warrendale, Pennsylvania

CAMBRIDGE UNIVERSITY PRESS
Cambridge, New York, Melbourne, Madrid, Cape Town,
Singapore, São Paulo, Delhi, Mexico City

Cambridge University Press
32 Avenue of the Americas, New York NY 10013-2473, USA

Published in the United States of America by Cambridge University Press, New York

www.cambridge.org
Information on this title: www.cambridge.org/9781107409224

Materials Research Society
506 Keystone Drive, Warrendale, PA 15086
http://www.mrs.org

© Materials Research Society 2004

First published 2004
First paperback edition 2012

Single article reprints from this publication are available through
University Microfilms Inc., 300 North Zeeb Road, Ann Arbor, MI 48106

CODEN: MRSPDH

ISBN 978-1-107-40922-4 Paperback

CONTENTS

LOW-k DIELECTRICS, PROCESSING, AND CHARACTERIZATION

*Invited Paper

METALLIZATION, BARRIERS, AND CAPPING

*Invited Paper

*Invited Paper

*Invited Paper

PREFACE

This proceedings volume contains 60 papers presented at Symposium F, "Materials, Technology and Reliability for Advanced Interconnects and Low-k Dielectrics," which was held April 13–15 at the 2004 MRS Spring Meeting in San Francisco, California. Over half of the contributions are from universities and national research institutes, testifying to the continuing recognition of the stimulating scientific and engineering challenges advanced interconnects brings to the microelectronics industry. Furthermore, the global importance of the field is apparent by the large number of contributions coming from outside the United States.

The issues addressed in this symposium cannot be dispelled as to simply selecting a low-k material and integrating it into a copper damascene process. The intricacies of the Back End for sub-100nm technology include novel processing of low-k materials, employing pore sealing techniques and capping layers, introducing advanced dielectric and diffusion barriers, and the development of novel integration schemes, in addition to the concerns of performance, yield, and reliability appropriate to nano-scaled interconnects. Although many challenges continue to impede progress along the ITRS roadmap, the contributions in this proceedings confront them head-on in order to provide a scientific understanding of the issues so that solutions may be achieved in the future.

The development of both the symposium and this proceedings volume would not have been possible without the support of the Materials Research Society and the assistance of the MRS staff. We thank them for their dedication and diligence. The financial support of the following organizations is also deeply appreciated:

AMD Saxony LLC & Co. KG
The Dow Chemical Company
Honeywell Electronic Materials
JSR Micro, Inc.
Intel Corporation
LSI Logic Corporation

Finally, we are indebted to the numerous university, industrial, government, and international colleagues who have contributed to the formulation and content of the symposium and proceedings.

R.J. Carter
C.S. Hau-Riege
G.M. Kloster
T.-M. Lu
S.E. Schulz

June 2004

MATERIALS RESEARCH SOCIETY SYMPOSIUM PROCEEDINGS

MATERIALS RESEARCH SOCIETY SYMPOSIUM PROCEEDINGS

Prior Materials Research Society Symposium Proceedings available by contacting Materials Research Society

MATERIALS RESEARCH SOCIETY SYMPOSIUM PROCEEDINGS

Volume 80X—Ion Beam Synthesis and Processing of Advanced Materials, XXXIII, V.M. Donnelly, C.D. Weber, 2004, ISBN: 1-55899-743-0

Volume 80X—Semiconductor Defect Engineering—Materials Science, and Technology III, 2004, K. Jonstad, B.L. Sopori, P. Stradins, J.C. Rojas, Sumanthenang, enonessessess-X

Volume 80X—High-Mobility Group IV Materials and Devices, M. Oehme, L. Colace, 2004, C.F. Caymax, P.E. Thompson, 2004, ISBN: 1-5899-756-6

Volume 81X—Group-III Nitride Materials for Generation Illumination—Low-dimensional Structure, R.Y. Nakamura, W. Wang, 2004, ISBN: 1-5899-760-4

Volume 81X—Semiconductor Mater... Topics... , L.C. Feldman, R.S. Oh, D.I. Sullivan, K. Wong, R.P. Wong, S. Zoles, 2004, ISBN: 2004, ISBN: 1-5899-761-2

Volume 81X—Electronic Packaging and Reliability for Advanced Interconnects and Low-k Dielectrics, 2004, J.L. Pan, A.J. Berger, P. Kohl, T.M. Lu, C.-U. Pinzer, 2004, ISBN: 1-5899-762-0

Volume 81X—Flexible ... Electronics ... , J. Hou, H.-S. Kwok, et al., 2004, ISBN: 1-5899-764-7

Volume 81X—Flexible ... Electronics—Materials and Device Technology, H.Q. Chiang, R.L. Chung, N. Pryds, X. Friederich, Jang, J.Y. Lee, 2004, ISBN: 1-5899-766-3

Volume 81X—Amorphous and Nanostructured Materials, M. Ichikawa, H. Phillips, B. Yan, 2004, ISBN: 1-5899-767-1

Volume 81X—Micro-and Nanosystems—Materials and Devices, D. Hesse, M. Kamins, W. Barnes, 2004, ISBN: 1-5899-768-X

Volume 82X—... Science ... Materials ... , D.M. Tanenbaum, M. Pan, H.I. Smith, 2004, ISBN: 1-5899-769-8

Volume 82X—Semiconductor Defect Engineering II, 2004, M. Stutzmann, R. Nemanich, Chen, H. Phillips, 2004, ISBN: 1-5899-771-X

Volume 82X—Thin-Film Engineering for Optoelectronics, Ill, Cosmabon, M. Kozicki, V. Fowler, C.B. Yarling, 2004, ISBN: 1-5899-772-8

Volume 82X—Semiconductor Materials for Sensing ... Technology, J.J.-L. Chen, 2004, L. Boarman, J. Schneider, et al., ... Springer ... Science, D.L. Kwong, 2004, ISBN: 1-5899-774-4

Volume 82X—Sensors, Actuators, and Microsystems: Advanced Materials, Processing, Assembling, M. Laudon, J.C. Bustillo, J.T. Borenstein, A.J. Bruno, P. Yang, 2004, ISBN: 1-55899-775-2

Volume 82X—Microelectromechanical Systems—Materials and Devices, D. LaVan, M. McNie, J. Bustillo, 2004, ISBN: 1-55899-781-7

Volume 82X—... , R.W. Gregory ... Materials ... , A. Aldawi, C.A. Ross, 2004, ISBN: 1-55899-784-1

Volume 82X—Synthesis, Characterization and Properties of Energetic/Reactive Materials, R.W. Armstrong, N.N. Thadhani, W.H. Wilson, J.J. Gilman, R.L. Simpson, 2004, ISBN: 1-55899-790-6

Volume 82X—Mechanisms of Surface and Microstructure Evolution in Deposited Films and Film Structures, J.G. Amar, G.H. Gilmer, et al., 2004, ISBN: 1-55899-791-4

Volume 82X—Structure and Mechanical Behavior of Biological Materials, P. Fratzl, W.J. Landis, R. Wang, F.H. Silver, 2004, ISBN: 1-55899-796-5

Volume 83X—Ferroelectric Thin Films XII, A.I. Kingon, S. Hoffmann-Eifert, et al., 2004, ISBN: 1-55899-797-3

Volume 83X—Materials and Devices for Smart Systems, Y. Furuya, J. Su, I. Takeuchi, V.K. Varadan, J. Ulicny, 2004, ISBN: 1-55899-798-1

Volume 83X—Ferroelectrics, Multiferroics, and Magnetoelectrics, A. Gruverman, V.A. Fonoberov, et al., 2004, ISBN: 1-55899-801-5

Prior Materials Research Society Symposium Proceedings available by contacting Materials Research Society

Low-k Dielectrics, Processing, and Characterization

Mat. Res. Soc. Symp. Proc. Vol. 812 © 2004 Materials Research Society

Molecular Caulk: A Pore Sealing Technology for Ultra-low k Dielectrics

Jay J. Senkevich[1], Christopher Jezewski[1,2], Deli Lu[1], William A Lanford[2], Gwo-Ching Wang[1] and Toh-Ming Lu[1]

[1]Rensselaer Polytechnic Institute, Dept. of Physics, Troy, NY 12180
[2]University at Albany, Dept. of Physics, 1400 Washington Ave., Albany, NY 12222

Abstract

Much effort has been undertaken to develop high performance ultra-low k (≤ 2.2) (ULK) dielectrics to improve the interconnect speed of ultra-large scale integrated devices. Metallization issues and their poor mechanical properties have plagued the successful integration of these porous ULK dielectrics. Both of these issues are exasperated by their open pore structure. We have developed a pore sealing technology, which may allow the successful integration of these materials. We have coined the term Molecular Caulking to describe the materials that use the parylene platform for the chemical vapor deposition of these polymers. They are very unique since they can be conformally coated on demanding geometries pin-hole free at a thickness of ~10 Å. We have shown that the 1st generation material, poly(p-xylylene), is selective with respect to copper and can completely seal porous-MSQ (methyl silsesquioxane) after 30 Å of deposition. This is roughly the pore diameter of this porous ULK dielectric. These materials exhibit very fast lateral growth rates and the penetration of the sealant into the ULK dielectric can be controlled. An overview will be give with respect to the Molecular Caulking technology.

Introduction

In future gigascale integrated circuits resistive-capacitive (RC) delay is an increasingly important issue [1]. Dense hybrid chemical vapor deposited dielectrics with a siloxane backbone are the current materials of choice [2]. These materials can be made nanoporous to reduce their dielectric constant further; however, the introduction of porosity results in a number of undesirable properties such as a reduction in mechanical strength and susceptibility to penetration of chemicals. Most important, during chemical vapor (CVD) or atomic layer (ALD) deposition of the barrier layer, the gas-phase precursors have a tendency to infiltrate the porous dielectric. Significant penetration of these metallic species will significantly damage the electrical properties of the ULK dielectric [3-6].

A recent review addresses some of the currently proposed strategies to seal porous dielectrics [7]. One method is to conformally deposit a dielectric layer to seal the porous dielectric. However, the primary drawback of this method is the lack of selectivity to the copper via. The dielectric material presented here, Molecular Caulk, retains selectivity over Cu and its penetration into the porous dielectric can be controlled to address fracture mechanics issues of the barrier layer/dielectric interface. Molecular Caulk is deposited via chemical vapor deposition at room temperature utilizing a free-radical polymerization mechanism. The approach taken was to measure the new sealant's ability to prevent penetration of metal precursors (copper, cobalt) during CVD. The depth distribution of deposited metals was measured by Rutherford backscattering spectrometry (RBS). In addition, changes in dielectric constant as a result of Molecular Caulk deposition were determined by metal insulator semiconductor (MIS) capacitance measurements. Deposited film thickness was determined by both ellipsometry and ion beam backscattering using the 4.28 MeV ^{4}He elastic nuclear resonance of ^{12}C. Finally, the

3

effect of Molecular Caulk deposition on surface topology was measured by atomic force microscopy (AFM).

Experimental

The porous MSQ films was used as-received and went through a series of baking stations before the final cure in an N_2 ambient at 420 °C before they were received. Two types of porous-MSQ films were studied here: $k = 2.2$ and $k = 2.0$, which had 29% and 37% porosity and an average pore size of 22 Å and 37 Å respectively.

Copper CVD experiments were undertaken via $Cu^{II}(tmhd)_2$ and H_2 in a vertical, low pressure, warm-wall reactor. The precursor bubbler was held at a constant temperature of 127.5 $\pm$ 0.6 °C and delivered with 15 sccm of argon carrier gas. The substrate was kept at 217 $\pm$ 5 °C and the chamber walls and precursor transfer lines all held at 150 $\pm$ 5 °C. The total pressure of argon, H_2, and precursor, was approximately 2 Torr. The deposition time was 30 min for all experiments. Bare MSQ and several MC/MSQ films of varying MC thickness were placed side-by-side on the substrate heater in each experiment. For further details on the copper CVD process employed in this study, we refer to an earlier publication [6].

Cobalt deposition experiments were undertaken in a vertical, low pressure, warm-wall reactor. $Co_2(CO)_8$ was sublimed at room temperature. The substrate was kept at 60 $\pm$ 2 °C and the deposition time was 2 minutes. No carrier or purge gas was used and deposition pressure was approximately 18 mTorr. Bare MSQ and several MC/MSQ films of varying MC thickness were placed side-by-side on the substrate heater in each experiment.

Molecular Caulk thin films were deposited using the Gorham method [8]. The reactor consisted of a sublimation furnace, a pyrolysis furnace, and a bell jar type deposition chamber. Base pressure in the deposition chamber was at mid 10^{-6} Torr. During growth the deposition chamber pressure was at 2.0 mTorr yielding deposition rates between 5-8.4 Å/min. A detailed description of the reactor and deposition process has been described elsewhere [9-10]. Briefly, the precursor [2.2] paracyclophane was sublimed at a temperature of 155 °C and the pressure controlled by a heated valve and measured by a heated capacitance in the deposition chamber. The sublimed precursor flew into a high temperature region (650 °C) of the reactor inlet where it was quantitatively cleaved into two p-xylylene monomers by vapor phase pyrolysis. These reactive intermediates were then transported to a room temperature deposition chamber where upon physisorption, a free radical polymerization took place. Linear chains of poly(p-xylylene) with un-terminated end groups were formed. Bulk poly(p-xylylene) has an average dielectric constant of 2.65 [11]. In this work, ultra-thin films (10-50 Å) were deposited. Si (100) 50 Ω-cm substrates were rinsed in ethanol, followed by de-ionized water, blown dry with nitrogen and then placed side-by-side with the porous MSQ in the deposition reactor. Ultra-thin poly(p-xylylene) films (10-50 Å) are low molecular weight and more oligomeric than polymeric. Molecular weight increases with film thickness. Annealing thin poly(p-xylylene) films has shown indications of conversion from low polymer to high polymer [12]. It is advantageous to have the pore sealant to be high polymer since it possesses more robust chemical bonds. In addition, unreacted monomer or larger oligomers that are volatile at higher temperatures will diffuse out of the film upon annealing. Therefore after deposition, samples were annealed in forming gas at 250 °C for 30 minutes.

The deposited film thickness was measured by a variable angle spectroscopy ellipsometer (VASE, J. A. Woollam, Lincoln, NB) on the silicon samples. VASE measurement interpretation is difficult on MSQ films, so thickness measurements were used from the silicon wafers and assumed similar growth on the MSQ films. The thickness of MC was determined by using the Cauchy coefficients of poly(p-xylylene) (A_n = 1.6, B_n = 0.01) or an index of refraction of 1.6658 at 634.1 nm.

Copper growth was characterized by RBS with the 4.0 MV Dynamitron accelerator at the Ion Beam Laboratory: Department of Physics, University at Albany. Measurements were made with 2.0 MeV ^{4}He particles. The RBS determined areal density was converted into an equivalent thickness by dividing by the bulk atomic density of copper 8.45 $\times 10^{22}$ atoms/cm^3. Spectra were collected with a 20 mm^2 area beam spot, 2-4 μC of charge, and with 3 nA of current. Ion beam backscattering, using the 4.28 MeV ^{4}He elastic nuclear resonance of ^{12}C, was undertaken at the same facility.

Two samples were chosen for extraction of the k value: 5400 Å MSQ/50 Å SiO$_2$/Si (control sample) and 250 °C annealed 37 Å MC/5400 Å MSQ/50 Å SiO$_2$/ Si. Top aluminum dots of 0.5, 1.0, and 1.5 mm diameter were electron beam evaporated via a shadow mask. In order to achieve a good Ohmic contact at the backside of the silicon wafer, 300 nm of aluminum was sputter deposited. The capacitance-voltage (C-V) characteristics of the Al /low-κ (stack) / 50 Å SiO$_2$ /Si structures were measured with a HP 4280A 1 MHz Capacitance meter / CV plotter. At least five measurements were performed for each capacitor size for each sample.

Surface morphology was measured using an atomic force microscope (AutoProbe CP) made by Park Scientific Instruments, TM Microscope. A triangular silicon cantilever with silicon conical tip (Veeco Metrology Group) was used in non-contact mode to measure the surface topography. The tip radius of curvature is <10 nm and had a half apex angle of 12 °.

Results and Discussion

From previous work, copper was shown to deposit inside the porous dielectric rather than on the surface during CVD [6]. The copper precursor penetrated the interconnected porous-MSQ dielectric and selectively deposited at the interface between the MSQ film and the silicon substrate. This was the result of a different interfacial chemistry from the use of adhesion promoters. This deposition was quantitatively measured by RBS and observed by scanning electron microscopy (SEM). RBS has monolayer sensitivity to Cu and Co precursors and therefore used to quantitatively measure the amount of metallorganic penetrant or metal deposited. Figure 1 shows the RBS spectra after copper deposition on bare MSQ/SiO$_2$/Si and a 11 Å Molecular Caulk/MSQ/SiO$_2$/Si sample. The arrows labeled "surface copper" and "surface silicon" show the kinematic energy of backscattering from copper and silicon at the surface.

The observed peak at ~1400 keV shows that copper is deposited at the interface between the MSQ and the SiO$_2$/Si substrate. Figure 2 is a plot of the RBS determined amount of copper deposited by CVD at this MSQ/SiO$_2$ interface as a function of Molecular Caulk (MC) thickness. Data from two separate experiments are shown in Fig. 2, which shows only 11 Å of Molecular Caulk resulted in a 96 % reduction in copper penetration during CVD. After 35 Å of deposition, the copper penetration goes below the 0.5 Å detection limit of RBS. Reactor modifications, including precursor and purge gas inlet lines, resulted in different growth rates of the two experiments, but the Molecular Caulk liner thickness sufficient to prevent penetration was consistent.

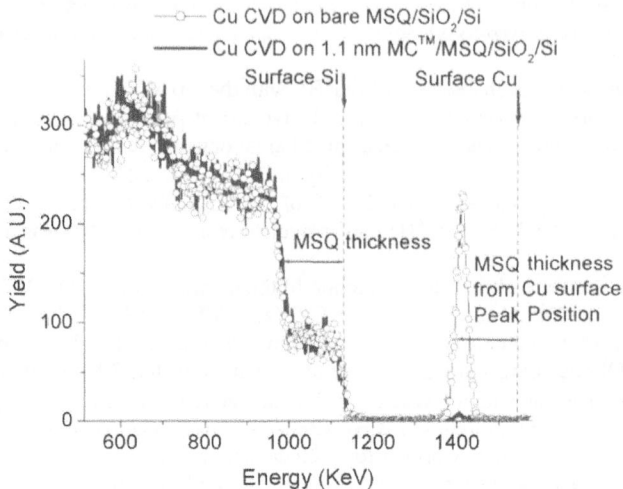

Figure 1 - RBS spectra showing penetration of copper into interface on bare MSQ/SiO₂/Si and 11 Å thick MC/MSQ/SiO₂/Si. The double arrows show the thickness of the MSQ film as determined by the width of the silicon signal in MSQ. Double arrow length also corresponds with the signal peak of penetrated copper.

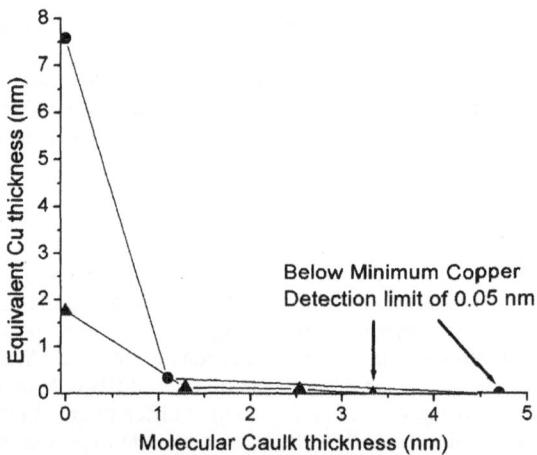

Figure 2 - RBS was used to determine the equivalent thickness of copper deposition on and in MC/MSQ/SiO₂/Si as a function of Caulk thickness. Two data sets: circles and triangles are from experiments before and after reactor alterations respectively. Data point size is larger than error bars.

If the Molecular Caulk deposition passivates the MSQ/SiO$_2$ interface, there would be no deposition to signify the penetration of the copper precursor. A non-selective metallorganic precursor Co$_2$(CO$_8$) was chosen to provide further evidence that Molecular Caulk blocks the pores. Cobalt CVD is non-selective and readily grows on top of the Molecular Caulk layer. Deposition was ~5 Å/min at 60 °C (precursor sublimation was room temperature), a temperature where cobalt deposition is surface-reaction controlled and thus expected to grow at the surface and penetrate the porous MSQ. Figure 3 shows RBS spectra of the three samples: Co/27 Å MC/SiO$_2$/Si, Co/27 Å MC/MSQ/SiO$_2$/Si, and Co/bare MSQ/SiO$_2$/Si. The bare MSQ shows penetration of cobalt supported by the long tail of the cobalt surface peak, while the other samples have only surface deposition. The width of penetrated cobalt is consistent with the silicon width (thickness) in MSQ and shows cobalt penetrates completely through the MSQ film without Molecular Caulk sealing.

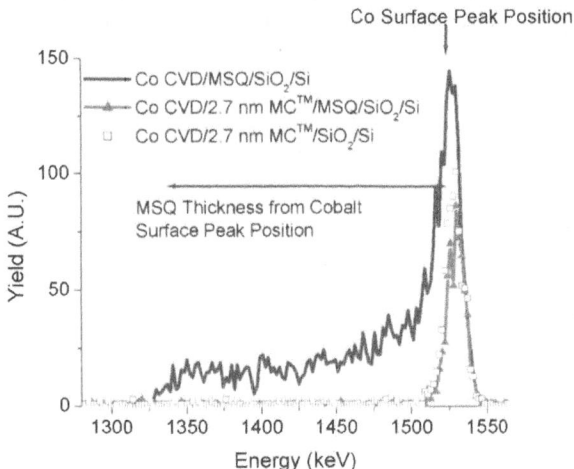

Figure 3 - RBS spectra of cobalt on 27 Å Molecular Caulk/Si, 27 Å MC/MSQ/SiO$_2$/Si, and bare MSQ/SiO$_2$/Si. Cobalt penetration is only observed on the bare MSQ/SiO$_2$/Si substrate. The 27 Å MC/Si and 27 Å MC/MSQ/SiO$_2$/Si samples show Cobalt deposition only at the surface.

Capacitance measurements showed that the dielectric constant of a MSQ(5000 Å)/SiO$_2$/Si film covered with 37 Å of Molecular Caulk had essentially the same dielectric constant as the bare MSQ/SiO$_2$/Si film. The effective dielectric constant of the low k stack was calculated from the slope of the measured accumulation capacitance vs. capacitor area. The measured dielectric constant of porous MSQ was 2.26, this increased to 2.30 after an annealed MC coated dielectric film. The ~ 1.7 % increase in dielectric constant after annealing at 250 °C was less than the sample-to-sample variation in capacitor area. Of course, this MSQ film was rather thick. However, the point to be made here is that the Molecular Caulk material itself is low k and the penetration has a minimal deleterious effect on k effective. As will be shown, the conditions above are the worse case scenario for the penetration of the MC into the ULK dielectric since they were undertaken at 2.0 mTorr.

Molecular Caulk thickness was determined by Variable Angle Spectroscopic Ellipsometry (VASE) on the native oxide of Si rather than on MSQ/SiO$_2$/Si since it is not possible to measure a 10 Å thick MC film on 2000-5000 Å MSQ. During CVD, precursor gases can readily diffuse into the porous material and therefore the penetration of the Molecular Caulk into the porous dielectric was expected. The quantitative measure of deposited film content in/on the porous MSQ/SiO$_2$/Si stack was determined by adding a stoichiometric percentage of hydrogen (C$_8$H$_8$) to the measured carbon content as determined by ^{4}He ion beam backscattering analysis at 4.28 MeV or 5.75 MeV. ^{12}C exhibits a strong (α,α) elastic scattering resonance in these energy regimes. In these energy regimes the cross-section may be more than 100 times the Rutherford cross-section. The resonance is sufficiently broad to enable depth profiling over depths on the order of microns. Essentially ion beam analysis at these energies is similar to RBS at 2 MeV, providing there is accurate absolute cross-section data available. For example, Davies *et al.* [13] measured the non-Rutherford cross-sections in the energy region of 5.5-5.8 MeV. A representative example of a Molecular Caulked porous-MSQ dielectric is shown in Figure 4 in terms of ^{12}C resonance.

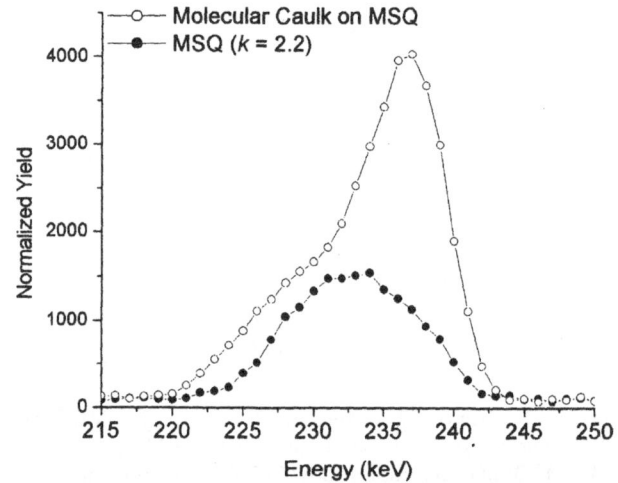

Figure 4 - ^{4}He 4.28 MeV ion beam backscattering spectra of MC/MSQ/SiO$_2$/Si and bare MSQ/SiO$_2$/Si samples. The MC was deposited at 6.4 mTorr and ~400 Å thick, which is observed as the high energy peak. View has been expanded to show only the carbon region of spectra and the silicon substrate contribution has been subtracted from both spectra to emphasize relative carbon content in each sample. The tail on the low energy side of the Caulked spectrum is due to penetration.

Measurements of carbon content were made on bare MSQ/SiO$_2$/Si and MC/MSQ/SiO$_2$/Si samples, where the MC was deposited at various pressures. Also, two different MSQ dielectrics were used, k = 2.2 and k =2.0. Amorphous carbon (H-Square, Sunnyvale, CA) was used as a standard. Amorphous carbon is dense and has a mirror smooth surface. No detectable bulk contamination was determined by RBS, and hydrogen depth profiling showed no bulk contamination of hydrogen at a detection limit of ~0.1 %.

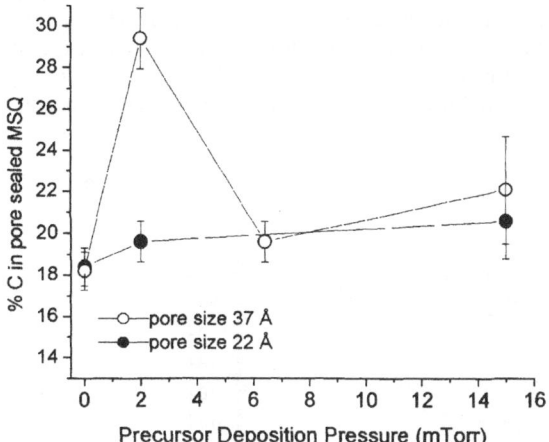

Figure 5 - ⁴He 4.28 MeV ion beam backscattering data for MC/MSQ/SiO₂/Si and bare MSQ/SiO₂/Si samples (0 mTorr) as a function of MC precursor pressure. The MC was deposited at 2 to 15 mTorr at ~400 Å of thickness. In real application the MC layer might only be 30-50 Å depending on the integration scheme.

It is important to be able to control the penetration of the Molecular Caulk itself into the porous-MSQ dielectric. Ideally, this would be undertaken from the side-walls of a reactive-ion etched dielectric; however, experimentally that is challenging to measure via the method presented here. The lateral growth rate of the parylene family of polymers can be controlled by deposition temperature or pressure. At higher pressures, the MC tends to exhibit very high lateral growth rates and at some point is not very conformal but the later only at much higher pressures than would be typically be undertaken. Figure 5 shows the carbon incorporation into the porous-MSQ films as a function of MC precursor pressure. Two porous-MSQ films were used, MSQ $k =$ 2.2 (average pore size 22 Å) and MSQ $k =$ 2.0 (average pore size 37 Å). A balance exists with respect to the pore size and the pressure needed to seal it assuming the penetration is kept constant. In other words, a larger pore size will necessitate the use of higher pressure so that no more MC is penetrated into the dielectric. Fig. 5 shows that at 2 mTorr significant MC is penetrated for the MSQ $k =$ 2.0 compared to the same film Caulked at higher pressure or with MSQ $k =$ 2.2 at the same pressure. Even with significant penetration of the MC into the porous-MSQ, still the k effective will not substantially increase since the dielectric of the MC itself is between 2.35 and 2.65. The later being the first generation material used as proof-of-concept.

To estimate the increase in dielectric constant, a uniform penetration of MC layer was assumed. One MSQ film was 5400 Å thick and contained 50 % porosity (relative to SiO₂), so an MC equivalent thickness of 300 Å evenly distributed through the MSQ film would fill 11.1 % of the porosity. An upper bound to calculation of the composite film dielectric constant can be found by adding the contribution of components in parallel [14]:

$$k_{tot} = k_1P_1 + k_2P_2 + k_3P_3, \qquad (1)$$

where k_{tot}, k_1, k_2, and k_3 are the dielectric constants of the total film, air, MC, and the dense MSQ, respectively. P_1, P_2, and P_3 are the respective fractions of air, MC, and the dense MSQ. An increase of approximately 4 % in the dielectric constant should be expected assuming a dielectric constant of 2.65 of the MC material. A capacitance-determined dielectric constant increase of 1.5 ± 3.3 % for a 37 Å thick MC/MSQ/SiO$_2$/Si is consistent with these results.

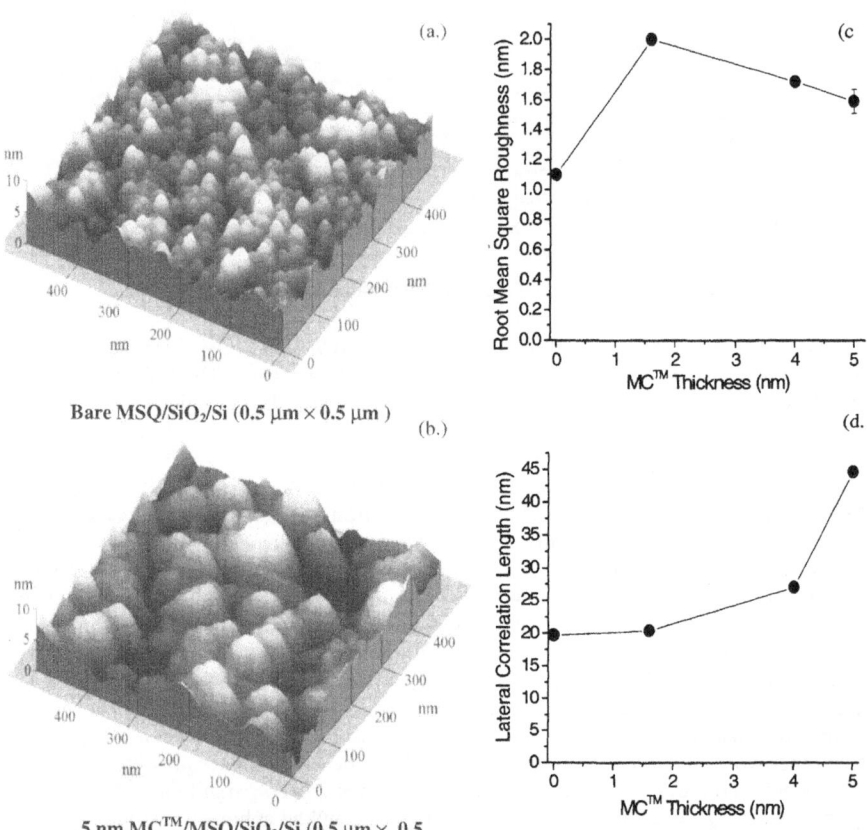

Figure 6 - AFM measurements: (a) bare MSQ/SiO$_2$/Si, (b) MC/ MSQ/SiO$_2$/Si, (c) Plot of rms roughness as a function of MC thickness, and (d) Lateral correlation length as a function of MC thickness.

Roughness was measured by AFM and quantitative information about surface morphology was extracted from a height-height correlation function defined elsewhere [15]. Figures 6 (a) and (b) show the AFM images of bare $MSQ/SiO_2/Si$ and 50 Å $MC/MSQ/SiO_2/Si$ stacks. 40 Å is a suitable MC thickness to prevent penetration and the RMS surface roughness is 6.2 Å greater than the bare $MSQ/SiO_2/Si$ sample. The morphology after deposition shows an initial increase in roughness followed by an apparent smoothing as deposition proceeds (Fig. 6(c)). The lateral correlation length continues to increase with deposition thickness and increases from an initial 200 Å for the bare $MSQ/SiO_2/Si$ to 446 Å for the 50 Å thick $MC/SiO_2/Si$ coated stack. An increase in the lateral correlation length indicates that as the film grows there is some longer range smoothing. The MC technology might be able to smooth rough side walls generated after the RIE of the porous dielectrics. The lateral correlation length is roughly the wavelength of fluctuation of the surface. The lateral correlation length changes are evident in the AFM images of bare MSQ (Fig. 6(a)) and 50 Å MC coated MSQ (Fig. 6(b)) by an overall increase in surface feature size.

Finally, we have shown [16] along with Vaeth *et al.* [17] that no deposition occurs on a air-exposed Cu surface. Vaeth *et al.* showed that 335 nm of polymer grew on a silicon substrate before any growth occurred on copper. Selectivity is a highly desired quality of the free radical polymerization process used here. For Dual Damascene structures in the back-end-of-the-line processes there would be no deposition on the copper via. Only the dielectric would be coated and there would be no series contribution to Ohmic contact from the MC film. This would alleviate the need to etch back the dielectric liner and therefore reduce the number of processing steps.

Conclusions

A polymer sealant, coined Molecular Caulk, as thin as 35 Å has been shown to prevent copper penetration during CVD. Ion beam backscattering at a ^{12}C resonance at 4.28 MeV proved to be a powerful technique to measure the penetration of the MC into the porous-MSQ dielectric. The first generation Molecular Caulk has a low dielectric constant ($k = 2.65$) and effectively seals the porous dielectric. The penetration of the Caulk could be controlled and resulted in a negligible increase in dielectric constant as determined by capacitance measurements. Higher precursor pressure favors a high lateral growth rate, which in turn favors minimal penetration into the porous-MSQ thin film. The $k = 2.0$ porous MSQ film has a larger pore diameter of 37 Å that necessitates the deposition of the MC at higher pressures to minimize the penetration into that film. AFM results show a 6.2 Å increase in RMS roughness of the sealed MSQ surface. The sealant is a selective CVD process and will not deposit on the copper via in Dual Damascene architectures. Selective growth is very important to eliminate the need to etch back the sealant before further metallization steps.

References

[1] S.P. Murarka, *Mat. Sci. Eng.* **R19**, 87-151 (1997).
[2] The National Technology Roadmap for Semiconductors, 2001 edition (Semiconductor Industry Association, San Jose, CA, 1997).
[3] M.E. Thomas, D.M. Smith, S. Wallace, and N. Iwamoto, *Proc. Inter. Interconnect Technol. Conf. (IITC)* p. 223, (2002).
[4] W. Besling, A. Satta, J. Schuhmacher, T. Abell, V. Sutcliffe, A.-M. Hoyas, G. Beyer, D. Gravesteijn, and K. Maex, *Proc. Inter. Interconnect Technol. Conf. (IITC)* p. 288 (2002).

[5] W. F. A. Besling, A. Satta, J. Schumacher, A. M., G. Beyer, K. Maex, O. Kilpella, and H. Sprey, *Proc. 3rd Inter. Conf. on Microelectron. and Interf., American Vacuum Society (AVS)* p. 52 (2002).

[6] C. Jezewski, W. Lanford, J.J. Senkevich, D. Ye, and T.-M. Lu, *Chem. Vapor Dep.* **9**(6) 305-7 (2003).

[7] K. Maex, M. R. Baklanov, D. Shamiryan, F. Iacopi, S. H. Brongersma. Z. S. Yanovitskaya, "Low dielectric constant materials for microelectronics" Appl. Phys. Rev.-Focused Review, *J. Appl. Phys.*, **93**(11) 8792 (2003).

[8] W. Gorham, *J. Polym. Sci., Part A-1*, **4** 3027 (1966).

[9] J.B. Fortin, and T.-M. Lu, *J. Vac. Sci. Technol. A*, **18**(5) 2459 (2000).

[10] J. B. Fortin and T.-M. Lu, *Chem. Mater.* **14** 1945 (2002).

[11] J.B. Fortin and T.-M. Lu, Chemical Vapor Deposition Polymerization: The Growth and Properties of Parylene Thin Films, p. 58, Kluwer Academic Publishers, MA (2004).

[12] J.J. Senkevich, G.-R. Yang, and T.-M. Lu, *Colloids and surfaces A: Physicochem. Eng. Aspects*, **216** 167 (2003).

[13] J.A Davies, F.J.D. Almeida, H.K. Haugen, R. Siegele, J.S. Forster, and T.E. Jackman (IBA-11 Balaton Fured, Hungary, 1993), *Nucl. Instr. and Meth.*, **B85** 28 (1994)

[14] D. E. Aspnes, *Thin Solid Films*, **89** 249, (1982).

[15] Y.-P. Zhao, G.-C.Wang, and T.-M. Lu, Experimental Methods in the Physical Sciences: Characterization of Amorphous and Crystalline Rough Surfaces: Principles and Applications, Vol. 37, p. 18, Academic Press (2000).

[16] J.J. Senkevich, C.J. Wiegand, G.-R. Yang, T.-M. Lu "Selective Deposition of Ultra-thin Parylene-N Films on Dielectrics versus Copper Surfaces" **In Press** *Chemical Vapor Deposition.*

[17] K.M. Vaeth and K.F. Jensen, *Chem. Mater.*, **12** 1305 (2000).

Repair of Porous Methylsilsesquioxane Films using Supercritical Carbon Dioxide

Bo Xie and Anthony J. Muscat[1]
Department of Chemical & Environmental Engineering
University of Arizona, Tucson, AZ 85721, U.S.A.
[1]email: muscat@erc.arizona.edu

ABSTRACT

Porous methylsilsesquioxane (p-MSQ) films (JSR LKD 5109) were treated with alkyldimethylmonochlorosilanes having chain lengths of one, four, and eight carbon atoms dissolved in supercritical carbon dioxide at 150-300 atm and 50-60°C to repair oxygen ashing damage. Fourier transform infrared (FTIR) spectroscopy showed that trimethylchlorosilane (TMCS), butyldimethylchlorosilane (BDMCS), and octyldimethylchlorosilane (ODMCS) reacted with silanol groups on the surfaces of the pores producing covalent Si-O-Si bonds. Self-condensation between alkylsilanols produced a residue on the surface, which was partially removed using a pure scCO$_2$ rinse. The hydrophobicity of the blanket p-MSQ surface was recovered after silylation treatment as shown by contact angles >85°. The initial dielectric constant of 2.4 ± 0.1 increased to 3.5 ± 0.1 after oxygen plasma ashing and was reduced to 2.6 ± 0.1 by TMCS, 2.8 ± 0.1 by BDMCS, and 3.2 by ODMCS.

INTRODUCTION

Porosity reduces the dielectric constant of insulating thin films below 2.4, which is needed in microelectronic devices for the 45 nm technology node and beyond to lower power consumption and minimize cross talk between copper metal lines. Pores not only compromise the structural integrity but also expose the interior of the film to the processing environment. Plasma ashing using O$_2$ and N$_2$/H$_2$ used to pattern Si-based low-k films induces damage in the form of silanol (SiOH) groups [1]. The silanols groups must be removed to reduce k and the pores of the films must be sealed to prevent contamination from subsequent processes. The goal of this study was to investigate the reactivity of a series of monochlorosilanes with surface silanol groups on porous methylsilsesquioxane (p-MSQ). The silanes were dissolved in supercritical carbon dioxide (scCO$_2$) because of the processing advantages over conventional solvents. The CO$_2$ and unreacted silanes or by-products are easily separated by reducing pressure lowering the environmental burden of processing. A wide range of silanes can be dissolved in scCO$_2$. Supercritical CO$_2$ has no surface tension so can wet any surface and is nonaqueous [2].

EXPERIMENTAL

The details of the experimental setup have been described previously [3]. P-type (100) orientated Si wafers with a minimum resistivity of 0.5 ohm·cm containing as deposited and O$_2$ ashed blanket p-MSQ films (JSR LKD 5109) were supplied by International Sematech. The silicon wafers were cleaved into 1.5 x 1.5 cm^2 pieces for processing and analyzed with FTIR. The chemistries chosen for this study were TMCS [$CH_3Si(CH_3)_2 Cl$] (99+%), BDMCS

$[CH_3(CH_2)_3Si(CH_3)_2Cl]$ (98%), and ODMCS $[CH_3(CH_2)_7Si(CH_3)_2Cl]$(97%). All chemicals were purchased from Sigma-Aldrich Co. and used as received. Samples were placed against the side of the reactor perpendicular to the bottom. The chlorosilanes molecules were added using a syringe to the bottom of the reactor. Direct contact between a sample and the chemistry occurred only when the reactor was filled with liquid CO_2. Processing was done batch using a mixture of 1 vol% TMCS and scCO$_2$ at 215 atm and 58°C, 0.025 vol% BDMCS at 181 atm and 54°C, and 0.025 vol%, ODMCS at 187 atm and 51°C. The ODMCS was followed by a pure scCO$_2$ rinse for 2 min at 254 atm and 55°C. After depressurizing and cooling the reactor for approximately two hours, a sample was taken out and post-process FTIR spectra were collected. Transmission FTIR spectroscopy (Nicolet Nexus 670 with a MCTA detector) was used *ex situ* to monitor chemical changes in the low-k films (200 scans at 4 cm^{-1} resolution co-added for each spectrum). The film thickness of p-MSQ films was measured using a variable angle spectroscopic ellipsometer (J. A. Woolam Co. M-2000). The hydrophobicity of p-MSQ films was determined by measuring the contact angle of DI water (18.2 MΩ·cm) with a volume of 10 μL using a goniometer (Ramé-Hart Inc. Model 100-00). Electrical measurements were made on metal-insulator-semiconductor (MIS) capacitors fabricated after processing in scCO$_2$. Capacitance-voltage (C-V) curves were measured at 1 MHz with an AC bias from –30V to +30V using an Agilent 4284A precision LCR meter at ambient conditions.

RESULTS AND DISCUSSION

The FTIR difference spectrum for TMCS addition to p-MSQ using scCO$_2$ is shown in Figure 1a. This difference spectrum is post minus pre scCO$_2$ treatment. The strong band at 1068 cm^{-1} indicates that TMCS is covalently bound to the p-MSQ matrix via surface siloxane Si-O-Si bonds. Tripp and coworkers have reported results for octadecyltrichlorosilane (OTS) $CH_3(CH_2)_{17}SiCl_3$ adsorption on silica, which show that the 1060 cm^{-1} band is indicative of covalent attachment to the surface [2,4,5]. Water is necessary for adsorption of chlorosilanes on silica, and hydrolysis occurs either in the fluid phase by reaction of water with a chlorosilane to form an alkylsilanol or on the surface by reaction of a surface silanol with chlorosilane to form siloxane. Condensation between alkylsilanols in solution forming polymeric compounds can also occur and produces FTIR vibrations at 1120 and 1020 cm^{-1} [4]. There is a shoulder at 1034 cm^{-1} in spectrum 1a but no feature near 1120 cm^{-1}. Moreover, the decreases in absorption at 3745, 3400, and 942 cm^{-1}, which are surface SiOH modes, are caused by reaction of TMCS with surface silanols. The bands at 1254, 841 and 793 cm^{-1}, which are various CH$_3$ rocking, Si-C, and Si-CH$_3$ modes, as well as the bands at 2964 and 2906 cm^{-1}, which are CH$_3$ asymmetric and symmetric stretches, respectively, show that TMCS adsorbed on p-MSQ.

The difference spectrum for BDMCS addition to p-MSQ using scCO$_2$ is shown in Figure 1b. This difference spectrum is post minus pre scCO$_2$ treatment. BDMCS contains a four-carbon alkyl chain attached to Si in addition to two methyl groups in contrast to the one carbon methyl groups on TMCS. The strong, sharp band at 1063 cm^{-1} is similar to TMCS but, in addition, there is a broad shoulder with a peak at 1134 cm^{-1} and a less distinct shoulder at 1016 cm^{-1}. These results suggest that both a covalent attachment of BDMCS to the surface occurred as well as self-condensation between hydrolyzed BDMCS molecules forming a polymeric residue on the p-MSQ surface. Loss of intensity in the 3745 cm^{-1} peak, which is the isolated and geminal surface silanol stretch, and the 3400 cm⁻¹ range supports the reaction of hydrolyzed BDMCS with surface silanol groups. The appearance of a peak at 3700 cm⁻¹ is due either to the O-H stretch of butyldimethylsilanol or to a shift in the isolated and geminal surface silanol groups at 3745 cm⁻¹ because of hydrogen-bonding with butyldimethylsilanol. This shows that hydrolyzed BDMCS was produced and that these molecules did not react completely with the silanol groups on the surface of the p-MSQ matrix as was observed for TMCS. The bands at 1468, 1271, 1254, 841 and 793 cm⁻¹ which are various CH₃ rocking, Si-C, and Si-CH₃ modes, the bands at 2958 and 2926 cm⁻¹, which are methyl (-CH₃) asymmetric and symmetric stretches, respectively, and the bands at 2875 and 2856, which are methylene (-CH₂-) asymmetric and symmetric stretches, respectively, indicate the presence of the alkyl moieties of BDMCS on p-MSQ. In contrast to the results of Tripp and coworkers, a band at 898 cm⁻¹ indicative of a Si-OH stretch of alkylsilanol was not present. There is a peak at 881 cm⁻¹ but this is likely a Si-CH₃ mode.

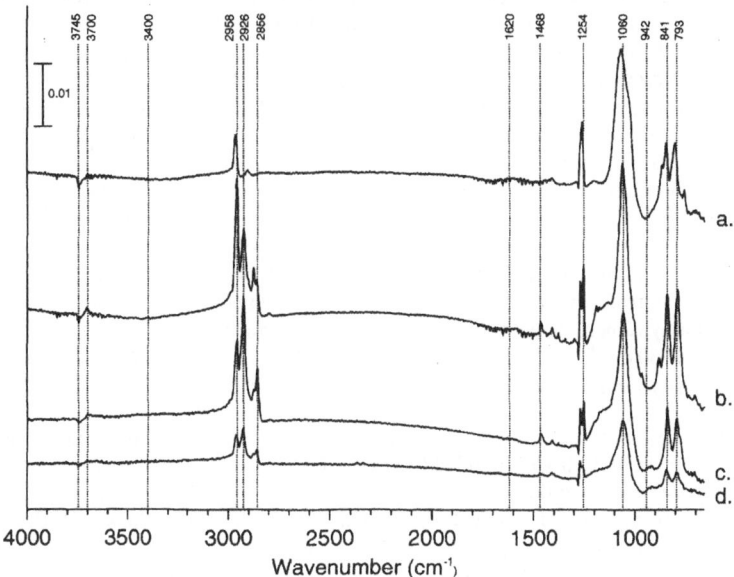

Figure 1. FTIR difference spectra for silylation of p-MSQ using mixtures of scCO₂ and (a) TMCS, (b) BDMCS, (c) ODMCS, (d) sample in (c) after scCO₂ rinse as described in the text.

The difference spectrum for addition of ODMCS, which contains an eight-carbon alkyl chain, to p-MSQ using scCO₂, is shown in Figure 1c. This difference spectrum is post minus pre scCO₂ treatment. The spectrum is similar to the one obtained after BDMCS treatment in that a strong band at 1060 cm⁻¹ and a broad shoulder starting at 1120 cm⁻¹ are present, however, there is no shoulder in the 1020 cm⁻¹ range. The hydroxyl band loss at 3745 cm⁻¹ and gain at 3700 cm⁻¹ occurred but are smaller than for BDMCS. There is a new band at 920 cm⁻¹ due to the Si-OH stretch of octyldimethylsilanol produced by the hydrolysis of ODMCS as well as a decrease in the band at 950 cm⁻¹ due to the condensation of surface silanols with octyldimethylsilanol. The same peaks in the 2920 cm^{-1} range are present as for BDMCS but the intensity of the methyl pair and the methylene pair of stretches is opposite. The characteristic bands for the alkyl frustrated translational modes and Si-alkyl stretches in the 1280-1250 and 850-790 cm^{-1} ranges indicate the presence of the octyl and methyl moieties attached to Si. These results show that ODMCS was covalently attached to the surface through Si-O-Si bonds and that hydrolyzed or nonhydrolyzed ODMCS was present either as free molecules or as a polymeric compound produced by self-condensation. The expected loss of hydrogen-bonded surface silanol intensity due to condensation was not observed, due perhaps to the production of free or polymerized octyldimethylsilanol molecules containing O-H stretches in the same frequency range.

Rinsing the sample treated in ODMCS with pure scCO$_2$ produced the difference spectrum shown in Figure 1d. This difference spectrum is post pure scCO$_2$ treatment minus the starting surface so peak changes are with respect to ashed p-MSQ. All of the bands are smaller, most notably the 1060 cm^{-1} peak, which shows that the scCO$_2$ rinse removed ODMCS moieties from the surface. The peak height of the 1060 cm^{-1} band decreased by approximately a factor of two, whereas the peak heights of the group of bands from 2960-2850 cm^{-1} and the two peaks at 841 and 793 cm^{-1} decreased by a factor of three. Moreover, the shoulder at 1120 cm^{-1} is not as steep after rinsing. These results show that the ODMCS (hydrolyzed or nonhydrolyzed) on the surface was removed by the scCO$_2$ rinse. This conclusion is supported by the behavior of the two Si-CH$_3$ deformation modes between 1280 and 1250 cm^{-1}; the surface Si-CH$_3$ band at 1271 cm^{-1} is smaller than the -CH$_3$ band of the ODMCS molecule at 1254 cm^{-1} after ODMCS treatment (Figure 1c) but the surface band is larger than the molecule band after rinsing (Figure 1d). Pure scCO$_2$ removed unreacted or hydrolyzed ODMCS from the surface preferentially to the ODMCS that condensed with the surface. It is not clear whether all of the unattached ODMCS was removed. The decrease in the 1060 cm^{-1} band is likely the result of the removal of the polymerized octyldimethylsilanol molecules from the surface, which have Si-O-Si stretches on either side of 1060 cm^{-1}, and not to the hydrolysis of siloxane bonds between the surface and octyldimethylsilyl, but this must be proven by further experiments.

The water necessary to hydrolyze a chlorosilane originated in the fluid phase or on the surface in the form of hydrogen-bonded O-H groups. Molecular water present on the surface of the ashed p-MSQ films would give rise to a characteristic signature produced by a bending mode at 1620 cm^{-1}. The absence of this peak shows that water dissolved in the scCO$_2$ fluid from the air present inside the reactor or incipient water in the form of surface silanol groups reacted with the monochlorosilanes yielding alkylsilanols. The amount of water from either source is relatively small and indicates that these molecules readily react with water as has been shown by previous work [4]. A portion of the alkylsilanols undergo a condensation reaction with surface silanols forming new siloxane Si-O-Si bonds as shown by the 1060 cm^{-1} band. The broad negative band at 3400 cm^{-1} in the difference spectra for TMCS and BDMCS was the result of reacting surface silanols, however, the 1060 cm^{-1} band shows that ODMCS also formed a covalent attachment to the surface even though a negative 3400 cm^{-1} band was not present. A small decrease in the 950 cm^{-1} peak indicative of the Si-OH stretch of surface silanols after ODMCS treatment and after scCO$_2$ rinsing of a ODMCS treated surface shows that surface silanols reacted but apparently much less for the eight carbon chain compared to the four or one carbon chain molecules. One interpretation of these results is that the condensation reaction, which occurs first at the top surface then percolates deeper into the film, deposits molecules that interfere with further reaction. Using ellipsometric porosimetry (EP) with toluene as adsorptive, pore sizes for the p-

MSQ are distributed between 1 and 6 nm [6]. Two BDMCS molecules bound to opposite sides of a pore would reduce the pore diameter by approximately 1 nm, but two ODMCS molecules bound in the same way would reduce the diameter by 2 nm or one-third of the diameter of the largest pores. Once these molecules are bound at the perimeter, the alkyl chains would be directed toward the center of the pore and are in a position to interact with the alkyl chains of free molecules. The interaction between alkyl chains would be relatively weak van der Waals forces, which combined with physisorption of polymeric molecules formed by self condensation, serve to limit mass transport of alkylsilanols into and out of pores. The implication for pore sealing is that a multistep sequence might be needed where the first step uses small molecules to repair patterning damage and the second step uses larger molecules to seal the pores of the film. The key to this approach is to balance the number of silanol groups reacted by the repair chemistry, in order to lower the dielectric constant to the desired range, with the number of silanol groups needed to serve as points of attachment for the sealing chemistry.

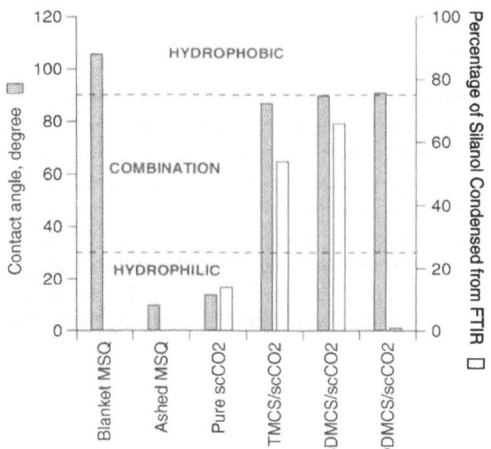

Figure 2. Contact angle and percent silanol condensed for unprocessed blanket and ashed p-MSQ films, and ashed p-MSQ films processed with scCO₂ containing TMCS, BDMCS, and ODMCS.

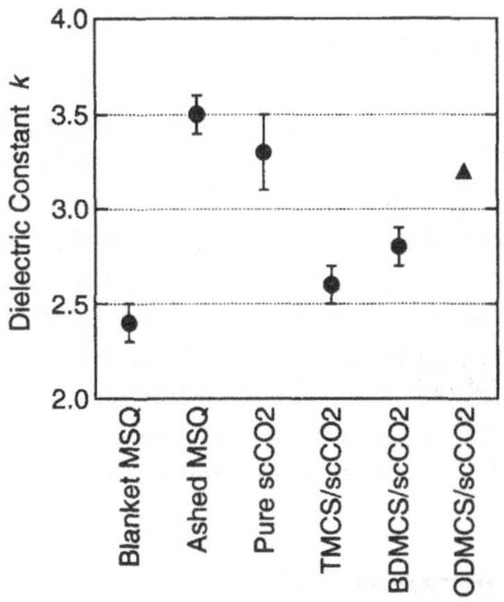

Figure 3. k and % silanol reacted for unprocessed blanket and ashed p-MSQ films, and ashed p-MSQ films processed with scCO₂ containing TMCS, BDMCS, and ODMCS after rinsing.

Contact angles

Figure 2 shows the contact angle and percent silanol reacted for different processes. The contact angle of blanket p-MSQ was over 100° indicating that the top surface of the film was hydrophobic. Ashing in an oxygen plasma converted a portion of the Si–CH$_3$ groups into silanol (SiO–H) groups in the pores of the film. The silanol moieties reduced the contact angle below 10° indicating that the ashed surface was hydrophilic. Reaction of TMCS and BDMCS removed over half of the silanol groups from the film and produced a hydrophobic surface as shown by contact angles of 87° and 90°, respectively. Addition of ODMCS did not reduce the silanol concentration as discussed but did produce a hydrophobic surface (91°).

Electrical measurements

Figure 3 shows the dielectric constant values corresponding to the processed surfaces. A k value of 2.4 ± 0.1 was measured on MIS devices made using the blanket films from CV curves in accumulation. Ashing in an oxygen plasma raised the k value to 3.5 ± 0.1 due the presence of polarizable O-H groups. The dielectric constant was reduced to 2.6 ± 0.1 and 2.8 ± 0.1 for TMCS and BDMCS chemistries, respectively, which are within 20% of the starting value before ashing. The dielectric constant was 3.2 for the sample processed in ODMCS and $scCO_2$ and rinsed with $scCO_2$. The lower value than the ashed wafer indicates that surface silanols reacted with ODMCS forming siloxane bonds. The relatively minor reduction, however, could be the result of a large fraction of unreacted surface silanol groups in the film or the presence of octyldimethylsilanol physisorbed to the surface. This silanol also contains polarizable silanol groups, which will increase the k value.

CONCLUSIONS

All three of the monochlorosilanes examined reacted with surface silanol groups in $scCO_2$ producing Si-O-Si bonds and converting the hydrophilic surface after ashing to hydrophobic surfaces. TMCS reacted the cleanest without forming a physisorbed layer and produced the largest reduction in the dielectric constant to 2.6 ± 0.1. TMCS is an excellent film repair molecule, however, owing to its short one carbon chain length is not expected to seal pores on the order of 5 nm in diameter. BDMCS in $scCO_2$ formed a physisorbed layer containing unreacted BDMCS, hydrolyzed BDMCS, and self-condensed silanol but reduced the dielectric constant to 2.8 ± 0.1. ODMCS also formed a physisorbed layer, which was partially removed by a $scCO_2$ rinse, but reduced the k value to only 3.2. ODMCS is a potential sealing chemistry for p-MSQ since its chain length is on the order of 1 nm.

ACKNOWLEDGEMENTS

This work was cofunded by International Sematech (306077-OF) and the NSF/SRC Engineering Research Center for Environmentally Benign Semiconductor Manufacturing (EEC-9528813/2001-MC-425). We are grateful to P. Josh Wolf, Steve Burnett, and Eric Busch at International Sematech for program guidance.

REFERENCES

1. K. Yonekura, *et al.*, J. Vac. Sci. Technol. B, **22**, 548 (2004).
2. C. P. Tripp and M. L. Hair, Langmuir, **8**, 1961 (1992).
3. B. Xie and A. J. Muscat, in *Eighth Intern. Symp. Cleaning Technol. in Semicond. Dev. Manufacturing*. Edited. by J. Ruzyllo, T. Hattori, R. Opila, and R.E. Novak, (Electrochemical Society Proceeding), **PV 2003-26**, (2004) pp. 279-288.
4. C. P. Tripp and M. L. Hair, Langmuir, **11**, 1215 (1995).
5. J. R. Combes, L. D. White, and C. P. Tripp, Langmuir, **15**, 7870 (1999).
6. A. Das, *et al.*, Microelectronic Engineering, **64**, 25 (2002).

Mat. Res. Soc. Symp. Proc. Vol. 812 © 2004 Materials Research Society

Processing damage and electrical performance of porous dielectrics in narrow spaced interconnects

F.Iacopi[1], Y.Travaly[1], M.Stucchi[1], H.Struyf[1], S.Peeters[1,2], R.Jonckheere[1], L.H.A.Leunissen[1], Zs.Tökei[1], V.Sutcliffe[1,3], O.Richard[1], M.Van Hove[1] and K.Maex[1,4]

[1]IMEC, Kapeldreef 75, B-3001 Leuven, Belgium; [2]affiliate from Lam Research, Fremont, Ca.; [3]affiliate from Texas Instruments, Dallas, Tx; [4]E.E. Dept., Katholieke Universiteit Leuven, Belgium

ABSTRACT

The damage induced in the low-k material upon exposure to dry etch and ash plasmas is a point of major concern in terms of preservation of the dielectric properties. There is urgent need to assess, classify and quantify the extent of such damage to allow the optimization of patterning processes and conditions. Meander-fork structures with spacings between 250nm and 70nm are used in this study as vehicle to compare trends in electrical performance for different dielectrics: SiO_2 and two SiOC:H low-k materials with pristine k values of 3.0 and 2.6. Here we demonstrate that the 'electrical equivalent damage' model is a valid and precise methodology for assessing dielectric damage upon processing from interline capacitance evaluation. This analysis allows to distinguish between bulk and sidewall modification and to quantify the extent of damage. Moreover, it provides an interpretation for the degradation of leakage current and breakdown field of the interline dielectric, revealing different trends whether due to only sidewall or total damage.

INTRODUCTION

The preservation of the electrical properties of low-k dielectrics throughout integration in damascene processes is an extremely challenging task.

As pointed out throughout recent literature, the electrical evaluation of interconnects structures embedded in (porous) low-k dielectrics shows considerable loss in performance and reliability as compared to the use of SiO_2. Moreover, the few attempts to quantify the 'effective' k for such structures have typically yielded values higher than those expected for low-k materials.

The electrical characterization of as-deposited blanket low-k films confirms a 'low-k' value and shows satisfactory *intrinsic* dielectric properties such as leakage current and breakdown field. There is clearly a considerable amount of degradation associated to dielectric processing and integration. Such degradation can occur at various different stages of the integration processes. The electrical issues linked to an incomplete sealing of pores by a diffusion barrier layer was already addressed in several studies [1,2]. Here we focus on the dielectric modification induced by patterning processes. Previous studies have shown that porous low-k dielectrics are extremely susceptible to damage upon exposure to plasma processes, due for instance to the diffusion of reactive species into the film through the pore network [3,4]. The consequences to electrical performance become even more critical as interconnects line spacing shrinks. There is urgent need for a full understanding of such problems. The goal of this study is to propose an efficient and general methodology for the evaluation and quantification of the extent of damage to the low-k films.

EXPERIMENTAL DETAILS

Meander-fork damascene structures 2 to 3 cm long, with spacing ranging from 250nm to 70nm were fabricated by means of electron beam lithography. SiO_2 and silica-based low-k materials, 150nm thick, are compared as Inter-Metal-Dielectric (IMD). Two microporous SiOC:H materials with as-deposited k value of 3.0 and 2.6 were chosen (SiOC:H −1 and −2, respectively), capped by 10nm SiO_2. Dry processing of the dielectric materials was performed using similar chemistries and plasma conditions in a medium density plasma chamber. Fluorine plasma mixtures were used for dielectric etch and O_2/CF_4 for photoresist ash. Line edge roughness of the lines was measured to be spacing independent and below 7nm (3σ). Care was taken to deposit a TaN/Ta diffusion barrier thick enough at the recess sidewalls to ensure an efficient sealing of the dielectric pores (about 10nm). The structures were passivated with a 800nm thick $SiC/SiO_2/Si_3N_4$ dielectric stack.

Inter-line capacitance measurements were performed at 100 kHz. The measured structures were analyzed by cross-sectional Transmission Electron Microscopy (TEM) micrographs combined to Energy Filtering TEM (EFTEM) analysis to retrieve compositional profiles. Measurements of breakdown and leakage current were performed at 100°C using an HP4063 semi-automatic wafer level tester. A voltage sweep from 0 to 100V of 2ms/V was applied to analyze the I-V behaviour and measure the breakdown field. Capacitance data were analyzed according to the 'electrical equivalent damage' model, based on the extraction and analysis of the parallel plate contribution of the low-k dielectric from the total interline capacitance data. For a more detailed description of the model, we refer to our work in [5].

RESULTS AND DISCUSSION

A. Capacitance and physical characterization

Experimental interline capacitance data for structures embedded in the different dielectrics were compared to the values computed through an electrostatic field solver assuming the dielectric preserved its pristine k value, as described in [6]. For the meander-forks embedded in SiO_2, a very good agreement was found between experimental and computed values in the whole spacing range investigated. In **Fig.1** data from structures embedded in SiOC:H −1 and SiOC:H −2 films are plotted versus meander-fork spacing. In both cases the experimental values tend to be higher than the computed ones for the smallest spaced structures. In particular, while for the SiOC:H −1 material considerable deviation is only observed for structures with spacing below 100nm, for the SiOC:H −2 dielectric there is substantial deviation already below 150nm spacing. From this analysis we can conclude that the effective k value of both low-k dielectrics once integrated is higher than their pristine value, and that this effect is more pronounced for the SiOC:H −2, but no quantification is possible at this stage.

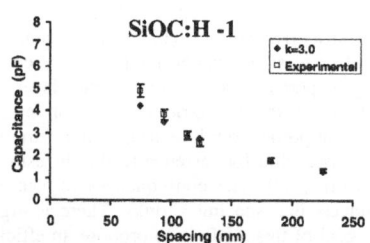 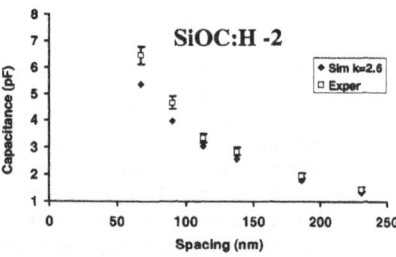

Fig.1 Experimental interline capacitance data are plotted against values computed by assuming a pristine k value for the SiOC:H -1 and -2 dielectrics. While for the SiOC:H -1 significant deviation is observed for spacings below 100nm, the experimental data for the SiOC:H -2 deviate considerably from the computed values already below 150nm spacing.

The capacitance data as analyzed according to the 'electrical equivalent damage' model are shown in **Fig.2** A and B for the SiOC:H −1 and SiOC:H −2, respectively. The experimental data from the SiOC:H −1 material can be fit with the slope corresponding to the pristine value of the material. The fitting line does not pass through the origin, which indicates the presence of damage at the sidewalls of the dielectric. The equivalent sidewall damage, estimated through the intersection point of the fitting line with the spacing axis, is estimated in about 10nm. The

equivalent damage $S' = 2t\left(1 - \dfrac{k_{lk}}{k_d}\right)$ is a combination of the thickness t and the ratio between the

pristine k (k_{lk}) and that of the damaged region at the sidewalls (k_d) [5]. As demonstrated in [5], this parameter can also be used to extrapolate to smaller spacings the increase of capacitance due to the presence of sidewall damage.

On the other hand, the same type of analysis on the capacitance from the integrated SiOC:H −2 (**Fig.2** B) indicates a total modification of the interline dielectric: the slope of the line fitting the experimental data corresponds to a k significantly higher than the pristine one (about 4 against 2.6). A total modification is observed up to the largest measured spacing, here 230nm. The capacitance calculations with the electrostatic solver using k=4 instead of the pristine k value yields indeed much better agreement with the experimental capacitances (not shown).

Compositional profiling by means of EFTEM of the integrated low-k materials in the proximity of the sidewalls are reported in **Fig.3**. Surprisingly, the profiles are comparable, both showing C depletion (typical effect of SiOC:H materials upon exposure through the sidewalls to oxidizing ash chemistries [7]) extending up to about 20nm. Also, a vertical profile of the interline SiOC:H −2 indicates no compositional gradient. This suggests that besides sidewall oxidation, structural material modifications not involving changes in composition can have significant impact onto the electrical film properties. In the case of the SiOC:H −2 material, such composition −independent modifications seem to dominate the electrical behaviour.

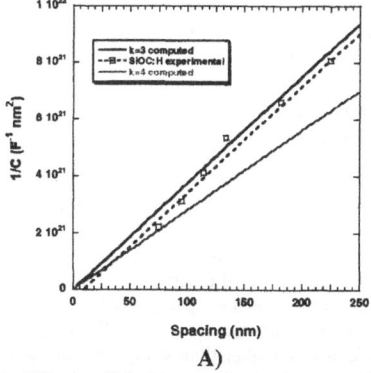
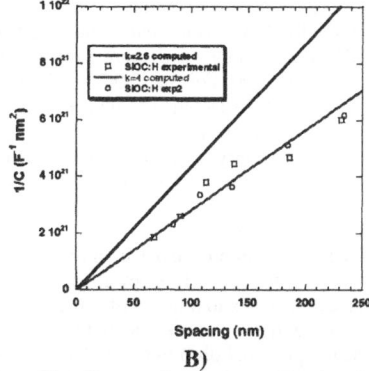

A)	B)

Fig.2 The parallel plate contribution is extracted from experimental interline capacitance data and is plotted versus spacing (1/C, normalized per unit area) for meander-forks embedded in SiOC:H −1 (A) and SiOC:H −2 (B). The SiOC:H −1 is found to have a 10nm equivalent sidewall damage: the fitting line has slope corresponding to the pristine k value of the dielectric and it intercepts the spacing axis at about 10nm. The data from the SiOC:H −2 are fitted with a linear trend with slope corresponding to k≅4 (the slope for k=2.6, pristine value, is also shown), indicating complete modification of the material.

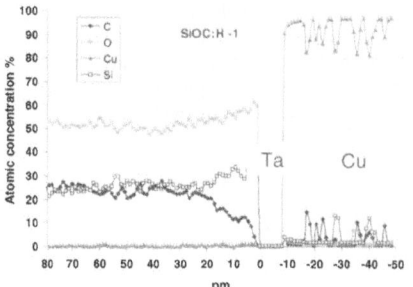 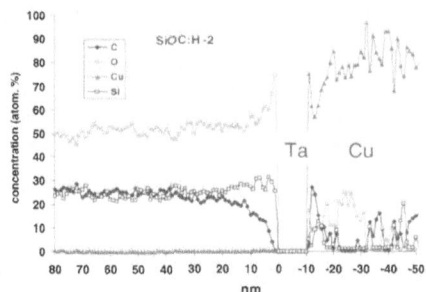

Fig.3 EFTEM compositional profiles of the meander-fork structures in proximity of the sidewalls. The dielectric sidewall is taken as zero position. The profiles appear similar for both dielectrics, both showing C depletion extending up to about 20nm from the sidewalls.

B. Leakage current and breakdown field

In **Fig.4** the breakdown field distributions for structures embedded in SiO_2, SiOC:H −1 and −2 are reported. The breakdown for the SiO_2 lies around 6MV/cm for all spacings (**Fig.4** A). For the SiOC −1 a strong reduction is observed for the structures with spacing below 100nm (**Fig.4** B). For the SiOC −2 breakdown field values are significantly lower, and this regardless of the spacing (**Fig.4** C).

A clear degradation of leakage current is observed from the I-V characteristics for the SiOC:H −1 as the spacing shrinks. The slope of current versus applied field becomes steeper indicating that for an equivalent applied field the narrow spaced structures show higher leakage than the larger ones. In **Fig.5** the curves for the largest (225nm) and the smallest (75nm) measured spacing are reported versus the square root of the nominal applied field. The leakage current at $E^{1/2}=2$ $(MV/cm)^{1/2}$ for the smallest spacing is about one order of magnitude higher than that for the largest spacing. For the larger spacings, the log of the current looks linear versus the square root of the field, thus consistent with either the Frenkel-Poole or the Schottky conduction mechanisms [8], while a more and more pronounced deviation from linearity is seen as interline spacing shrinks (**Fig.5**). For the SiOC:H −2 material there is no significant dependence of the curves on the measured spacing. The leakage current is substantially higher than for the SiOC:H −1 and all curves deviate in a similar way from linearity, as shown in **Fig.6**.

These reliability data are consistent with the conclusions drawn on the basis of the equivalent damage analysis of the interline capacitance data. The integrated SiOC:H −2 shows a reduced breakdown, relatively high leakage and no spacing dependence, thus in agreement with uniform dielectric damage.

The degradation of leakage and breakdown as spacing shrinks for SiOC:H −1 is consistent with the presence of only sidewall damage. The presence of modified dielectric regions along the sidewalls leads to a non-uniform dielectric constant within the interline space, and particularly to a concentration of electric field in the undamaged portion, with lower k value. This way, the actual potential drop in the undamaged part of the dielectric is higher than what it would be in a homogeneous dielectric. This effect is shown in **Fig.7**, where a comparison of the electrostatic field computed for an interline dielectric with and without sidewall damage (A and B, respectively) is given. As the spacing shrinks the difference between the applied and the actual field becomes larger, causing breakdown to occur at lower applied voltage. Also, the I-V ((or I-E) behaviour is expected to become steeper for the same reason (the actual local field is different and higher than the nominal one), which explains also an increasing deviation from linearity for narrow spaces as in **Fig.5**.

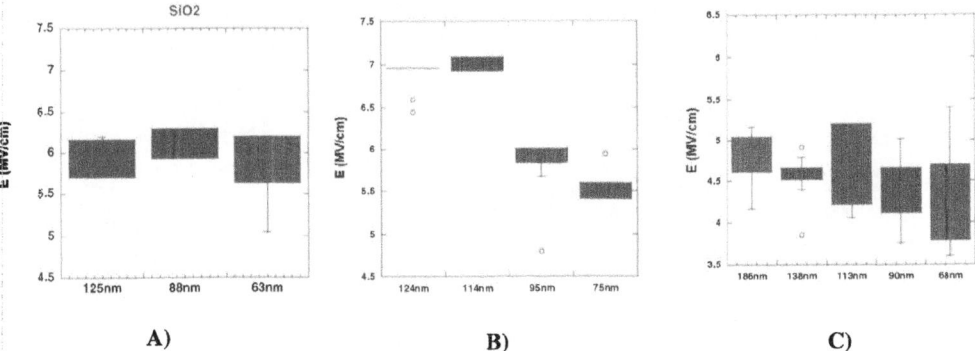

A)   B)   C)

Fig.4 Breakdown field distributions upon fast voltage sweep at 100C for lines embedded in SiO_2 (A), SiOC:H -1 (B) and SiOC: -2 (C). The SiOC:H is the only material showing pronounced degradation as line spacing shrinks.

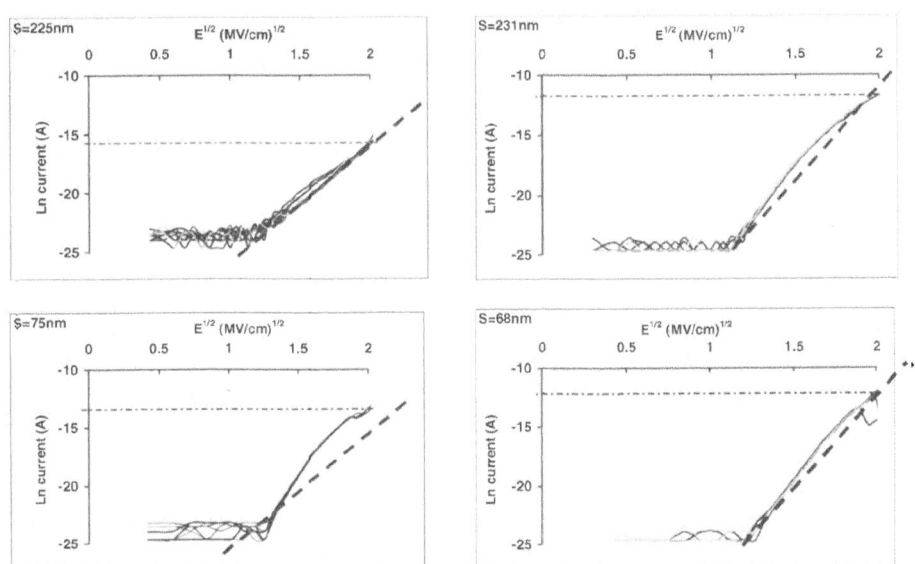

Fig.5 I-E curves for lines embedded in the SiOC:H -1 material with 225nm and 75nm spacing. As spacing shrinks the leakage current for the same applied field tends to increase and to deviate from linear behaviour.

Fig.6 I-E curves for lines embedded in the SiOC:H -2 material. All curves deviate from linearity in the same way, regardless of line spacing.

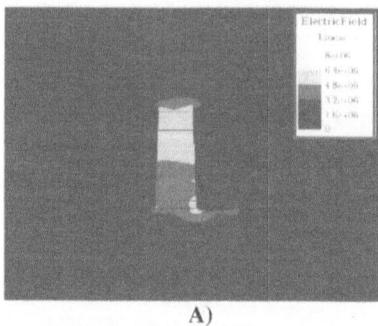 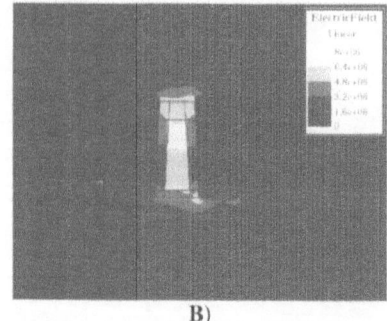

A)   B)

Fig.7 Electric field distributions when the same ΔV=1V is applied across an interline dielectric with k=2 when homogeneous (A) and when an extent of damage of about ¼ of the total interline spacing was considered, with a k value close to SiO_2 (k=4.2). In the homogeneous case the electrical field in the low-k dielectric reaches about 6×10^6 V/m, while more than 8×10^6 V/m are reached in presence of sidewall damage.

CONCLUSIONS

Porous low-k films are extremely susceptible to damage upon patterning processes. This leads to degraded electrical performance of the dielectrics integrated in damascene structures. Here we demonstrate that the 'electrical equivalent damage' model is a valid methodology for assessing process-induced dielectric damage. It allows the distinction between bulk and sidewall modification and the quantification of the extent of damage. It is shown that compositional analysis alone is not sufficient to characterize the actual modification of the dielectrics and does not necessarily show a correct correlation to the consequences in terms of electrical behaviour. It is also shown that the presence of processing damage to the dielectric corresponds invariably to a degradation of leakage current and breakdown field, and this in different ways whether due to only sidewall or total damage.

ACKNOWLEDGMENTS

The authors gratefully acknowledge J.Moonens, I.Bovie, G.Potoms and H.Costermans for their contribution in the e-beam lithography work.

REFERENCES

[1] F.Iacopi, Zs.Tőkei, M.Stucchi, F.Lanckmans, and K.Maex, IEEE Electron Device Letters 24 (3), pp.147-149, 2003.
[2] Zs.Tőkei, M.Patz, M.Schmidt, F.Iacopi, S.Demuynck, K Maex, Materials for Advanced Metallization Conference, 8-10 March 2004, Brussels, Belgium; Zs.Tokei, V.Sutcliffe, S.Demuynck, F.Iacopi, P.Roussel, G.P.Beyer, K.Maex, proc. of the International Reliability Physics Symposium, April 2004, Phoenix, (AZ).
[3] F.Iacopi, M.R.Baklanov, E.Sleeckx, T.Conard, H.Bender, H.Meynen and K.Maex, J.Vac.Sci.Technol. B20 (1), 109 (2002).
[4] D.Ernur, F.Iacopi, L.Carbonell, H.Struyf and K.Maex, Microelectron.Eng. 70 (2-4), pp.285-292, 2003.
[5] F.Iacopi, K.Maex, M.Stucchi, O.Richard, Electrochemical and Solid State Letters 7 (4), G79-G82, 2004.
[6] K.Maex, M.R.Baklanov, D.Shamiryan, F.Iacopi, S.Brongersma, Z.S.Yanovitskaya, Applied Physics Focused Review, J.Appl.Phys.93 (11), pp.8793-8841, 2003.
[7] M.Lepage, D.Shamiryan, M.Baklanov, H.Struyf, G.Mannaert, S.Vanhaelemeersch, K.Weidner, H.Meynen, proc.of International Interconnects Technology Conference 2001, pp.174-176.
[8] T.C.Chang, Electrochem.Solid St. Lett. 6(4), F13-F15, 2003; S.M.Sze, *Physics of semiconductor devices*, Wiley, New York, 1981.

Mat. Res. Soc. Symp. Proc. Vol. 812 © 2004 Materials Research Society

New hybrid low-k dielectric materials prepared by vinylsilane polymerization

Jung-Won Kang, Byung Ro Kim, Gwi-Gwon Kang,
Myung-Sun Moon, Bum-Gyu Choi, Min-Jin Ko,
LG Chem. Ltd./Research Park, Corporate R & D,
Daejon Korea

Abstract

Spin-on Low-K materials are potentially very attractive as interconnection materials in a wide range of semiconductor structures. In this work, new organic-inorganic hybrid materials synthesized by vinylsilane polymerization were proposed. According to compositions and additional fabrications, dielectric constants of these materials were evaluated to be 2.3~3.1. The hardness was 2.0GPa after 430°C curing. These materials had good adhesion strength such that fracture toughness on silicon wafer was 0.22 $MPa*m^{0.5}$ without any adhesion promoters. This result indicates that these organic-inorganic hybrid materials are very promising candidates for low-K dielectrics.

Introduction

Reducing interline capacitance and line resistance is required to minimize RC delays, reduce power consumption and crosstalk in 100nm node technology [1]. For this purpose, various inorganic- and organic polymers have been tested to reduce dielectric constants in parallel with the use of copper as the metal line. However, requirements related to integration of low-k materials are so diverse and stringent that it is difficult to satisfy the whole requirements. Lowering the dielectric constants, in particular, causes the detrimental effect on mechanical properties and then leads to film damage and/or delamination during chemical-mechanical planarization(CMP) or repeated thermal cure cycles. To overcome this issue, new carbon-bridged hybrid materials synthesized through vinyl polymerization and sol-gel reaction were proposed. Organic-inorganic hybrid was selected because of the potential of combining distinct properties of both organic and inorganic components within a single molecular composite. While organic components offer structure flexibility, inorganic ones have an advantage in thermal and mechanical stability. Polymerization with other materials was attempted to improve mechanical properties. Porogen loading inside the materials was carried out to lower dielectric constant.

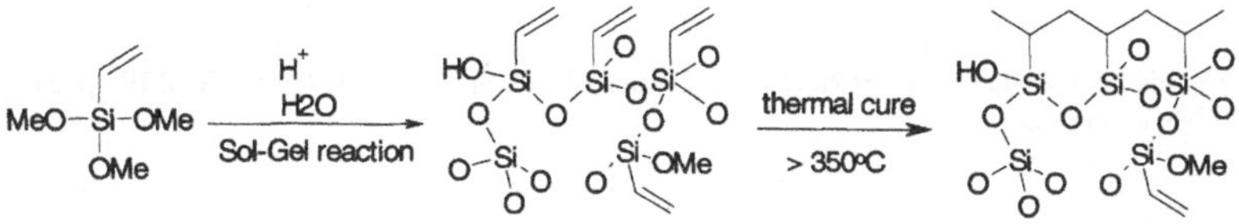

Figure 1 General process scheme preparing PVSSQ gels by the sol-gel reaction followed by thermal curing.

Result and Discussion

Sol-gel polymerization of VTMS(vinyltrimethoxysilane) monomer proceeded by a series of hydrolysis and condensation reaction under acidic condition (Figure 1). A starting mixture containing VTMS, water, catalyst and solvent was heated at 60 ℃ for overnight. Under this condition the VTMS were partially hydrolyzed and condensed. To prepare dielectric films, a sol-gel product was spin-coated at spinning rate of 2000rpm for 20s onto a silicon(001) substrate.

After deposition, the coatings were dried for 1 min at 120 ℃ on a hot plate. Subsequent annealing was performed for 1 hr at 430℃ in N2

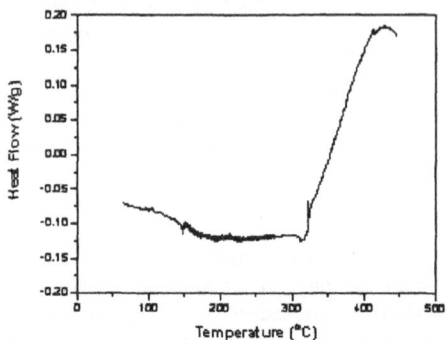

Figure 2 DSC measurement of PVSSQ homopolymer

atmosphere, leading to the further condensation of the hybrid siloxane polymer. This procedure produced thin films with uniform thickness of approximately 700nm. During the curing process, carbon-carbon bond formation reaction also proceeded spontaneously at the high temperature. DSC result in Figure 2 reveals that carbon-carbon bond formation between vinyl groups started at 300°C roughly.

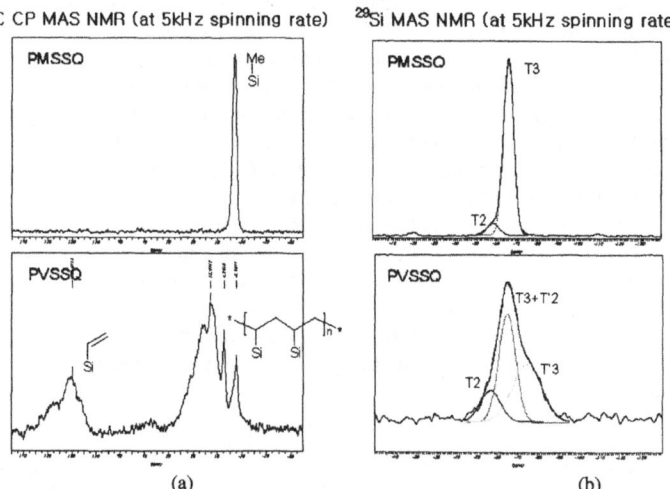

Figure 3 NMR spectra of PMSSQ and PVSSQ after 430°C curing. (a) ^{13}C CP MAS NMR spectra. (b) ^{29}Si MAS NMR spectra.

26

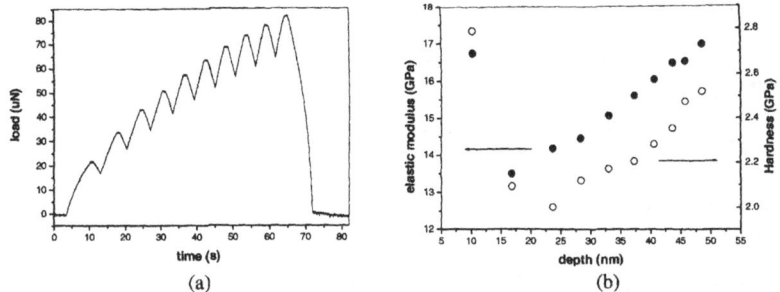

Figure 4 (a) The loading function for nanoindentation technique (b) Reduced modulus and hardness of PVSSQ film as a function of indented depth.

Figure 3 represents NMR of PMSSQ(polymethylsilsesquioxane) and PVSSQ after 430°C curing. The process synthesizing PMSSQ from MTMS monomer is the same as that of PVSSQ. ^{13}C CP MAS NMR spectrum represents the direct evidence for the formation of carbon-carbon bonds and ^{29}Si MAS NMR spectrum also proves the existence of 2 types of Si, unreacted vinylsilane T species and reacted carbon bridged silane T' species, after 430°C curing.

The carbon-carbon bond formation can be confirmed indirectly by measuring the mechanical strength of PVSSQ film with the nanoindentation test [3]. The load and the indented depth were continuously monitored while the films were indented using the load function as shown in figure 4(a). Figure 4(b) plots the modulus and hardness of PVSSQ film as a function of indented depth. Hardness is defined by the ratio of the applied load and the contact area, and reduced modulus is extracted using the projected contact area and the slope of the unloading curve. Both mechanical properties versus indented depth profiles have concave shapes and the minimum values are normally regarded as the properties of the thin film. While high values observed at large indented depth are definitely attributed to substrate effect, it is not clear so far why both properties in shallow depth are increasing. Modulus of PVSSQ film is 13.5 GPa at 15nm depth and hardness is 2.0 GPa near 25nm.

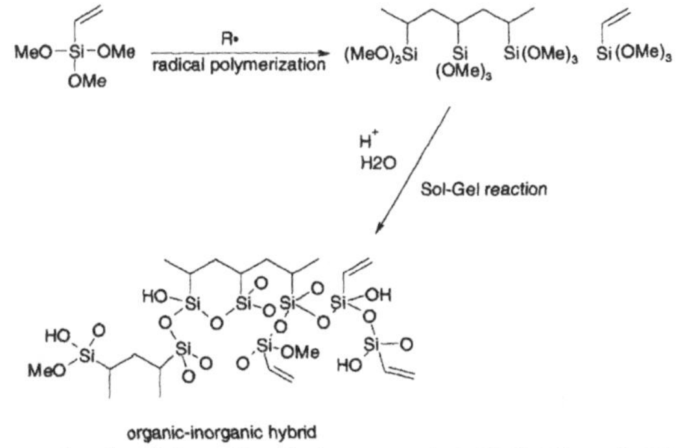

organic-inorganic hybrid

Figure 5 Carbon-carbon bond formation first process *via* radical polymerization.

initiator	equivalent	conversion	Er (GPa)	H (GPa)
none	-	0%	13.5	1.9
AIBN	0.01	14%	15.1	2.1
AIBN	0.03	19%	14.4	2.1
AIBN	0.10	31%	11.6	1.7
AIBMe	0.01	27%	14.6	2.1
AIBMe	0.03	44%	13.7	2.0
AIBMe	0.10	62%	9.9	1.4

Table 1 Radical polymerization of VTMS : AIBN (2,2'-Azobisisobutyronitrile), AIBMe (Dimethyl 2,2'-Azobisisobutyrate).

In addition to the characterization of the mechanical properties, the dielectric constant and refractive index were measured. The dielectric constant was obtained with an area capacitor which consisted of a metal/dielectrics/Si(MIS) structure and refractive index was measured by ellipsometry method at 632.8nm wavelength [4,5]. Dielectric constant at 1 MHz and refractive index of the PVSSQ film were 2.87 and 1.45.

Radical polymerization was also attempted in the presence of azo-type initiator because hydrophobic vinyl end groups should be suppressed to improve the miscibility with porogen. Radical polymerization proceeded uneventfully to give linear carbon chained oligomer, subsequently, which was converted into silicate sol *via* hydrolysis and condensation with catalytic amount of acid and excess water(Figure 5).

Table 1 shows variation of conversion ratio with the type of initiator and the amount of the equivalent of initiator. An increase in the equivalent of initiator caused the increase of conversion ratio and the decrease of mechanical properties. Maximum value of the mechanical strength came out at 10%~20% conversion of vinyl groups. This result indicates that the carbon-carbon bond formation of radically polymerized PVSSQ with a high conversion is significantly suppressed during a curing process and final conversion of the films are much influenced by the formation of carbon-carbon bond at curing rather than in prepolymer. The dielectric constant of radical polymerized PVSSQ polymer varied from 3.0 to 3.4 respectively, which was tunable by controlling its compositions and additional fabrications.

PVSSQ-PMSSQ copolymer was tested to restrict the dielectric constant to the PMSSQ level maintaining the higher mechanical properties. Figure 6 illustrates the strong increase in elastic modulus as the portion of vinyl monomer increases.

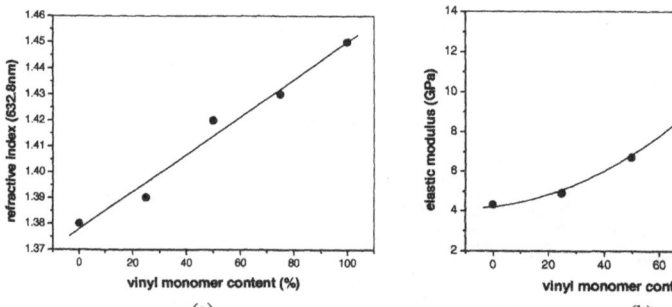

(a)   (b)

Figure 6 (a) Refractive index of PVSSQ-PMSSQ copolymer (b) elastic modulus of PVSSQ-PMSSQ copolymer, 0% vinyl monomer content represents PMSSQ homopolymer and 100% vinyl monomer represents PVSSQ homopolymer.

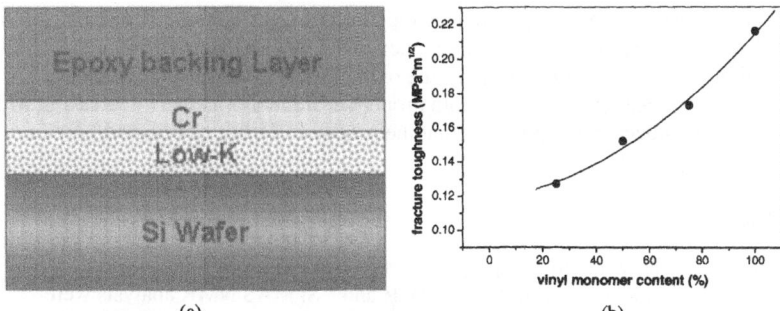

(a)   (b)

Figure 7 (a) MELT test structure (b) fracture toughness of PVSSQ-PMSSQ copolymer with adhesion promoter, 100% vinyl monomer represents PVSSQ homopolymer.

Fracture toughness of PVSSQ, PVSSQ-PMSSQ copolymers and PMSSQ was measured using the modified-edge liftoff test (MELT) [6]. Figure 7(a) illustrates the schematics of the MELT that utilizes to measure adhesive fracture toughness. Epoxy backing layer was applied in order to increase the stored energy in the system and to delaminate low-k films from the substrates without going to the extremely low temperature. Small amount of Cr coated on low-k film improved adhesion between backing layer and low-k film, and induced delamination to occur only at the interface between thin film and Si wafer. If the backing layer is much thicker than low-k films, the fracture toughness can be calculated with the backing layer thickness and the temperature that delamination starts. Figure 7(b) is a plot of fracture toughness as a function of vinyl monomer content. As PMSSQ homopolymer had poor interfacial adhesion and was debonded at room temperature, it had a difficulty in obtaining reliable data. Fracture toughness of the copolymer film increased with an increase in the vinyl monomer content and, finally, the toughness of PVSSQ homopolymer film was determined to be 0.22 MPa*m$^{0.5}$ without adhesion promoters. The PVSSQ film had a much higher toughness than the conventional PMSSQ film.

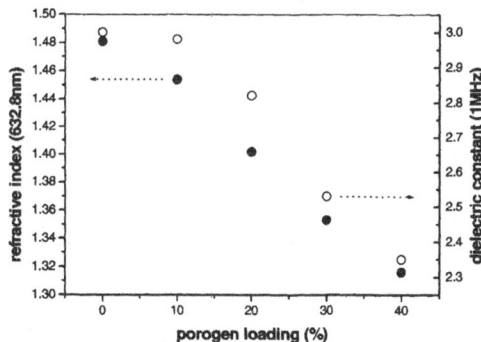

Figure 8 Refractive index and dielectric constant as a function of porogen loading for a PVSSQ matrix polymer.

Increasing porosity is one way to reduce the dielectric constant of an existing material. Figure 8 shows the variation of refractive index and dielectric constant when monomethyl terminated PEO (poly(ethylene oxide), Aldrich) linear polymer was employed as a porogen. The fact that both properties are monotonically decreasing with the increase in porogen loading proves that the pore is uniformly distributed pore in PVSSQ matrix polymer.

Conclusion

Novel organic-inorganic hybrid polymers were synthesized by vinylsilane polymerization. Solid-state ^{13}C CP MAS NMR and ^{29}Si MAS NMR analysis were conducted for the structural characterization of hybrid dielectrics. The C-C bond formation of vinyl group induced by thermal and radical process resulted in the excellent mechanical property of the hybrid dielectric films. It has been also demonstrated that electrical and mechanical properties of the hybrid films can be tailored by copolymerization with PMSSQ and through the introduction of porogen. To verify the adaptability for integration, further studies for other properties are needed.

Reference

1. Semiconductor Industry Association, The National Technology Roadmap for Semiconductor, 1994.

2. Y. Abe, J. Non-crystalline Solid, **261**, p39, 2000

3. W.C. Oliver, and G.M. Pharr, J. Mat. Res. **7**, p1564, 1992.

4. R.F. Pierret, *Semiconductor Device Fundamentals*, 1995.

5. H.G. Tompkins, *A User's guide to Ellipsometry*, 1993.

6. E.O. Shaffer II, F.J. Mcgarry, and L. Hoang, Polymer Eng. Sci., **36**, p2375, 1996.

7. D. Zhao et al., Science **279**, p548, 1998.

Mat. Res. Soc. Symp. Proc. Vol. 812 © 2004 Materials Research Society

NOVEL EPOXY SILOXANE POLYMER AS LOW-K DIELECTRIC

Pei-I Wang*, Jasbir S. Juneja*, Shyam Murarka*, Toh –Ming Lu*, Ram Ghoshal**, and Rajat Ghoshal**

*Center of Integrated Electronics, Rensselaer Polytechnic Institute, Troy, NY
**Polyset Co. Inc., Mechanicville, NY

ABSTRACT

This paper introduces a low-k dielectric material, a novel epoxy siloxane polymer, made by Polyset Co. Inc, which has promising properties. The polymer was spin-deposited, and thickness and optical properties were measured using variable-angle spectroscopic ellipsometry (VASE). Fourier transform infrared (FTIR) spectra of as deposited and cured polymers showed that the polymer is fully cured at 165 °C. The low curing temperature of the polymer lowers stress in back-end-of-line (BEOL) stack and thus improves the reliability. The polymer is thermally stable up to 400 °C. The polymer has Young's modulus of ~5 GPa and hardness of greater than 0.4 GPa. After multiple stress cycles up to 300 °C, the residual stress in the polymer at room temperature is less than 60 Mpa. The polymer has good adhesion with semiconductor and dielectrics such as Si, SiC, and SiO_2, metals such as Al, Cu, Co, and W, and barrier materials such as TaN. The bulk dielectric constant of the polymer is 2.4 - 2.7. The leakage current density in the polymer at the applied electrical field of 1 MV/cm is in 10^{-9} A/cm^2 range and the breakdown field of the polymer is ranging from 5 to 7 MV/cm. The polymer when subjected to bias-temperature stress (BTS) conditions of 150 °C and 0.5 MV/cm shows no C-V shift for up to 100 min indicating that the polymer resists Copper diffusion. The current density under stress conditions of 150 °C and 0.5 MV/cm was less than 10^{-9} A/cm^2 for up to 7 hrs.

INTRODUCTION

The speed of the logic devices is governed by the transistor gate delay, which is proportional to the size of the transistor, and the interconnect signal delay, characterized by the resistance-capacitance (RC) time constant. With the reductions in the dimensions of the devices below 0.18 micron, the interconnect delay has become the limiting factor.[1] The drive in the industry is to replace the aluminum-SiO_2 interconnects by copper-low-k. The industry has shifted to Cu technology but is still struggling to implement inter-layer dielectric (ILD) using low-k.[2]

The electrical, mechanical and chemical property requirements of the candidate ILD low-k materials are very stringent.[3,4] At the same time, thermal budget is of concern for the semiconductor industry leading to the requirement of low k dielectric materials that can be processed at low temperatures. In addition, high temperature stability in the range of 400 °C of processed dielectric films is necessary to ensure circuit reliability.[5]

In addition to the inherent properties of the dielectric, compatibility with the current copper technology is a must for any candidate low k material. Copper ion injection in dielectric results in high leakage and premature failure.[6] A diffusion barrier is required in between metal and dielectric to avoid metal ion injection in the dielectric. With the scaling down of the interconnect dimensions, there is a requirement of an effective ultra thin barrier or, ultimately, of a dielectric that inherently resists metal ion injection.

In this paper a Polyset spin-on low k polymer, which has a dielectric constant of 2.4 – 2.7 has been studied for its inherent mechanical, electrical and thermal properties. Our results show that this polymer has promising properties to be an ILD, and moreover, this polymer resists Cu diffusion. According to International Technology Roadmap for Semiconductors (ITRS), the effective k value requirement by 2007 is 2.3 – 2.7. Since this polymer relaxes the need for diffusion barrier, lower effective dielectric constant than that of most other non-porous existing low k dielectrics can be achieved. Thus this Polyset epoxy siloxane polymer is a potential candidate low k material for advanced interconnect technology.

EXPERIMENTAL

The polymer films were deposited on N type, 4-inches silicon wafers with low resistivity of $\leq$ 0.02 ohm-cm for MIM (metal-insulator-metal) structures, 25 nm thermally oxidized P type silicon wafers for MIS (metal-insulator-semiconductor) structures, and bare p type silicon wafers for mechanical property measurements. The dielectric films were spin-deposited at 3000 rpm for 100 sec and then baked in the vacuum of 10^{-3} torr for 1 hr. Curing processes were preformed in a furnace with hydrogen gas flow. The final film thickness and refractive index were measured using variable-angle spectroscopic ellipsometry (VASE, J.A. Wollam Co., Inc.). The chemical bond structure was studied using FTIR spectroscopy performed on a Mattson Galaxy Series 3000 spectrometer.

BTS C-V measurements were made on a HP 4280A 1 MHz capacitance meter. I-V measurements were performed on the MIM samples at room temperature using 4140 pA meter. Current vs. time (I-t) measurements were performed on the MIM samples that were subjects to BTS conditions of 0.5 MV/cm and 1 MV/cm at 150°C and 200°C up to 7 hrs.

Adhesion test of Polyset polymer film on various materials was performed using Scotch tape peeling test. Adhesion of Polyset polymer on copper was measured using 4 point bending technique on FSM AquaFlex – 4 Point Bend adhesion Tester.

RESULTS AND DISCUSSION

Polyset material is an epoxy alkoxy siloxane oligomer with many unique properties in contrast with other available epoxy-siloxanes.[7] The multifunctional nature of the molecule holds great advantages in regards to both the speed and the extent of cure, leading to a highly cured, highly cross-linked three-dimensional structure with enhanced thermal and mechanical properties. The repeating units of the polymer have low polarizability, which allow the k value of the Polyset material to be significantly lower than those of existing siloxanes.[7]

Figure 1a shows ex situ FTIR spectra of the spin on polymer films at various stages of processing, namely as-spun, baked at 100°C, cured at 165°C, 250°C, and 300°C. It can be seen that the peak associated with the solvent at 1200 cm^{-1} in the FTIR spectrum of as-spun film disappears after the baking step. In conjunction with the loss of ~ 9% in the corresponding film thickness at this stage in Fig. 1b, our results indicate that all the solvent evaporates by 100°C. This decrease of the film thickness is also partially attributed to the cross-linking process which the linear chains of the polymer are linked together to form the three-dimensional network structure. Furthermore, an increase in the intensity of the low-frequency component of IR band associated with stretching mode of the Si-O-Si chain structure is observed. This intensity of the low-frequency further increases at the curing temperature of 165°C while the film thickness decreases additional 2% as a result of the continuous cross-linking upon curing. The refractive index seems to increase slightly during curing process due to the formation of a denser structure.

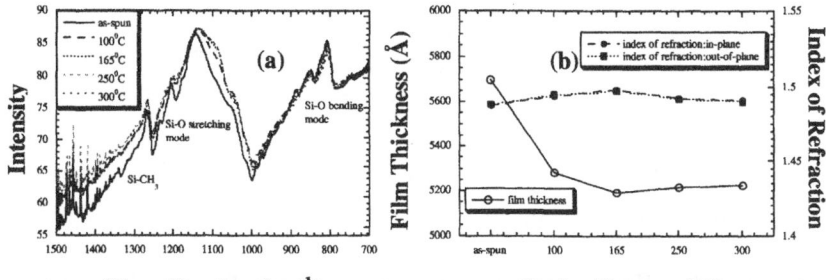

Wave Number (cm⁻¹)   Curing Process (°C)

Figure 1(a) FTIR spectra of as-deposited, baked, and cured polymer films and (b) their corresponding ellipsometric thicknesses

No additional increase of the intensity of the low-frequency is observed as the films were subjected to higher curing temperatures. Because the low-frequency component is assigned to the long chains and the high-frequency is assigned to the short chains, our result clearly shows that the cross-linking process starts at the baking process and equilibrates at a low temperature at 165°C. The corresponding film thickness and refractive index show no change at higher curing temperatures (see Fig. 1b) further suggesting the completion of the structural transformation of the dielectric film.

    Figure 2 provides the result of a thermogravimetric analysis (TGA) analysis of the as spun polymer film. This TGA curve consists of three distinctive regions. A significant mass loss is observed in the temperature range of 100°C to 160°C, corresponding to the loss of solvent from the material. The curve then reaches a plateau in the temperature range of 160°C to 400°C, indicating this polymer is thermally stable in this temperature range. A second significant mass loss thus follows as the temperature goes beyond 400°C. Table 1 shows the thermal stability of the polymer film at 400°C for 30 min in hydrogen and Ar-3%H₂ ambient, respectively. No significant change of polymer film thickness is observed, thus the polymer film is shown to be thermally stable for at least 30 min under these annealing conditions.

    The hardness and Young's modulus of polymer film on Si substrate were determined using nanoindentation. Figures 3a and 3b present the comparison of the hardness and Young's modulus of Polyset polymer with SiLK. Despite the artifact attributed to displacement near the Si

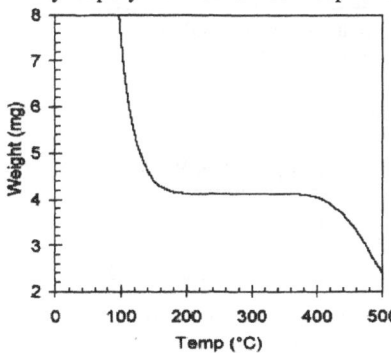

Annealing process	As-prepared	→	High temp. anneal
Annealing Ambient	nitrogen		nitrogen
Annealing Temp.-time	250°C-1 hr		400°C-0.5 hr
Film Thickness (Å)	1506		1493
Annealing Ambient	nitrogen		Ar-3%H₂
Annealing Temp.-time	250°C-1 hr		400°C-0.5 hr
Film Thickness (Å)	1592		1541

Figure 2 TGA of as-spun polymer   Table 1 Thermal stability of polymer at 400°C

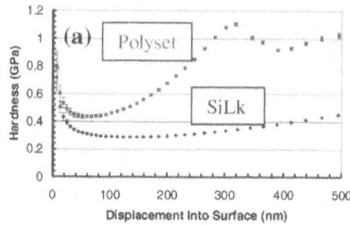 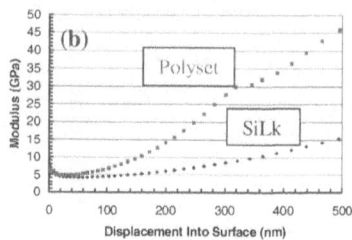

Figure 3 Comparison of (a) Hardness and (b) Young's modulus of the Polyset polymer with SiLK

substrate, the Polyset polymer clearly has favorable Young's modulus of ~5 GPa and hardness of greater than 0.4 GPa. Figure 4 shows a plot of the stress in the polymer film on a Si substrate during heating to 300°C. The sample had previously been cured at 250°C. During the first cycle, the stress drops linearly as the sample is heated. On cooling, stress recovers, leading to a high tensile stress at room temperature. The higher stress is attributed to the difference of the curing temperature and the stress temperature. On subsequent heat cycles, the linear path is reversible during repeated cycling. It illustrates that after multiple stress cycles up to 300 °C, the residual stress in the polymer at room temperature is less than 60 Mpa. Adhesion of the polymer film on various substrates has been performed using Scotch tape peeling test. Table 2 shows that the polymer has good adhesion on a series of materials that have been commonly used in the semiconductor industry. Adhesion strength of polymer on Cu measured using 4-point bending is 12 N/m.

Figure 5 shows the film thickness dependency of the dielectric constant of polymer. The dielectric constant is ≤ 2.4 for film thickness less than 100 nm and equilibrates to a value of 2.7 for greater film thickness. This phenomenon is attributed to the intrinsic property of the material. The dielectric constant of the thick film, however, can be maintained in the range of 2.5 if the polymer film consists of multiple thin layers.

In the experiments of Fig. 6, leakage current curves for the polymer are presented. Capacitors with three different areas, with diameters 0.5, 1, and 1.5 mm, were used to show the validity of the extreme value statistics. It can be clearly observed that the intrinsic breakdowns at high

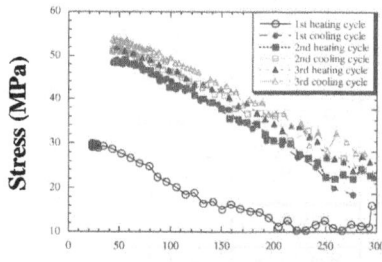

Substrates that can be adhered by the polymer (pass peeling test)	
Semiconductor	Si
Metal	Al, Cu, Co, W, Ta
Dielectrics	SiO$_2$, SiC, SiN, TaN

Figure 4 Stress temperature plot of a polymer Film on Si substrate

Table 2 A list of materials can be adhered by Polyset polymer

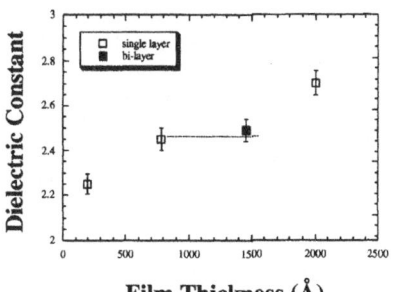

Film Thickness (Å)

Figure 5 Dielectric constant

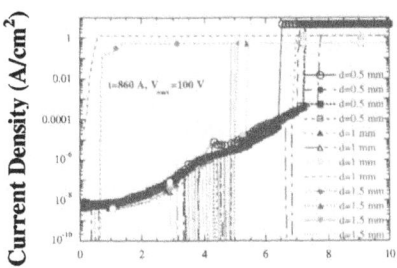

Applied Field (MV/cm)

Figure 6 Leakage current and breakdown distributions of the polymer film

electric field values while the defect-related breakdowns occur with a wide distribution at low fields. The intrinsic leakage current curves show that the leakage current densities of the polymer are in the range of 10^{-9} A/cm^2 at the applied field of 1 MV/cm and the average breakdown field for the polymer is in the range of $5 - 7$ MV/cm.

Figure 7 shows various C-V plots of the MIS capacitors with Cu electrodes that were subjected to BTS at 150°C and 0.5 MV/cm for 10, 20, 30, 40, and 100 minutes in sequence. The flat-band voltage shift of less than 2 V was observed to the left after BTS for 10 min. No significant flat-band voltage shift was observed as the capacitors were subjected to further BTS treatments suggesting that the C-V curves reached the equilibrium state. Continuous Cu ion injection in a dielectric film would manifest itself by progressive shifting of the flat band voltage of C-V curves. Since the flat-band voltage shift observed is not significant, it can be interpreted that the Cu diffusion in the dielectric is negligible. Also, the initial shift in the C-V curves has been attributed to annealing out or movement of trace alkali contaminants[6] or to charging at the silicon oxide polymer interface.[8] It has been shown earlier that the availability of oxygen at the interface promotes metal ion formation and these ions move under the applied bias.[9] The absence of free oxygen at the polymer-metal interface helps to contain Cu diffusion.

The effects of temperature and electrical field on Cu transport in polymer film with time were studied by current vs. time (I-t) response of four samples that were subjected to the BTS

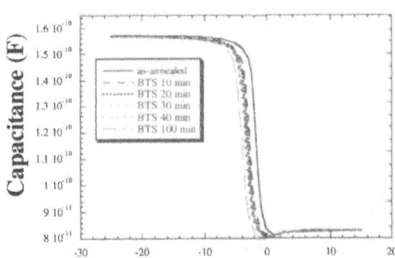

Voltage (V)

Figure 7 C-V curves of MIS capacitors of Cu/polymer/SiO$_2$/Si after BTS at 150°C and 0.5 MV/cm

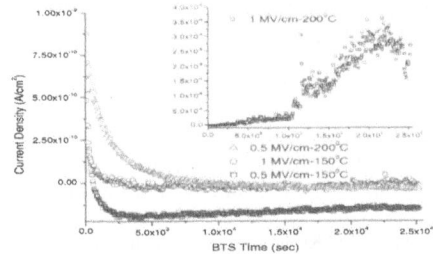

Figure 8 I-t curves of MIM capacitors of Cu/polymer/conductive Si subjected to BTS at 150°C and 200°C and applied field of 0.5 and 1 MV/cm for 7 hrs

conditions of applied field of 0.5 and 1 MV/cm while temperatures were held at 150°C and 200°C. Figure 8 shows that in the first three BTS conditions the I-t curves can be divided into two distinct regions. At short time scales, the positively charged Cu ions build up in the Cu/dielectric interface and are unable to penetrate the dielectric. The accumulated positive Cu ions then set up an electric field opposing further injection of Cu ions. The increase of the opposing field leads to a reduction in injected current and thus results in the sharp current descent in the initial region. The ionic current continues to reduce until it matches the electron leakage current in the dielectric film. The current density for these measurements at BTS condition of 0.5 and 1 MV/cm at 150°C and 0.5 MV/cm at 200°C remained low in the range of 10^{-10} A/cm^2 for up to at least 7 hrs, indicating that under these test conditions, no further defects are created in the film. This is a strong indication that diffusion of Cu ions into the dielectric is not taking place at a rate significant enough to cause leakage. The stress-induced leakage current is not observed until the dielectric film is subjected to a higher BTS of 1 MV/cm at 200°C, at which point though tremendous charge injection into the dielectric occurs. However, it should be noted that this threshold to degrade the dielectric characteristics is essentially a severe stress condition, despite the fact that the dielectric has not yet reached the final break down stage even after 7 hrs under the ultimate test. These results reinforce the barrier property of this dielectric to the diffusion of Cu ions.

CONCLUSIONS

The Polyset polymer exhibits appealing mechanical properties such as good adhesion with metals and various dielectrics, and have dielectric constants as low as 2.4-2.7. The Polyset polymers show excellent reliability with respect to their dielectric properties when subjected to standard test conditions of thermal and electrical bias. Furthermore, the low-k dielectric polymers studied herein require low processing temperatures, exhibit low leakage currents, and remain stable at temperature of 400 °C making it particularly attractive for use in the semiconductor industry as low k dielectric.

REFERENCES

1. S. P. Jeng, R. H. Havemann, and M.-C. Chang, *Mater. Res. Soc. Symp. Proc.*, **337,** 25 (1994).
2 M. Morgen, E. T. Ryan, J.-H. Zhao, C. Hu, T. Cho, and P. S. Ho, *Annu. Rev. Mater. Sci.*, **30**, 645 (2000).
3. S. P. Murarka, *Solid State Technol.*, **39**, 83 (1996).
4. M. Morgan, E. T. Ryan, J.-H. Zhao, Chuan Hu, T. Cho, and P. S. Ho, *Annu. Rev. Mater. Sci.*, **30**, 645 (2000).
5. S. P. Murarka, *Mater. Sci. Eng.*, **R 19**, 88 (1997).
6. R. Tsu, J. W. Mcpherson, and W. R. McKee, *IEEE Inter. Reliability Phys. Sym. Proc.*, 348 (2000).
7. C. H. Kim, and J. S. Shin, *Bull. Korean Chem. Soc.*, **23**, 413 (2002).
8. S. Rogojevic, A. Jain, F. Wang, W. N. Gill, P. C. Wayner, J. L. Plawsky, T.-M Lu, G. R. Yang, W. A. Lanford, A. Kumar, H. Bakhru, and A. N. Roy, *J. Vac. Sci. technol.*, **B 19**, 354 (2001).
9. A. Mallikarjunan, C. Wiegand, Jay J. Senkevich, G.-R. Yang, E. Williams, and T.-M Lu, *Electrochemical and Solid-State Letters*, **6**, F28 (2003).

Mat. Res. Soc. Symp. Proc. Vol. 812 © 2004 Materials Research Society

Supercritical CO2 Treatments for Semiconductor Applications

S. Gangopadhyay[a], J.A. Lubguban[a], B. Lahlouh[b], G. Sivaraman[b], K. Biswas[a], T. Rajagopalan[b], N. Biswas[b], H. -C. Kim[c], W. Volksen[c] and R. D. Miller[c]

[a]Dept. of Electrical and Computer Engg., University of Missouri, Columbia, MO, 65211
[b]Nano-Tech Center, Texas Tech University, Lubbock, TX 79409
[c]IBM Almaden Research Center, San Jose, CA 95120

Supercritical fluids (SF) have been used in a wide variety of applications: in industrial processes, analytical, waste detoxification, etc. Recently, its usefulness extends to the semiconductor industry. Researches have shown that supercritical CO_2 ($SCCO_2$) can be used to remove photoresists and significantly reduce the amount of waste from solvents in comparison to conventional stripping techniques. SF will also find its usefulness in cleaning high aspect ratio vias and deep trenches as semiconductor features shrink to sub-micron levels. We will report here the use of supercritical CO_2 treatments in extraction of porogens from a nanohybrid film fabricated via templated-porogen approach. Its use as a medium to repair the damage in porous films from plasma ashing will also be presented. The ability to tune the solvation and diffusion power of $SCCO_2$ and to swell the film matrix make it a good medium for silylation to restore hydrophobicity and functionalize the film.

Introduction

The further development of ultra-large-scale-integration microelectronics devices will depend on the introduction of new materials and the development of new processing techniques and procedures. The reasons are: using current materials to fabricate devices will ultimately offset the speed and performance and using current processing techniques will create reliability problems that come with scaling to dimensions below 100 nm. New materials for low-k and high-k gate dielectrics, new method of cleaning residues and new deposition techniques in high aspect ratio vias and deep trenches are just few examples. Among the new processes for microelectronics processing is the emerging supercritical CO_2, $SCCO_2$.[1-3] This technique has been used extensively for various applications like industrial, pharmaceutical, analytical and waste detoxification.[4,5] The recent interest in $SCCO_2$ process lies on its versatile characteristics which includes gas-like diffusivity, liquid-like dissolving power and dynamic properties.[4,5] In high aspect ratio vias and deep trenches, these properties of $SCCO_2$ can be exploited in a variety of ways: as a solvent carrier to clean residues, as a medium for chemical reaction, as a deposition medium, and as solvent for removal of certain species. The efficient and minimal use of co-solvents in $SCCO_2$ is a strong motivator to use this process. The use of $SCCO_2$ with a co-solvent in the removal of photoresists has been reported and resulted in minimal waste products.[1] This means the availability of an environmentally friendly and cost-cutting process which make it attractive. Another group has reported the use of $SCCO_2$ as toughening agent in the fabrication of nanoporous films.[3]

In this paper, we will report the various applications of $SCCO_2$ that our research group has been investigating. This includes generating porosity on spin-coated inorganic-organic hybrid films and surface modification of dielectrics. Introducing porosity to dielectric films seems to be the obvious way of further lowering the dielectric constant below 2.0, since there is no dense material that has a k-value below 2.0. Surface modification is needed to cure plasma-damaged surfaces during photoresists ashing and also to restore the hydrophobic character of the film.

Experiment

A supercritical CO_2 system has been assembled in this study. The system consists of a high pressure reactor with operating pressures up to 10,000 psi, an oven operating up to 200°C to control the reactor and fluid temperatures, an air-driven gas booster, a solvent pump that can inject liquids at supercritical pressures of CO_2, and a vacuum system to evacuate the ambient before the experiment. We used here static and pulsed modes of $SCCO_2$ treatments. A diagram of our system is given elsewhere.[6] Most films studied here were prepared via spin coating by IBM and TEL.

Results and Discussion

A. Porogen extraction in nanohybrid organosilicate films fabricated via spin coating.[7,8]
As stated above, future dielectric materials are geared towards the introduction of porosity and the popular approach is incorporating porogens (sacrificial material) in a thermally stable matrix. Current process of pore generation involves thermolysis of porogens from the thermosetting matrix. The problem here is that the process window can be narrow, since the porogen decomposition must occur well below the glass transition (T_g) of the matrix to prevent pore collapse. Since many organic polymers have relatively low T_gs compared to the degradation temperatures of porogens, this processing constraint can lead to incomplete porogen decomposition often resulting in char formation. Other issues in this approach are high temperature thermolysis of many porogens usually greater than 400°C and longer times of processing. There is a need for a low temperature processing and we have reported $SCCO_2$ has the capability of low temperature processing. Hybrid films with open (55wt% porogen loading) and closed (25wt% porogen loading) pore structures with poly(propeleneglycol) (PPG) porogens and MSSQ matrix were processed for $SCCO_2$ PPG extraction.[7,8] We have succeeded to extract 92 and 85% of PPG respectively at temperatures below 200°C.[7,8] Complete results from this experiments were reported in references [7,8].

Representative 3-dimensional porous morphologies were generated from SAXS data using an analysis method that builds on work by Cahn, Berk, and others.[8] While the morphologies are only representative (i.e., not a reconstruction), they do statistically reproduce the pore morphology. Details of this method are shown in ref [8]. Fig. 1 shows representative morphologies for different porogen loadings for a 200nm cube obtained using the method described above. Yellow and red correspond to the PMSSQ matrix and the pore interior as seen through the cube side, respectively. As shown in Fig. 1, the average pore size increases as the loading increases for porous films prepared by both thermal decomposition and supercritical CO_2 extraction (SCF). It would be quite valuable to state the average pore sizes obtained from this method. For 55wt%, the morphology becomes bi-continuous (see below) and one cannot easily define an average pore size. Fig. 2 presents another visualization of the pore morphology that permits determination of the extent of pore interconnection. The PMSSQ matrix is transparent and the pores are shown as seen from outside the cube. The largest pore is colored red (left side), and this pore is removed on the right side of the figure. For 55wt% porogen loadings, the elimination of the largest pore results in the removal of most of the pore volume. This shows that at 55wt% loading the pore morphologies are bi-continuous for both pore generation methods. Moreover, at 25wt% loading Fig. 2 demonstrates that the morphologies are highly interconnected, but not bi-continuous.

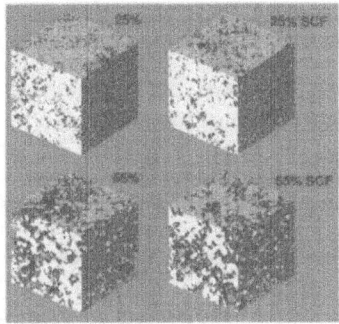

Fig. 1. Visualization of representative microstructures for 200 nm cubes. Yellow and red correspond to the PMSSQ matrix and the pore interior as seen through the cube side, respectively.

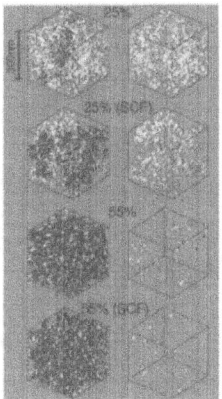

Fig. 2. Images of microstructures for a cube side of 300 nm. The PMSSQ matrix is transparent and the pores are seen from the outside of the cube side. On the left, the largest pore is shown in red, while this is removed on the right side.

B. Surface modification using SCCO$_2$ as a medium for silylation.[9] Potential dielectric materials contains significant amount of methyl groups like MSSQ. This is a good starting material for the incorporation of porosity due to the steric hindrance of CH$_3$ that results in a lower dielectric constant. However, during device processing (particularly O$_2$ ashing of photoresists), the surface methyl groups are also etched and become hydrophilic. This results in an increase in refractive index, dielectric constant, and leakage currents to values greater than that of conventional SiO$_2$.[10,11] This creates reliability problems and is unacceptable. Hence, SCCO$_2$ was used as a medium of reaction for silylation. This is to recover the hydrophobic surface by removing acidic H in Si-OH in the surface. SCCO$_2$ is a better alternative to either liquid or vapor silylation because of the enhanced diffusity and solvation properties. Trimethylchlorosilane (TMCS) is used as the silylating agent to facilitate the reaction in the SCCO$_2$ medium. We have succeeded in incorporating alkylsilyl group and restore the hydrophobic character of the two films above. Figs. 3(a) and (b) show the FTIR absorption spectra of the open pore sample (55wt% loading) before plasma damage, after plasma damage, and after SCCO$_2$/Butanol, TMCS treatments. Fig. 3(a) shows the changes in the total absorption intensity of the CH groups in the 2800 to 3000 cm^{-1} region as it undergoes ashing and SCCO$_2$/Butanol, TMCS treatments. The effectiveness of SCCO$_2$/ Butanol, TMCS treatments in reversing the damage is shown by the re-incorporation of most CH groups lost during ashing. The -CH$_3$ lost in oxygen ashing is replaced by -Si-(CH$_3$)$_3$. Fig. 3(b) shows the silanol groups (3000 to 3800 cm^{-1}) of the film. It shows that silanol groups are effectively removed by SCCO$_2$/Butanol, TMCS treatments.

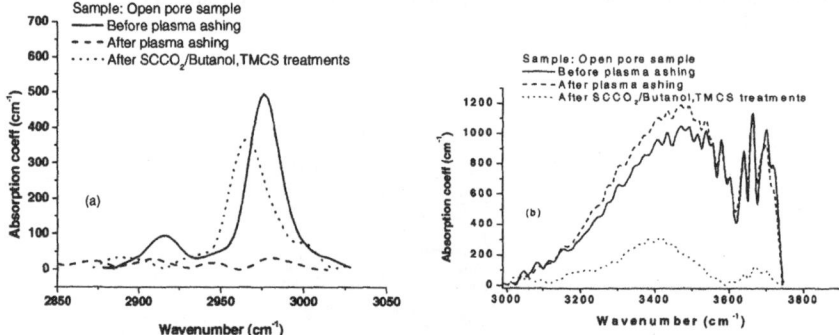

Fig.3 (a) FT-IR of CH$_x$ and (b) -OH absorption bands for open pore film before oxygen ashing, after ashing and after SCCO$_2$/Butanol, TMCS treatments

Figs. 4(a) and (b) show the CH and OH absorption bands of the closed pore sample (25wt% loading) before plasma damage, after plasma damage and after SCCO$_2$/Butanol, TMCS treatments. The results here show identical trend with results shown in Figs. 3 (a) and (b). These results imply that supercritical fluid treatments proposed in this investigation is effective in repairing the plasma damage not only on open pore structure but also, on closed pore morphologies. However, the effectiveness of supercritical fluid in re-incorporating methyl groups is better on open pore films compared with closed pore structure.

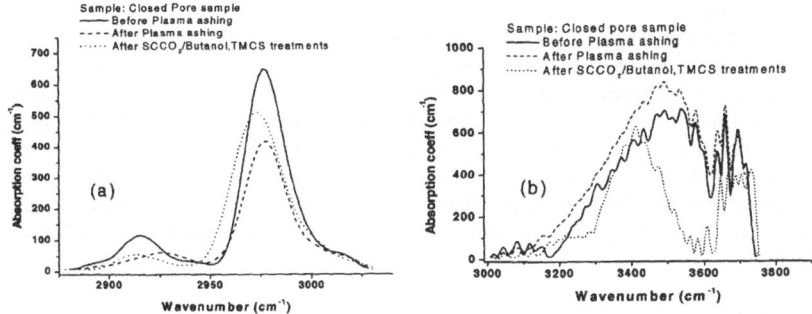

Fig. 4(a). FT-IR of CH and (b) –OH absorption bands for closed pore film before oxygen ashing, after ashing and after $SCCO_2$/Butanol, TMCS treatments

We have also used hexamethyldisilazane (HMDS) in $SCCO_2$ as silylating agent to cure plasma-damaged nanoporous films. The results shows that $SCCO_2$ as a medium for silylation with HMDS is not effective in incorporating CH-groups as shown in Fig. 5. Another problem in using HMDS is the inability in removing all the water content in the films. We observed that 45% OH remained in the films. J.R Combes et al.[12] have shown that HMDS reacts with CO_2 at 3,000 psi to form ammonium carbamate and some of this by-product is then adsorb on the surface hydroxyl groups. This blocks the silylation reactions sites and may be the reason for incomplete water removal and no incorporation of CH-groups.

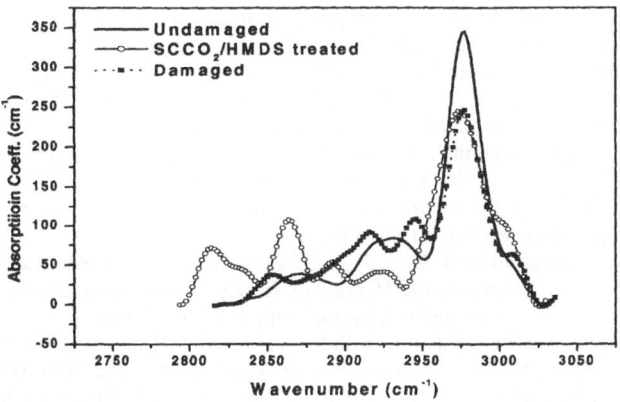

Fig. 5 CH-absorption band of HMDS treated film

Conclusion

We have shown various studies using $SCCO_2$ for applications in microelectronics. $SCCO_2$ treatments on spin coated samples shows that pores can be generated in order to decrease the dielectric constants. We have also shown the ability of $SCCO_2$ to be used as reaction medium for silylation in order to modify dielectric surfaces. $SCCO_2$ has shown its versatility in various applications in semiconductor device fabrication and may become an indispensable tool in the future. Our future investigations will focus in using different solvents for removing different porogens and decrease the extraction time. Our group is also looking at the possibility of using $SCCO_2$ as a deposition tool for dielectric materials

Acknowledgement

We would like to acknowledge J. Sun, D.H. Huang and S.L Simon in helping us in some experiments. The SRC and NSF are acknowledged for providing funding to conduct this research.

References

1. Dateline Los Alamos, Fall 2001.
2. M.A. Biberger, P. Schilling, D. Frye and E. Mills, Semiconductor FabTech 12th ed. p. 239-243.
3. S. Ogawa, T. Nasuno, M. Egami and A. Nakashima, "Formation of Mechanically Strong Low-k Film using Supercritical Fluid Dry Technology", International Interconnect Technology Conference, June 3-5, 2002.
4. M.D. Loque de Castro, M. Valcarcel and M.T. Tena, Analytical Supercritical Fluid Extraction, Springer-Verlag, Berlin Heidelberg (1994).
5. L.T. Taylor, Supercritical Fluid Extraction, John Wiley & Sons, Inc, USA (1996)
6. B. Lahlouh, T. Rajagopalan, J. A. Lubguban, N. Biswas, S. Gangopadhyay, J. Sun, D. Huang, S. L. Simon, H. C. Kim, W. Volksen and R. D. Miller, "Creating Nanoporosity By Selective Extraction of Porogens Using Supercritical Carbon Dioxide/Co-solvent Processes", Proceeding for Materials Research Society Spring 2003
7. T. Rajagopalan, B. Lahlouh, J. A. Lubguban, J. Sun, D. H. Huang, N. Biswas, S. L. Simon, S. Gangopadhyay, A.Mallikarjunan, H. -C. Kim, W. Volksen, M. F. Toney, E. Huang, P. M. Rice, E. Delenia and R. D. Miller, Appl. Phys. Lett. 82, 4328 (2003)
8. J. A. Lubguban, B. Lahlouh, T. Rajagopalan, N. Biswas, S. Gangopadhyay, J. Sun, D. H. Huang, S. L. Simon, A. Mallikarjunan, H.-C. Kim, J.Hedstrom, W. Volksen, R. D. Miller, and M. F. Toney, *Supercritical CO_2 Extraction of Porogen Phase: An Alternative Route to Nanoporous Dielectrics*, to be submitted
9. B. Lahlouh, G. Sivaraman, R. Gale, J. A. Lubguban and S. Gangopadhyay, "*Silylation using Supercritical Carbon Dioxide Medium to Repair Plasma- damaged Porous Organosilicate Films*", submitted to Electrochemical Society Letters
10. T.C. Chang, C.W. Chen, P.T. Liu, Y.S. Mor, H.M. Tsai, T.M. Tsai, S.T. Yan, C.H. Tu, T.Y. Tseng and S.M. Sze, *Electrochem. and Solid State Lett.*, **6,** F13 (2003).
11. T.C. Chang, Y.S. Mor, P.T. Liu, T.M. Tsai, C.W. Chen, Y.J. Mei, and S.M. Sze, *J. Electrochem. Soc.*, **149,** F81 (2002).
12. J.R. Combes, L.D. White, and C.P. Tripp, Langmuir, 15 (1999) 7870

Mat. Res. Soc. Symp. Proc. Vol. 812 © 2004 Materials Research Society

Comparative Studies of Ultra Low-*k* Porous Silica Films with 2-D Hexagonal and Disordered Pore Structures

Nobutoshi Fujii[1], Kazuhiro Yamada[1], Yoshiaki Oku[1], Nobuhiro Hata[2], Yutaka Seino[2], Chie Negoro[2] and Takamaro Kikkawa[2, 3]
[1]MIRAI-ASET, Tsukuba, Japan
[2]MIRAI-ASRC-AIST, Tsukuba, Japan
[3]RCNS, Hiroshima Univ., Higashi-Hiroshima, Japan

ABSTRACT

Periodic 2-dimensional (2-D) hexagonal and the disordered pore structure silica films have been developed using nonionic surfactants as the templates. The pore structure was controlled by the static electrical interaction between the micelle of the surfactant and the silica oligomer. No X-ray diffraction peaks were observed for the disordered mesoporous silica films, while the pore diameters of 2.0-4.0 nm could be measured by small angle X-ray scattering spectroscopy. By comparing the properties of the 2-D hexagonal and the disordered porous silica films which have the same porosity, it is found that the disordered porous silica film has advantages in terms of the dielectric constant and Young's modulus as well as the hardness. The disordered porous silica film is more suitable for the interlayer dielectrics for ULSI.

INTRODUCTION

In order to reduce interconnect delays for high-speed ULSIs, it is necessary to develop ultra-low-*k* materials whose dielectric constants are less than 2.5. The mesoporous silica films using a sol-gel self-assembly technique with a surfactant as the template are formed for the interlayer dielectrics (ILD). The pore structures have been controlled by the sol-gel self-assembly technique using various surfactants for forming the 2-D hexagonal, the cubic, the 3-D hexagonal and disordered phases [1-5]. A 2-D hexagonal periodic structure has been developed for ILD, however, it is difficult to control the orientation of pore channels in 2-D hexagonal porous silica films, resulting in the possibility of anisotropic dielectric constant.

Therefore, a disordered mesoporous silica film, which has no periodic pore structures, was formed by controlling the static electrical interaction between the micelle and the silica oligomer. In this study, the 2-D hexagonal and disordered porous silica films were produced and compared in terms of the dielectric constant and the Young's modulus to reveal the potential of the material as ultra-low-*k* dielectrics.

EXPERIMETNTAL PROCEDURE

The precursor of the 2-D hexagonal and the disordered porous silica films were synthesized based on the raw materials such as tetraethyl-orthosilicate (TEOS), ethanol, H_2O, HNO_3 as acid catalyst and the tri-block copolymer $((EO)_x(PO)_y(EO)_x$; where (EO) is polyethylene oxide and (PO) is polypropylene oxide block. These surfactants, Pluronic^{TM}, such as P45 $[(EO)_{13}(PO)_{20}(EO)_{13}]$, F485 $[(EO)_{74}(PO)_{20}(EO)_{74}]$, and P103 $[(EO)_{15}(PO)_{55}(EO)_{15}]$ were employed in this study (Table 1). F485 ascertains to form the periodic pore structure by the (EO) blocks, which periodically assemble via the interaction with the silica oligomer. On the other hand, the (EO) blocks of P45 makes difficult to form the periodic pore structure when the ratio of surfactant/Si decreased to obtain the desired k value. However, P45 and F485 form the same pore diameters because the pore diameter does not depend upon the (EO) length but the numbers of the (PO) block under the same condition of the silica oligomer [6]. The mixtures were agitated at 298 K for 24 hours. The molar ratio of surfactant/Si in each solution was varied from 0.017 to 0.1, depending on the desired k-value of the disordered and 2-D hexagonal porous silica films. In order to control the static electrical interaction of assembling disordered pore structure, dimethyl(diethoxy)silane (DMDEOS) was added in the precursor solutions.

The precursor as described above was spin-coated on a Si substrate to form the uniform thickness. Then, the substrates were calcined using a vertical furnace at 673 K in the dry air. To make the porous silica films hydrophobic, hexamethyldisilazane (HMDS) vapor was injected into the chamber of the furnace at the highest calcination temperature and treated for 30 minutes.

In order to investigate the properties of these porous silica films, spectroscopic ellipsometry (SE), X-ray diffraction, nanoindentation and transmission electron microscopy (TEM) were used for determining film thicknesses and refractive indices, pore sizes, mechanical strength and cross sectional view, respectively.

Table 1. Data of molecular weights, the numbers of EO and PO, and the surface tensions for Pluronic^{TM} P45, F485 and P103.

Surfactant	Molecular Weight	(EO)	(PO)	Surface Tension (mN/m) (0.1wt%)
P45	2400	13	20	46.3
F485	8000	74	20	53.3
P103	4950	17	60	34.4

RESULTS AND DISCUSSION

Figure 1 shows the X-ray diffraction patterns from 2-D Hexagonal pore structured silica film formed with F485 and P103. However, no diffraction peaks were observed by X-ray diffraction measurement from the disordered porous silica films. Figure 2 shows a typical pore diameter distribution for the disordered porous silica film produced with using P45 as determined by small angle X-ray scattering (SAXS) [7]. The cross sectional TEM images of the P45 and P103 disordered porous silica films are shown in Fig. 3. The periodic structures are not confirmed except the top surface and the bottom interface of the P103 film. In these images, the black points show the images of the pores colored by ruthenium oxide (RuO_4), which was induced in the pores to observe the pores by using TEM. The results of the XRD and the TEM images are consistent with each other. No ordered structures were observed in the porous silica film. SAXS measurements indicated that these pores sizes were not random but rather homogeneous. The disordered pore structure is considered as a wormhole-like network having uniform pore diameter.

The electrical and mechanical measurements were carried out to estimate the values of the dielectric constant, Young's modulus and hardness. The observed values of the periodic and disordered porous silica films were compared in Table 2.

The result shows that the dielectric properties of sample (A) with 2-D hexagonal pore structure and sample (B) with disordered structure are very similar. On the other hand, the samples (C) and (D) were produced by using the same surfactant P103 but had different structures such as 2-D hexagonal pore structure and disordered structure, respectively. The ratios

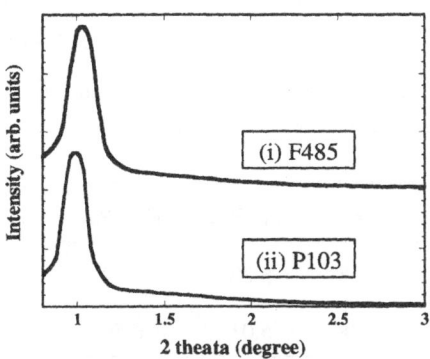

Figure 1. X-ray diffraction patterns from 2-D hexagonal silica films formed with F485(i) and P103(ii).

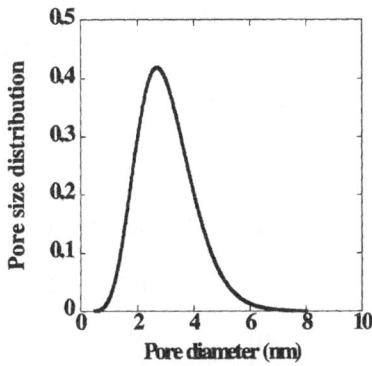

Figure 2. Pore size distribution of a P45 disordered porous silica film determined by small angle X-ray scattering.

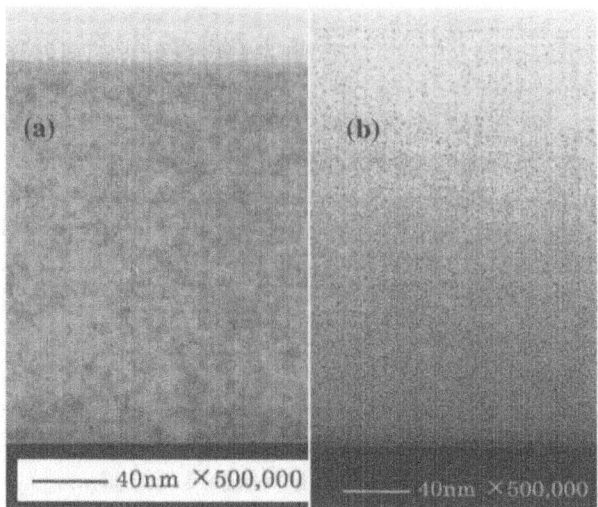

——— 40nm ×500,000   40nm ×500,000

Figure 3. Transmission electron microscopy (TEM) images of disordered porous silica films. P45 (a) and P103 (b) were used as templates to form pores, respectively. There are no remarked periodic images except for the top surface and the bottom interface in (b).

of the surfactant/Si for the samples (C) and (D) were equal, while the observed values of k and E were different. At this point, it should be noted that disordered structures are considered to have a wormhole-like structure with vertical pore channels. Then, the origin of the difference in the mechanical strength could be attributed to the fact that the disordered porous structure includes vertical pore channels, which makes the sample (D) mechanically harder against the vertical stress than the sample (C). Another reason such as siloxane networks should be discussed. Siloxane networks of sample (D) grow faster than sample (C) due to doped (CH_3). It is known that DMDEOS affects the reactivity of the monomers and makes it more significant [8], so that the reactivity of DMDEOS in the sol-gel solution, is much higher than that of TEOS. In fact, the change of the network type of [-O-Si-O-] in the mixture of TEOS and DMDEOS was observed by means of

Table 2. Values of the dielectric constant (k), the Young's modulus (E) and the hardness (H) for 2-D Hexagonal and disordered mesoporous silica films formed by different surfactants.

Sample	Surfactant	Pore structure	k	E (GPa)	H (GPa)
(A)	F485	2-D Hexagonal	2.25	5.20	0.52
(B)	P45	Disorder	2.28	5.01	0.50
(C)	P103	2-D Hexagonal	2.22	3.15	0.38
(D)	P103	Disorder	2.14	4.14	0.40

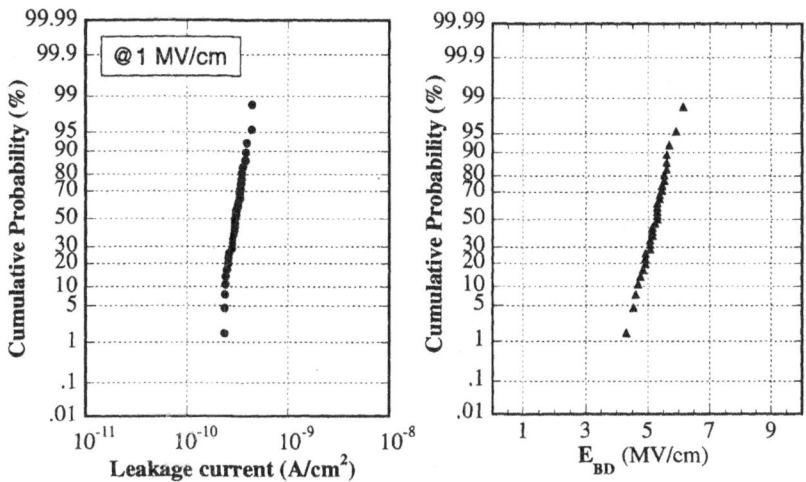

Figure 4. Leakage current density at 1MV/cm and dielectric breakdown field strength for disordered porous silica films formed by P45.

^{29}Si NMR [8]. Thus, the DMDEOS doped sol-gel solution makes the density of the silica wall higher than the case of TEOS solution. Moreover, the hydrophobicity was given by CH_3 and the dielectric constant of the disordered porous silica film is kept unchanged without uptake of moisture. These results suggest that the mechanical strength of porous silica films depend on the pore structure and the siloxane network formed through the reaction of the hydrolyzed TEOS and DMDEOS.

Figure 4 shows the cumulative probability of the leakage current density at an electrical field of 1 MV/cm and the dielectric breakdown field strength for the disordered porous silica films formed by P45. Both of them show excellent insulating properties. The integration of the disordered porous silica film with Cu damascene was successfully demonstrated [9]. We conclude that our disordered porous silica film is one of the most promising ultra-low-*k* materials in 45 nm technology node and beyond.

CONCLUSION

The mechanical strength of two kinds of pore structures such as 2-D hexagonal and disordered porous silica films was investigated. For the same molar ratios of the surfactant/Si, the mechanical strength is not always necessarily the same. The disordered porous silica was harder than the 2-D hexagonal although they were formed by the same surfactant P103. It is considered that the mechanical strength of the porous silica is influenced by the pore structure and the degree of cross-linkage of the siloxane networks. The disordered porous silica films with the methyl-group have advantages not only in the dielectric constant but also in the mechanical strength.

ACKNOWLEGEMENTS

This work was supported by NEDO (New Energy and Industrial Technology Development Organization). The Japan Science and Technology Agency is acknowledged for the Cooperative System for Supporting Priority Research.

REFERENCES

[1] C. Jeffrey Brinker, Yunfeng Lu, Alan Sellinger and Hongyou Fan, Adv. Mater. **11,** 579 (1999).

[2] K. Yamada, Y. Oku, N. Hata, S. Takada and T. Kikkawa, Jpn. J. Appl. Phys. **42,** 1840 (2003).

[3] Q. Huo, D. Margolese, P. Feng, T. Gier, P. Siegr, R. Leon, P. Petroff, F. Schuth and G. D. Stucky, Chem. Mater. **6,** 1176 (1994).

[4] R. Ryoo, J.M. Kim, C. H. Koo and C. H. Shin, J. Phys. Chem. **100,** 17718 (1996).

[5] T. Linssen, K. Cassiers, P. Cool and E. F. Vansant, Adv. Colloid and Int. Sci. **103,** 121 (2003).

[6] K. Flodström and V. Alfredsson, Microporous and Mesoporous Mater. **59,** 167 (2003).

[7] N. Hata, C. Negoro, S. Takada, Y. Oku and T. Kikkawa, Mater. Res. Soc. Symp. Proc. **766,** 191 (2003).

[8] J. Brus, J. Dybal, Polymer **40,** 6933 (1999).

[9] Y. Oku, K. Yamada, T. Goto, Y. Seino, A. Ishikawa, T. Ogata, K. Kohmura, N. Fujii, N. Hata, R. Ichikawa, T. Yoshino, C. Negoro, A. Nakano, Y. Sonoda, S. Takada, H. Miyoshi, S. Oike, H. Tanaka, H. Matsuo, K. Kinoshita and T. Kikkawa, IEDM 2003, 139.

Mat. Res. Soc. Symp. Proc. Vol. 812 © 2004 Materials Research Society   F5.2

3-dimensional evaluation of nm-pores in porous low-k films using TEM stereoscopic / electron tomographic observation method

J. Shimanuki, and Y. Inoue
NISSAN ARC, LTD., 1 Natsushima-cho, Yokosuka 237-0061, Japan
M. Shimada, and S. Ogawa
Semiconductor Leading Edge Technologies, Inc., 16-1 Onogawa, Tsukuba 305-8569, Japan

ABSTRACT

Stereoscopic and electron tomographic observation methods using Transmission Electron Microscope (TEM) were examined to characterize three-dimensionally a shape and size of pores and spatial distribution in porous low dielectric constant (low-k) films. In a case of TEM observation, nm size pores in an amorphous film are difficult to be imaged since contrast from the amorphous layer affect the pore imaging. An optimum image capture method at a modified electron beam condition was studied in which the amorphous contrast from the porous low-k film is weakened to enhance scattering contrast from the pores and a matrix. As a result, a stereoscopic observation and 3-D reconstruction images clarified that the shape of pores was not spherical but distorted and almost pores were partially connected. For measuring pore size and spatial distribution, connected pores were segmented into smaller pores by separating at the narrowest part. From images after segmentation, it is indicated that there existed more and larger pores at interfaces than a center area. It was found that the pores did not homogeneously distribute in the film but concentrated at the interfaces.

INTRODUCTION

Amorphous porous low-k films in which nm size pores are introduced have been expected to be a next generation ultra low-k materials as a dielectric interlayer for multilevel Cu interconnects below a 65nm node [1]. It is notable that shape and size of pores and spatial distribution in the films may affect mechanical and electrical properties of the interconnects during BEOL processes such as cleaning, metallization, and CMP, and packaging processes [2]. Several methods for characterizing the pore distribution have been eagerly studied, e.g. high-resolution specula X-ray reflectivity, small-angle neutron scattering, and gas-absorption [3], however these methods are not able to detect a pore shape and spatial distribution. On the other hand, since contrast from an amorphous layer affects pore imaging, pores are difficult to be imaged in an amorphous film. Thus only a few works on the TEM observation of the pores have been reported [4], while TEM has a high potential ability to observe nm size structures directly. In general, TEM images are 2-dimensional transmitted, so it is difficult to obtain sufficient information of pores in the amorphous layer for the three-dimensional imaging.

In this work, two methods of the three-dimensional TEM observation techniques were investigated for the three-dimensional characterization of nm-pores in porous low-k films.

One is a stereoscopic observation method and the other is an electron tomography. In the former method, the three dimensional image was taken by superposing only two images, which were obtained by tilting a specimen in two different angles. The shape and size of pores and spatial distribution in the films were characterized qualitatively. In the latter method, several tens images were taken by tilting the sample from low to high angles using a special TEM holder. 3-D reconstruction and image quantitative analysis were done by a commercial software.

EXPERIMENTAL

Porous low-k films spin-coated in a Si-substrate / SiO2 (500nm) / SiC (50nm) / porous low-k film (250nm) / SiC (50nm) stack with commercially adequate cure treatment were evaluated. TEM samples were prepared by using a SMI9200 Focused Ion Beam Microscope (Seiko Instruments Inc.), and TEM observations were carried out using two electron microscopes. One was H-9000UHR high-resolution electron microscope (Hitachi Co.) operated at 300kV, and the other was TECNAI G2 F20 (FEI Co.) operated at 200 kV. For a TEM stereoscopic observation, H-9000UHR was used to obtain a series of stereo-pairs. Through a preliminary experiment, it was found that optimum angles for tilt to get stereoscopic images were +/- 5 degrees. Obtained TEM micrographs were digitized using a Film Scanner LS-4500AF (Nikon Co), and images were analyzed using an image editing software. For the analysis, if necessary, the contrast of TEM images was enhanced using software. For electron tomography, TECNAI G2 F20 was used. In this method, several tens images have been collected with tilt angles from -65 degree to +65 degree (2 degree steps) by using GIF imaging filter (Gatan Co.). Automatic data collection was quickly done in order to avoid damage generation into the porous low-k film by electron beam. An optimum image capture method at a modified electron beam condition was developed in which the amorphous contrast from the porous low-k film is weakened to enhance scattering contrast from the pores and matrix. A 3-D reconstruction and quantitative analysis of obtained images were done by a 3-D structure analysis algorithm software.

RESULTS AND DISCUSSION

TEM stereoscopic observation

In a case of TEM observation, nm size pores in the amorphous film are difficult to be imaged since contrast from the amorphous layer affect the pore imaging. Accordingly, an optimum image capture method at modified electron beam condition was developed in which the amorphous contrast from the porous silica film is weakened to enhance scattering contrast from the pores and matrix.

Figure 1 shows a pair of stereoscopic micrographs of porous low-k films after 400 degree C, 0.5 hours cure treatment. High density of white granules is recorded in the micrographs. By superposing two micrographs (a) and (b), it is clearly seen that these white granules are three dimensionally distributed within the film. This fact indicates that the white granules are pores in the amorphous silica matrix, and are not surface damages (or artifact) produced by the FIB sample preparation. It is worthy to note that the stereoscopy makes it possible to examine not only shape of individual pores but also spatial connection between (or among) pores in detail. The obtained micrographs showed shapes of pores were not spherical, but random, and that they were connected in place. Further, spatial distributions of pores in the films were uneven. This type of 3-D analysis would be not possible by a conventional projected images observation.

The TEM stereo observation is simple and convenient method, since this method does not need complicated transformation and correction in comparing with an electron tomography, which will be discussed next. Furthermore, in this work, the stereoscopic micrographs were employed as a reference to create the 3-D image at the same time.

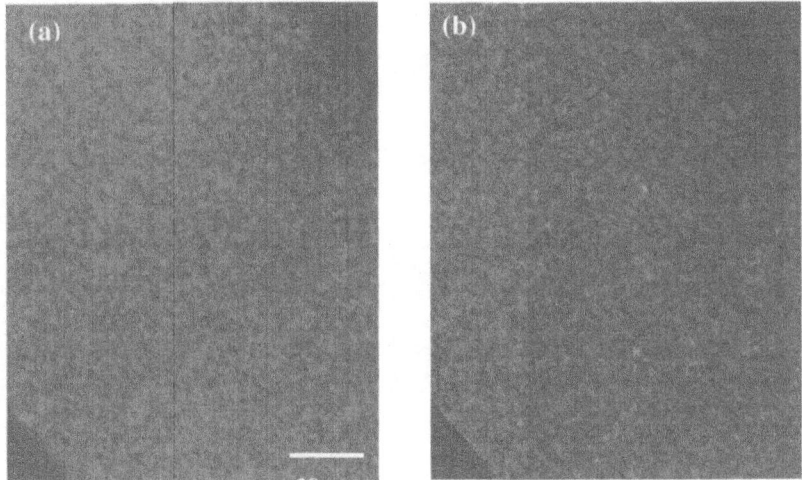

Fig.1 A pair of stereoscopic micrographs of the porous low-k films. (a) Tilt angle -5 degree, (b) tilt angle +5 degree

Electron tomography

In an electron tomography, complicated processes to obtain images are required, but it has various merits, such as visualization using P.C. (rotation image, virtual 2-D slice image, etc), and quantitative evaluation.

Figure 2 shows volume rendering of the 3-D reconstruction of the pores in the porous low-k films using the electron tomographic technique. In fig.2, white contrast regions show lower mass density in the sample; namely, it was recognized as pores in the film.

Shape of pores was distorted similar to stereoscopic observation results. Quality of images is restricted because of a 2-D display here, but a reconstructed 3-D animation image shows shape of pores clearly.

To examine size of pores and spatial distribution in the film quantitatively, the pores were modeled by extracting pores from the 3-D image. Figure 3 shows volume rendering of the 3-D reconstruction of the modeled pores in the porous low-k films using electron tomographic technique, in which modeled region: 100pixel*100pixel*100pixel, 1pixel=0.6nm. Connected pores were displayed by the same colors and isolated pores were displayed by the different colors. Total volumes of pores were supposed to be 30% in the modeled region.

In fig.3, shape of pores was distorted similar to fig.2, and almost pores were connected.

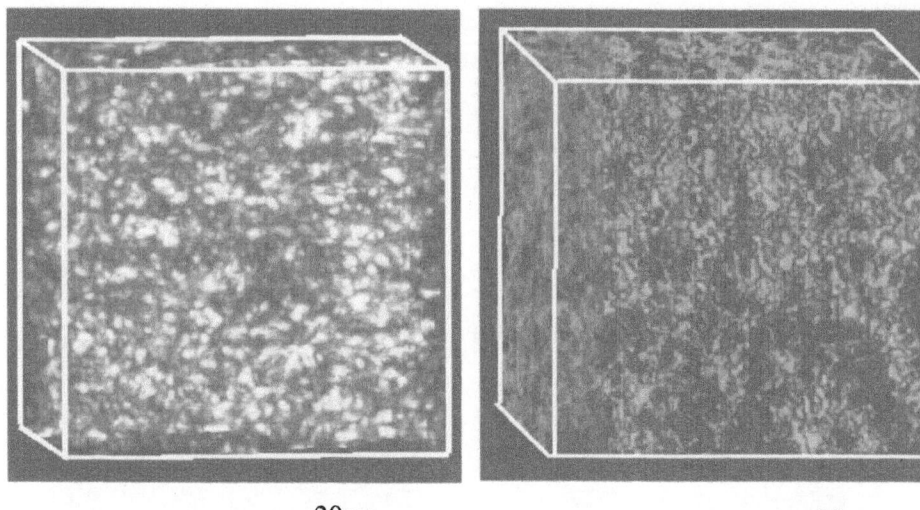

20nm   20nm

Fig. 2 Volume rendering of the 3-D   **Fig. 3** Volume rendering of the 3-D
reconstruction of the pores in the   reconstruction of the modeled pores
porous low-k films   in the porous low-k films

Figure 4 shows a low magnification image of the 3-D reconstruction of pores in the porous low-k film. In fig.4, almost pores are homogeneously connected. Since the software recognize connected pores as one huge pore, we introduced a 3-dimensional segmentation method and characterized the distribution of pore size and spatial distribution in the films to compare the results from other method such as SXR.

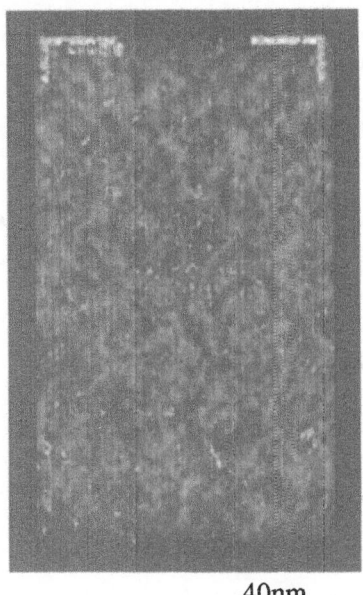

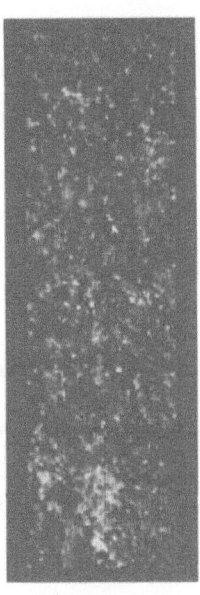

40nm   40nm

Fig.4 Volume rendering of the 3-D reconstruction of the modeled pores in the porous low-k films (before segmenting)

Fig.5 A part of Volume rendering of the 3-D reconstruction of the modeled pores in the porous low-k films (after segmenting)

Figure 5 shows a part of volume rendering of the 3-D reconstruction of the modeled pores in the porous low-k film after segmenting. In fig.5, connected pores were segmented into smaller pores by separating at the narrowest part. Since isolated pores were displayed by the different colors, it was possible to confirm that segmentation was successful.

Figure 6 shows distribution of volume of pores in the porous low-k film calculated from the 3-D reconstruction image in Fig 5. Quantitative analysis was performed using a 3-D structure analysis algorithm for the segmented connected pores. In fig.6, it is impossible to evaluate smaller pores exactly, because it is difficult to remove influence from amorphous contrast and image noise due to image processing. In fig.6, it showed remarkable peak around $4nm^3$, using spherical approximation, diameter of pore is equivalent to 2nm nearly. In this fact, it has been suggested that it is possible to characterize a size distribution of nm-sized pores.

Figure 7 shows spatial distribution of pores in the porous low-k film calculated from the 3-D reconstruction image. Number of pores (open squares) and volume of pores (solid circles) tended to increase at interfaces. It has been indicated that there existed more and larger pores at the interfaces than the center area. It was found that the pores did not homogeneously distribute in the film but concentrated at the interfaces.

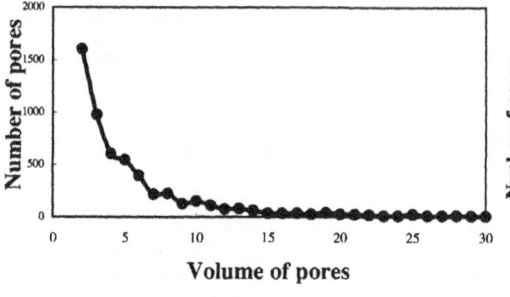

Fig.6 Distribution of volume of pores in the porous low-k film calculated from the 3-D reconstruction image.

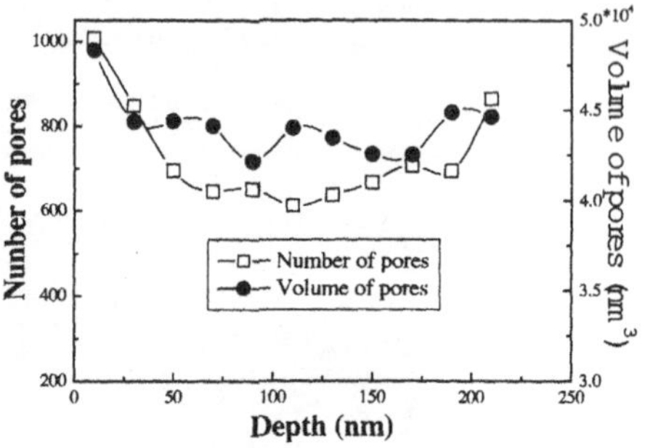

Fig.7 Spatial distribution of pores in the porous low-k film

CONCLUSION

Stereoscopic and electron tomographic observation methods using TEM were successfully applied to characterize three-dimensionally the shape and size of pores and spatial distribution in the porous low-k films for the first time. As the results, stereoscopic observation and the 3-D reconstruction images clarified that the shape of pores was not spherical but distorted shape and almost pores were partially connected. For measuring pore size and spatial distribution, the connected pores were segmented into smaller pores by separating at the narrowest part. From obtained images after segmenting, it is indicated that there existed more and larger pores at the interfaces than the center area. It was found that the pores did not homogeneously distribute in the film but concentrated at the interfaces. The more and larger pores at the interfaces would weaken mechanical properties of the interconnect structure.

It was clarified that novel knowledge of pores in the porous low-k films was obtain by 3-dimensional TEM observation technology.

REFERENCES

[1] L.W. Hrubesh et al., J. Mater. Res. 8, 1736 (1993)
[2] M. Rasco et al., in proc. 2002 IITC, P.74 (2002)
[3] W. Wu et al., J. Appl. Phys. 87, 1193 (2000)
[4] S. Ogawa et al., in proc. 2003 IITC, P.100 (2003)

Mat. Res. Soc. Symp. Proc. Vol. 812 © 2004 Materials Research Society   F5.4

Ellipsometric Porosimetry of Porous Low-k Films with Quazi-Closed Cavities.

Mikhail R.Baklanov, Konstantin P.Mogilnikov[1], Jin-Heong Yim[2]
IMEC, Kapeldreef 75, B-3001 Leuven, Belgium
[1] SOPRA, Bois Colombes, France
[2] Samsung Advanced Institute of Technology (SAIT), Kyungki-do, South Korea

ABSTRACT

Evaluation of quasi-closed cavities connected with air through narrow necks is discussed. These cavities behave as closed pores when they are studied by Positron Annihilation Lifetime Spectroscopy (PALS). The reason is a short lifetime of o-positronium (Ps) and energy barrier that exist for Ps diffusion from large pores (d>3 nm) to small ones (d<3 nm). It is shown that more comprehensive information can be obtained using adsorption porosimetry. Standard adsorptives used in adsorption porosimetry have infinite lifetime allowing complete penetration and filling all the cavities during the measurement. Calculation of the neck and cavity sizes is based on the theory of metastable adsorption phases developed by Derjagin, Broekhoff and de Boer (DBdB). Results of evaluation are in good agreement with data obtained by SEM and TEM.

INTRODUCTION

Introduction of porous low dielectric constant (low-k) films into ultralarge scale integrated (ULSI) technology meets number of controversial requirements [1]. Increase of porosity decreases the k-value but reduces strength/stiffness properties. Decrease of the pore size facilitates their sealing but the materials with small pore size and high porosity have thin wall between the pores that deteriorates mechanical properties and their stability. For this reason, several companies have been developing porous films with relatively large embedded voids. Although there is a hope that these voids/cavities might be closed, most of these materials have narrow necks, connecting the voids with air. The "necks" might be a constitutive property of the matrix material and/or might form during the porogen removal. Evaluation of the pore structure of these materials is quite complicated. The techniques based on Small Angle Neutron and X-ray scattering have problems to resolve the micropores/necks because of the not well-defined density contrast between the walls and pores. PALS is extremely sensitive to small pores and shows bi- or multimodal porosity (voids and necks). However, Ps diffusion from large voids to narrow "necks" has energetic limitations and the cavities connected through narrow necks behave as "closed" for the Ps diffusion [2]. Therefore, the percolation threshold and the diffusion length determined by PALS reflect Ps properties and the pores closed for Ps still can facilitate gas, solvent, and Cu diffusion.

The k-value is only parameter that is directly defined by porosity. The pore size and structure are important for feasibility of the films in ULSI technology because they define the sealing ability and their behavior during different technological steps. Therefore, it is important to evaluate the pore structure using more "realistic" probes with properties similar to chemistries used in technology. EP uses stable organic adsorptives like toluene, isopropyl alcohol etc. Therefore, the measurement time can be long enough allowing the adsorptive permeation through the narrow necks and adsorption/condensation in the cavities. In this paper, we demonstrate evaluation of such films with quazi-closed cavities by EP with toluene. Special samples with the same matrix material and porosity equal to 30% but different pore structure have been prepared by Samsung Advanced Institute of Technology (SAIT) using controlled change of functional groups in the cyclodextrin (CD) molecules [3].

EXPERIMENTAL DETAILS.

The porous low-k films have been prepared by the following way. The precursor solutions were prepared by properly mixing the siloxane-silsesquioxane hybrid polymer as a matrix with various kinds of cyclodextrin compounds as a porogen and propylene glycol methyl ether acetate (PGMEA) as a solvent. The mixture was spin cast at 3000 rpm, 30 sec onto silicon wafers. The wafers were then subjected to a series of soft baking on a hot plate, including 1 minute at 150°C and another minute at 250°C to remove the organic solvent. The porous thin films were then produced by curing the wafers in a cylindrical furnace at 420°C for 60 min under vacuum [3]. The pore structure of the porous films is strongly related to interaction between the functional groups in the CD molecules.

The pore structure was investigated by FE-SEM (Field Emission Scanning Electron Microscopy) and TEM (Transmission Electron Microscopy: Hitachi H9000NA) at 300kV for the porous film powder specimens. SEM images were recorded on a Hitachi S4500 operating at 20kV. The film porosity and pore radius distribution were analyzed by EP [4].

MODELS FOR ADSORPTION OF FILM WITH SPHERICAL CAVITIES.

Pore geometry significantly affects thermodynamic properties of fluids, their adsorption/desorption and diffusion behavior. In the simple case of infinite cylindrical pores the capillary condensation is described by Kelvin equation for the cylindrical interface between the adsorbed layer and vapor,

$$RT \ln(P_a / P_0) = -\frac{\gamma V_L}{R_p - h_a} \tag{1}$$

while evaporation/desorption is related to the formation of a hemispherical meniscus between the condensed fluid and vapor,

$$RT \ln(P_d / P_0) = -\frac{2\gamma V_L}{R_p - h_d} \tag{2}$$

where P_0 is equilibrium vapor pressure of the adsorptive, P_a and P_d are pressures corresponding to the adsorption and desorption, respectively, γ and V_L are the surface tension and the molar volume of the liquid adsorptive. R_p is the pore radius, h_a and h_d are the thicknesses of the adsorbed layers on pore wall at the relative pressures P_a/P_0 and P_d/P_0. When h_a and h_d are much smaller than R_p, the width of hysteresis loop is defined by equation

$$(P_d / P_0) = (P_a / P_0)^2 \tag{3}$$

Condensation and desorption in spherical cavities connected through narrow cylindrical necks (ink-bottle pores) can be described by the DBdB theory (Derjaguin [5], Broekhoff and de Boer [6]). The equilibrium thickness of the adsorbed layer h in a spherical pore of radius R_p is determined from the balance of capillary and disjoining pressures [7]:

$$\Pi(h)\,V_L + \frac{2\gamma V_L}{R_p - h} = RT\ln(P_0/P) \qquad (4)$$

where $\Pi(h)$ is the disjoining pressure of the adsorbed layer. According to this equation, the capillary condensation in a spherical cavity, which is connected with the neighboring cavities through narrow necks is always associated with a hysteresis larger than expected by equation (3). During the adsorption, the isotherm traces a sequence of metastable states of the adsorption layer, and the capillary condensation occurs spontaneously when the layer thickness approaches the limit of stability. The critical thickness of metastable layers is determined from the condition

$$-\left(\frac{d\Pi(h)}{dh}\right)_{h=h_{cr}} = \frac{2\gamma}{(R_p - h_{cr})^2} \qquad (5)$$

Desorption from a cavity is delayed until the vapor pressure is reduced below the equilibrium desorption pressure from the neck, which leads to larger hysteresis effect in comparison with cylindrical pores. Thus, while the capillary condensation condition is determined by the size of cavity, the desorption condition is related to the neck size.

However, a different behavior can be observed if the neck size is much smaller than the cavities. The decrease in the vapor pressure may cause the fluid in the cavity to become thermodynamically unstable at a higher pressure than the equilibrium desorption pressure for the pore neck. In this case, the confined fluid may cavitate because of huge negative pressure [8] and evaporate spontaneously before the neck has opened. This effect would cause a sharp knee on the desorption isotherm. The neck size calculated from this slope is apparent and doesn't reflect the real necks. To establish the reality of the neck, it is useful to study the same material using different adsorptives with different molecular characteristics (Gurvitch test). The slopes appeared as a result of cavitation have to disappear or to show a different behavior.

Quantitative scheme of the processes occurring during the adsorption and desorption is demonstrated on Figure 1. The open triangles show adsorption/desorption hysteresis in ideal cylindrical pores. In this case the hysteresis loop is defined by equation (3). If the film contained cavities connected through micro-necks, the hysteresis loop is larger (dark circles) because reflects the size difference between the necks and voids.

First, all the pores and cavities are emptied in vacuum (1). When the adsorptive vapor is introduced into the chamber, the micropores/necks are filled by adsorptive at P/Po ≤ 0.4 (2). However, the pressure is still not enough for the condensation in the cavities and they started to be filled at the critical pressure for spherical voids (P/Po=0.75) (3). When we succeed the same P/Po=0.75-0.8 during the pumping down, the cavities can't be emptied because they are blocked by the liquid plug in the microporous necks that can be emptied only at P/Po=0.4-0,45. As soon as the necks are opened (P/Po<0.45), the adsorbate evaporates from the cavities very fast (4). If the neck size is very small, the cavities can be emptied even when the necks are closed because of effects of cavitation. Finally, all pores are completely emptied at zero pressure.

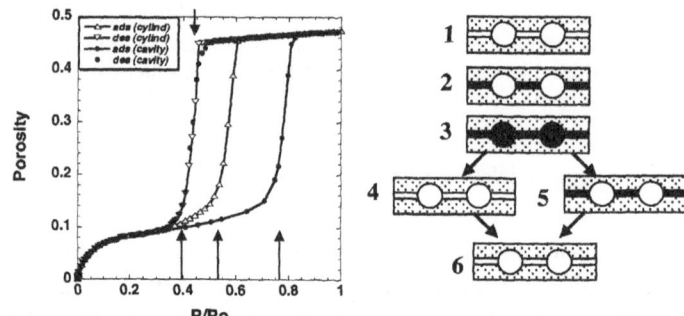

Figure 1. Adsorpton/desorption isotherms in a porous film with cylindrical pores (open triangles) and in a film containing quasi-closed cavities (dark circles). The necks in the film with quazi-closed cavities have the same size as in the film with cylindrical pores. The size of the cavities is calculated from the adsorption branch.

RESULTS AND DISCUSSIONS.

Figure 2 shows SEM pictures of the porous CSSQ films. The micropores are not visible in the film #12. Films #14 and #17 have internal cavities/voids. The void diameters are in the range between 10-40 nm in the film #14 and 20-100 nm in the film #17.

The same films have also been evaluated by EP. Figure 3 shows adsorption/desorption isotherms of toluene. One can see that the porosity of all three films is close to 30%. The presence of embedded cavities drastically changes the shape of hysteresis loop as it has been described in the previous section. All three samples have micropores/necks with radius 0.85 nm (#12), 0.8 nm (#14) and 0.75 nm (#17). Sample #17 has additional necks with radius 1.7 nm that probably form during the porogen removal. Samples 14 and 17 have cavities with quite wide radius distribution with max at 7 nm and 40 nm, respectively. The size of the cavities calculated from the adsorption branch well correlates with the results of SEM and TEM analysis. Similar isotherms are observed in porous SiLK films prepared by Dow Chemical. The calculated cavity sizes in p-SiLK are in good agreement with results obtained by Small Angle X-ray Scattering (SAXS) and PALS [9,10].

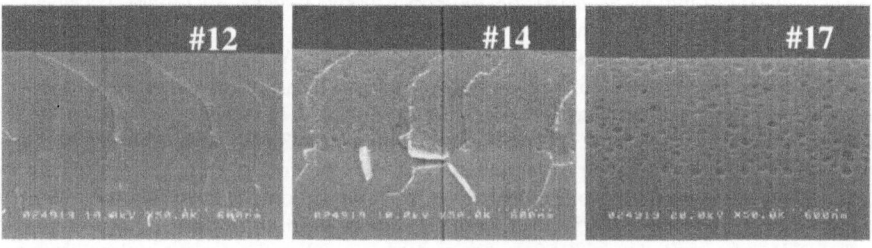

Figure 2. SEM pictures of CSSQ based porous films with quazi-closed cavities.

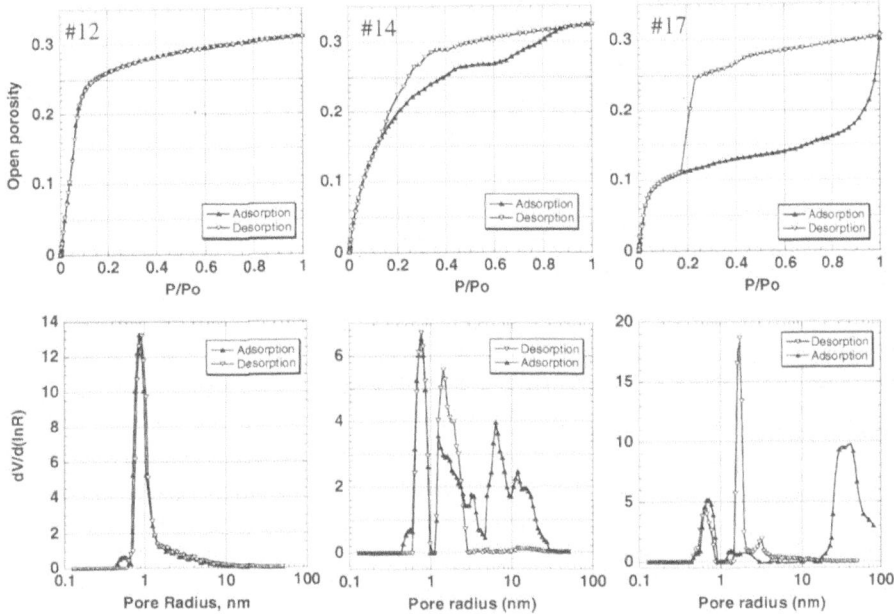

Figure 3. Adsorption/desorption isotherms of toluene and pore radius distribution in low-k films with micropores (#12) and cavities (#14, #17). The size of the cavities calculated from adsorption curves well correlates with the results of analysis of SEM images. The neck sizes in samples 12 and 14 are in good agreement with the results obtained by PALS [3].

The specific features of condensation lead to different responses of the film thickness to the capillary forces during the adsorption and desorption. Figure 4 shows change of the film thickness during the toluene adsorption in the film #12 and isopropyl alcohol (IPA) adsorption in the sample #17. Change of the film thickness in the case of sample #12 has a typical shape related to capillary shrinkage during the adsorption and desorption that is described by Young-Laplace equation and depends on Young Modulus of the film (E) and molar volume of the adsorptive [8] :

$$d = d_0(1 - \pi_c / E) = d_0 - k \cdot \ln \frac{p}{p_0}; \quad E = \frac{d_0 \cdot RT}{kV_L}, \qquad (6)$$

where π_c is capillary pressure, d_0 is the film thickness when the pores are completely filled by adsorptive, d is the current film thickness between the minimum point and maximum. In the case of a porous film with cavities, the capillary shrinkage happens only during the filling of the cavities (P/Po $\rightarrow$ 1) because of the stress relaxation due to the air in the cavities. Inversely, relaxation of the film thickness to the initial value happens at the pressure corresponding to the cavity empting (P/Po = 0.4-0.5). The slop at P/Po $\approx$ 0.4-0.5 observed during the IPA desorption from sample #17 corresponds to the neck with R $\approx$ 1.9 nm (1.7 nm for toluene, Figure 3). This fact suggests that the appearance of this peak is not related to the solvent cavitation.

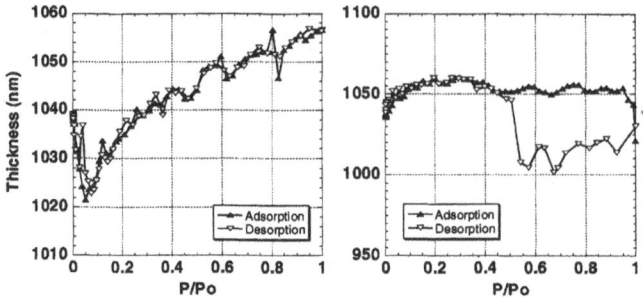

Figure 4. Typical changes of the film thickness during the adsorption and desorption in a film with cylindrical pores (a) and in the film with cavities (b). In this figure: toluene adsorption in the microporous sample #12 and IPA adsorption in the sample with cavities #17.

CONCLUSIONS

It is shown that Ellipsometric Porosimetry can be successfully used for evaluation of porous films with quasi-closed cavities. These cavities behave as closed pores when PALS is used for evaluation. The reason is the limitation of Ps diffusion in pores with complicated shape and short lifetime of positronium. Adsorption porosimetry provides a more realistic picture useful for ULSI technology because the adsorptives have properties similar to solvents used in technology.

ACKNOWLEDJEMENTS

It is our pleasure to thank Hironobu Shirataki of Asahi Kasei for preparation of low-k films with ordered cylindrical pores, Mihail Petkov of California Institute of Technology and K.Maex of IMEC for fruitful discussions.

REFERENCES

1. K.Maex, M.R.Baklanov, D.Shamiryan, F.Iacopi, S.Brongersma and Z.Sh.Yanovitskaya. *J.Appl.Phys.*, **93**, 8793 (2003).
2. K.P.Mogilnikov, M.R.Baklanov, D.Shamiryan and M.P.Petkov. *Jpn.J.Appl.Phys.* **43**, 243 (2004).
3. Jin-Heong Yim, Jong-Baek Seon, Hyun-Dam Jeong, Lyong Sun Pu, Mikhail R.Baklanov and David W.Gidley. *Adv.Funct.Mat.*, **14**, 277 (2004).
4. M.R.Baklanov, K.P.Mogilnikov, V.G.Polovinkin and F.N.Dultsev, *J.Vac.Sci.Technol.* **B18**, 1385 (2000);
5. B.V.Derjaguin. *Acta.Phys.-Chim.*, **12**, 181 (1940).
6. J.C.P.Broekhoff and J.H.de Boer. *J.Catal.*, **10**, 153 (1968).
7. P.I.Ravikovitch and A.V.Neimark. *Langmuir*, **18**, 1550 (2002).
8. K.P.Mogilnikov and M.R.Baklanov. *Electrochem.Sol.St.Lett.*, **5**, F29 (2002).
9. D.W.Gidley. *SEMATECH Porosimetry Workshop*, Santa Clara, 2003.
10. E.Huang. *SEMATECH Porosimetry Workshop*, Santa Clara, 2003.

Mat. Res. Soc. Symp. Proc. Vol. 812 © 2004 Materials Research Society

Effect of Mode-Mixity and Porosity on Interfacial Fracture of Low-k Dielectrics

Caroline C. Merrill[1] and Paul S. Ho
Interconnect and Packaging Laboratory, The University of Texas at Austin
10100 Burnet Road, PRC MRB 160, Austin, TX 78758, USA
[1]Intel Corporation 4100 Sara Road, F09-610, Rio Rancho, NM 87124, USA

ABSTRACT

In this study, we developed a system allowing interfacial adhesion measurements as a function of mode-mixity, from pure tension to pure shear. Results show that the debonding energy increases, by a factor of 3 to 10, as the amount of shear stress increases, approaching mode II conditions. For low k dielectrics, the debonding energy was found to decrease with increasing porosity and increase with increasing plasticity. The crack propagation is also dependent on mode-mixity.

INTRODUCTION

With device scaling continuing beyond the 130 nm node, low-k interlevel dielectrics (ILD) are being implemented to replace oxide in Cu interconnects. Their low modulus and high coefficient of thermal expansion can cause interfacial debonding during thermal cycling or CMP. The dual-damascene structure of the Cu interconnect can give rise to complex stress states at the Cu/ILD interface. Therefore, it is important to study interfacial adhesion as a function of mode-mixity, from pure tension to pure shear. In this study, we developed a system allowing interfacial adhesion measurements as a function of mode-mixity. This system was evaluated by measuring the adhesion of a porous MSQ-based spin-on-glass to different cap layers and comparing the results with previous results obtained with the four-point bending method. The effects of mode-mixity, porosity and plasticity were also investigated. Additionally, the crack propagation path as a function of mode-mixity was examined using analytical tools including SEM, FIB and AFM.

EXPERIMENTAL DETAILS

Mixed-mode loading technique

The mixed-mode loading apparatus, shown in figure 1, was developed using an approach originally conceived by G. Fernlund and J.K Spelt [1]. This design utilizes a double cantilever beam (DCB) geometry. The instrument allows interfacial fracture measurements for phase angles ranging from 0° (pure tension) to 90° (pure shear). By changing the positions of the different links, the forces, F_1 and F_2, applied respectively on the upper and lower beams, vary in the following manner:

$$F_1 = F\left(1 - \frac{s_1}{s_3}\right) \tag{1}$$

$$F_2 = F_1 \frac{s_1}{s_2} \frac{1}{\left(1 + \frac{s_3}{s_4}\right)} \tag{2}$$

[1] methylsilsesquioxane
[2] organo-silicate glass

Where s_1, s_2, s_3 and s_4 are the distances between two links as labeled in figure 1.

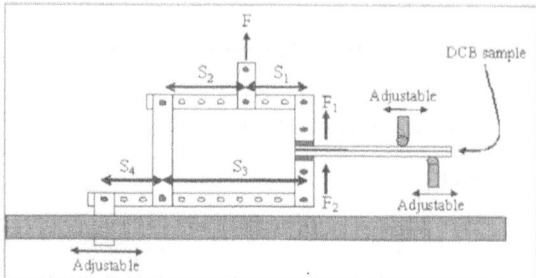

Figure 1. Mixed-mode loading fixture

The energy release rate per unit crack length can then be defined as:

$$G = \frac{(F_1 a)^2}{2D}\left[1+\left(\frac{F_2}{F_1}\right)^2 - \frac{1}{8}\left(1+\left(\frac{F_2}{F_1}\right)^2\right)\right] \tag{3}$$

$$D = \frac{Eh^3}{12} \text{ (flexural rigidity per unit width)} \tag{4}$$

The phase angle, Ψ, varies as function of the ratio F_1/F_2 as shown in equation (5):

$$\psi = \arctan\left[\frac{\sqrt{3}}{2}\frac{\left(\frac{F_1}{F_2}+1\right)}{\left(\frac{F_1}{F_2}-1\right)}\right] \tag{5}$$

s_1, s_2, s_3 and therefore F_1 can only be positive. s_4 and F_2 can be positive or negative. s_4 is negative if the connection between the lower drilled bar and the base plate is closer to the specimen than the connection between the two drilled bars.

Load is applied on the specimen at a constant displacement rate. As shown in figure 2, initially, the sample is pre-cracked so load increases linearly with displacement. This linear relationship corresponds to the body stiffness for the specific crack length of this sample. At a sufficiently high load, the crack starts propagating. Since G_c is constant, load decreases with increasing crack length. Every point in the crack extension region corresponds to a specific crack length and can therefore be used to calculate G_c.

Sample Preparation

The first step consists in preparing four-point bending samples [3]. One edge of each sample is sputtered with aluminum with a target thickness of 400 Å. A thin layer of gold ($\approx$50 Å) is immediately sputtered on the edge of each sample to avoid oxidation of the Al film. A small crack is introduced at the weakest interface using the 4-point bending technique [3,4,5]. Each sample is subsequently cleaved in half to obtain DCB samples. In order to measure the crack length, a piece of Si_3N_4 coated wafer is also coated with aluminum and gold along with every

[1] methylsilsesquioxane
[2] organo-silicate glass

batch of samples. It is diced with a wafer saw to replicate the edges of actual DCB samples with different crack lengths. Each replica is used to measure resistance as a function of crack length to obtain a calibration curve. A potential drop method based on a Wheatstone bridge circuit [2],

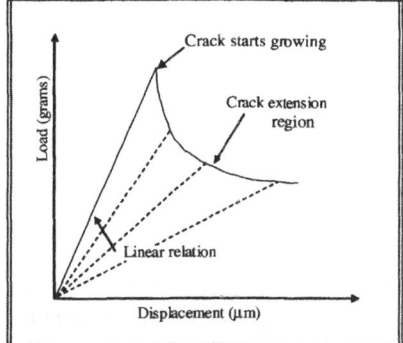

Figure 2. Characteristic load vs. displacement curve for DCB geometry

shown in figure 3, is used to measure the aluminum film resistance changes for each sample. The resistance change can be correlated with a specific crack length using the calibration curve. Using a two-part silver epoxy, two holders are glued on each side of the samples to facilitate their loading in the mixed-mode test fixture. Figure 4 summarizes the sample preparation process flow.

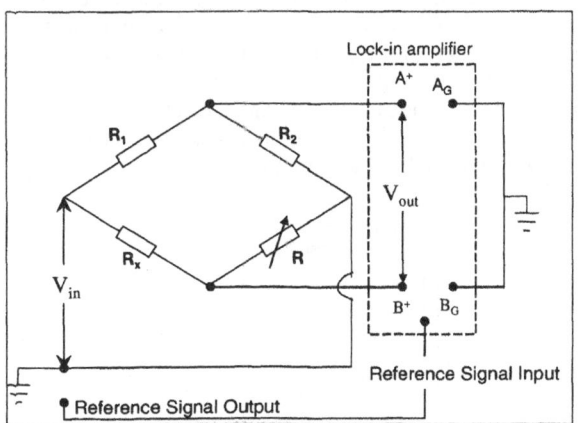

Figure 3. Wheatstone bridge set up to measure crack length

RESULTS

The four-point bending method [4], which is now widely used in industry and academia, was used as a reference to evaluate the reliability and accuracy of the mixed-mode fixture as a tool to measure interfacial adhesion. Nanoglass, a porous MSQ-based[1] spin-on OSG[2] was used for

[1] methylsilsesquioxane
[2] organo-silicate glass

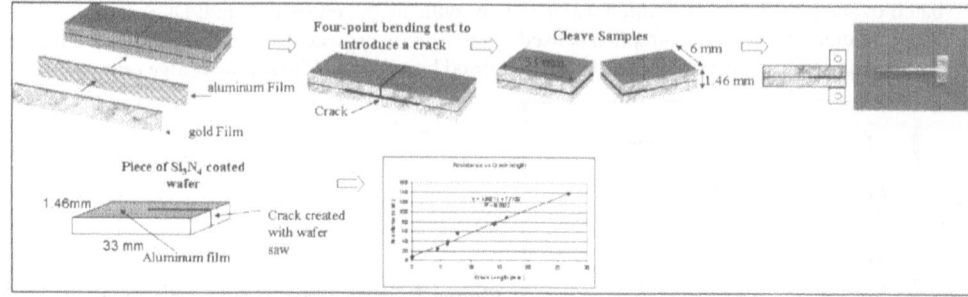

Figure 4. Sample Preparation

this study. Samples were tested with the four-point bending method ($\Psi \approx 42°$) and the mixed-mode fixture technique ($\Psi = 44.70°$). For each configuration, twenty-two samples were tested using the four-point bending technique. Because it's very time consuming, only five samples were tested using the mixed-mode fixture method. Both methods yielded almost identical results in all three cases studied, as shown in figure 5. This proved that the mixed-mode fixture method was adequate for measuring interfacial adhesion energy.

Effects of mode-mixity and porosity were studied with the same 3 configurations used to compare four-point bending and mixed-mode loading techniques. Results are summarized in figure 6. The results obtained with configuration (a) show that as the shear stress component increased, the critical energy release rate increased (G_c)by a factor of 10. Likewise, with configuration (b), the critical energy release rate increased by a factor of 3.

NGk1.9 and NGE are different versions of the same material. A high dielectric constant indicates a low porosity and vice versa. Nanoglass NGE is stronger than Nanoglass NGk1.9 because the mechanical properties of low-k materials degrade with increasing porosity. The

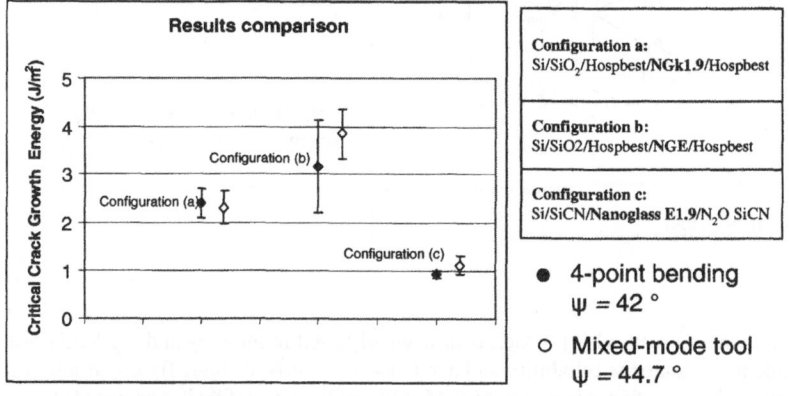

Figure 5. Comparison of results from four-point bending and mixed-mode tests

[1] methylsilsesquioxane
[2] organo-silicate glass

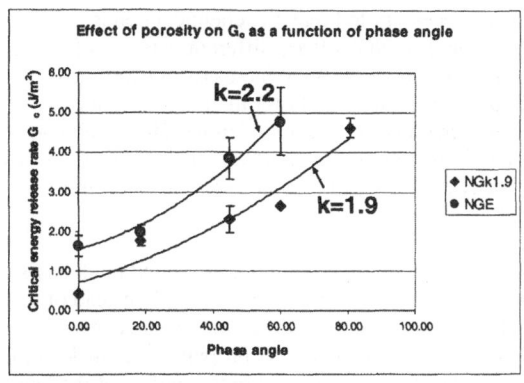

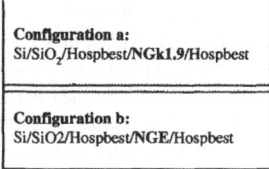

Configuration a:
Si/SiO$_2$/Hospbest/NGk1.9/Hospbest

Configuration b:
Si/SiO2/Hospbest/NGE/Hospbest

Figure 6. Effect of mode-mixity and porosity on interfacial adhesion energy

two curves in figure 6 are almost parallel. This is due to the fact that the surrounding layers are identical in both cases and therefore the contribution of plastic deformation is identical for both configurations. G_c values for configuration (b) are twice as high as G_c values for configuration (a). This is a direct result of the mechanical properties of the materials due to different porosities.

The contribution of surrounding layers in the film stack was also investigated. The results are summarized in figure 7. In this case, the low-k material is the same in both thin film stacks. However, the thin films surrounding the low-k materials are different. At $\Psi = 0°$, G_c values are very similar. In pure tension, the macroscopic energy release rate measured is close to the intrinsic work of adhesion G_0. In this case, since failure occurred cohesively inside the low-k

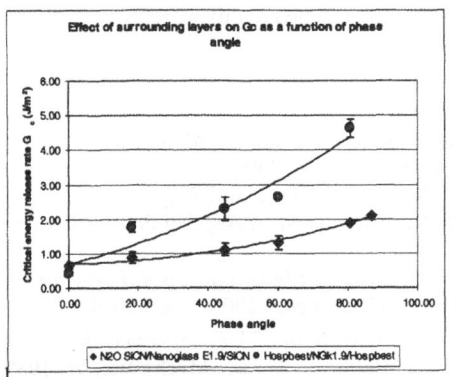

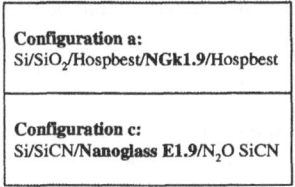

Configuration a:
Si/SiO$_2$/Hospbest/NGk1.9/Hospbest

Configuration c:
Si/SiCN/Nanoglass E1.9/N$_2$O SiCN

methylsilsesquioxane
[2] organo-silicate glass

Figure 7. Effect of surrounding layers on Gc as a function of mode-mixity

film, G_c provides a direct measure of the energy necessary to break the chemical bonds composing the material. Nanoglass E1.9 and Nanoglass NGk1.9 are different versions of the same material, which explains why the G_c values are so close. As Ψ increases, a gap forms between the curves. The G_c values for configuration (a) increase faster than the G_c values for configuration (b). At $\Psi = 80.63°$, the curves are separated by a factor of three. This result can be attributed to the contribution from the plastic deformation of the materials surrounding the low-k. Hospbest is a polymer, which deforms plastically during delamination and contributes more to the macroscopic work of adhesion than SiCN.

Using analysis tools such as FIB, SEM and AFM, cohesive failures were found in all cases examined. The critical energy G_c is therefore a measure of the material's strength instead of a measure of its ability to adhere to another material. This result is not surprising since low-k materials have extremely weak mechanical properties. A crack always propagates at the location that minimizes the energy released. At $\Psi=42°$, cohesive failure occurs in the middle of the low-k material. As Ψ increases to $80.63°$ or decreases to $\Psi = 0°$, the crack deviates and the failure still occurs cohesively but very close to the upper interface [5].

CONCLUSION

We found that as porosity increases, the critical energy required for crack propagation decreases regardless of the phase angle. Additionally, the plasticity effect of layers surrounding the low-k is greater as the phase angle increases. Failure analysis showed that, in all three porous low k materials studied, cohesive failure occurred at $\psi=42°$. Failure analysis revealed that as the phase angle approaching a pure mode ($\psi=0°$, $\psi=80.63°$), the crack moves up closer to the upper interface.

ACKNOWLEDGEMENT

The Semiconductor Research Corporation (SRC) supported this work through the SRC graduate scholars program. International Sematech also contributed to this work by providing processed wafers.

REFERENCES

1. Fernlund, G. & Spelt, J.K., *"Mixed mode fracture characterization of adhesive joint"s*, Composites science and technology, Vol. 50, No. 4 (1994) 441-449.
2. Hu, C, Morgen, M., Ho, P. S., Jain, A., Gill, W. N., Plawsky, J. L., and Wayner, P. C., *"Thermal conductivity study of porous low-k dielectric material"s*, Applied Physics Letters, Vol. 77, No.1 , p 145-7 (2000)
3. Ho, P.S., Miller, M.R., *"Interfacial Adhesion Study for Low-k Interconnects in Flip-chip Packages"*, Electronic Components and Technology Conference (2000) 1089-1094
4. P.G. Charalambides, J. Lund, A.G. Evans and R.M. McMeeking, *"A TestSpecimen for Determining the Fracture Resistance of Bimaterial Interfaces,"*J. App. Mech., Vol. 56, 1989, pp. 77-82
5. Merrill, C.C., *"Effects of Mode-Mixity and Porosity on Interfacial Adhesion of Porous Low-k Materials"*, Master's thesis, SRC website

[1] methylsilsesquioxane

[2] organo-silicate glass

Mat. Res. Soc. Symp. Proc. Vol. 812 © 2004 Materials Research Society

Anisotropic Elastic Properties of Low-k Dielectric Materials

A.A. Maznev[1], A. Mazurenko[1], G. Alper[1], C.J.L. Moore[1], M. Gostein[1],
Michelle T. Schulberg[2], Raashina Humayun[2], Archita Sengupta[2], Jia-Ning Sun[2]
[1]Philips Advanced Metrology Systems, Natick MA 01760
[2]Novellus Systems, Inc., San Jose, CA 95134

ABSTRACT

A non-contact optical technique based on laser-generated surface acoustic waves (SAWs) was used to characterize elastic properties of two types of thin (150-1100 nm) low-k films: more traditional non-porous organosilicate glass PECVD films (k=3.0) and novel mesoporous silica films fabricated in supercritical CO_2 (k=2.2). The acoustic response of the non-porous samples is well described by a model of an elastically isotropic material with two elastic constants, Young's modulus and Poisson's ratio. Both parameters can be determined by analyzing SAW dispersion curves. However, the isotropic model fails to describe the SAW dispersion in the mesoporous samples. Modifying the model to allow a difference between in-plane and out-of plane properties (i.e., a transversely isotropic material) results in good agreement between the measurements and the model. The in-plane compressional modulus is found to be 2-3 times larger than the out-of plane modulus, possibly due to the anisotropic shape of the pores. Elastic anisotropy should therefore be taken into account in modeling mechanical behavior of low-k materials.

INTRODUCTION

While low-k dielectrics are being pursued by the semiconductor industry for their electrical characteristics, their mechanical properties such as hardness and elastic modulus are equally critical to provide structural integrity to the interconnect structures [1]. Techniques currently accepted in the industry for measuring mechanical properties (e.g. nanoindentation) are based on the assumption that film properties are isotropic. In thin film deposition, however, the substrate presents an asymmetric condition that may lead to structural anisotropy resulting from one-dimensional shrinkage, preferential alignment of polymer chains or ordering or orientation of the pores. Consequently, elastic properties of thin film materials may be different for the directions parallel and perpendicular to the plane of the substrate. Nanoindentation is not adequate for characterization of such anisotropic materials. Methods based on surface acoustic waves (SAWs), on the other hand, have proven capable of determining anisotropic elastic properties of thin films [2]. In this work, we used a non-contact technique utilizing laser-generated SAWs to measure the elastic properties of two classes of low-k films: PECVD organosilicate glass (k~3.0) and a novel mesoporous silica material (k~2.2). We will show that in the former case the measurements are well described by the isotropic model of the elastic properties of the low-k material. In the latter case, the material is found to exhibit strong elastic anisotropy, with the in-plane compressional modulus 2-3 times larger than the out-of-plane modulus. We will discuss the relationship between the elastic anisotropy and the structure of the films, as well as the potential impact of a higher in-plane stiffness on the film integration characteristics.

EXPERIMENTAL DETAILS

Both sets of low-k samples were fabricated on 200 mm Si wafers. Samples of High Mechanical Strength (HMS) Coral^{TM}, an organosilicate glass [3], with thickness varying between 150 and 800 nm were deposited in a Novellus Sequel^{TM} PECVD tool. This material typically has a dielectric constant of ~3.0 [3].

Samples of mesoporous silica with thicknesses 550-1100 nm were fabricated by a novel process involving infusion of a solution of methyl triethoxysilane (MTEOS) and tetraethyl orthosilicate (TEOS) in supercritical CO_2 into a block copolymer template [4,5]. Following depressurization, the hybrid film is subjected to a plasma treatment to remove the porogen, and the resulting porous oxide maintains the ordered structure established by the phase-segregated block copolymer. The process, described in more detail in Ref.[5] as "Process 1", yields a dielectric constant k=2.25.

To characterize the elastic properties of the samples, we used a technique based on the laser generation and detection of acoustic waves and referred to as transient laser-induced gratings or Impulsive Stimulated Thermal Scattering (ISTS) [6]. The technique uses two sub-nanosecond laser pulses to create a spatially periodic pattern of laser intensity at the sample surface. Absorption of laser light by the material results in a temperature rise and impulsive thermal expansion that launches counter-propagating SAWs with the wavelength Λ equal to the period of the optical interference pattern, typically in the range 2-10 μm. Surface "ripples" caused by SAWs are then detected via diffraction of a CW laser beam. A detailed description of the compact automated measurement set-up enhanced by optical heterodyne detection is presented in Ref. [7]. The measurement head is integrated into a fully automated platform capable of measuring 200 and 300 mm wafers.

Figure 1 shows a typical signal waveform obtained from a mesoporous silica film and the corresponding frequency spectrum. The peaks in the spectrum represent surface acoustic modes excited in the film/substrate structure at the wavelength Λ defined by the excitation intensity pattern. By varying the wavelength, we measure SAW dispersion curves, i.e. the dependencies of the acoustic phase velocity, obtained by dividing the measured frequency by Λ, vs. the wavenumber $q=2\pi/\Lambda$. The dispersion curves are analyzed with the help of model calculations of SAW propagation in a layered structure to determine the elastic properties of the film material.

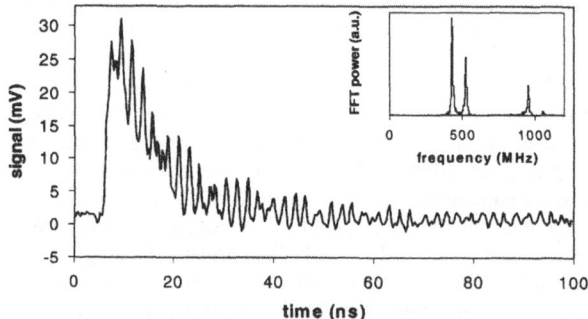

Figure 1. Surface wave signal waveform and frequency spectrum (inset) for 780 nm mesoporous silica film, measured at acoustic wavelength Λ=5 μm.

RESULTS AND DISCUSSION

Non-porous PECVD samples

Fig 2 presents SAW dispersion curves measured on five HMS Coral™ samples plotted versus the product of the film thickness h and the wavenumber q, which is a convenient way to present the dispersion data for samples of different thickness. One can see that the dispersion curves for the different samples match almost perfectly, indicating that all of the samples have the same elastic properties. Solid lines in Figure 2 show results calculated by modeling the acoustic response of an isotropic film on a cubic crystal substrate. The three curves correspond to acoustic modes with different elastic displacement patterns [6]. In the limit $qh=0$, the lowest mode turns into the Rayleigh wave in silicon with velocity 5080 m/s.

An elastically isotropic solid is described by three parameters: density ρ and two elastic constants, Young's modulus E and Poisson's ratio σ. An alternate representation of the elastic constants uses two independent components of the elastic stiffness tensor C_{ij}: compressional modulus C_{11} and shear modulus C_{44}. SAW dispersion curves for a soft light film on a hard substrate are mainly dependent on two parameters of the film material: the ratio E/ρ and σ (or, alternately, C_{11}/ρ and C_{44}/ρ). Therefore, in order to determine the elastic constant values, the density should be known.

In our analysis, we used a value of 1.31 g/cm^3 for the density of Coral™ [8] while E and σ were varied in order to provide the best fit to the data. The calculated curves in Figure 2, corresponding to E=11.3 GPa and σ= 0.205, simulate the experimental data very well. The values of E and σ determined by fitting the data for each sample individually are listed in Table I. Note that for the two thinnest samples, where only the lowest acoustic mode appeared in the data, only Young's modulus was determined by the dispersion curve fitting while Poisson's ratio was set to the average of the values determined for the three thicker samples.

The sample-to-sample variation of the elastic constants does not exceed 2%. We have also measured a sample of a previous version of Coral™ material which yielded a modulus value of ~6 GPa. Thus the modulus of the HMS Coral™ had increased by almost a factor of two compared to the previous material. Nanoindentation measurements of the modulus yield ~15 GPa for the HMS Coral™ and ~9 GPa for the previous version of the material [3]. The fact that

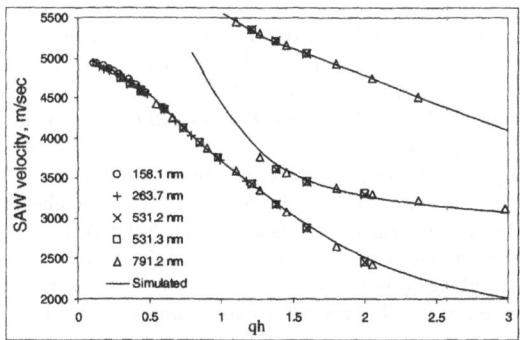

Figure 2. SAW dispersion curves for HMS Coral™ samples of different thickness plotted vs. the product of the wavenumber q and film thickness h.

Table I. Properties of HMS Coral™ samples. Assumed values (not measured) are marked by an asterisk.

Sample #	Thickness, nm	Young's Modulus E, GPa	Poisson's Ratio σ	Compressional Modulus C_{11}, GPa	Shear Modulus C_{44}, GPa
1	158.1	11.15	0.205*	12.47	4.63
2	263.7	11.15	0.205*	12.46	4.63
3	531.2	11.15	0.206	12.49	4.62
4	531.3	11.15	0.205	12.47	4.63
5	791.2	11.33	0.202	12.63	4.71

nanoindentation yields higher modulus values than SAW spectroscopy is typical for low-k materials [9]. We estimate that the absolute error of the SAW method, which includes the error in the SAW velocity measurements, the error in film thickness measurements by a spectroscopic ellipsometer, and the error in the density, is of the order of several percent. The repeatability of the modulus measurements was <0.3%, as determined by repeating the measurements at the same site with unloading the sample from the system between the measurements.

Novel mesoporous material

One advantage of the SAW method is that the dispersion curve analysis tests the validity of the model used to describe the elastic behavior of the film, and the measurement of mesoporous samples demonstrates this. Figure 3(a) shows dispersion curves measured on three samples of different thickness. The dispersion curves for the three samples do not overlap, indicating sample-to-sample variations in elastic properties. However, all three sets of data share a common feature: an "avoided mode crossing" at $qh\sim0.7$ where the 1st and 2nd acoustic modes become very close to each other. Model calculations with isotropic properties for the film material fail to reproduce this feature. For example, gray lines in Fig 3(b) show the best fit with the isotropic model for the 550 nm sample. Clearly, the isotropic model does not work well for this material.

An elastically anisotropic model was therefore employed. A possible explanation for the anisotropy lies in the film fabrication process: when the template is extracted, the film shrinks in the direction normal to the wafer such that the final thickness is ~30% smaller than the thickness following infusion. The resulting pores are therefore isotropic within the plane of the film, but compressed along the axis perpendicular to the film, as shown in the TEM micrograph in Figure 4. It is natural to assume that the elastic properties are also isotropic within the plane of the film, but may be different in the out-of-plane direction. Materials with this kind of symmetry of elastic properties are referred to as transversely isotropic materials.

For elastically anisotropic materials, Young's modulus and Poisson's ratio become direction-dependent and are no longer used to describe the elastic properties. Instead, the elastic constant tensor Cij is used. For a transversely isotropic material, this tensor has 5 independent components. SAW measurements are only sensitive to 4 of those (defined with the z axis perpendicular to the plane of the film): C_{11} (in-plane compressional modulus), C_{33} (out-of-plane compressional modulus), C_{44} (shear modulus for xz or yz shear deformation) and C_{13}. Table II lists the best fit values for each sample obtained using an experimentally determined density value of 1 g/cm^3.

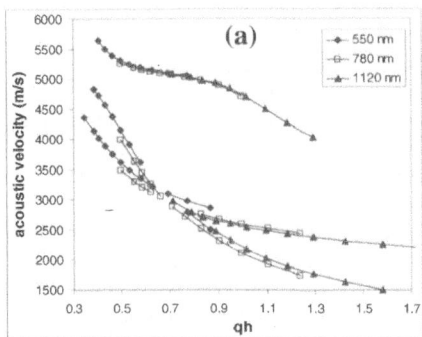 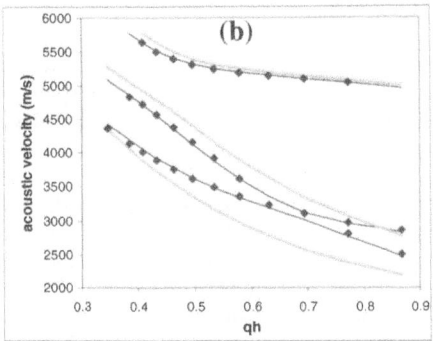

Figure 3. (a) Measured dispersion curves for three mesoporous samples of different thickness plotted vs. the product of the wavenumber q and film thickness h. (b) Modeling the data for the 550 nm sample (symbols) with the isotropic model (gray lines) and transversely isotropic model (black lines).

The black lines in Figure 3(b) show the best fit achieved with the elastic constants from Table II. The transversely isotropic model fits the data very well. Notably, the model yields a much larger in-plane compressional modulus C_{11} compared to the out-of-plane modulus C_{33}. Results for the two other samples presented in Table II show a similar trend. The fact that this material successfully withstood chemical mechanical planarization [5] is possibly attributed to the high in-plane compressional modulus. Nanoindentation measurements of the mesoporous silica material yielded a modulus of 4.4 GPa for a 1μm-thick film. However, nanoindentation data are analyzed based on the model of an elastically isotropic material. A proper comparison between SAW measurements and nanoindentation would require a further study.

Table II. Elastic properties of mesoporous samples.

Sample #	Thickness, nm	C_{11}, GPa	C_{33}, GPa	C_{44}, GPa	C_{13}, GPa
1	546	5.58	1.92	0.85	0.77
2	782	5.03	1.76	0.85	0.82
3	1120	4.72	2.04	0.97	0.89

Figure 4. TEM micrograph of the mesoporous silica ultra low-k material deposited by supercritical infusion.

CONCLUSIONS

Elastic properties of two sets of low-k samples have been characterized with surface acoustic wave spectroscopy. The elastic response of non-porous PECVD organosilicate glass is described well by the isotropic model for the elastic properties of the material. In contrast, the novel ultra-low-k mesoporous silica material is strongly anisotropic, with the in-plane compressional modulus exceeding the out-of-plane modulus by a factor of 2 to 3. Elastic anisotropy, which cannot be detected by traditional mechanical testing techniques such as nanoindentation, must be taken into account in modeling the mechanical behavior of the material and understanding its performance at critical process steps such as chemical mechanical planarization and packaging. Non-contact laser-based surface acoustic wave measurements that can be performed by commercially available instruments provide a convenient tool for characterization of elastic properties of both isotropic and anisotropic low-k materials.

REFERENCES

1. K. Maex, M.R. Baklanov, D. Shamiryan, F. Iacopi, S.H. Brongersma, Z.S. Yanovitskaya, *J. Appl. Phys.* **93**, 8794 (2003).
2. J.A. Rogers, L. Dhar and K.A. Nelson, *Appl. Phys. Lett.* **65**, 312 (1994).
3. W. Liu, Y.K. Lim, A. See et al., in *Proceedings of IEEE International Reliability Physics Symposium*, April 2004, to be published.
4. R.A. Pai, R. Humayun, M.T. Schulberg, A. Sengupta, J.-N. Sun, and J.J. Watkins. *Science* **303**, 507 (2004).
5. M.T. Schulberg, R. Humayun, A. Sengupta, and J.-N. Sun, *Proceedings of 2004 MRS Spring Meeting*, this volume.
6. J.A. Rogers, A.A. Maznev, M.J. Banet and K.A. Nelson, *Annu. Rev. Mater. Sci.* **30**, 17 (2000).
7. A.A. Maznev, A. Mazurenko, L. Zhuoyun, and M. Gostein, *Rev. Sci. Instrum.* **74**, 667 (2003).
8. Haiying Fu, private communication.
9. C.M. Flannery, C. Murray, I. Streiter, S.E. Schulz, *Thin Solid Films* **388**, 1 (2001)

Mat. Res. Soc. Symp. Proc. Vol. 812 © 2004 Materials Research Society

Scanning Near-Field Microwave Probe for In-line Metrology of Low-K Dielectrics

Vladimir V. Talanov, Robert L. Moreland, André Scherz, Andrew R. Schwartz, and Youfan Liu[1]
Neocera, Inc., 10000 Virginia Manor Road, Beltsville, MD 20705 USA
[1] Intel Assignee, International Sematech, Austin, TX 78741 USA

ABSTRACT

We have developed a novel microwave near-field scanning probe technique for non-contact measurement of the dielectric constant of low-k films. The technique is non-destructive, non-invasive and can be used on both porous and non-porous dielectrics without any sample preparation. The probe has a few-micron spot size, which makes the technique well suited for real time low-k metrology on production wafers. For dielectrics with k<4 the precision and accuracy are better than 2% and 5%, respectively. Results for both SOD and CVD low-k films are presented and show excellent correlation with Hg-probe measurements. Results for k-value mapping on blanket 200mm wafers are presented as well.

INTRODUCTION

The *International Technology Roadmap for Semiconductors* (ITRS) [1] projects rapid transitions over the next decade to new materials for use in the interconnects: from aluminum to copper to reduce the resistance of the metal wires, and from SiO_2 to dielectrics with lower dielectric constant k, the so-called 'low-k' materials, to reduce the delay times on interconnect wires, and to minimize the crosstalk between such wires. "The introduction of new low-k dielectrics ... provide[s] significant process integration challenges," and "...will greatly challenge metrology for on-chip interconnect development and manufacture" [1]. In addition, "design of interconnect structures requires measurement of the high frequency dielectric constant of low-k materials. High frequency testing of interconnect structures must characterize the effects of clock harmonics (5× to 10× base frequency), skin effects, and crosstalk" [1]. The desired metrology should also be non-destructive, non-contaminating, and provide real time/rapid data collection and analysis [2].

In response to this need for a fab-compatible low-k dielectric metrology we have developed a novel technique that is based on a scanning near-field microwave probe. The main feature of the near-field approach is that the spatial resolution is defined by the probe tip geometry rather than by the wavelength of the radiation used [3–6]. Therefore, even for microwave frequencies with wavelengths of centimeters, a near-field probe can achieve spatial resolution down to the sub-micron scale. Here we present quantitative dielectric constant measurements on a variety of low-k dielectric films and show excellent correlation with mercury probe measurements on the same films. The microwave measurements were made with a probing area on the order of a few microns, and therefore the measurements could easily be made over a test region in the scribe line on a device wafer.

EXPERIMENTAL SETUP

To fabricate our near-field microwave probe a quartz bar with ~1×1 mm^2 cross-section is tapered down to a sub-micron size using a laser-based micropipette puller. Aluminum is deposited onto two opposite sidewalls of the bar and the very end of the tapered portion is cut off around 1 μm in size using a focused ion beam (FIB) technique similar to [7], which provides a well-defined tip. The whole structure forms a tapered balanced parallel-plate transmission line (Fig. 1) with open tip end and no cut-off frequency.

Numerical simulations using Ansoft's High Frequency Structure Simulator (HFSS™), a 3D finite element package, confirm that the fringe E-field at the end of such a tip forms a "bubble" with typical diameter on the order of the tip size. Unlike the apertureless STM-like [5] or AFM-like [6] schemes that have been previously employed, this "apertured" approach allows us to perform truly quantitative measurements on a few-micron length scale where the result is insensitive to the material property outside this probing area (Fig. 2). The balanced line geometry virtually eliminates the parasitic stray fields and reduces the amount of microwave power radiated from the tip by a few orders of magnitude when compared to unbalanced coaxial probes [3, 5, 6].

When the probe tip is placed in close proximity to the sample under test its fringe capacitance C_t is governed by the tip geometry, the sample permittivity, and the tip-sample distance. The complex refection coefficient from the tip can be found as follows:

$$\Gamma \cong \exp(-i\,2\omega Z_0 C_t) \tag{1}$$

where ω is the operating frequency, Z_0 is the line characteristic impedance, and $\omega Z_0 C_t \ll 1$. In order to increase the measurement sensitivity the transmission line is formed into a half-lambda resonator, which is done by etching the back end of the aluminum strips to the appropriate length. The resonator has a resonant frequency $F\sim4$ GHz and an unloaded quality factor $Q\sim100$. A conventional magnetic loop is used to couple the microwave signal into it. The resonator is packaged inside a metallic enclosure with the tapered portion protruding out through a clear hole in the enclosure wall. The probe resonant frequency F is experimentally determined from the minimum in the probe reflection coefficient S_{11} using a microwave reflectometer (Fig. 3) with resolution down to 100 Hz.

From Eq. (1) we find the relative shift in the probe resonant frequency F versus change in the tip capacitance C_t:

$$\frac{\Delta F}{F} = -\frac{Z_0}{L\sqrt{\varepsilon_0\mu_0\varepsilon_{\text{eff}}}}\Delta C_t \tag{2}$$

where L is the resonator length, ε_0 is vacuum permittivity, μ_0 is vacuum permeability, ε_{eff} is the transmission line effective dielectric constant, and Z_0 is the line characteristic impedance. An estimate for the tip capacitance in air is $C_{t0}\sim\varepsilon_0 a_t$, where a_t is the tip size; for $a_t \sim 1\mu m$ $C_{t0}\sim10$aF. For typical probe parameters ($L\sim25$ mm, $Z_0\sim100\Omega$, $\varepsilon_{\text{eff}}\sim2.5$) and a 100 Hz precision in ΔF, Eq.(2) yields sensitivity to changes in the tip

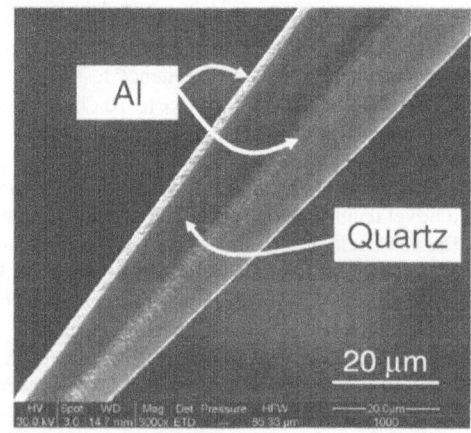

Figure 1: SEM image of a typical probe with two metal strips on the quartz dielectric support

74

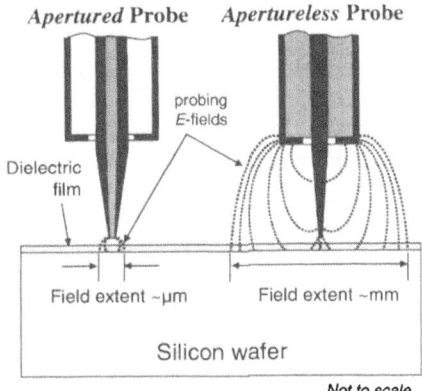

Apertured **Probe**   *Apertureless* **Probe**

probing
E-fields

Dielectric
film

Field extent ~μm   Field extent ~mm

Silicon wafer

Not to scale

*Figure 2: Comparison of apertured vs.
apertureless probes with the same tip cross-
section or curvature ~micron. The apertured
probe has much high probing field confinement
and thus quantitative spatial resolution.*

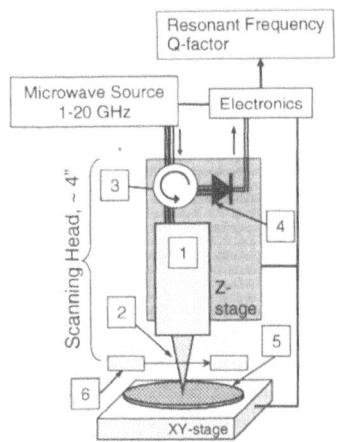

*Figure 3: Microscope schematic showing (1)
resonator, (2) probe, (3) circulator, (4) detector,
(5) wafer, and (6) distance control system.*

capacitance on the order of 3×10^{-20} F = 30zF.

A modified shear-force method similar to the one typically employed in near-field scanning optical microscopy (NSOM) applications [8, 9] is used to actively control the tip-sample separation with precision down to a few nanometers. The tapered quartz bar forms a mechanical resonator with the fundamental transverse mode frequency on the order of a few kHz. The probe package is dithered at this frequency with sub-nm amplitude using a piezo tube, and the tapered portion of the tip is illuminated with a laser beam projecting onto a photo-detector (Fig. 3). The ac portion of the photo-detector output is proportional to the tip vibration amplitude, which is a strong function of the tip-sample distance d for $d<100$ nm. The photo-detector output is fed into a lock-in amplifier and then into a PID controller, which in turn controls the amplifier for the piezo z-stage and therefore the tip-sample separation. From experimental data and theoretical estimates, the typical operating tip-sample separation is found to be on the order of 50 nm. The shear-force distance control provides typical MTBF of a tip ~10^3 hours.

The probe enclosure and the microwave and the shear-force circuits are installed into a measurement head that is mounted onto a highly accurate piezo-driven z-stage. The piezo stage is attached to a mechanical coarse z-stage, which is mounted onto a gantry bridge. A 100 – 300mm wafer under test is placed onto a vacuum chuck that is scanned by the 350-mm-travel xy-stage (Fig. 3). The whole setup is mounted onto a vibration-isolation platform and placed inside an enclosure to improve acoustic isolation and thermal stability.

DATA ANALYSIS

One of the most challenging tasks in thin film metrology is to deconvolve the film and substrate contributions. It is a necessary procedure for our probe as well, since the measured tip capacitance C_t depends on both the film and the substrate properties. We have developed an approach for the probe calibration, which removes the substrate contribution from C_t.

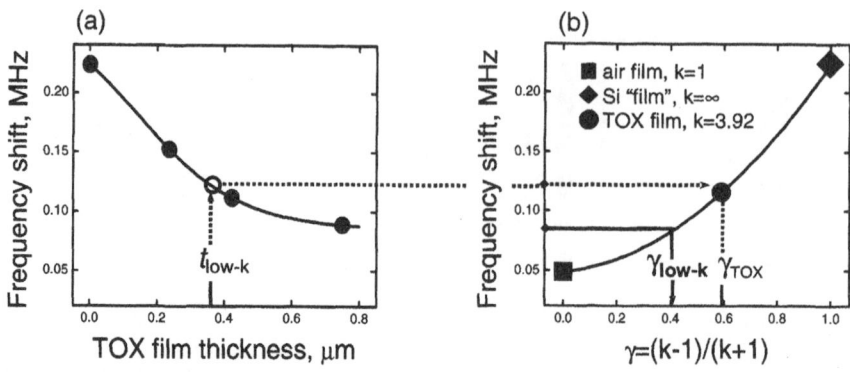

Figure 4: Analysis plots showing (a) the measured frequency shift for thermal oxide films as a function of film thickness with the empirical fit discussed in the text, and (b) the expected frequency shift as a function of γ at a given film thickness.

The probe is calibrated by measuring resonant frequency shift $\Delta F = F_e - F$ for three "standard" films of nominally the same thickness and on nominally the same substrate as the low-k film under test: $\Delta F_{air}[t_f]$ for an air film with dielectric constant k=1, $\Delta F_{TOX}[t_f]$ for a thermal oxide film (TOX) with well-known dielectric constant $k_{TOX}=3.92$, and ΔF_{Si} for a low-resistivity silicon wafer, which effectively exhibits infinitely large dielectric constant k>>1 for our probe. Here F_e is the probe resonant frequency with no sample, and F is the frequency measured at some distance d_0 above the film surface. In this study, all the low-k films under test and the TOX calibration films were fabricated on Si substrates with bulk resistivity $\rho<0.01$ $\Omega\cdot$cm. Under the condition $\omega C_t \rho/a_t << 1$ the material appears as a perfect metal for the probe. For $\rho\sim0.01$ $\Omega\cdot$cm and 200-nm-thick film with k=4, $\omega C_t\rho/a_t<10^{-3}$, and therefore all the films effectively had a perfect conductor backing.

In order to create a calibration curve (Fig. 4b), which is generally dependent on the probe geometry and the low-k film thickness, we plot the three calibration data points for frequency shift versus the parameter $\gamma=(k-1)/(k+1)$ and fit them to some empirical form. In order to determine the dielectric constant of the low-k film under test we measure the probe resonant frequency shift F_{low-k} for it, and then using the calibration fit function convert it into the k-value. It turns out that $\Delta F(\gamma)$ has highest sensitivity to k around k = 2 ($\gamma= 0.33$), which makes the technique perfectly-suited for measurements on low-k materials.

It would be impractical, of course, to have a TOX calibration film for each low-k film thickness. Instead we measure ΔF_{TOX} vs. thickness for a set of TOX films with variable thickness ranging around the film thickness t_f and use the interpolated $\Delta F_{TOX}[t_f]$ value (Fig. 4a), where $\Delta F_{TOX}[t=0]=\Delta F_{Si}$. While the calibration curve shown in Fig. 4b is for one particular film thickness ~400 nm, similar curves can be obtained for any thickness using the procedure above.

It is important to note that for our calibration procedure there is no need to know the absolute value of the tip to film surface separation d_0, however it must be nominally the same for all the measurements, which is provided by the virtually material independent shear-force distance control. The typical slope for resonant frequency F vs. distance is on the order of 100Hz per 1nm at typical operating tip to sample separation ~50 nm. For ΔF vs. γ dependence (see Fig. 4b) a practical limit of ±1nm precision in the tip-to-sample distance would provide ±100Hz deviation in ΔF, which in turn yields less then ±0.07% deviation in the γ value. For typical k~2.3

this corresponds to less then ±0.2% variation in the film dielectric constant. We also estimated less than 0.2% error in the k-value due to 1 nm inaccuracy in the film thickness for typical low-k film ~400-nm-thick.

RESULTS

Using the instrument and technique described above we have quantitatively measured the dielectric constant of 13 low-k films on Si substrates with $\rho<0.01$ $\Omega\cdot$cm. These samples were both SOD and CVD films with thicknesses ranging from 0.35 to 1.4 μm and dielectric constants ranging roughly from 2.1 to 3.1. For each film we compared the average dielectric constant across the wafer as measured by our technique to the values obtained by the conventional mercury probe. The correlation plot between the two data sets is shown in Fig. 5. The correlation is excellent with a correlation coefficient $R^2 = 0.97$. Although these data were all obtained on blanket wafers, the few-micron spot size and non-contact nature of the measurement means that this technique could easily be applied to test regions in the scribe line on a product wafer.

In Fig. 6 we show a 49-point map of one of the low-k films on a 200mm wafer. Each measured point is represented as a + or − indicating whether they are above or below the mean of 2.27. The solid contour line indicates the position of the mean and the dashed lines are at plus and minus one standard deviation of 0.034.

The sample-to-sample repeatability of these measurements $\Delta k/k$ is currently ~2%. It is controlled primarily by the repeatability of the tip-sample separation. Estimates show that the practical limit can be as low as 0.2%, and improvements are underway to achieve this. The accuracy of the measurements is currently ~5% and depends mainly on the accuracy of the empirical fit for the frequency shift vs. γ. We are studying the theoretical form to use for this fit, which should improve accuracy down to ~1%.

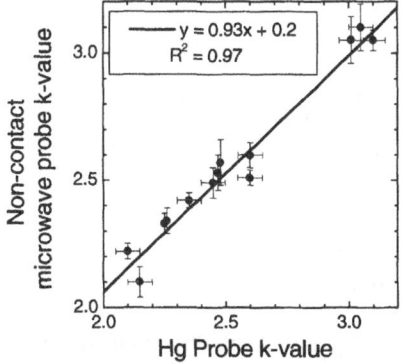

Figure 5: Correlation between k-values measured by the novel microwave probe and by Hg probe for thirteen low-k films. The samples were a mix of SOD and CVD films with thicknesses ranging from 0.3 to 1.4 μm.

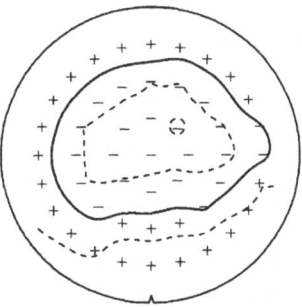

Figure 6: Wafer map for a low-k film on a 200mm wafer. The mean k was 2.27 and is shown by the solid contour line. The standard deviation was 0.034 and is shown by the dashed lines. The + and − symbols are the measured points and indicate values above and below the mean.

CONCLUSIONS

We have developed a novel technique for measuring the dielectric constant of low-k films on Si or metallic substrates. The advantages of this new technique are as follows: a) it is non-contact, non-invasive and non-contaminating; b) it has high spatial resolution (on the order of a few microns in this example, but potentially can be scaled below one micron); and c) the measurement is made at microwave frequencies (4GHz in this example, but can be scaled up), which is well matched to the operating frequencies of the devices in which these materials are used. The non-contact nature and high spatial resolution of this technique make it particularly well suited for measurements on patterned wafers, and the high-frequency measurement provides information about material properties at device frequencies. With the current probe geometries, inhomogeneities in dielectric constant can be quantitatively imaged on a few microns scale, and efforts are underway to scale the probing area down to sub-micron dimensions.

ACKNOWLEDGMENTS

This work was partially supported by NSF-SBIR Award No. 0078486 and NIST-ATP Award No. 70NANB2H3005. In addition we would like to acknowledge the support of the following people: Dr. Hans Christen of ORNL for his involvement in the early stages of this project at Neocera; Dr. Igor Smolyaninov and Prof. Chris Davis of the University of Maryland for help with the shear-force distance control; Prof. Steve Anlage of the University of Maryland for useful discussions; Alex Krechmer of FEI and Dr. Bin Ming of Neocera for help with the FIB cutting and probe fabrication; Mike Bangham and Mukund Shridharan of Neocera for the software development; Don Kuechle and Joe Zihmer of Neocera for help with the mechanical design; and Prof. Venky Venkatesan and John Matthews of Neocera for invaluable attention paid to this work.

REFERENCES

1. Semiconductor Industry Association. *International Technology Roadmap for Semiconductors.* Austin, TX: International SEMATECH, 1999.
2. J. Iacoponi, Presented at the Characterization and Metrology for ULSI Technology conference, Austin, TX, 2003 (unpublished).
3. C.P. Vlahacos, R.C. Black, S.M. Anlage, A. Amar, and F.C. Wellstood, Appl. Phys. Lett. **69**, 3272 (1996).
4. A. Lann, M. Golosovsky, D. Davidov, and A. Frenkel, Appl. Phys. Lett. **75**, 3133 (1999).
5. C. Gao and X.-D. Xiang, Rev. Sci. Instr. **69**, 3846 (1998).
6. K. Ohara, Y. Cho, Jpn. J. Appl. Phys. **41**, 4961 (2002).
7. S. Pilevar, K. Edinger, W. Atia, I. Smolyaninov, and C. Davis, Appl. Phys. Lett. **72**, 3133 (1998).
8. E. Betzig, P.L. Finn, and J.S. Weiner, Appl. Phys. Lett. **60**, 2484 (1992).
9. R. Toledo-Crow, P.C. Yang, Y. Chen, and M. Vaez-Iravany, Appl. Phys. Lett. **60**, 2957 (1992).

Mat. Res. Soc. Symp. Proc. Vol. 812 © 2004 Materials Research Society

Deposition and Integration of a Novel Ultra-Low k (2.2) Material

Michelle T. Schulberg, Raashina Humayun, Archita Sengupta, and Jia-Ning Sun
Novellus Systems, Inc., San Jose, CA 95134

ABSTRACT

Increasing demands for faster chip speed and reduced power consumption are driving the semiconductor industry to develop insulating layers with lower dielectric constants. As the dielectric constant of a material is reduced, however, it becomes increasingly difficult to achieve the mechanical strength required to manufacture a multilevel interconnect. A new route to the synthesis of mesoporous silica has been demonstrated on 200 mm wafers. Silicate precursors dissolved in supercritical CO_2 are infused into a block copolymer film. The polymer is then removed, but the resulting porous SiO_2 replicates its ordered structure, enhancing the strength of the network. Incorporation of alkyl silicates further improves the film properties. Post-treatment to cap residual silanol groups renders the surface of the film hydrophobic and stabilizes it to air exposure. By appropriate choice of the block copolymer and other process parameters, the pore size and density can be varied and k values as low as 1.8 can be achieved. For a film with a dielectric constant of 2.25, the pore size is ~4 nm. The hardness and modulus are 1.1 GPa and 7.8 GPa, respectively, as measured by nanoindentation. Four-point bend measurements yield fracture energies of 9.8 J/m^2. More importantly, the film can withstand chemical mechanical planarization (CMP) using standard oxide polishing conditions.

INTRODUCTION

One of the greatest challenges in the semiconductor industry has been to develop materials with a lower dielectric constant (k) than SiO_2 (~4) for use as insulators in multi-level interconnect structures. Lowering the dielectric constant shortens the RC delay of the circuit, decreases power consumption, and reduces cross-talk between adjacent conductors. Doping SiO_2 with fluorine can lower k to ~3.5, while addition of carbon to SiO_2 can create films with k as low as ~2.5. To achieve k < 2.5, though, it is necessary to incorporate porosity.

A viable low k material must also be able to withstand all of the processes involved in building an integrated circuit. Both the CMP step required for a dual damascene integration scheme and the packaging process to turn the wafer into a finished chip exert enormous physical stresses on the interconnect structure. The low k material must therefore have a high mechanical strength. Unfortunately, porous materials are inherently weaker than fully dense ones. To achieve the strongest porous material, the pores should be ordered and uniform in size. The ordered porous structure of mesoporous silica is therefore a good candidate for a low k material. Although the IUPAC definition of mesopores is 2 to 50 nm, only materials with pores sizes in the range of 1 to 10 nm are appropriate for interconnects.

Mesoporous silica has traditionally been made by evaporation induced self-assembly (EISA) [1-3]. To deposit a film, a solution containing a templating agent, one or more silicate precursors, a catalyst, and one or more solvents is spun on to a wafer. Several processes take place simultaneously: solution spreading, self-assembly, precursor reaction, and solvent evaporation. The conflicting requirements of these various processes may make it difficult to control the reaction in a repeatable manner. For this reason, a new approach is introduced: supercritical fluid infusion of a silicate precursor into a block copolymer template [4].

The supercritical fluid infusion process sequence is shown in Figure 1. A solution containing an organic block copolymer, a catalyst, and a solvent or solvent mixture is spun on to the wafer. The polymer self-assembles to create hydrophilic and hydrophobic domains, and the catalyst segregates to the hydrophilic blocks. The wafer is then introduced to a high pressure chamber and exposed to a silicate precursor dissolved in supercritical CO_2. The CO_2 swells the polymer and carries the precursor throughout the film. The precursor reacts to form SiO_2 only in the hydrophilic regions where the catalyst is found. After the pressure is released, the polymer is removed either by calcination or by plasma processing. The resulting porous SiO_2 has the ordered structure of the hydrophilic block of the polymer. This approach is preferable to EISA because the self-assembly and precursor reaction steps have been separated, allowing the process conditions for each to be tailored independent of the other. Also, the precursor and the catalyst mix only in the film itself, so the reaction cannot begin prematurely before the solution is applied to the wafer.

EXPERIMENTAL DETAILS

Mesoporous silica films were grown on 200 mm wafers using a series of standard and novel deposition tools. The template solution contained a polyethylene oxide (PEO) – polypropylene oxide (PPO) – polyethylene oxide (PEO) block copolymer (BASF Pluronic™ F108), para-toluenesulfonic acid (pTSA) as a catalyst, and an ethanol-water solvent mixture. This solution was deposited onto a wafer using an integrated spin track. After drying, the template film was loaded into a custom-designed high pressure reactor. The chamber was pressurized to a maximum of 1800 psi with a solution of either tetraethyl orthosilicate (TEOS) or a TEOS/methyl triethoxysilane (MTEOS) mixture in humidified supercritical CO_2. After depressurization, the wafer was transferred to a Novellus Concept 2 Sequel™ plasma-enhanced chemical vapor

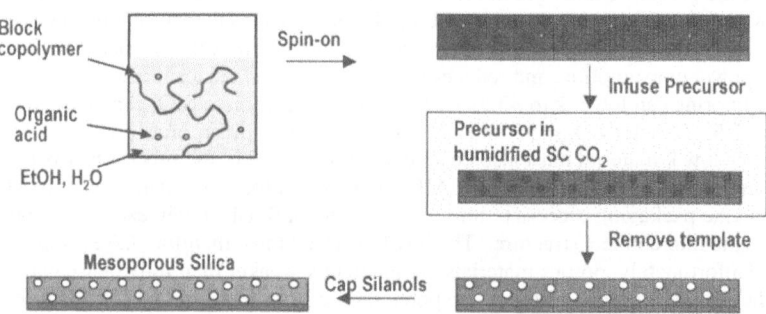

Figure 1: Process sequence for deposition of mesoporous silica by supercritical fluid infusion of a block copolymer template.

deposition (PECVD) tool and exposed first to a H_2/N_2 plasma to remove the template and then to hexamethyldisilazane (HMDS) vapor to restore the hydrophobicity of the film. Films for k measurements were also capped with a thin (~100 nm) layer of silicon oxide or silicon nitride.

RESULTS

The evolution of the film on the wafer can be followed with Fourier Transform Infrared Spectroscopy (FTIR), as shown in Figure 2. The bottom spectrum shows the block copolymer template as spun onto the wafer. The largest peaks are at 1115 cm^{-1}, representing the CO ether linkages in the PEO and PPO, and at 2890 cm^{-1}, due to the CH_x stretch of the hydrocarbons. Following the TEOS infusion, the middle spectrum shows a combination of the polymer peaks with those of the silicate network. New peaks appear at 1040 cm^{-1} (SiO_2), 950 cm^{-1} (H_2O), and 3400 cm^{-1} (H_2O). Finally, the plasma treatment removes the polymer and the water and the top spectrum shows only the features of an SiO_2 film. The small peaks at 1275 cm^{-1} ($SiCH_3$) and 2980 cm^{-1} (CH_3) are due to the methyl addition from the HMDS exposure. Films infused with MTEOS/TEOS mixtures have much larger $SiCH_3$ and CH_3 peaks, as the methyl groups are incorporated into the silicate backbone. The organic groups increase the hydrophobicity of the MTEOS/TEOS films, resulting in a much lower k drift. TEOS-based films exposed to ambient for two weeks show an increase in k of 1.7, while k of the organosilicate films increases by <0.2 in the same period. In the latter case, half of the k drift can be recovered by annealing the films.

The pore structure of both TEOS-only and MTEOS/TEOS-based films with a dielectric constant of 2.25 was examined by Positronium Annihilation Lifetime Spectroscopy (PALS) [5,6]. The TEOS-only film had an average pore diameter of >3 nm, while the pores in the

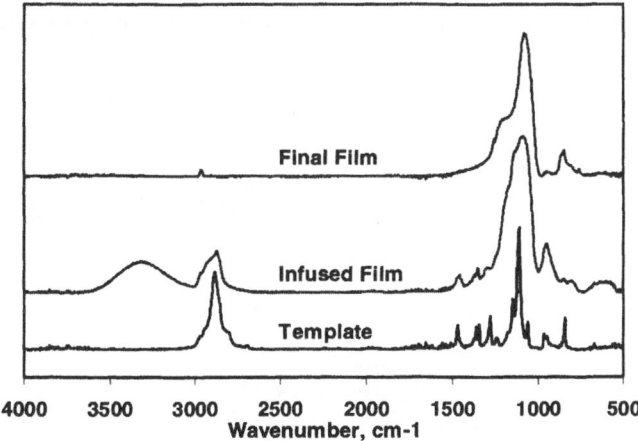

Figure 2: FTIR spectra during growth of mesoporous silica by supercritical fluid infusion of a block copolymer. Bottom: the block copolymer template. Middle: the composite film following TEOS infusion into the polymer. Top: the final mesoporous silica film.

MTEOS/TEOS-based film measured 4 nm. The MTEOS/TEOS film was also analyzed with Ellipsometric Porosimetry (EP) [7] and Transmission Electron Microscopy (TEM), as shown in Figure 3. EP reveals a monodisperse pore size distribution with a maximum intensity at a pore radius of 1.7 nm, or pore diameter of 3.4 nm. There is a sharp cut-off at the high end of the distribution, indicating the absence of larger "killer pores." The total porosity is 37%. In the TEM image, the pores appear evenly spaced and of uniform size, roughly 5 nm in diameter. The supercritical fluid infusion process has therefore succeeded in transferring the well-ordered structure of a phase-separated block copolymer film to the carbon-doped porous oxide network.

The density of the film, and therefore its k, hardness, refractive index, and other properties, is governed by the ratio of pore volume to wall volume. This is determined in part by the volume ratio of the hydrophilic and hydrophobic blocks of the polymer. Even with a single polymer template, however, variation of the process conditions can control the amount of precursor that reacts within the template and therefore the final film density. Figure 4 shows the variation of the film's hardness versus dielectric constant for films formed from Pluronic F108 infused with a MTEOS/TEOS mixture and processed with two different plasma conditions for template removal. Further variation of the precursor uptake allows dielectric constants as low as 1.8 to be achieved. Process II yields exceptionally strong films for this range of k values, with a hardness of 1.1 GPa and a modulus of 7.8 GPa for k=2.25.

Further tests with Process I also indicate the superior mechanical properties of the mesoporous carbon-doped silica. The cracking threshold of a film with k=2.25 was determined to be >1.5 μm. The adhesion of the mesoporous film to a silicon carbide layer was evaluated with the four-point bend technique [8]. The following film stack was prepared: Si/SiO_2 (1000 nm)/SiC (70 nm)/mesoporous carbon-doped silica (600 nm)/SiO_2 (50 nm). This stack was glued to a bare Si wafer backing plate. The interface fracture energy, G_c, was 9.8 J/m^2, well above the 5 J/m^2 threshold recommended for CMP. The failure mechanism appeared to be delamination at the low k/SiC interface, indicating that the cohesive strength of the mesoporous film was even greater.

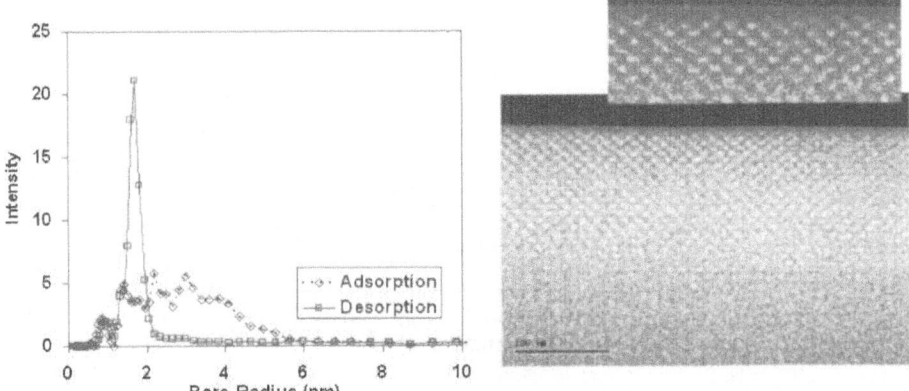

Figure 3: Pore size measurements of the mesoporous silica film. Left: Ellipsometric porosimetry. Right: Transmission electron microscope image, and expanded view. The scale bar represents 100 nm. (TEM image by Cerium Laboratories)

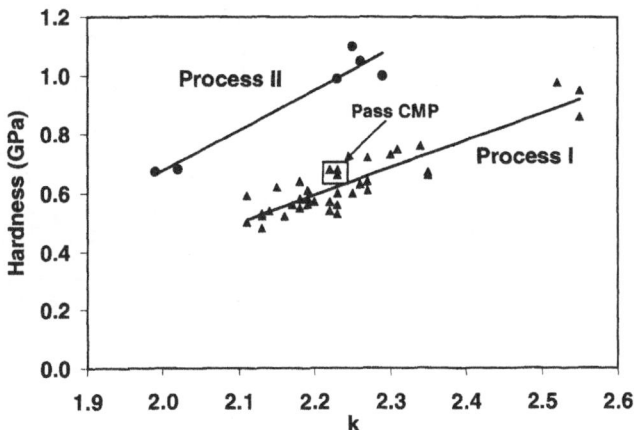

Figure 4: The relationship of dielectric constant to hardness as measured by nanoindentation for mesoporous carbon-doped silica films treated with two different template removal processes. The films highlighted in the box survived CMP processing.

Finally, MTEOS/TEOS-based mesoporous films with k=2.25, as indicated in Figure 4, were subjected to the same CMP process normally used for oxide films. The dielectric stack was similar to that used for the 4-point bend measurements. Metal barrier and seed layers were sputter-deposited onto the dielectrics, followed by a thick layer of electroplated copper. Some of the wafers were polished only through the copper, leaving behind the metal barrier and SiO_2 cap, while others were polished all the way to the low k layer. Examples of each are shown in Figure 5. No delamination was observed in any of the tests. The polished surface of the mesoporous film is smooth with no indication of cohesive failure.

CONCLUSIONS

A new process, supercritical fluid infusion of a block copolymer, has been described for depositing carbon-doped mesoporous silica films for use as a low-k material for interconnects. An ordered porous structure is predicted to have the highest mechanical strength for a given porosity, and indeed films with a dielectric constant of 2.25 have proven to be robust enough to withstand the chemical mechanical planarization processing required to build a dual damascene interconnect structure. Process variations allow control of the film's density, and therefore its dielectric constant, such that k can be extended down to 1.8.

ACKNOWLEDGEMENTS

We gratefully acknowledge the contributions of Professor James J. Watkins and Rajaram Pai at the University of Massachusetts. We thank IMEC for providing ellipsometric porosimetry data, and BASF for supplying the Pluronic polymers. The assistance of Patrick Van Cleemput, Feng Wang, Dayton Cheatham, and Tom Mountsier at Novellus is also appreciated.

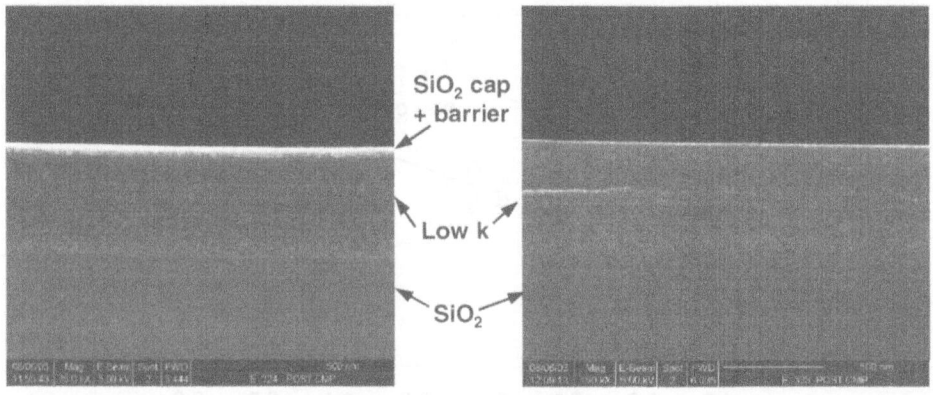

Figure 5: Scanning electron microscope images of mesoporous low k films that have undergone CMP. The wafer on the left was polished only through the copper layer while the wafer on the right was polished down to the mesoporous film.

REFERENCES

1. C. T. Kresge, M. E. Leonowicz, W. J. Roth, J. C. Vartuli, and J. S. Beck, *Nature* **359**, 710 (1992).
2. D. Zhao, J. Feng, Q. Huo, N. Melosh, G. H. Fredrickson, B. F. Chmelka, and G. D. Stucky, *Science* **279**, 549 (1998).
3. C. J. Brinker, Y. Lu, A. Sellinger, H. Fan, *Adv. Mater.* **11**, 579 (1999).
4. R. A. Pai, R. Humayun, M. T. Schulberg, A. Sengupta, J.-N. Sun, and J. J. Watkins, *Science* **303**, 507 (2004).
5. D. W. Gidley, W.E. Frieze, A.F. Yee, T.L. Dull, H.-M. Ho, and E.T. Ryan, *Phys. Rev. B. Rapid Comm.* **60**, R5157 (1999).
6. D. W. Gidley, W.E. Frieze, T.L. Dull, J. Sun, A.F. Yee, C. V. Nguyen, and D.Y. Yoon, *Appl. Phys. Lett.* **76**, 1282 (2000).
7. M. R. Baklanov, K. P. Mogilnikov, V. G. Polovinkin, and F. N. Dultsev, *J. Vac. Sci. Technol.* **B18(3)**, 1385 (2000).
8. R. H. Dauskardt, M. Lane, Q. Ma, and N. Krishna, *Eng. Fracture Mechanics* **61**, 141 (1998).

Mat. Res. Soc. Symp. Proc. Vol. 812 © 2004 Materials Research Society

A Novel Organosiloxane Vapor Annealing Process for Improving Elastic Modulus of Porous Low-k Films

Kazuo Kohmura[1], Shunsuke Oike[1], Masami Murakami[1], Hirofumi Tanaka[1], Syozo Takada[2],
Yutaka Seino[3], and Takamaro Kikkawa[3,4]
[1] MIRAI-ASET, Tsukuba, Japan
[2] ASRC-AIST, Tsukuba, Japan
[3] MIRAI-ASRC-AIST, Tsukuba, Japan
[4] RCNS, Hiroshima Univ., Higashi-Hiroshima, Japan

ABSTRACT

A novel organosiloxane-vapor-annealing method has been developed for improving the mechanical strength of porous silica films with a low dielectric constant. Treatment of a porous silica film with 1,3,5,7-tetramethylcyclotetrasiloxane (TMCTS) under atmospheric nitrogen above 350 °C significantly enhanced the mechanical strength (i.e., elastic modulus and hardness) of the film. Results of Fourier transform infrared spectroscopy (FT-IR) and thermal desorption spectroscopy (TDS) suggested the formation of cross-linked poly(TMCTS) network on the porous silica internal wall surfaces by the TMCTS treatment. Such TMCTS cross-linked network is thought to enhance the mechanical strength of the low-k film.

INTRODUCTION

Considerable efforts have been devoted to the development of ultralow dielectric constant ($k \leq 2.0$) films with a relevant mechanical strength because of the great demand for low-k/Cu interconnects in ultra large scale integrated circuits (ULSIs). Porous silica films possess considerable potential for ultralow-k values and high mechanical strengths. The drawbacks of these porous films include the presence of hydrophilic silanol groups on the internal wall surfaces and the lack of mechanical strength (e.g., elastic modulus, hardness). Although the silanol groups can be transformed to the hydrophobic moieties by treatment with hexamethyldisilazane (HMDS) [1-4], the mechanical strength of the resulting films is insufficient as ULSI interconnects.

The objective of this work is to develop porous silica films that possess both ultralow dielectric constant and mechanical strength.

EXPERIMENTAL

Porous silica films were prepared using a sol-gel method based on the self-organization of surfactant templates [5-7]. A precursor solution was prepared by mixing 4.8mmol of tetraethoxysilane (TEOS), 5mmol of H_2O and 0.1mmol of HCl in 10 ml ethanol at 50 ºC, being kept for 90 min under agitation. Then 2.9g of Pluronic P123(BASF Corp.) as a surfactant template and 40ml of ethanol solution were added in 21ml of the precursor solution at 30ºC and further left for 2 hours. Next, 8 ml of H_2O and 20ml of N,N-dimethylacetoamide were added to it at a molar ratio of $TEOS:P123:H_2O:HCl = 1:0.01:10:0.02$.

The resulting solution was spin-coated onto Si substrates and calcined at 400ºC for 3h in air to form a porous silica film.

The film was treated with TMCTS at temperature from 250 to 400 ºC for 30 sec – 180 min under atmospheric nitrogen. For comparison, the film was also treated with HMDS under identical conditions. The mechanical properties of the resulting films were evaluated by nanoindentation at the shallow indentation depth compared to the film thickness. FT-IR and TDS analyses were performed for thus obtained films. Note that k-values of TMCTS treated porous silica films remained unchanged over 3days.

RESULTS AND DISCUSSION

Improvement of Film Properties by Vapor Annealing Method

Figure 1 summarizes the elastic modulus and the hardness of the porous silica films treated with TMCTS or HMDS. It is shown that both the elastic modulus and hardness increased significantly when TMCTS treatment was performed at 350ºC or above. For example, the TMCTS treatment at 400ºC enhances the elastic modulus and the hardness by a factor of about 2.0 compared to non-TMCTS-treated films with an elastic modulus of 4.3 GPa and hardness of 0.34 GPa.

The effect of TMCTS treatment time on the elastic modulus was studied as shown in Figure 2. The elastic modulus sharply increased with the treatment time up to 10 min and tends to saturate over 10min. It is likely that structural changes of the films, which are associated with the increase in the mechanical strength, occurred at the early stages of the treatment. In contrast, the HMDS treatment gave rise to no significant increase in the elastic modulus.

A cross-sectional TEM image of porous silica film after TMCTS treatment shown in Figure 3 reveals that the pore structure in silica film is unchanged before and after TMCTS treatment. The treated films have a k-value as low as 2.2, which was comparable to those treated with conventional HMDS.

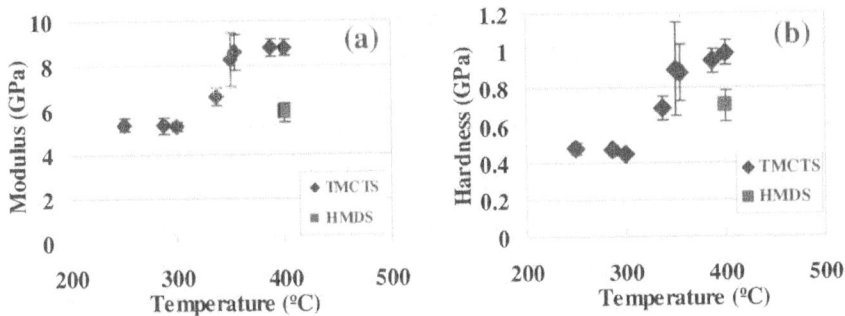

Figure 1. Effects of annealing temperature on (a) elastic modulus and (b) hardness of porous silica films.

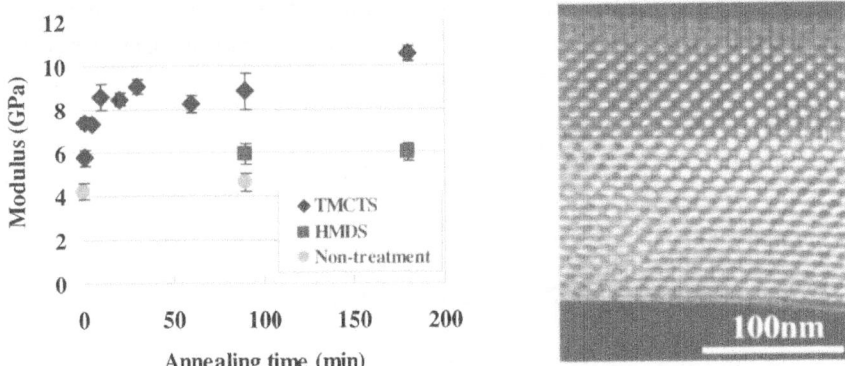

Figure 2. Effects of annealing time on elastic modulus of porous silica films.

Figure 3. Cross-sectional TEM image of porous silica film after TMCTS treatment.

Possible Mechanism for Mechanical Strength Enhancement

In order to get insights in the improvement mechanism of the mechanical strength of the TMCTS-treated films described above, FT-IR and TDS analyses were performed on both TMCTS-treated and non-treated films. Figure 4 shows the FT-IR spectra of the TMCTS-treated films, showing that the decrease in the ratio of the Si-H stretching peak intensity to C-H one above 350ºC results in the enhancement of the elastic modulus of the films. It is well-known that polymerization of TMCTS can provide two types of structurally different polymers (i.e., linear and cross-linked structures) [8]. The decrease in the Si-H bonds absorbance by maintaining C-H absorption strength suggests that TMCTS was transformed to cross-linked polymers by reducing Si-H bonds above 350ºC. It should be noted that H_2O is essential for the cross-linking polymerization of TMCTS. The supply of H_2O needed for the polymerization presumably comes from adsorbed water on the silica film surface. This is supported by the fact that the elastic modulus and the hardness of the TMCTS-treated films reached the maximum at the desorption temperature of the chemically adsorbed water (Figure 5). Furthermore, X-ray photoelectron spectroscopy (XPS) analysis was performed on TMCTS-treated film in its transversal direction (results not shown). The carbon content was found to be constant from the top to the bottom of films. Since TMCTS has smaller molecule size than the pore diameter, it diffuses and reacts uniformly with the pore wall surface. Based on these results, we have concluded that the treatment of the porous silica films with TMCTS leads to the formation of the cross-linked poly(TMCTS) network, which enhances the mechanical strength of the films. A possible structural model for the cross-linked poly(TMCTS) network is illustrated in Figure 6, where pore wall surfaces are covered with a mechanically stable layer.

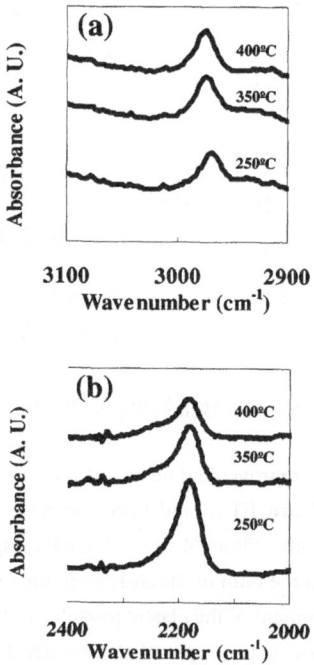

Figure 4. FT-IR spectra for porous silica films; (a) C-H stretching peak intensity and (b) Si-H stretching peak intensity after TMCTS treatment at 250, 350 and 400ºC.

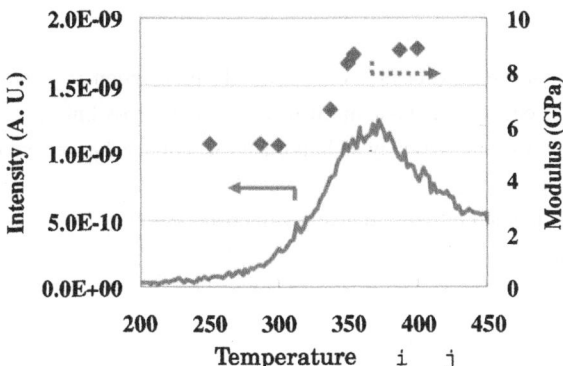

Figure 5. TDS spectra of porous silica films before TMCTS treatment (solid curve) and elastic modulus of the films after TMCTS treatment (   ).

Cross-linking poly(TMCTS) network

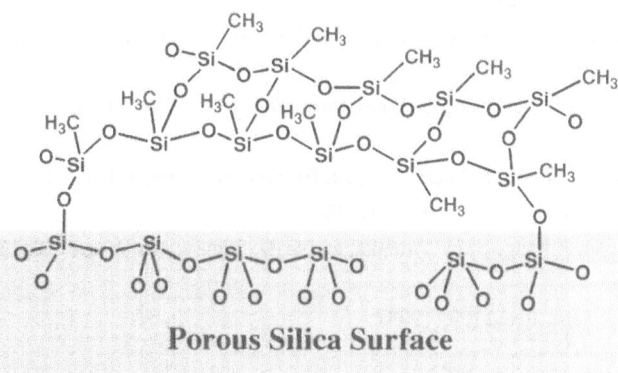

Porous Silica Surface

Figure 6. A structural model for a cross-linked poly(TMCTS) network formed on porous silica wall surfaces.

CONCLUSIONS

The TMCTS-vapor-annealing method has been developed for the preparation of porous silica films having an ultralow dielectric constant and mechanical strength. Cross-linked poly(TMCTS) network on the pore wall surfaces play an important role in the enhancement of the mechanical strength of the films.

The TMCTS-vapor-annealing method introduced herein will open a pathway for synthesizing interlayer dielectrics.

ACKNOWLEDGMENT

This research was supported by NEDO of Japan.

REFERENCES

1. J. Y. Chen, F. M. Pan, A. T. Cho, K. J. Chao, T. G. Tsai, B. W. Wu, C. M. Yang, and Li Chang, *J. Electrochem. Soc.* **150,** F123 (2003).
2. D. M. Smith, T. Ramos, K. H. Roderick, S. Wallace, J. Drage, H. –J. Wu, N. Viernes and L. B Brungardt, *US Patent 6395651* (2002).
3. Y. Lu, H. Fan, N. Doke, D. A. Loy, R. A. Assink, D. A. LaVan, and C. J. Brinker, *J. Am. Chem. Soc.* **122,** 5258 (2000).
4. C. –Y. Ting, D. –F. Ouyan and B. –Z. Wan, *J. Electrochem. Soc.* **150,** F126 (2003).
5. M. Ogawa, *Chem. Commun.* 1149 (1996).
6. Y. Lu, R. Ganguli, C. A. Drewien, M. T. Anderson, C. J. Brinker, W. Gong, Y. Guo, H. Soyes, B. Dunn, M. H. Huang, and J. I. Zink, *Nature* **389,** 364 (1997).
7. D. Zhao, P. Yang, N. Melosh, J. Feng, B. F. Chmelka, and G. D. Stucky, *Adv. Mater.* **10,** 1380 (1998).
8. H. Fukui, *J. Soc. Cosmet. Chem. Jpn.* **27,** 3 (1993).

Mat. Res. Soc. Symp. Proc. Vol. 812 © 2004 Materials Research Society

Probing Effects of Etching Plasmas on the Properties of Porous Low-k Dielectrics

L. Wang, J. Liu, W.D. Wang, and D.Z. Chi[*]
Institute of Materials Research & Engineering, 3 Research Link, Singapore 117602

D. W. Gidley
Department of Physics, University of Michigan, Ann Arbor, Michigan 48109

A. F. Yee
Department of Chemical Engineering & Materials Science, University of California, Irvine, CA 92697-2800

Abstract

The application of porous low-k interlayer dielectrics is needed for reducing the parasitical capacitance, especially at 65-nm node and beyond. The understanding of process-induced modifications to material properties is crucial for a successful integration of these low-k dielectrics. The dry etching processes of porous low-k materials are important modules in ULSI fabrication. In this study, the interaction between MSQ-based JSR LKD-5109 films (shown by PALS to have interconnected 2.8 nm size pores) with CF_4/O_2 plasma has been investigated. Various ratios of O_2 content were designed to characterize its effects on the etch rate, formation of polymerization layer, and properties of the LKD-5109 film. Composition analysis was conducted by SIMS and FTIR. Moisture absorption and fluorine diffusion into low-k films after etch process are observed, along with carbon depletion near the surface region. The influence of etching chemistries on the morphological characteristics of thin Ta barrier layers (8-nm in thickness) deposited on etched low-k films were further investigated by SEM, and it is found that oxygen concentration has significant influences on the morphological characteristics of thin Ta barriers.

1. Introduction

The size scaling in IC technology poses great challenges for back-end-of-line (BEOL) interconnect as the resistance-capacitance (RC) delay becomes a major limitation for the device performance. To address this issue, major changes must be implemented at each generation: copper metallization was first introduced to reduce interconnect line resistance, and low-k dielectrics were developed to reduce the parasitic capacitance [1].

To further meet the interconnect technology requirement outlined in International Technology Roadmap for Semiconductors 2003 (k<2.4) [2], the application of porous low-k materials is required to significantly reduce the dielectric constant, especially for 65-nm node and beyond. However, a number of integration challenges must be solved before these low-k materials can be implemented successfully, including the compatibility of the porous low-k films with various plasma processes. In particular, the damage caused by dry etching process for trench/via formation is one of major integration concerns for advanced Cu damascene interconnect scheme [3] as reactive plasma may cause depletion of carbon component, film densification, dangling bonds and defects [4]. Therefore, a detailed understanding of the effects of plasma dry etching process becomes important to minimize the integration damage.

Fluorocarbon plasma is normally used for the etching of inorganic low-k dielectrics with oxygen added to facilitate the process by effectively removing the passivation layer formed during etching. In this study, the conventional fluorocarbon CF_4/O_2 plasma was used and the interaction between the plasma and JSR's LKD-5109, a MSQ-based porous low-k film with k value of 2.2, was investigated.

*Corresponding author email: dz-chi@imre.a-star.edu.sg

2. Experimental

Commercially available JSR LKD-5109 film with thickness of 400 nm was used for this study. The film's dielectric constant is 2.2, and refractive index is 1.248. Beam-positron annihilation lifetime spectroscopy (PALS) measurement shows that LKD-5109 film has interconnected pores with mean pore size of 2.8 nm. The plasma etch process was carried out in Oxford Plasma Lab80 Plus Reactive Ion Etch (RIE) system using fluorocarbon feed gas CF_4 and O_2 on blanket LKD-5109 film for 1 minute. The frequency, RF power, chamber pressure were maintained at 13.56 MHz, 100 W, and 50 mTorr, respectively. The oxygen flow rate was varied from 0 to 60 sccm with total gas flow rate fixed at 100 sccm. The specific combinations of CF_4/O_2 flow ratios in this study were 100/0, 95/5, 90/10, 80/20, 60/40, and 40/60 sccm. The etched samples were immediately treated with 15s oxygen plasma flush in the same chamber in order to remove polymer layer and any residue on the surface. The etch rate was measured using KLA-Tencor P-10 surface profiler on patterned structures. 8 nm Ta films were deposited on processed samples by Denton magnetron sputtering system using Ar plasma to evaluate the effect of dry etching process on Ta morphological characteristics.

PerkinElmer FTIR spectrometer was employed to investigate the chemical bonding and structure under transmission mode, and barrier morphology was characterized using JEOL JSM-6700F field emission SEM. Time-of-flight secondary ion mass spectroscopy (TOF-SIMS) was used to analyze the compositional change of low-k films. The TOF-SIMS profiles were measured by an ION-TOF IV instrument using the dual beam configuration. The sputtering ion beam was Cs (3keV, 50 nA, 200 x 200 μm^2 raster) and the analysis beam was Ga (25keV, 3 pA, 100 x 100 μm^2 raster). All profiles were recorded in negative mode, using the non-interlaced mode of the spectrometer. The adhesion property of Ta to low-k film was assessed qualitatively using tape test method.

3. Results and Discussions

3.1. Etch rate

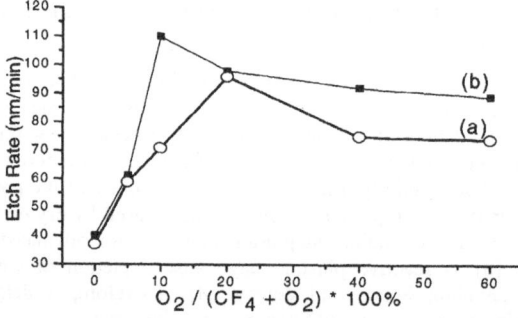

Fig. 1. Etch rate of LKD-5109 film with function of O_2 partial pressure in CF_4/O_2 plasma. (a) RF power: 100W, (b). RF power: 150W.

Fig. 1 schematically shows the etch rate of LKD-5109 film as a function of oxygen concentration. It is well known that the fluorocarbon deposition takes place on the surface during plasma processing, and controls the etch rate by suppressing the delivery of activation energy and transport of reactants [5,6,7]. The slow etch rate with the absence of O_2 can be attributed to the presence of a thick C_xF_y passivation layer formed on the film surface [7].

Etch rate increased with the increase of oxygen content, reaching a maximum etch rate of 96 nm/min with 20% O_2 content (compared with 40 nm/min when O_2 is not added). The enhancement of etch rate can be explained in terms of the thinning of polymer layer under O_2 exposure which facilitate the chemical reaction at the surface, as oxygen plasma is effective in removing the polymer layer. On the other hand, the slightly decreased etch rates at higher O_2 contents are due to the reduced reactant concentration at the etch front, resulting from further thinning of the polymer layer by O_2 plasma.

Fig. 1 also shows the RF power has a significant impact on etch rate. It is noted that at a higher RF power (150 W), the maximum etch rate with a higher value is reached at a lower O_2 concentration, as compared to that under 100 W RF-power. This is mainly because that the higher power regulates the ion energy through substrate bias, and the ionized plasma contains more energetic radicals for ion activation of surface sites. Such higher energy ion activation can inhibit in polymer formation and promote etching process [5], and hence, the function of oxygen can be compensated.

3.2. FTIR

The change in the chemical bonding structure in LKD-5109 as a function of oxygen content in the plasma was studied using transmission mode FTIR, as shown in Fig. 2. The spectrum of the as-deposited LKD-5109 film reveals a typical MSQ structure with Si-CH$_3$ peak at 1270 cm^{-1}, Si-O-Si stretching cage-like peak near 1130 cm^{-1}, Si-O-Si stretching network peak near 1070 cm^{-1}, and the C-H peak at 2970 cm^{-1}. It can be clearly seen that the intensity of the network peak related to Si-O-Si bonds was reduced when oxygen was increased, and disappeared when O_2 reached 60% in CF$_4$/O$_2$ mixture. Meanwhile, the intensity of the cage-like structure peak is enhanced, corresponding to the reduction of network structure signal. This observation indicates that the porous low-k film lost its intrinsic network Si-O-Si characteristic, and converted to cage-like structure under O_2 plasma exposure during etch process.

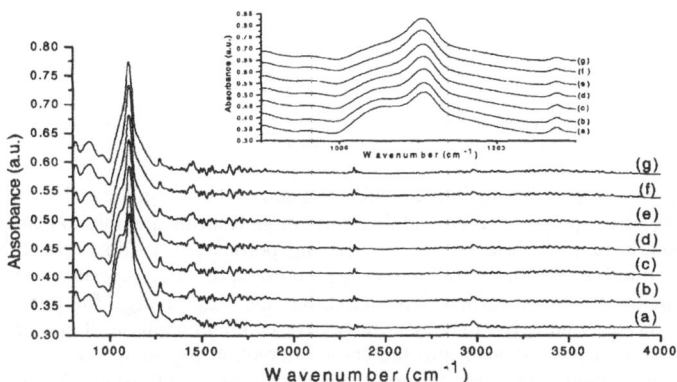

Fig.2. FTIR spectra of LKD-5109 films after etch process with different combinations of gas flow rate (a). As-deposited film, (b). CF$_4$/O$_2$ = 100/0, (c). CF$_4$/O$_2$ = 95/5, (d). CF$_4$/O$_2$ = 90/10, (e). CF$_4$/O$_2$ = 80/20, (f). CF$_4$/O$_2$ = 60/40, (g). CF$_4$/O$_2$ = 40/60 (unit: sccm).

The FTIR also shows that the CH$_3$ peak intensity was reduced with the increasing of the oxygen, and absorption peaks related to moisture absorption appeared within the 3000-3600 cm^{-1} interval. This indicates that the LKD film was degraded by the oxidizing ambient caused by dry-etch chemistries. Such degradation is due to the break-up of Si-C bonds through reacting with oxygen radicals diffused into the porous network of LKD film, which generates dangling bonds. The dangling bonds may be easily converted into Si-OH bonds when the film is exposed to the environment. The presence of Si-OH bonds makes low-k film hydrophilic, and consequently results in moisture absorption.

3.3. Composition Analysis

Fig. 3 shows a depth profile of the as-deposited LKD-5109 film as well as those of etched samples measured by TOF-SIMS. The carbon profile in Fig. 3(a) reveals the depletion of carbon at the top surface region. As expected, higher O$_2$ concentrations caused more severe carbon depletion, 50 nm depletion width for zero O$_2$ content as compared to 150 nm for 40% O$_2$. Meanwhile, fluorine penetration was also observed after etching process, which is shown in Fig 3(b), and the amount of diffused fluorine increases with increasing O$_2$ content.

Interestingly, the fluorine in the LKD-5109 film exhibits a two-step concentration distribution profile. It is proposed that the penetration of fluorine is mainly driven by the diffusion mechanism when the polymer layer is formed during the etch process. The polymer layer serves as a buffer layer to reduce the incident F energy, and also prevents more

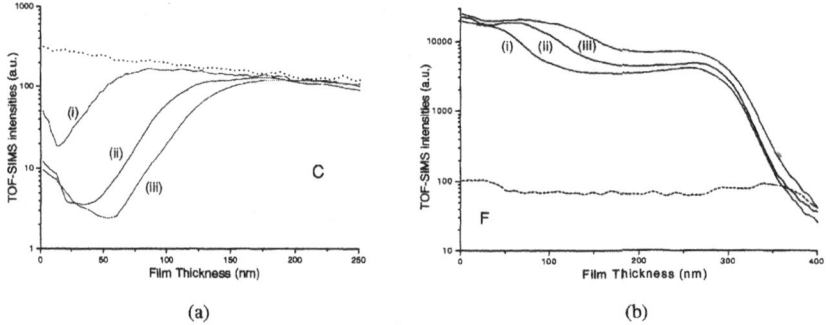

| (a) | (b) |

Fig.3. TOF-SIMS depth profile of carbon (a), and fluorine (b) for low-k films exposed to different etch process. (i). CF$_4$/O$_2$ = 100/0, (ii). CF$_4$/O$_2$ = 80/20, (iii). CF$_4$/O$_2$ = 60/40.

fluorine species from diffusing into the bulk. The addition of O$_2$ in the etching plasma causes the thinning of the polymer layer, and therefore, more fluorine are able to penetrate through the polymer layer to diffuse into the underlying porous low-k films, which explains the increase of fluorine content with adding O$_2$ partial pressure. It is important to note from Fig. 3 that both the width of carbon depletion layer and that of high fluorine concentration plateau are correspondingly same under respective etching condition. It can be inferred that fluorine may be easily trapped into the carbon depletion region where a large amount of dangling bonds readily available to form chemical bonds with fluorine, due to the break-up of Si-C bonds. On the other hand, the penetration of fluorine throughout the film is due to the fast diffusion of F in the highly interconnected porous matrix, which was revealed by the PALS characterization.

3.4. Barrier Integrity

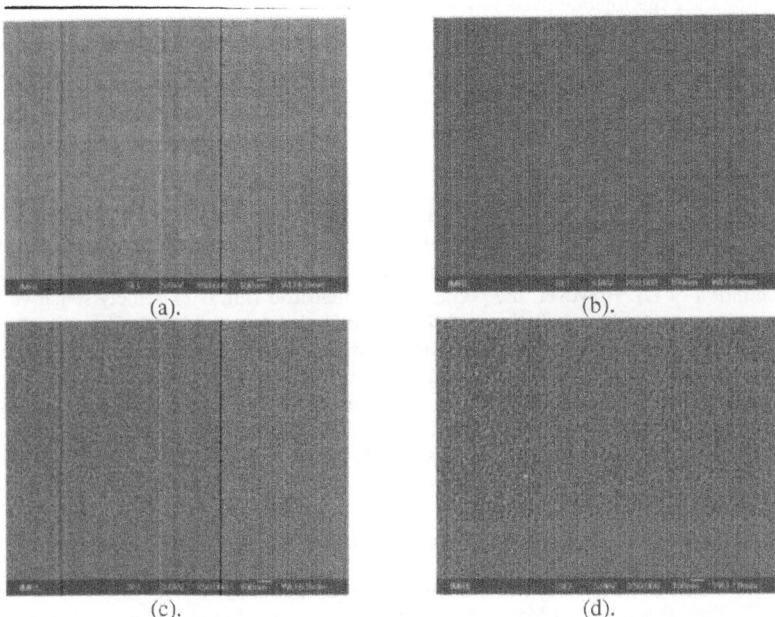

(a).   (b).

(c).   (d).

Fig.4. SEM micrographs showing the 8 nm Ta morphology deposited on (a). As-deposited LKD film, (b). LKD film with $CF_4/O_2 = 100/0$ etch, (c). LKD film with $CF_4/O_2 = 90/10$ etch, and (d). LKD film with $CF_4/O_2 = 60/40$ etch

Fig.4 shows the SEM images of 8 nm Ta morphology on LKD-5109 films with different etch treatments. Both Ta barrier films on as-deposited low-k and etched low-k using CF_4-only chemistry show smooth and continuous surfaces without voids, as shown in Fig.4(a) and 4(b). However, addition of O_2 in the etching plasma causes a dramatic change in Ta surface morphology, and worm-like features are observable in the Ta barrier layers (shown in Fig.4(c) and 4(d)), which indicates a bad wetting of Ta film on O_2 exposed surface [8]. Furthermore, the adhesion property of barrier to the low-k surface was qualitatively characterized using the standard tape test method, and the results are listed in table 1. The barrier layer deposited on the etched film surface without O_2 plasma exposure has acceptable adhesion strengths while the rest involving O_2 exposure failed the test.

It could be deduced that the composition of the surface has significant influences on barrier integrity. Several studies have found that the formation of Ta-C bonds at the interface when Ta compounds are deposited on substrates containing carbon [9-11]. It is also reported that the interaction between the C of the SiC:H and the Ta barrier layer occurs based on XPS analysis [12], and the chemical reaction at the interface could enhance the properties of Ta thin film. F. Iacopi et al. suggested that the presence of Ta, coupled with ion bombardment from the PVD deposition, result in weakening, breaking, and local cross-linking of carbon bonds [12,13], which may seal the surface of porous matrix, and promote barrier integrity. However, Carbon depletion takes place at the low-k surfaces when O_2 is added to etching plasma as previously discussed. The carbon-depleted surface may inhibit the chemical reaction with Ta, and leads to the formation of weaker bonding and poor adhesion, thus explaining the experimental observations.

Table 1. Tape test results showing the adhesion quality of Ta film to low-k surface

CF_4/O_2	As-deposited	100 / 0	95 / 5	90 / 10	80 / 20	60 / 40	40 / 60
Test	Pass	Pass	Fail	Fail	Fail	Fail	Fail

4. Conclusions

This study shows that the conventional CF_4/O_2 mixture plasma poses great challenges for the patterning of JSR LKD-5109 porous low-k dielectric. The addition of O_2 helps to effectively remove polymer layer to promote etching rate, with a maximum etch rate at an optimal concentration of O_2. However, the porous low-k dielectric film is extremely sensitive to the plasma exposure. Carbon depletion and fluorine diffusion into the low-k film was observed. O_2 addition in the etching process adversely modified the intrinsic structures of low-k film. Such influences lead to moisture absorption, and conversion of molecular bonding. The morphology and adhesion property of thin Ta diffusion barrier are also significantly influenced by the addition of O_2 in the plasma, which causes discontinuous formation of Ta film, as well as deterioration of adhesion strength.

References

1. M. Fayolle, G. Passemard, O. Louveau, F. Fusalba, J. Cluzel, Microelectronic Engg., 70, 255 (2003)
2. International Technology Roadmap for Semiconductors 2003.
3. E. Todd Ryan, Jeremy Martin, Kurt Junker, Jeff Wetzel, David W. Gidley, Jianing Sun, J. Mater. Res., 16, 3335 (2001)
4. E. Kondoh, M.R. Baklanov, H. Bender, and K. Max. Electro-chem. Solid-State Lett. 1, 224 (1998).
5. Da Zhang, Mark J. Kushner, J. Vac. Sci. Technol. A 19, 524 (2001)
6. N. R. Rueger, J. J. Beulens, M. Schaepkens, M. F. Doemling, J. M. Mirza, T. E. F. M. Standaert, and G. S. Oehrlein, J. Vac. Sci. Technol. A 15, 1881 (1997)
7. K. Miyata, M. Hori, and T. Goto, J. Vac. Sci. Technol. A 14, 2083 (1996)
8. D. Ernur, F. Iacopi, L. Carbonell, Herbert Struyf, K. Max, Microelectronic Engg. 70, 285 (2003)
9. G.R.Yang, Y.P.Zhao, B.Wang, E.Barnat, J.McDonald, and T.M.Lu, Appl. Phys. Lett. 72, 1846 (1998)
10. J.W.Nah, W.S.Choi, S.K.Hwang, and C.M.Lee, Surf. Coat. Technol. 123, 1 (2000)
11. T.Hara, K.Sakamoto, F.Togoh, H.Yang, and D.R.Evans, Jpn. J. Appl. Phys., Part 2, 39, L506 (2000)
12. F.Iacopi, Zs.Tokei, Q.T.Le, D.Shamiryan, T.Conard, B.Brijs, U. Kreissig, M.Van Hove, and K.Max, J. Appl. Phys. 92, 1548 (2002)
13. K.W.Gerstenberg and M.Grischke, J. Appl. Phys. 69, 736 (1991)

Mat. Res. Soc. Symp. Proc. Vol. 812 © 2004 Materials Research Society   F6.6

Dry Etch and Wet Clean Process Characterization of Ultra Low-k (ULK) Material Nanoglass®E

B. Ramana Murthy[1], C.K. Chang [1], Ahilakrishnamoorthy[1], Y.W. Chen[1] and Ananth Naman[2]
[1]Institute of Microelectronics, 11 Science Park Road, Singapore Science Park 2, Singapore 117685.
[2]Honeywell Electronic Materials, 1349 Moffett Park Drive, Sunnyvale, CA 94089, USA

ABSTRACT

NANOGLASS®E (NGE) ultra low-k (ULK) dielectric material, with a k-value of ~2.2, was integrated for 130 nm Cu/ULK interconnect process technology. This work deals with the characterization of reactive ion etching (RIE) and wet chemical processing of this film. Blanket films were characterized for etch rate, surface roughness, k-value change and chemical compatibility. Trench etching and post etch wet clean processes were developed and optimized enabling process integration for single damascene structures. Trench etch processes were evaluated for two etch schemes viz., etching under - photo resist and etching under hardmask. The details of each scheme will be described and advantages observed will be discussed. To evaluate effect of wet clean processes three different formulations were used. After formation of single damascene wafers, metal comb and serpentine structures were measured for metal continuity and bridging. Electrical continuity was achieved for long serpentine structures with 0.18µm/0.18µm line width/spacing. Based on voltage ramp test results the film was found to be sensitive to certain plasma etch conditions.

INTRODUCTION

It is well known that reduction in k-value of the films is principally achieved by reducing the film density, which implies that materials are to be formulated with more porosity [1]. The ability of porous Inter Metal Dielectric (IMD) layer to satisfy the stringent integration requirements depend on the material type and level of porosity [2]. Moreover, it is well known that if the materials are porous, they tend to be more susceptible to processes, mainly plasma etch and wet clean. In view of this, it is essential to evaluate the materials to find various effects of key processes such as plasma etching and post etch wet cleaning, well before integrating these materials for interconnect technology development. In the present context, firstly we have carried out material characterization of the ULK, after exposing to dry plasma etch and wet chemical processes on blanket films. Blanket films were characterized for the following parameters: etch rates, surface roughness, k-value change. Subsequently, RIE etch and wet clean processes were developed and evaluated for 130nm technology node Cu/ULK single damascene interconnects.

* This paper is a result arising from IME Wafer Technology Consortium (WTC). Member companies of the consortium are Honeywell Electronic Materials,USA, ASM, Japan, The Dow Chemical,USA, System on Silicon Manufacturing Co.Pte.Ltd.,Singapore, Cabot Microelectronics Corporation, USA

EXPERIMENTAL DETAILS

Initially, material spun-on 8" silicon wafer was cured and analyzed using Fourier Transform Infrared (FTIR) scan, X-Ray Photoelectron Spectroscopy (XPS) and Energy Dispersive X-Ray (EDX) analysis techniques. Typical material composition showed C, O and silicon as prime elements and XPS results showed 16% of C, 57% of O and 26% of silicon in the film.

Figure 1(a) shows a typical integration stack built. The ULK film was spin coated and Silicon Carbide (SiC) and USG (un doped silicate glass) dual hard mask layers were deposited by Plasma Enhanced Chemical Vapor Deposition (PECVD) [3]. Trench patterning was done on Nikon S203 model scanner which uses 248 nm laser source after coating bottom anti reflection coating (BARC) and 4800Å of UV-210 photo resist. Trenches were resolved at 180nm critical dimension (CD) at several locations. Wet chemical processing was carried out in a Semitool Magnum Spray Processor. Two types of chemical formulations viz., semi-aqueous fluoride mixture (Chemical A) and semi-aqueous amine mixture (Chemical B) were used in this study.

Trench etch process was evaluated for two schemes viz., trench etching of ULK under photo resist (scheme-1) and trench etching under USG hard mask (scheme-2). Trench etch process flow in both the schemes is depicted schematically in figure 1 and figure 2 respectively. Hitachi CD scanning electron microscope (CD-SEM), was used for qualitative analysis of top view images. JEOL Field Emission Scanning Electron Microscope (FE-SEM), was used to view the side walls of the trench. Plasma dry etch process was performed in a Dipole Ring Magnet (DRM) Unity-II dielectric etcher supplied by Tokyo Electron Limited.

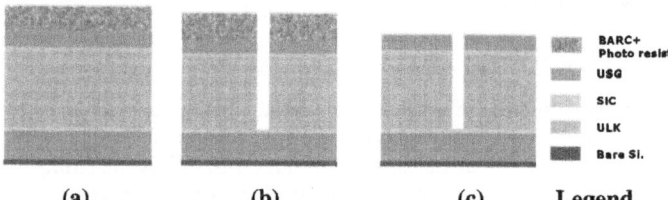

(a)   (b)   (c)   Legend

Figure 1: Stack details and etch process flow (scheme-1): (a): Stack; (b): Etching hard mask layers and ULK trench; (c): Resist removal.

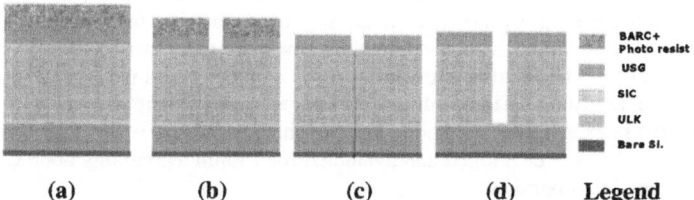

(a)   (b)   (c)   (d)   Legend

Figure 2: Stack details and etch process flow (scheme-2): (a): Stack; (b): Etching USG hard mask; (c): Resist removal; (d) Trench etching in second hard mask layer and ULK

This is a dual chamber single RF source medium density plasma etcher, where the dielectric film/ultra low k film etching was performed in the anodized aluminum type chamber and ashing was performed in the ceramic type chamber. Blanket film, trench etching and photo resist ashing was done using various conventional gas mixtures as described in the table 1.

RESULTS AND DISCUSSION

Blanket Film Characterization:

ULK film was etched with conventional $C_4F_8/CO/Ar/N_2$ gas mixture and moderate etch rate of 3000 Å/min. A low strip rate of 170 Å/min with O_2 strip plasma was obtained. A significant increase in surface roughness of the film was observed after etching. However, roughness had shown a 30% reduction after exposing to O2 plasma stripping compared to post etch roughness. Thus, in case of schemes with no hard mask, the strip process may not cause physical damage to the surface in terms of roughness. It was found that when film was exposed to plasma without CO, surface roughness could be reduced in the range of 30-50%. Also FTIR spectra of pre and post etched films showed no compositional changes.

The effect of wet chemical treatment was evaluated by measuring etch rates of ULK film when treated with chemicals A and B at room temperature. Etch rate on blanket films was measured to be <2Å/min. However, etch rate slightly increased to ~6Å/min at about 35-45°C and with longer treatment time. Dielectric Constant (k-value) measurements showed a 30% increase after treating with chemical A and 15% with chemical B. Subsequently baking was done at 200°C for 30 minutes and k-values showed a recovery of 50% and 5% respectively.

Trench etch process characterization-scheme-1:

Integration film stack was built as per shown in figure 2 and litho patterning was done using Shipley photo resist (PR) of 4800Å to resolve 180nm trenches. Etch process was characterized for etch rate, selectivity to PR and profile at SEM bar for dense and isolated trenches. An average etch rate of 4000Å/min was obtained at the trenches for ULK film. At the trenches, an etch selectivity (to photo resist) of ~4 for NGE film and ~3 for etching full stack including ULK, was obtained. However, at the dense trenches selectivity has dropped by 25% compared to isolated areas.

Table 1: Layers etched and typical plasma etch gas mixtures used

Etch Process	Layer Thickness (Å)	Gas Mixture
BARC etch (Trench)	600	CF_4/O_2
USG hard mask	2000	$C_4F_8/Ar./O_2$
SiC hard mask	500	$CH_2F_2/Ar./O_2$
NGE film	5000	$C_4F_8/Ar./CO/O_2$
Resist Ashing	------	Pure O_2

As in figure 3(a), Post etch CD (FICD) measurements showed a monotonic decrease across the pitch i.e. 1:1 to 1:5 for 180nm trenches, as shown in figure 3(a). This was the same as post litho trend which can be off-set by OPC in the mask design. However, optimized etch process has provided FICD within 10% limit of CD after litho patterning. Figure 3(b) shows RIE lag effect (ratio of etch rate inside trench/pad area) in NGE film studied after trench etching with three different etch gas combinations viz., C_4F_8/CO (base-line), C_4F_8 (no CO) and CHF_3. RIE lag was found to be more in case of CHF3 etch process and minimum in case of C_4F_8 (no CO), as shown in figure3(b). This type of RIE lag data will be useful to evaluate the etch process margin, if structures of different size and geometry are to be characterized. The plasma gas mixture flow ratio of C_4F_8/CO and C_4F_8/N2 was same (0.1) at a pressure of 35mT using a plasma density of 3 W/cm^2 for about 75 sec. The post etching O2 plasma stripping was done with a softer power density at 0.3 W/cm^2 where as H_2/N_2 stripping was done at 1.5 W/cm^2 at a flow ratio of 1:1. The chuck temperature (lower electrode) for the entire etch and strip processing was 40 ^{0}C.

Figure 4(a) and (b) shows typical cross-section of the 180nm trenches and close to vertical profile was obtained with flat trench bottom. However, a slight undercut was seen at the top of the trench in ULK film. The undercut was formed after the photo resist strip using pure O_2 plasma. However, undercut could be eliminated by performing photo resist strip using H_2/N_2 gas combination which was further confirmed by cross-section.

Trench etch process characterization-scheme-2:

An alternative trench etch process flow was evaluated which was different from scheme-1. In this, trench etching in ULK film was performed under USG hard mask after removing the photo resist, which was schematically shown in figure 3. After trench etching, severe under etch, corner rounding was observed at dense trenches and mildly at iso. This is found to be due to poor selectivity of ULK film to USG (hard mask) and selectivity being worse at dense areas compared to iso. Subsequently, based on the calculations in view of the other integration requirements, the USG hard mask thickness was increased to 5000Å from 2000Å and etching was done. Substantial improvement in the etch rate and profile was found after trench etching. However, post etch CD and profile was not within the acceptable limits as the aspect ratio was more in this case. This process flow completely eliminated the exposure of ultra low-k to strip plasmas (O_2 plasma in this case). This could possibly reduce the strip plasma damage to the ULK film [4, 5], which has been a serious concern in the low-k interconnects.

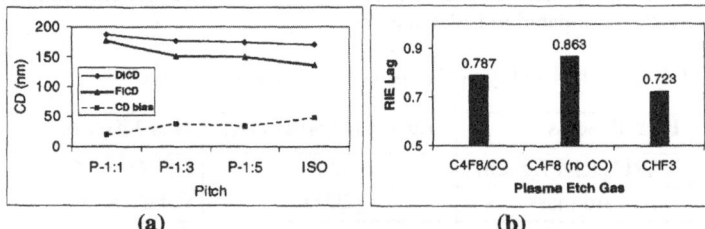

(a)   (b)

Figure 3: (a) Typical CD variation across pitch; (b) Typical comparison of RIE lag in ULK film after trench etching.

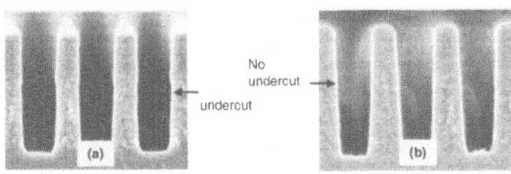

Figure 4: Trench etch profile after PRS with (a) O_2 plasma ; (b) H_2/N_2 plasma

Post trench etch cleaning:

Post trench etch clean process was evaluated using chemical A and B. Some delamination of trench structures was found after wet cleaning. This was attributed to the chemical susceptibility of the ULK film towards the formulations used. It is speculated that though the etch rates of ULK is insignificant on blanket films, after plasma etch and strip in the patterned structures might have possibly modified the sidewalls of the trench. This in turn become weak zone for chemical attack when exposed leading to instability of patterns. Alternatively, post etch cleaning process was optimized with DI water cleaning and performance was found to be acceptable for process integration, as shown in figure 5.

Electrical and voltage ramp tests:

The resistance of serpentine metal lines of width and space of 0.18 µm and length 0.192 m was measured from 42 dies of a wafer fabricated using etch under PR scheme and plotted in Figure 6(a). The electrical yield on the basis of resistance was 95%.

The leakage current through dielectric between interspersed comb metal lines of width and space of 0.18 µm and tine length of 400 µm was measured as a function of potential at a ramp rate of 1 V/s and the results are shown in Figure 6(b). The leakage current was high for the samples subjected to scheme 1 and O_2 PRS at any given potential (not shown here). On voltage ramping, these samples failed at very low electric fields as shown by the median curve in Figure 1. This is attributed to trench sidewall damage by O_2 PRS resulting in poor coverage of Ta. For scheme 1 with H_2/N_2 based PRS recipe, median breakdown field is higher of the order of 1.5 MV/cm, but this value is still far below that of the bulk (3.5 to 4 MV/cm) measured on blanket film. It can be noticed that when etching was done under hardmask (scheme 2), the dielectric breakdown is ~ 2.5 MV/cm higher than the other two split conditions. The sidewall dielectric damage is the lowest in this split condition as in this scheme the trench sidewall was never exposed to strip gas plasmas such as O_2 or H_2/N_2.

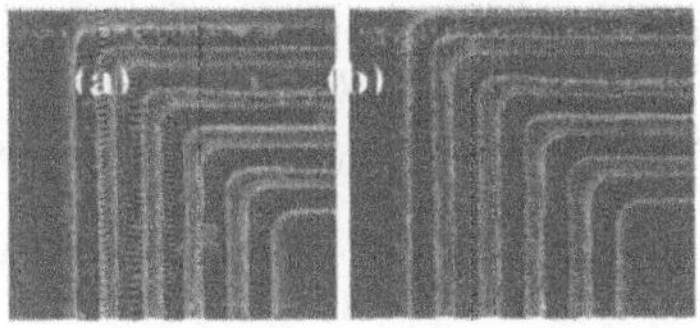

Figure 5: 180nm trenches before (a) and after (b) DI water cleaning

(a) 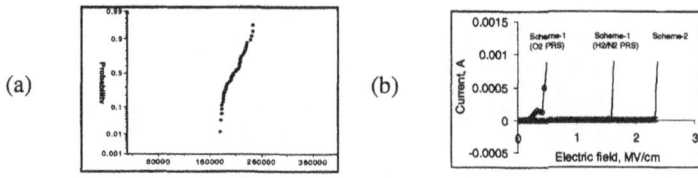 (b)

Figure 6: (a) - Resistance probability plot of serpentine structures; (b) – Leakage current as a function electrical field for different etch splits

CONCLUSIONS

The Nanoglass®E ultra low-k film was characterized for RIE and wet chemical clean processes. Trench etching and post wet cleaning processes were developed and optimized enabling successful integration for single damascene interconnects. Using various plasma etch gas combinations, surface roughness could be controlled on blanket films and wet chemical treatments showed insignificant etch rates at room temperatures. While chemical treatments showed significant impact on k-value the change was negligible with DI water treatment. Of the two trench etching schemes evaluated, etching under resist scheme gave better results in terms of electrical yield. Voltage ramp test has confirmed that when the film was not exposed to strip gas plasmas, breakdown field was closer to the bulk value.

ACKNOWLEDGEMENTS

Authors would like to acknowledge the wafer technology consortium, Rakesh Kumar, Moitreyee, Babu Narayanan, Samuel Lim, Chen Xian Tong for their support in this work.

REFERENCES

1. S.Purushothaman, S.V.Nitta, J.G.Ryan, C.Narayan, M.Krishnan, S.Cohen, S.Gates, S.Whitehair, J.Hedrick, C.Tyberg, S.Greco, K.Rodbell, E.Huang, T.Dalton, R.DellaGuardia, K.Saenger, E.Simonyi, S-T. Chen, K.Malone, R.Miller and W.Volksen, IEEE-2001 Proceedings of the International Electron Devices Meeting2001 pp. 529-532
2.Y.Y.Cheng, L.C.Chao, S.M.Jang, C.H.Yu and M.S.Liang. Proceedings of the IEEE 2000 International Interconnect Technology Conference, pp 161-163
3. K.Mosing, T.Jacobs, P.Kofron, M.Daniels, K.Brennan, A.Gonzales, R.Augur, J.Wtzel, .Havemann and A.Shiote, Proceedings of the IEEE 2001 International Interconnect Technology Conference, pp. 292-294.
4. T.Gao, W.D.Gray, M.Van Hove, E.Rosseel, H.Struyf, H.Meynen, S. Vanhaelemeersch, K. Maex, Proceedings of the IEEE 1999 International Interconnect Technology Conference1999, PP 53-55
5. Ching-fa Yeh, Yueh-chuan Lee, Yuh-ching Su, Kwo-hau Wu, Chein-hsin Lin, American Vacuum Society, 2000, pp 81-84

Modification of Nanoporous Silica Structures by Fluorocarbon Plasma Treatment

Woojin Cho, Ravi Saxena, Oscar Rodriguez, Ravi Achanta, Manas Ojha, Joel L. Plawsky, and William. N. Gill
Department of Chemical and Biological Engineering, Rensselaer Polytechnic Institute, Troy, New York 12180, USA
Mikhail R. Baklanov
IMEC,
Leuven, Belgium

ABSTRACT

Polymerization occurring during fluorocarbon plasma treatment as a potential method for pore sealing was investigated. CHF_3 was used as a reactant gas to expedite the rate of polymerization due to the presence of hydrogen and the low C/F ratio. The reactor pressure was varied from 30mTorr to 90mTorr to change the number of neutrals that act as the polymerizing species. The films were exposed to the plasma for times of 1min, 3min, and 5 min to observe the penetration depth of neutrals and the thickness of modified layer as a function of time. Dielectric constants were measured before and after plasma treatment. The film morphology was investigated by scanning electron microscopy before and after plasma treatment and a featureless surface morphology was observed at 90mTorr on a 56% porosity film. After plasma treatment, the average pore neck size decreases which may help reduce metal precursor penetration during metallization.

INTRODUCTION

Future integrated circuits require a low dielectric constant (k) material as an interlayer dielectric (ILD) [1, 2]. The incorporation of porosity is the most plausible way of reducing the dielectric constant of a material below 2. Nanoporous silica or silica xerogels are one promising alternative dielectric material as they can be made hydrophobic and the porosity and thickness can be tailored to desired values [2-4]. However, porous materials have several shortcomings such as metal precursor penetration into the pores during CVD, ALD and/or PVD processes [5, 6], poor mechanical strength [7], and low thermal conductivity [8]. The sealing of surface pores has been proposed as one way of preventing metal precursor penetration during metallization, without affecting (or even improving) the dielectric properties of the porous material. In this work, we study the formation and use of the polymer layer that forms during exposure to a fluorocarbon plasma for pore sealing.

EXPERIMENTAL DETAILS

Nanoporous silica films were prepared by a spin-on, binary solvent technique and the film porosity was controlled by changing the amount of glycol as the pore former [9]. The effect of plasma treatment in the nanoporous silica films was studied in a capacitively coupled plasma reactor (Plasmatherm 73, RPI). The plasma power was fixed at 247 W, the reactant gas (CHF_3) flow rate was 50 sccm, and the RF was 13.56 MHz. The films were exposed to plasma for 1 minute at three different pressures: 30, 60 and 90 mTorr. The films were also exposed to plasma at different times under 90 mTorr.

The refractive index and thickness of the films before and after plasma treatment were measured by ellipsometry. The wavelength of the light was 632.8 nm and the angle of incidence was 65°. A Cauchy model was used to model the dispersion of the refractive index as a function of wavelength.

To measure the dielectric properties of the films before and after plasma treatment, a metal-insulator-metal structure was used. A highly doped [N-type, Silicon Quest International, resistivity < 0.004 ohm-cm] Si substrate was used as one metal layer and 1 μm thick Cu layer was sputtered on the nanoporous film for use as the other metal layer. The capacitance of the nanoporous film was measured at 1MHz and 30 mV in a HP4280 impedance analyzer.

The surface morphology before and after plasma treatment was observed by field emission scanning electron microscopy (FESEM, JEOL-JSM840). Rutherford backscattering (RBS) depth profiling was used to determine the amount and depth of fluorine penetration into the dielectric. The average pore size of the films before and after plasma treatment was obtained by ellipsometric porosimetry.

DISCUSSION

Figure 1 shows the surface morphology of a 56% porous film before and after plasma treatment as a function of the reactant gas (CHF_3) pressure. As the reactant gas pressure increased we observed an increase in the contrast in the SEM images, which indicates that etching during plasma treatment occurs. After etching, the surface becomes rougher. The 90 mTorr case shows a featureless surface.

The plasma-etching rate at 30 mTorr was 14.4 nm/min. The 90 mTorr sample did not show net etching behavior. The overall film thickness actually increased slightly (Figure 3). Many of positive ions that take part in the etching process lose their energy at higher reactor pressures due to an increased collision rate. The number of radicals (CF and CF2), that serve as precursors to polymerization are proportional to the reactor pressure, thus at higher pressures there is more polymer deposition.

Another remarkable result from the 90 mTorr plasma treated film was that the film was composed of two layers. The upper layer is a mix of deposited polymer and the porous silica backbone, with a higher refractive index than the as deposited film. The lower layer is the as-deposited porous silica, whose properties (porosity and refractive index) are the same before and after plasma treatment. More details about this will be mentioned later.

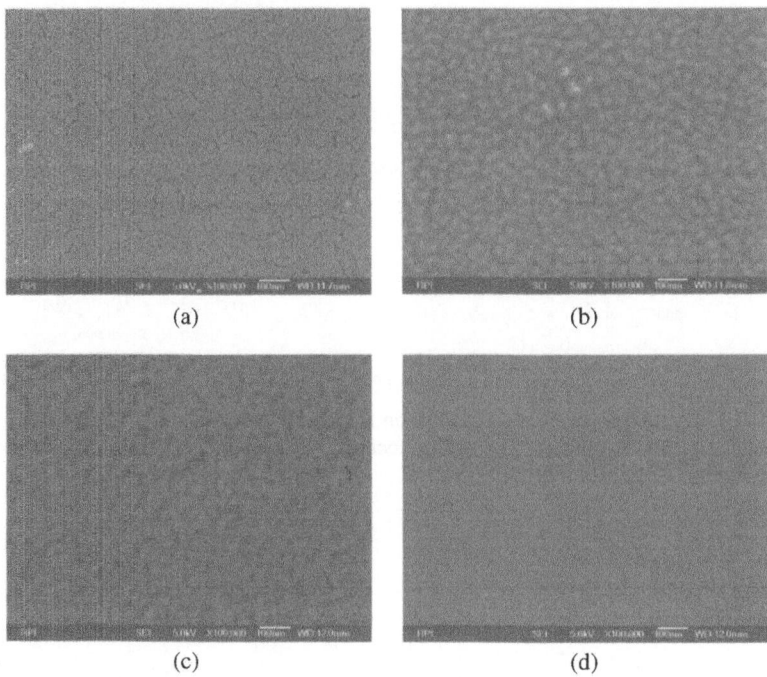

(a)   (b)

(c)   (d)

Figure 1. Effects of plasma pressure on polymerization as a pore sealing method:
(a) as prepared (56%), (b) 30 mTorr, (c) 60 mTorr, and (d) 90mTorr.

Figure 2 shows the fluorine penetration profile as a function of film porosity and
reactant gas (CHF_3) pressure, obtained by RBS. The depth of penetration of flourine is
significant so that if the feature size is smaller than 100 nm, we will have completely
modified the dielectric material. Figure 2(a) shows fluorine uptake vs. depth as a function
of film porosity for 60 mTorr pressure and an exposure time of 1min. Figure 2(b) shows
fluorine uptake vs. depth as a function of reactant gas pressure for a 58% porosity film and
an exposure time of 1 min. The fluorine penetration increases with porosity and fluorine
uptake increases with reactant gas pressure as expected.

Knudsen diffusion is the dominant mechanism for fluorine motion through the
nanoporous material in the pressure range of our plasma experiments. We have reported
that the average pore size of nanoporous silica is proportional to the porosity and varies
from 1 nm to 12 nm when the porosity changes from 20% to 70% [10]. Since the Knudsen
diffusion coefficient is linearly proportional to the radius of the pore the higher penetration
depth at higher porosities is expected. Since the concentrations of neutrals and radicals
increases with pressure, the fluorine uptake should increase with pressure and that is what
we observe. We assume that the detected fluorine atoms correspond to fluorine in the
deposited polymer.

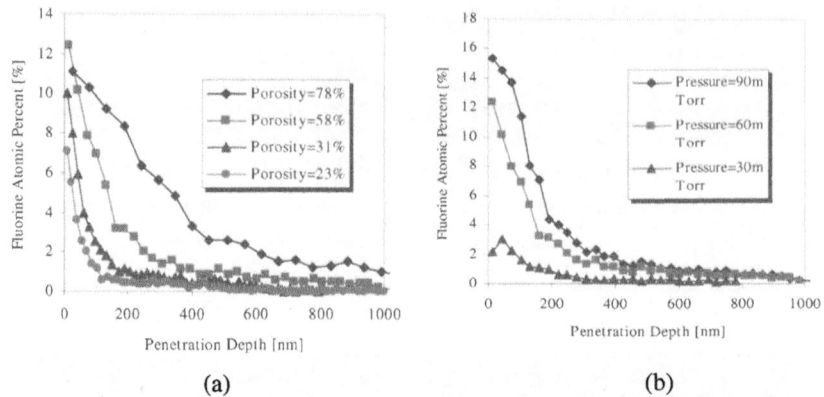

(a)   (b)

Figure 2. Effects of porosity and pressure on fluorine penetration in CHF_3 plasma using Rutherford backscattering depth profiling technique. (a) 60 mTorr, 1min plasma and (b) 58% porosity, 1min plasma

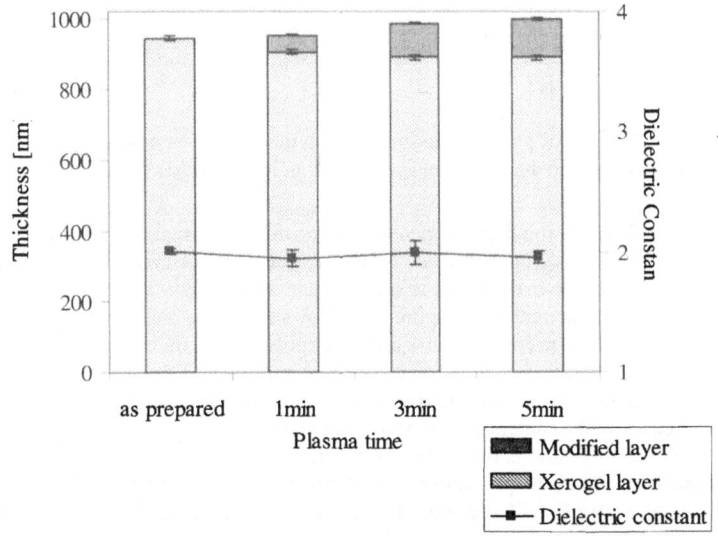

Figure 3. The effect of plasma time to the film on film properties (thickness and dielectric constant).

Figure 3 shows the effect of plasma exposure time on the film thickness (mixed and as-deposited layers) and the dielectric constant. The thickness of each layer was pretty uniform and their standard deviation was 3 ~ 7 nm. The mixed layer, with deposited polymer, shows a higher refractive index than that of the as-deposited layer. The refractive index (R.I.) of mixed layer increases (R.I. of the as-deposited film is 1.21, 1min CHF_3 plasma is 1.28, 3 min CHF_3 plasma 1.30 and 5 min CHF_3 1.35) with increasing plasma exposure time. As shown in Figure 3, the thickness of the mixed layer increases as the plasma exposure time increases. The dielectric constant does not change after plasma treatment and the dielectric constant is not affected by the deposited polymer layer. Therefore this polymer layer could be used as a protective or barrier layer.

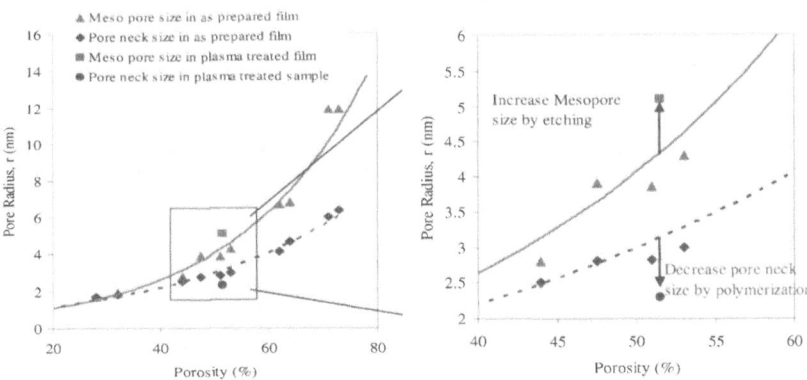

Figure 4. Pore (meso and neck) size as a function of porosity and comparison before and after plasma treatment [11, 12].

Figure 4 shows the average pore size of as-prepared films as a function of porosity [10] and the change in average pore size after plasma treatment (CHF3 90mTorr, 5min). In Figure 4, the graph on the left shows the overall average pore size as a function of porosity and the graph on the right is a magnification of that graph for the porosity range 40~60%. One can notice that the size of pore body increases (red square) and the size of the pore neck (blue circle) decreased in the plasma treated film. We think that the increase in the size of pore body is due to etching and the decrease in the size of pore neck is due to polymer deposition in that region.

We speculate that the modified structure by plasma treatment at higher pressures has the potential of making the dielectric more resistant to the diffusion of metal precursors during the metallization process due to the narrowing of pore neck (micro pore) size since the diffusivity in Knudsen regime is linearly proportional to pore size.

CONCLUSIONS

A deposited polymer layer produced during plasma treatment may be a potential solution to pore sealing of nanoporous materials. We observed a featureless surface morphology in SEM micrographs due to polymer deposition at a reactor pressure of 90mTorr on 56% porosity film. The polymer layer does not affect the dielectric constant of the film though it does have a higher refractive index than the underlying nanoproous silica. RBS depth profiling shows higher penetration for higher porosities and higher fluorine contents at higher reactant pressures. These results will be useful for correlating the Knudsen diffusivity with average pore size in nanoporous media. Ellipsometric porosimetry results show a narrowing of the pore neck and widening of the pore body following plasma treatment.

ACKNOWLEDGEMENT

This work is supported by the Semiconductor Research Corporation under task 995.015.

REFERENCES

1. International Technology Roadmap for Semiconductor, 2001.
2. K. Maex, M. R. Baklanov, D. Shamiryan, F. Lacopi, S. H. Brongersma, and Z. S. Yanvitskaya, *Applied Physics Reviews*, **93**, 8793 (2003).
3. P. S. Ho, J. Leu, W. W. Lee (Eds.), *Low Dielectric Constant Materials for IC Applications*, Springer (2002).
4. J. L. Plawsky, W. N. Gill, A. Jain, and S. Rogojevic, *Interlayer Dielectrics for Semiconductor Technology, Chapter 9*, S. P. Murarka, M. Eizenberg, and A. K. Sinha (Eds), Elsevier Inc., (2003).
5. M. Ritala and M. Leskela, *handbook of thin film materials*, H. S. Nalwa (Eds), Academic, (2002)
6. F. Iacopi, M. R. Baklanov, E. Sleeckx, T. Conard, H. Bender, H. Meynen, and K. Maex, *J. Vac. Sci. Technol*. **20**, 109 (2002).
7. A. Jain, S. Rogojevic, W. N. Gill, J. L. Plawsky, I. Matthew, M. Tomozawa, E. Simonyi, S. T. Chen and P. S. Ho, *J. Appl. Phys.*, **90**(11), 5832 (2001).
8. A. Jain, S. Rogojevic, S. Ponoth, W. N. Gill, J. L. Plawsky, E. Simonyi, S. T. Chen and P. S. Ho, . *J. Appl. Phys.*, **91**(5), 3275 (2002).
9. S. V. Nitta, V. Pisupatti, A. Jain, P. C. Wayner, Jr., W. N. Gill, and J. L. Plawsky, *J. Vac. Sci. Technol. B*, **17**, 205(1999).
10. R. Saxena, O. Rodriguez, W. Cho, W. N. Gill, J. L. Plawsky, M. R. Baklanov, and K. P. Mogilnikov, *J. Non-crystalline Solid*, accepted (2004).
11. M. R. Baklanov, K. P. Moginikov, V. G. Polovinkin, F. N. Dultsev, "Determination of pore size distribution in thin films by ellipsometeric porosimetry", *J. Vac. Sci. Technol. B*, 18(3), 1385 (2000).
12. M. R. Baklanov, K. P. Mogilnikov, "Non-destructive characterization of porous low-k dielectric films", *Microelectronic Engineering*, 64, 335 (2002).

Mat. Res. Soc. Symp. Proc. Vol. 812 © 2004 Materials Research Society   F6.18

Chemical Routes to Improved Mechanical Properties of PECVD Low K Thin Films

S.M. Bilodeau, A.S. Borovik, A.A. Ebbing, D.J Vestyck, C. Xu, J.F. Roeder, T.H. Baum
ATMI, Inc., 7 Commerce Drive, Danbury, CT 06810. Email: sbilodeau@atmi.com

ABSTRACT

Increasing the elastic modulus and hardness of low K films is one of the key challenges towards integration of these materials into future integrated circuits. Several approaches are explored for increasing the hardness of carbon doped oxide (CDO) dielectrics. Several low K precursors and their mixtures specifically chosen to enhance the hardness (H) and modulus (E) of CDO films through chemically induced cross-linking. Composition and FTIR measurements suggest the presence of C-C and C-Si cross-linking with concurrent observation of improved film hardness and modulus at relatively low deposition temperatures. Films deposited at 373°C using diethoxy-methyl-oxiranyl have a hardness and modulus of 2.5 GPa and 18.1 GPa respectively. Films deposited at 180°C using tetramethylcyclotetrasiloxane (TMCTS) and 25% hardener have hardness and modulus of 1.5 GPa and 9.4 GPa, respectively. These film properties are significantly higher than those observed for TMCTS alone under similar deposition conditions. Based on these results a low temperature process with 25% hardener and 75% TMCTS combined with a porogen was used to produce a porous film with a k<2.5 and a hardness of 0.72GPa.

INTRODUCTION

The most common chemical precursors used for the deposition of CDO dielectrics are silanes or siloxanes with hydride and alkoxide functionalities to yield network structures with stable methyl terminating groups. TMCTS, trimethylsilane, dimethyldimethoxysilane and methylsilane all fall into this group. There have been reports in the literature that increased elastic modulus and hardness can be obtained by processing these precursors under conditions that encourage the formation of bridging carbon structures (Si-C-Si) [1]. Alternatively, there have been reports of enhanced mechanical properties using precursors that have bridging carbon within the precursor structure [2], or for those expected to form carbon cross-linked structures such as vinyl silanes. Here we report on PECVD using two new low k precursors that contain the oxiranyl functional group, dimethyl-dioxiranyl silane (DMDORS) and diethoxy-methyl-oxiranyl silane (DEOMORS) and mixtures of DMDORS with TMCTS.

In addition to high hardness and modulus it is also desirable to have the ability to reduce the dielectric constant of these films by incorporation of pores. For PECVD low k materials this can be accomplished by incorporation of a sacrificial material that is thermally removed in subsequent processing.[3] This processing route requires a somewhat lower deposition temperature due to the need to remove the porogen at a lower temperature than it is incorporated at. In this paper we focused on low deposition temperature processes to allow a larger processing temperature window for porogen incorporation and removal.

EXPERIMENT

Deposition

Films were deposited in an Applied Materials P5000 reactor onto 8" wafers using the deposition conditions summarized in Table 1. Generally a range of properties is achievable for each of these precursors with a tradeoff between higher hardness/modulus and lower k. Deposition conditions were optimized for each precursor with the goal of producing the highest hardness and modulus consistent with a dielectric constant near 2.8.

Table 1. Deposition conditions

Deposition Parameter	Value
Precursor Flow Rate (ml/minute)	0.13 - 0.81
Process CO_2 Flow (sccm)	0 - 430
Substrate Temperature (°C)	130 - 375
RF Power (W)	100 - 250
Pressure (torr)	2.5 - 6.0
Showerhead Spacing (mil)	225 - 460
CO_2 Carrier Flow (sccm)	100 - 200
O_2 Flow (sccm)	0 - 150
Precursor	TMCTS, DMDORS, TMCTS+DMDORS, DEOMORS, TMCTS+DMDORS+porogen

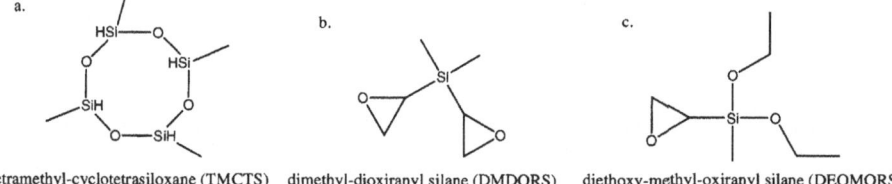

tetramethyl-cyclotetrasiloxane (TMCTS)   dimethyl-dioxiranyl silane (DMDORS)   diethoxy-methyl-oxiranyl silane (DEOMORS)

Figure 1. Chemical structure of precursor compounds a) TMCTS b)DMDORS c)DEOMORS

Post Deposition Processing

After deposition, films were annealed in a tube furnace in a 4%H_2/96%Ar mixture (referred to subsequently as a forming gas anneal). A ramp rate of 1°C/minute was used from 150°C to 400°C, with a soak at 400°C for 12 hours and a furnace cool to below 150°C. These films were then subjected to a 100W hydrogen plasma in the P5000 with a hydrogen flow of 200sccm and a showerhead spacing of 400mil at a pressure of 3.5torr for 420s Films were characterized in the as-deposited state as well as after the anneal and after the hydrogen plasma exposure.

Film Characterization

Dielectric constants were measured using a MDC 802-150 mercury probe connected to an HP 4192A impedance analyzer. The dielectric constants for the film deposited using the porogen and the film deposited using the TMCTS/25% DMDORS mixture were verified to be within 1% of the mercury probe's measured value using e-beam evaporated aluminum dots, patterned using a shadow mask. Spectroscopic ellipsometry (Sopra GES5) was used to measure the film thickness, as required for the dielectric constant calculation, and the film refractive index. Elastic modulus and hardness were measured using a MTS Nano Indenter SA2 in continuous stiffness measurement mode. The reported modulus and hardness are average values of 10 measurements, taken at indentation depths 50nm and 100nm for modulus and hardness, respectively. These indentation depths usually correspond to the minimum in the hardness or modulus versus indentation depth curves. FTIR transmission spectra were measured using a Nicolet Magna-IR 760 ESP spectrometer. Composition was measured on selected films using a combination of Rutherford backscattering spectrometry (RBS), hydrogen forward scattering spectrometry (HFS) and nuclear reaction analysis (NRA). Porosity of the porous low k films were characterized by nitrogen porosimetry (Micromeritics ASAP 2405). The Horvath-Kawazoe method [4] has been used extensively for microporous materials and was used to calculate a pore size distribution from the adsorption isotherms.

RESULTS and DISCUSSION
The dielectric constant and hardness for films deposited from TMCTS and DEOMORS are shown as a function of temperature in Figures 2 and 3. For both materials, the hardness after annealing and hydrogen plasma exposure is significantly higher for higher deposition temperatures. This is even more pronounced in the as-deposited state as is shown for films deposited from TMCTS in Figure 2.

The dielectric constant, indentation hardness and indentation modulus are compared for low temperature (180°C) deposited films using DEOMORS, DMDORS, TMCTS, a 25%DMDORS / 75%TMCTS mixture and the DMDORS/TMCTS mixture with a porogen in Figure 4. FTIR of these films in the as deposited condition, after the forming gas anneal, and after hydrogen plasma exposure are shown in Figures 5 and 6. The composition of selected films is shown in Table 2.

An increase in modulus and hardness was observed after forming gas annealing for the films deposited from DEOMORS, DMDORS, TMCTS and the 25% DMDORS / 75% TMCTS mixture. Examination of the FTIR spectra (Figures 5 and 6) show a significant increase in the absorbance of the network Si-O-Si network bands near 1040cm^{-1} for these films. This is consistent with the observed increase in hardness and modulus. No significant change is observed in the Si/O ratio, suggesting that the increase in network Si-O-Si is related to reorganization of oxygen present in the film after deposition.

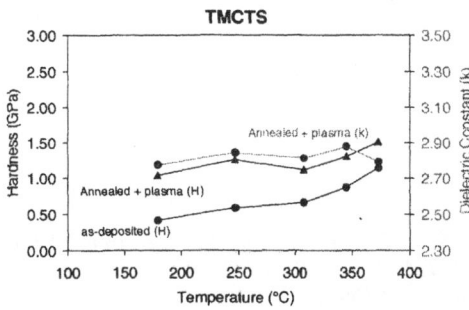

Figure 2. Hardness and K of films deposited from TMCTS at temperatures between 170 and 370 °.

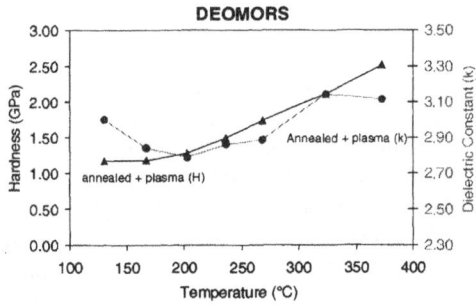

Figure 3. Hardness and K of films deposited from DEOMORS at temperatures between 170 and 370 °C

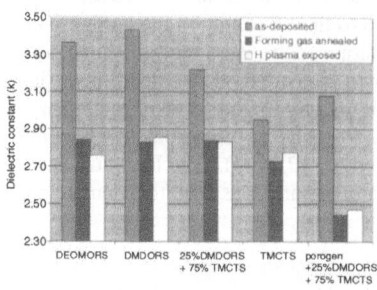

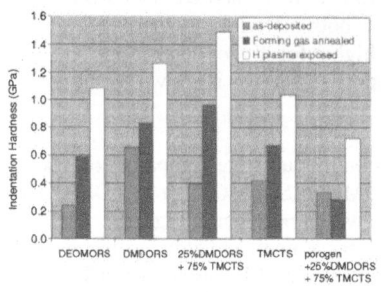

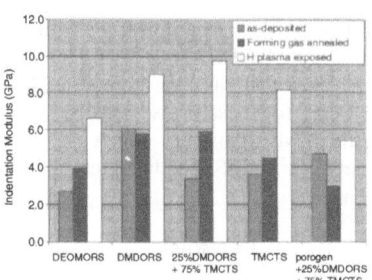

Figure 4. *(a) Dielectric constant, (b) Hardness and (c) modulus for films deposited from TMCTS, DEOMORS, DMDORS and TMCTS/25% DMDORS deposited at 180 °C, as-deposited, after a forming gas anneal and after hydrogen plasma exposure.*

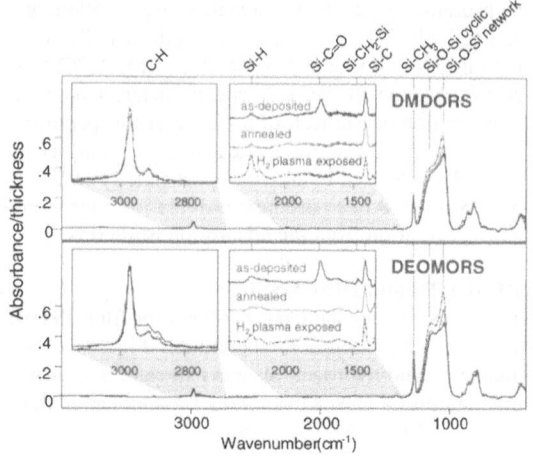

Figure 5. *FTIR spectra of a film deposited from DMDORS and DEOMORS deposited at 180°C as deposited, forming gas annealed, and after hydrogen plasma exposure*

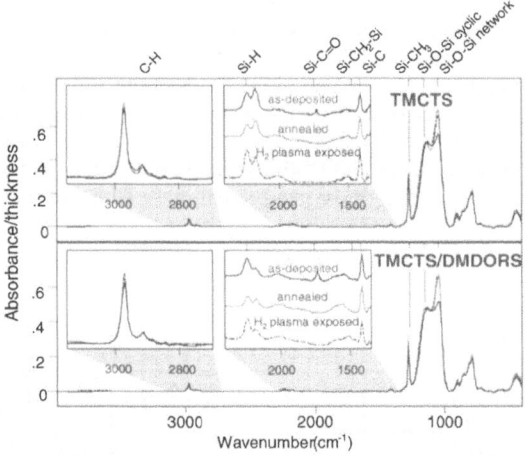

Figure 6. *FTIR of low k films deposited from TMCTS and 75%TMCTS/25%DMDORS at a deposition temperature of 180°C.*

The hardness and modulus decrease slightly after annealing for the film deposited from the mixture of DMDORS/TMCTS and porogen. We believe that in this case the stiffness of the backbone is increased by the forming gas anneal, but this is more than offset by the introduction of porosity (see Table 3).

For all of the films deposited at 180°C there was a substantial reduction in dielectric constant (k) during the forming gas anneal. For the porogen containing film this is attributed to the introduction of porosity. For all of the films deposited at 180°C we have observed the presence of the carbonyl functionality in the as deposited film FTIR centered near 1725cm^{-1} that is removed by a forming gas anneal. The carbonyl functionality is highly polarizable and its removal contributes to the reduction in k upon forming gas anneal. The observation of carbonyl consumption, network Si-O-Si formation and loss of carbon suggest the decomposition of carbonyl to yield a siloxane bridge and hydrocarbon species.

There are also contributions to the reduction in k from removal of some hydrocarbon species as evidenced by loss of FTIR absorbance near 2900cm^{-1} and reductions in the C/Si ratio as observed by RBS/HFS/NRA.
The hardness and modulus of all of the films deposited at 180°C exhibited increases in hardness and modulus after hydrogen plasma exposure. The hardness increases ranged from a 50% increase for the film deposited using DMDORS to a 150% increase for the film deposited using TMCTS+DMDORS+porogen. Examination of the network Si-O-Si region in the FTIR shows some additional network formation for the case of DMDORS and DEOMORS but very little for the case of TMCTS and TMCTS+DMDORS. The only other significant change in the FTIR spectra after hydrogen plasma exposure is a small increase in the Si-H absorbance (near 2200cm^{-1}). This would be expected to either decrease modulus/hardness or not change it significantly depending on whether network groups or terminal groups were consumed in the formation of Si-H. There is a drop in C/H ratio upon hydrogen plasma exposure. This might indicate the removal of H from C with the formation of additional C-Si or C-C cross-linking.

Figure 8 displays a comparison of the thickness normalized FTIR spectra of each of the low temperature deposited films after annealing and hydrogen plasma exposure. DMDORS resulted in a higher modulus and hardness than either DEOMORS or TMCTS even though it has less absorbance in the network region. We speculate that the additional strength derives from C-C or C-Si cross-linking.

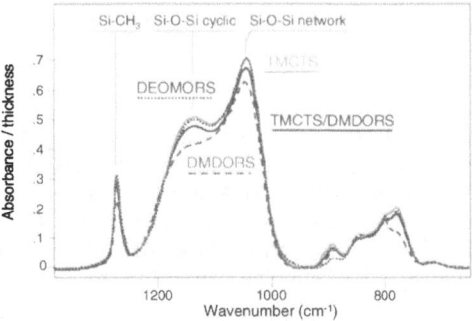

Figure 8. *Comparison of FTIR spectra of fully processed films deposited from TMCTS, DEOMORS, DMDORS and 25%DMDORS/75%TMCTS.*

113

Table 2. *Composition of selected films*

Precursor	H	C	O	Si	C/Si	O/Si	H/C
DEOMORS as-deposited	50.0	13.2	22.8	14	0.94	1.63	3.79
DEOMORS FG400C	44.6	13.9	25.6	16	0.87	1.61	3.21
DEOMORS FG400C + H plasma	35.0	15.7	30.7	19	0.84	1.65	2.23
TMCTS/DMDORS as-dep	38.7	17.1	27.9	16	1.05	1.71	2.26
TMCTS/DMDORS FG400C	37.6	15.9	29	18	0.91	1.66	2.36
TMCTS/DMDORS FG400C + H plasma	36.3	17	29	18	0.96	1.63	2.14

Table 3. *Properties of a porous low k film deposited using a mixture of 25%DMDORS, 75% TMCTS and a porogen.*

Property	Value
K	2.47
Hardness	0.72 GPa
Modulus	5.4 GPa
Median pore diameter	8.5 Å
Total nanopore volume	21% of film volume

CONCLUSIONS

The use of oxiranyl containing silanes for PECVD deposition of low k films can result in significant increases in indentation hardness and modulus when compared to films deposited from TMCTS. Films deposited with these functional units in the precursor had higher carbon content and low C/H ratios suggesting the presence of carbon cross-linking. Post deposition processing was found to both reduce the dielectric constant and increase film indentation hardness and modulus. This is partially due to an increase in the Si-O-Si network bonding as observed by FTIR.

DMDORS was particularly effective at producing high hardness low k films at low deposition temperatures. We have investigated mixtures of DMDORS and TMCTS and found that the hardness of mixtures was higher than that of either DMDORS or TMCTS alone. A 25%DMDORS/75% TMCTS mixture was deposited with a porogen to demonstrate a material with k~2.5 hardness of 0.72GPa and modulus of 5.4GPa.

ACKNOWLEDGEMENTS

The authors thank Donald Carruthers for his help with the nitrogen porosimetry measurements and their analysis.

REFERENCES

1 . G. Xu, J. He, E. Andideh, J. Bielefeld and T. Scherban, 2002 IEEE International Interconnects Technology Conference Proceedings, 11.4 (2002).
2 K. Usami, S. Sugahara, T. Kadoya and M. Matsumura, Proceedings of the 7th International Symposium on Quantum Effect Electronics, November 21, 2000, Tokyo (2000).
3 . A. Grill and V. Patel, Mat. Res. Soc. Symp. Proc., **612**, pD2.9.1 (2000).
4 Horvath, G., Kawazoe, K., J. Chem. Eng. Japan, **116** (6), 470 (1983).

Metallization, Barriers, and Capping

Mat. Res. Soc. Symp. Proc. Vol. 812 © 2004 Materials Research Society

Optimization of Dielectric Cap Adhesion to Ultra-Low-k Dielectrics

Greg Spencer, Alfred Soyemi, Kurt Junker, Jason Vires, Michael Turner, Stuart Kirksey, David Sieloff and Narayanan Ramani
Semiconductor Products Sector, Motorola, Inc.
3501 Ed Bluestein Blvd
Austin, Texas 78721

ABSTRACT

In this work, the adhesion of CVD dielectric caps to ULK MSQ spin-on dielectric materials with k values of 2.2 and 2.0, and a ULK CVD material with a k value of 2.7 is presented. A substantial improvement in cap adhesion to both the k2.2 ULK MSQ and the k2.7 ULK CVD material is demonstrated. The improvement is obtained using a low-k CVD glue material between the ULK dielectric and the subsequent cap material and/or by optimizing the CVD cap film deposition. Four-point bend measurement of adhesion strength is used to quantify the improvement in interface adhesion. The improvement in CVD cap adhesion is demonstrated to be strongly dependent upon both the glue layer film and the cap deposition conditions. While optimization of the CVD cap materials results in adequate adhesion for the k2.2 ULK MSQ, these improvements are demonstrated not to extend to the k2.0 ULK MSQ film.

INTRODUCTION

In order to meet low RC-delay requirements of future technologies, low-dielectric constant (low-k) back-end integrations are required. Future low-k integrations will likely require the use of ultra-low-k (ULK) dielectric materials with dielectric constant (k) below 2.7. The 2003 International Technology Roadmap for Semiconductors indicates k values of less than 2.7 will be required for the 90 nm technology node with a reduction to k values of less than 2.4 for the 65 nm node, and further reduction to less than 2.1 for the 45 nm node [1]. While the integration of materials with k value near 3.0 has been demonstrated [2], the integration of ULK materials poses many challenges due to low mechanical strength and high porosity of the dielectrics. In addition, ULK dielectric interface adhesion strength, both to underlying substrates and to capping materials, is known to be an issue. This current work focuses on characterizing and improving the adhesion strength of chemical vapor deposition (CVD) oxide caps to the ULK material as this is the weakest interface for the materials studied, i.e. adhesion failures during 4-point bend testing consistently occur at this interface. Cap materials are required due to the relatively low mechanical strength of ULK materials, and since the cap material must provide adequate strength to act as a chemical-mechanical polish (CMP) stop, adhesion of the cap is critical to prevent delamination during polishing.

While improving the adhesion of cap materials to ULK materials is critical, the need for improved adhesion must be balanced with the need for a ULK solution. Techniques to improve adhesion, such as plasma pre-treatments, while often capable of enhancing adhesion energies also can significantly degrade the ULK benefit of the materials. The goal of this study is to improve the adhesion of CVD dielectrics caps while having no or minimal impact on the ULK material or film stack k value.

EXPERIMENTAL DETAILS

This work investigates the adhesion of CVD dielectric caps to ULK methylsilsesquioxane (MSQ) based spin-on dielectric (SOD) materials with k values of 2.2 and 2.0, and an octamethylcyclotetrasiloxane (OMCTS) based ULK CVD material with a k value of 2.7. Throughout this paper, these materials will be referred to as k2.2 ULK MSQ, k2.0 ULK MSQ and k2.7 ULK CVD, respectively. Some of the properties of these materials are shown in Table 1. Note that while the measured k values of the SOD materials are lower than anticipated, the values are within expected tolerances of the k value measurement. Film thicknesses quoted in Table 1 and subsequent thickness data are determined using spectroscopic ellipsometry.

The CVD cap materials presented here are limited to oxides of silicon. The silicon oxides are deposited in single-wafer plasma-enhanced CVD chambers using tetraethyl-orthosilicate (TEOS), trimethylsilane (TMS), and OMCTS precursors. Although each of these precursors is carbon containing, the deposition conditions are established to minimize carbon incorporation into the film resulting in a silicon-oxide.

A dense low-k CVD SiCOH material with a dielectric constant of 3.0, subsequently referred to as low-k CVD, is also deposited in this study. This film is deposited in single–wafer plasma-enhanced CVD chamber using the TMS precursor.

This study is logically divided by the three ULK materials. For the k2.2 ULK MSQ, the use of the low-k CVD film as a glue layer between the ULK material and the oxide cap material is investigated. This dielectric stack is graphically represented in Figure 1. The thickness of the glue layer is varied between 20 nm and 100 nm. Two oxide cap precursors are investigated: TEOS and TMS. In addition, the effect of continuous deposition plasma is determined.

For the k2.7 ULK CVD material, the optimal solution for the k2.2 ULK MSQ is applied. In addition, the use of the OMCTS-based oxide cap is investigated with and without continuous deposition plasma.

To investigate the extendibility of the k2.2 ULK MSQ results to the k2.0 ULK MSQ film, experiments like those performed on the k2.2 ULK MSQ are conducted on the k2.0 ULK MSQ film. In the case of the k2.0 ULK MSQ investigation, the low-k CVD glue layer is varied from 20 nm to 50 nm.

Property	Units	Technique	k2.2 ULK MSQ	k2.0 ULK MSQ	k2.7 ULK CVD
Thickness	nm	Ellipsometry	300	300	560
k-value		Hg Probe	2.13	1.89	2.70
RI		Ellipsometry	1.256	1.217	1.365
Stress	MPa	Curvature	14	13	50

Table 1. Material Properties for ULK films included in this study.

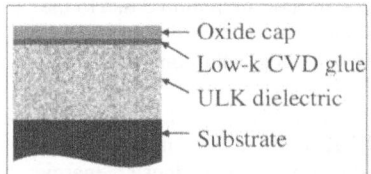

Figure 1. Graphical representation of dielectric stack

Adhesion of the CVD dielectric cap oxide to the underlying ULK material is quantified using the four-point bend technique. Four-point bend is a widely used technique to measure interface adhesion [3,4]. Samples are prepared either in a symmetric configuration or using a silicon bending beam from cleaved samples 5 to 8 mm wide. Samples are bonded using a two part epoxy with a 1 hour, 180 °C thermal cure. Five samples are evaluated for each dielectric stack using a four-point bend delamination load cell. The average and standard deviation of the adhesion energy is reported. While not reported here, repeatable measurement of adhesion is observed wafer to wafer. Although integration may be possible with a lower adhesion energy, a generally accepted robust adhesion requirement is 5 J/m^2. [5] Verification of the failing interface is performed visually and/or with time-of-flight secondary-ion mass spectroscopy.

In order to quantify the impact of the adhesion improvement techniques on the dielectric stack k values, the stack is modeled by assuming simple infinite parallel-plate capacitors in series for the ULK / low-k CVD glue layer/ oxide cap stack. Representative metal layer dimensions are modeled and held at a constant total thickness with the ULK dielectric layer thinning as the glue layer thickens. A post-CMP cap oxide thickness of 20 nm is assumed for the modeling.

DISCUSSION

Table 2 shows the oxide cap adhesion energies for multiple capping processes on the k2.2 ULK MSQ. The adhesion energies for a CMP compatible dielectric stack range from a low of 2.4 J/m^2 for a TEOS-based oxide cap with the low-k CVD glue layer to a high of 15.8 J/m^2 for a 100 nm low-k CVD glue layer with a TMS-based oxide cap. Also shown is the impact of the presence of the glue layer on the stack k value. As shown, the TEOS-oxide cap adhesion directly to the k2.2 ULK MSQ material is poor with an adhesion energy of 3.0 J/m^2. The adhesion energy of the TMS-based low-k CVD dielectric to the k2.2 ULK MSQ is, however, excellent with the adhesion failure occurring at the cap to epoxy interface with an energy of 30.0 J/m^2. (See 80nm glue layer thickness data in Table 2.) Although, like the ULK materials, this low-k CVD material does not provide an adequate CMP stop, it will prove useful as a glue layer to improve the adhesion of the oxide cap material.

Glue Material	Glue Thickness (nm)	Cap Oxide Precursor	Cont. Plasma	Oxide Cap Thickness (nm)	Adhesion Energy (J/m^2)		Stack k Value	k value increase (%)
					Avg.	%1σ		
N/A	0	TEOS	N/A	100	3.0	2.0	2.27	Ref.
TMS low-k	80	N/A	N/A	0	30.0	1.3	N/A	N/A
TMS low-k	20	TEOS	No	100	2.4	0.5	2.31	1.7
TMS low-k	20	TMS	No	100	4.5	1.0	2.31	1.7
TMS low-k	20	TMS	Yes	100	5.8	1.3	2.31	1.7
TMS low-k	30	TMS	Yes	100	7.2	0.4	2.33	2.6
TMS low-k	50	TMS	Yes	100	11.4	1.3	2.37	4.5
TMS low-k	100	TMS	Yes	100	15.8	3.4	2.48	9.4

Table 2. Dielectric Cap adhesion energy for various glue and oxide cap thicknesses on the k2.2 ULK MSQ. Sample stacks are Substrate / k2.2 ULK MSQ / Low-k CVD glue / Oxide cap. A "Yes" in the "Cont. Plasma" column signifies that the deposition plasma is maintained between the glue and cap layer depositions. Modeled stack k values are shown.

As indicated by the data in Table 2, the adhesion energy of 2.4 J/m^2 for a stack utilizing 20 nm of the low-k CVD material as a glue layer capped with 100 nm of TEOS oxide is significantly lower than that of the glue layer itself, showing no benefit to the presence of the glue layer at all. The failing interface is observed to be the low-k CVD glue layer to k2.2 ULK MSQ which was previously excellent. If instead of a TEOS-based oxide cap, a TMS-based oxide cap is deposited after the TMS-based low-k CVD glue layer is deposited, but without breaking vacuum between the depositions, the adhesion is significantly improve to 4.5 J/m^2. The adhesion can be further improved by maintaining a continuous deposition plasma between the low-k CVD glue layer and the TMS-based oxide. In this case, an additional improvement in the adhesion is observed with an adhesion energy of 5.8 J/m^2, exceeding the 5 J/m^2 goal.

The effect on the adhesion energy of increasing the thickness of the glue layer shown in Table 2 is graphically represented in Figure 2. As the glue layer thickness is increased from 20 nm to 100 nm, the adhesion energy also increases from 5.8 J/m^2 to 15.8 J/m^2. A possible explanation for this improvement in adhesion with increasing low-k CVD glue layer thickness is that film growth may be discontinuous for thinner films. The k value of the low-k CVD glue material of 3.0, while low, is significantly higher than that of the k2.2 ULK MSQ. Therefore, increasing the glue layer thickness negatively impacts the overall k value of the stack as seen by the modeling results in Table 2. However, as shown, the adhesion energy may be tailored to meet integration requirements by balancing it with stack k value needs for a given technology. For all but the thickest glue layer evaluated, the degradation in the stack k value is less than 5%.

Table 3 shows the adhesion results for the k2.7 ULK CVD film for various capping strategies. The adhesion energy varies from 3.6 J/m^2 for the TEOS-based oxide and OMCTS-based oxide with discontinuous deposition plasma, to 8.0 J/m^2 for an OMCTS-based oxide with continuous deposition plasma, all without the low-k CVD glue layer. As shown, the adhesion improvements observed on the k2.2 ULK MSQ film are also seen on the k2.7 ULK CVD film. The TEOS oxide on k2.7 ULK CVD adhesion at 3.6 J/m^2 is a bit higher than, but similar to, that of the k2.2 ULK MSQ film. Also like the k2.2 ULK MSQ, the adhesion is improved using the TMS-based low-k CVD glue layer and TMS-based oxide cap with plasma maintained through

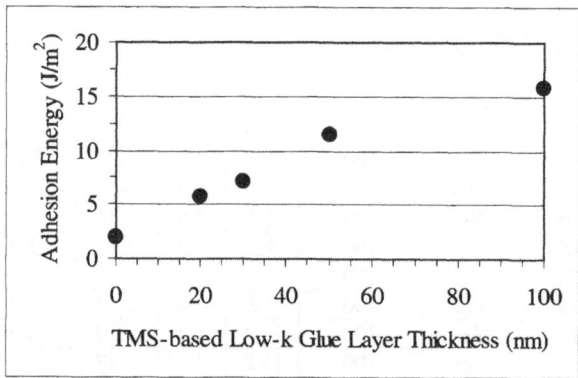

Figure 2. Adhesion energy as a function of glue layer thickness for a stack consisting of Substrate / k2.2 ULK MSQ / low-k CVD glue layer / oxide cap with TMS-based deposition plasma maintained throughout the glue and cap layers deposition. Note that the adhesion is tunable with glue layer thickness.

Glue Material	Glue Thickness (nm)	Cap Oxide Precursor	Cont. Plasma	Oxide Cap Thickness (nm)	Adhesion Energy (J/m^2)		Stack k Value	k value increase (%)
					Avg.	%1σ		
N/A	0	TEOS	N/A	100	3.6	0.5	2.76	Ref.
TMS low-k	80	N/A	N/A	0	36	1.3	N/A	N/A
TMS low-k	20	TMS	Yes	100	6.8	1.3	2.78	0.6
N/A	0	OMCTS	No	100	3.6	0.6	2.76	0.0
N/A	0	OMCTS	Yes	100	8.0	0.9	2.76	0.0

Table 3. Dielectric Cap adhesion energy for various glue and oxide cap thicknesses on the k2.7 ULK CVD. Sample stacks are Substrate / k2.7 CVD ULK / Low-k CVD glue / Oxide cap. A "Yes" in the "Cont. Plasma" column signifies that the deposition plasma is maintained between the glue and cap layer depositions. For the OMCTS-oxide cap, plasma is maintained between the k2.7 ULK CVD and oxide depositions. Modeled stack k values are shown.

the depositions. In this case the adhesion energy improves from 3.6 J/m^2 to 6.8 J/m^2. Due to the similar k values of the k2.7 ULK CVD and low-k CVD films, the effect on k-value is negligible.

Based upon the results of combining the low-k CVD glue layer with a oxide cap having the same precursor, one might expect similar results from the OMCTS-based k2.7 CVD ULK material with a OMCTS-based oxide cap and without any glue layer. As observed in Table 3, while the OMCTS-based CVD oxide shows poor adhesion of 3.6 J/m^2 when the deposition plasma is not maintained throughout the k2.7 ULK CVD and cap depositions, adhesion is improved significantly to 8.0 J/m^2 when the plasma is maintained. The use of a OMCTS-based oxide cap for the OMCTS-based k2.7 CVD ULK is more desirable than using the TMS-based low-k glue/TMS-based oxide solution due to reduced process complexity, cost, and cycle time.

A natural question is: do these cap solutions extend to lower k value regimes? While a lower dielectric constant OMCTS-based ULK CVD material does not exist at the time of study, lower dielectric constant ULK MSQ SOD materials do. Application of the TMS-based low-k CVD glue and TMS-based oxide cap with continuous deposition plasma to the k2.0 ULK MSQ material has been evaluated. The results are shown in Table 4 with the adhesion energies ranging from a low of 1.3 J/m^2 to a high of 2.2 J/m^2 for a CMP compatible stack. Unfortunately,

Glue Material	Glue Thickness (nm)	Cap Oxide Precursor	Cont. Plasma	Oxide Cap Thickness (nm)	Adhesion Energy (J/m^2)		Stack k Value	k value increase (%)
					Avg.	%1σ		
N/A	0	TEOS	N/A	100	1.8	0.0	2.07	Ref.
TMS low-k	80	N/A	N/A	0	8.3	1.2	N/A	N/A
TMS low-k	20	TEOS	No	100	2.2	0.4	2.11	2.2
TMS low-k	20	TMS	Yes	100	1.3	0.1	2.11	2.2
TMS low-k	30	TMS	Yes	100	1.8	0.2	2.14	3.3
TMS low-k	50	TMS	Yes	100	2.1	0.6	2.19	5.7

Table 4. Dielectric Cap adhesion energy for various glue and oxide cap thicknesses on the k2.0 ULK MSQ. Sample stacks are Substrate / 300 nm k2.0 ULK MSQ / Low-k CVD glue / Oxide cap. A "Yes" in the "Cont. Plasma" column signifies that the deposition plasma is maintained between the glue and cap layer depositions. Modeled stack k values are shown.

the adhesion improvements are not shown to extend to the lower k material with the adhesion of each of the k2.0 ULK MSQ / low-k CVD glue layer / oxide cap stacks having an adhesion energy below 2.3 J/m^2. An indication of the source for the significantly worse adhesion results even with the low-k CVD glue-layer is the lower adhesion of the low-k CVD glue layer itself with an adhesion energy of 8.3 J/m^2 for the k2.0 ULK MSQ as compared to 30.0 for the k2.2 ULK MSQ. While it is possible to obtain adhesion energies exceeding the 5 J/m^2 target for cap materials to the k2.0 ULK MSQ, these studies are beyond the scope of this work.

CONCLUSIONS

The adhesion of CVD dielectric caps to ULK MSQ SOD materials with k values of 2.2 and 2.0, and a ULK CVD material with a k value of 2.7 has been investigated utilizing four-point bend analysis. A substantial improvement in cap adhesion to both the k2.2 ULK MSQ and the k2.7 ULK CVD material is demonstrated. These improvements are obtained using a low-k CVD glue material between the ULK dielectric and the subsequent cap material. While adding the glue layer alone is not sufficient to improve the adhesion, utilizing an oxide cap with the same precursor as the low-k material (TMS) and maintaining the deposition plasma through the glue layer and cap layer depositions significantly improves cap adhesion. Measurements indicate a nearly 100% improvement in the cap to ULK film adhesion, exceeding the targeted 5 J/m^2 adhesion energy. This improvement in CVD cap adhesion is demonstrated to be strongly dependent upon the presence of the glue layer film, the cap oxide precursor and the deposition conditions. In addition, the improvement in cap adhesion is obtained without significant degradation (<2%) in the k value of the dielectric stack. Furthermore, it is demonstrated that the use of a same-precursor oxide cap for the OMCTS k2.7 ULK CVD material with the deposition plasma maintained through the ULK/cap deposition provides good adhesion without the need for a glue layer. Finally, studies to determine whether the k2.2 ULK MSQ solutions are extendable to the k2.0 ULK MSQ show that while the k2.0 and k2.2 materials are similar, the improvement in adhesion does not convey to the k2.0 ULK MSQ material. These results indicate that as the k values of ULK materials continue to decrease, additional effort will be required to maintain sufficient adhesion.

ACKNOWLEDGEMENTS

The authors would like to acknowledge the historical efforts of Todd Ryan on the improvement of dielectric cap adhesion for SOD materials with k values greater than 2.2. In addition, the efforts of the Motorola Physical Analytical Laboratories in Austin and Phoenix are greatly appreciated.

REFERENCES

1. International Technology Roadmap For Semiconductors, Interconnect, 5-6 (2003).
2. K.C. Yu, et al; International Interconnect Technology Conference Proceedings, 9-11 (2002).
3. Q. Ma; Journal of Materials Research, **12**, 840 – 845 (1997).
4. R.H. Dauskardt, et al; Engineering Fracture Mechanics, **61**, 141-162 (1998).
5. T Sherban, et al; International Interconnect Technology Conference Proceedings, 257-259 (2001).

Mat. Res. Soc. Symp. Proc. Vol. 812 © 2004 Materials Research Society

Self-Assembled Monolayers as Model Substrates for Atomic Layer Deposition

Caroline M. Whelan[1], Anne-Cécile Demas[1], Jörg Schuhmacher[1], Laureen Carbonell[1] and Karen Maex[1,2]
[1]IMEC, Kapeldreef 75, B-3001 Leuven, Belgium.
[2]Department of Electrical Engineering, Katholieke Universiteit Leuven, Kasteelpark Arenberg 1, B-3001 Heverlee, Belgium.

ABSTRACT

Our understanding of the role of the initial surface on atomic layer deposition (ALD) of Cu diffusion barrier materials is limited by the complexity of the sequential reactions and the heterogeneous nature of typical dielectric substrates. The atomically controlled surface chemistry of self-assembled monolayers (SAMs) provides a means of creating model substrates for ALD. Here we report on ALD of WC_xN_y films on SAMs derived from bromoundecyltrichlorosilane adsorbed on silicon dioxide. The as-prepared SAM is macroscopically ordered with the expected Br-termination and has a well-defined chemical composition as determined by contact angle measurements and X-ray photoelectron spectroscopy, respectively. Temperature programmed desorption spectroscopy confirms that the SAM is stable to 550°C. It survives multiple cycles of ALD at 300°C as evidenced by the detection of mass fragments characteristic of the alkyl chain and supported by the persistence of a Br 2p peak at 71 eV. X-ray fluorescence, ellipsometry and atomic force microscopy reveal that the underlying SAM influences WC_xN_y film coverage, thickness, and morphology.

INTRODUCTION

The trend towards increasing functional density in integrated circuits cannot be supported by traditional materials [1]. Device performance dictates a transition from SiO_2 to an insulator with lower dielectric constant (k) and Cu instead of Al for lower resistance wiring. Such interconnect metallization requires the introduction of a barrier layer to prevent Cu diffusion under electrical bias. It is, however, difficult to obtain conformal barriers of the thicknesses (< 5 nm) foreseen in future device architectures without resorting to unconventional methods. To this end, the unique self-limiting and inherently conformal method of atomic layer deposition (ALD) of films from W, Ti, and Ta compounds is being investigated [2]. Based on sequential saturated gas phase-surface reactions, the growth rate in ALD is in theory layer-by-layer, controlled by the number of deposition cycles. In practice, the early stages of film formation may be non-linear, involving three-dimensional growth depending on substrate reactivity [3].

Due to the heterogeneous nature of dielectric materials, information on the role of the initial surface on ALD-mediated growth mechanisms is limited. In particular, our understanding of the surface reactions relevant to ALD of WC_xN_y diffusion barrier is in its infancy. One approach is to simplify the substrate. Self-assembled monolayers (SAMs), molecular assemblies that are formed spontaneously by the exposure of an appropriate substrate to an active surfactant in an organic solvent or in the gas phase, have been extensively explored as model substrates for studying interactions at metal or metal oxide/organic interfaces [4]. To this end, we have investigated various organosilane-derived SAMs as model substrates for WC_xN_y film growth. In this report, we focus on SAMs derived from bromoundecyltrichlorosilane adsorbed on silicon dioxide. The Br-termination acts as a convenient label to monitor SAM composition before and after ALD. The composition and thermal stability of the SAM are characterized using contact angle measurements, X-ray photoelectron spectroscopy (XPS) and temperature programmed desorption spectroscopy (TDS). X-ray fluorescence (XRF), ellipsometry, and atomic force microscopy (AFM) applied to determine the influence of the underlying SAM on WC_xN_y film coverage, thickness, and morphology.

EXPERIMENTAL DETAILS

SAMs were prepared by immersion of a SiO_2 substrate, < 1 nm chemical oxide on Si(100) prepared by an IMEC-clean [5], in 10^{-1}, 10^{-2}, or 10^{-3} M solution of bromoundecyltrichlorosilane (Br-UTS) used as received from Gelest at ambient temperature between 15 min. and 24 h. The modified wafer was then rinsed with copious amounts of toluene, then, acetone and, finally, ethanol before being dried under nitrogen flow [6]. The deposition of WC_xN_y (with cycle numbers ranging from 50 to 1500) was performed using an $ALCVD^{TM}$ Pulsar$^{®}$ 2000 reactor integrated with an automated wafer handling platform (ASM Polygon^{TM} 8200). A precursor pulse sequence of triethylborane ($(C_2H_5)_3B$), tungsten hexaflouride (WF_6), and ammonia (NH_3) represents one deposition cycle producing WC_xN_y. The precursors were mixed with a nitrogen carrier gas flow. The $(C_2H_5)_3B$ was evaporated by flowing nitrogen over the liquid at 170 kPa and 18°C. Excess precursor gas was removed by flowing nitrogen for 2 s after each precursor pulse. The temperature during deposition was 300°C and the maximum pressure was 2 hPa.

Static contact angles of deionised water were measured using a software-controlled Video Contact Angle System OCA-20 (DataPhysics). X-ray photoelectron spectroscopy (XPS) measurements were performed with an SSX-100 (Surface Science Instruments) photoelectron spectrometer with a monochromatic Al K_α X-ray source hv = 1486.6 eV) and concentric hemispherical electron energy analyser using an energy resolution of 0.92 eV. Thermal desorption spectroscopy (TDS) measurements were carried out in a commercial single wafer rapid thermal processing tool (RTP AST SHS2800) equipped with an atmospheric pressure ionization mass spectrometer (APIMS, VG Trace +) [7]. The amount of W deposited was determined from the W L_α feature of X-ray fluorescence (XRF) spectra (Energy Dispersive Spectro2000). A Sentech automatic multi-wavelength SE-801 ellipsometer was used for thickness measurements. AFM characterisation was carried out under ambient conditions using a Nanoscope III Dimension 3000 in tapping mode.

DISCUSSION

SAM preparation was optimized with a view to developing a method suitable for 8-inch wafer level processing in a clean room environment. The preparation method was evaluated in terms of solution concentration and immersion time using contact angle measurements to monitor the macroscopic film order. We found that immersion in a 10^{-3} M solution of Br-UTS for 1 h. produces SAMs of reproducibly high quality with a contact angle of 86.6 ± 1.5° which compares favorably with literature values [8].

Detailed quantitative and qualitative evaluation of the SAMs was performed using XPS. The as-prepared surfaces were characterized by C 1s photoemission at 285.1 eV attributed to the aliphatic methylene units with broadening towards higher binding energies due to a contribution at ~286.5 eV from the most electron deficient carbon atom bonded to bromine. The observation of Br $3d_{3/2}$ photoemission at 71 eV is in agreement with a previous study of a Br-terminated SAM [8].

Br-UTS adsorption on SiO_2 is predicted to take place through the hydrolysis of the Si-Cl bonds to form Si-OH groups. In core level scans from 190 to 210 eV, we find no evidence of a Cl 2p peak. Its absence is an indication that all Si-Cl bonds have been hydrolysed, thus favoring monolayer formation via Si-O-Si bonding resulting from a condensation reaction between the Br-UTS OH groups and the oxidized Si surface. The absence of Cl also implies that precursor polymerization on the surface, a consequence of excess water, has not occurred.

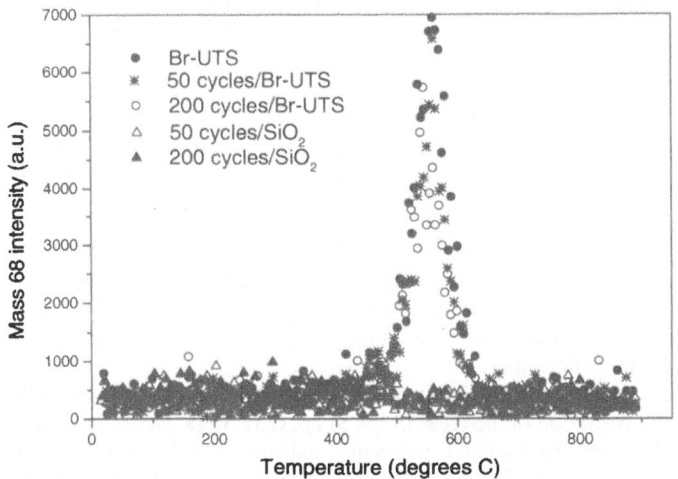

Figure 1   Thermal desorption spectra for mass 68 from Br-UTS SAMs before (●●●) and after deposition of 50 (***) and 200 (OOO) cycles of WC_xN_y ALD. The corresponding data for 50 (ΔΔΔ) and 200 (▲ ▲ ▲) cycles of ALD on SiO_2 are shown for comparison.

The thermal stability and decomposition pathway of Br-UTS SAM were investigated using TDS. While only selected data (mass 68 desorption) is shown in Fig. 1, our interpretation is based on the analysis of masses from 11 to 100 a.m.u. The decomposition process starts at 450°C as evidenced by desorption of high molecular weight hydrocarbon fragments created by alkyl chain C-C bond cleavage. A maximum in desorption intensity occurs at 550°C. After annealing to 650°C the monolayers have completely desorbed. These results are in excellent agreement with a previous study of the decomposition of alkyltrichlorosilane-derived monolayers on SiO_2 [9] and indicate a thermal stability well within the limits required for ALD processing. It has been suggested that above 540°C only methyl groups remain adsorbed implying that desorption of the SAMs is primarily the result of C-C rather than Si-C bond cleavage [9]. While methyl groups can be considered to decompose via methyl radical or methane desorption, our analysis of masses 15 and 16 failed to reveal evidence of such reactions, suggesting decomposition to surface carbon and hydrogen.

AFM images of the Br-UTS surface are featureless, with a root-mean-square (RMS) roughness of 0.272 nm, which is slightly larger than the RMS of 0.106 nm for SiO_2, in agreement with literature values for similar SAM surfaces [8].

Table 1   XPS composition analysis of Br-UTS SAMs before and after WC_xN_y ALD.

ALD cycles	O %	C %	Si_{ox}%	$Si_{substrate}$%	Br %	W %	N %
0	26.77	40.35	11.9	18.8	2.0		
50	33.15	42.2	9.8	13.5	0.46	0.79	
100	41.95	44.08	3.5	4.2		6.23	
200	41.75	41.29				11.13	5.80
500	43.17	36.15				13.21	7.37

Having established the expected chemical composition, suitable thermal stability, and smooth surface morphology of the Br-UTS SAM, it was then used as a substrate for ALD. XPS was used to monitor core level spectra as a function of increasing number of ALD cycles. A decrease in Br 3d intensity was observed to coincide with the appearance of W 4f and N 1s peaks assigned to formation of a WC_xN_y film. The fact that Br is still observed following 50 ALD cycles (Table 1) unambiguously confirms the compatibility of the SAM to the processing conditions. This is supported by TDS analysis showing desorption characteristic of the Br-UTS as-prepared sample is maintained following 50, and even 200, cycles of ALD (Fig. 1). Obviously, the metal overlayer eventually attenuates photoemission from the SAM to the extent that the Br signal can no longer be detected.

The atomic percentage (±0.5%) ratio of silicon, oxygen, carbon, bromine, nitrogen, and tungsten were calculated by integration of the areas of the Si 2p, O 1s, C 1s, Br 3d, N 1s, and W 4f peaks after application of a linear background subtraction and correction for atomic sensitivity factors including analyzer transmission. The results are summarized in Table 1. Quantification yields an average carbon to bromine ratio of approximately 20:1 from the as-prepared SAM. This deviation from the theoretical stoichiometry of 11:1 is explained by the known susceptibility of halide-terminated SAMs to X-ray-induced damage during XPS measurements [8].

The growth of WC_xN_y on the Br-UTS surface was monitored by determining the W content as a function of number of ALD cycles using XRF. In Fig. 2 we compare the growth on Br-UTS with SiO_2, a cyanoundecyltrichlorosilane-derived SAM (CN-SAM), and an octadecyltrichlorosilane-derived SAM (CH$_3$-SAM). It is clear that the W coverage varies depending on the substrate. For example, 200 cycles of ALD on Br-UTS results in 14.1 CPS (3.8 $\times 10^{16}$ W atoms) compared with 20.1 CPS (5.4 $\times 10^{16}$ W atoms) on the cyano-terminated SAM. The corresponding thicknesses of 10.3 and 14.2 nm, respectively, estimated by ellipsometry, imply the growth of a lower density film on the Br- versus the cyano-terminated surface. These differences translate to material properties that influence the electrical performance of the film as evidenced by sheet resistances (R_s) of 489 and 313 Ohms/square, respectively. Following 500 cycles of ALD WC_xN_y films of similar thicknesses, 32.1 and 35 nm, are found on Br-UTS and CN-SAM, respectively, and give rise to similar R_s values of 112 and 99 Ohms/square, respectively. This implies that substrate-induced differences in the early stages of film growth eventually disappear as the coverage increases.

Further, the morphology of the diffusion barrier layer must be considered, in particular in terms of its impact on subsequent processing steps such as Cu seed layer growth. For Br-UTS, the RMS increases from 0.272 nm to 0.879 nm following 200 cycles of ALD. In contrast, the RMS value increases only slightly from 0.455 nm for the CN-SAM substrate to 0.497 nm following 200 cycles of WC_xN_y growth demonstrating that WC_xN_y film growth varies depending on the substrate.

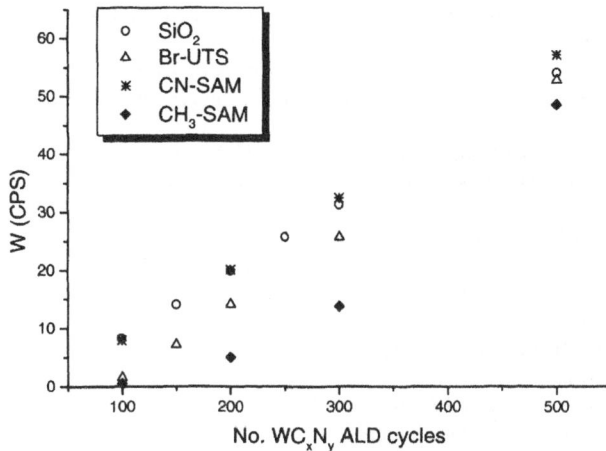

Figure 2   W content following ALD of WC_xN_y on Br-UTS SAMs (ΔΔΔ) determined by XRF. The corresponding data for SiO_2 (OOO), CN-SAM (***) and CH_3-SAM (♦ ♦ ♦) are shown for comparison.

CONCLUSIONS

The role of the initial surface on ALD of WC_xN_y can be investigated using SAMs as model substrates. Br-UTS SAMs provide films of known chemical composition and macroscopically well-ordered surfaces. With thermal stability extending to 550°C, such substrates withstand multiple cycles of ALD at 300°C. Comparison of WC_xN_y grown on Br-UTS with films grown on SiO_2 and SAMs with different terminal groups shows differences in W coverage, film thickness, density, and resistance and surface morphology. This work demonstrates that, via tunable structure and surface chemistry, SAMs provide a novel approach to diffusion barrier engineering.

ACKNOWLEDGEMENTS

The authors gratefully acknowledge T. Conard, D. Vanhaeren, and B. Eyckens for assistance with XPS, AFM, and ellipsometry measurements, respectively. The authors also wish to thank A. Martin Hoyas and V. Sutcliffe for insightful discussions. This work has been carried out within IMEC's Industrial Affiliation Cu/low-k BEOL Program.

REFERENCES

1. International Technology Roadmap for Semiconductors, Semiconductor Industry Association, San Jose, CA. For more details see http://public.itrs.net/.
2. M. Ritala and M. Leskela, "Atomic Layer Deposition," *Handbook of Thin Film Materials*, edited by H.S. Nalwa, *Volume 1: Deposition and Processing of Thin Films* (Academic Press, 2002) pp. 103-159.
3. A. Satta, J. Schuhmacher, C. M. Whelan, W. Vandervorst, S. H. Brongersma, G. P. Beyer, K. Maex, A. Vantomme, M. M. Viitanen, H. H. Brongersma, and W. F. A. Besling, *J. Appl. Phys.* **92**, 7641 (2002).
4. G.C. Herdt, D.R. Jung and A.W. Czanderna, *J. Adhesion* **60**, 197 (1997).
5. M. Meuris, P. W. Mertens, A. Opdebeeck, H. F. Schmidt, M. Depas, G. Vereecke, M. M. Heyns, and A. Philipossian, *Solid State Technol.*, 38, 109 (1995).
6. H. Brunner, T. Vallant, U. Mayer, and H. Hoffmann, *Langmuir* 12, 4614 (1996).
7. G. Vereecke, E. Kondoh, P. Richardson, K. Maex and M. M. Heyns, IEEE Trans. Semicond. Manuf., 13, 315 (2000).
8. M.-T. Lee and G.S. Ferguson, *Langmuir* **17**, 762 (2001).
9. G.J. Kluth, M.M. Sung, and R. Maboudian, *Langmuir* 13, 3775 (1997).

Mat. Res. Soc. Symp. Proc. Vol. 812 © 2004 Materials Research Society

Ruthenium Sputter Deposition on Organosilicate Glass and on Paralyne: an XPS Study of Interfacial Chemistry, Nucleation and Growth

X. Zhao, N.P. Magtoto, and J. A. Kelber
Department of Chemistry, University of North Texas, Denton, TX 76203

ABSTRACT

The interactions of sputter-deposited ruthenium with organosilicate glass (OSG) at 300 K have been studied by x-ray photoelectron spectroscopy (XPS) for Ru coverages from ~ 0.1 monolayer to several monolayers, using *in-situ* sample transfer between the deposition and analysis chambers. The results indicate Stranski-Krastanov (SK) type growth, with the completion of the first layer of Ru at an average thickness corresponding to 1 monolayer average coverage. Ru(0) is the only electronic state present. XPS core level spectra indicate weak chemical interactions between Ru and the substrate. A less pronounced tendency towards SK growth was observed for Ru deposition on parylene. Deposition of Ru on OSG followed by electroless deposition of Cu resulted in the formation of a shiny copper film that failed the Scotch Tape test. Results indicate failure mainly at the Ru/OSG interface.

INTRODUCTION

Ruthenium is of growing interest as a diffusion barrier for Cu in low-k integration, primarily because the semi-noble nature of the metal allows for facile electrodeposition of Cu films, while the high melting point and immiscibility with Cu are favorable for microelectronics applications[1]. In diffusion barrier applications, the adhesion of Ru films to dielectric substrates is also important. The semi-noble nature of Ru is detrimental for adhesion and conformal growth on dielectric substrates, and one would intuitively expect that Ru would exhibit much less interfacial interaction with a Si-O-C low k dielectric than would, e.g., Ta[2]. XPS has been demonstrated[2-4] to be an excellent method for characterizing the formation of metal-substrate chemical bonds (or lack of same) during gradual metal deposition under controlled conditions. In such experiments, it is usually helpful to transfer the sample from the deposition chamber to the analysis chamber without exposure to atmosphere. This report documents results of XPS core level spectra for sputter-deposited Ru on OSG and on parylene, at coverages up to several monolayers. OSG is a leading low-k candidate for replacing SiO_2 as an interlayer dielectric, while parylene is a candidate for pore sealing of porous ultra-low k materials.[5] The XPS results confirm generally weak Ru-substrate interactions, and this is corroborated by results for a Scotch Tape test of an electrolessly deposited Cu film on Ru sputter-deposited onto OSG, showing failure at the Ru/OSG interface.

EXPERIMENTAL

Experiments were carried out in an ultrahigh vacuum (UHV) system with a sputter deposition system equipped with dual magnetron sputter systems. This system has been described previously in some detail [3, 4]. Briefly, the XPS system included a hemispherical analyzer and dual anode x-ray source, as well as an ion gun for sample cleaning. XPS spectra presented in this report were acquired using Mg Kα radiation with the analyzer operated in the constant pass energy mode (25 eV). Methods for the treatment of XPS spectra, including background subtraction, determination of film stoichiometry, and the estimation of overlayer thickness, were as previously described.[2, 4] Since XPS on insulating samples can result in sample charging that shifts the apparent photoelectron energies, spectra were calibrated by referencing the peak from adventitious carbon at 284.5 eV.

An elemental Ru sputter target was used, with target/sample distance adjusted so that a minimum sputter deposition rate of 0.004 Å-sec^{-1} was observed, as determined by the attenuation of core-level XPS intensity during Ru deposition on clean Cu foil. Sample transfer between deposition and analysis chambers occurred under UHV ($< 10^{-8}$ Torr). The ~ 1 cm^2 samples consisted of either commercially prepared OSG deposited on Si, or 1000Å thick parylene deposited on ~20Å SiO$_2$ on a p-type Si(100) substrate. Sample temperature was monitored by a type-K thermocouple mounted between the sample and sample holder, and the temperature could be varied between 100 K and 1200 K by a combination of liquid nitrogen cooling and resistive heating of the sample holder.

RESULTS

XPS core-level spectra are displayed in figure 1 for the OSG sample after annealing in UHV at 500 K for 30 minutes, which removed some adventitious carbon.

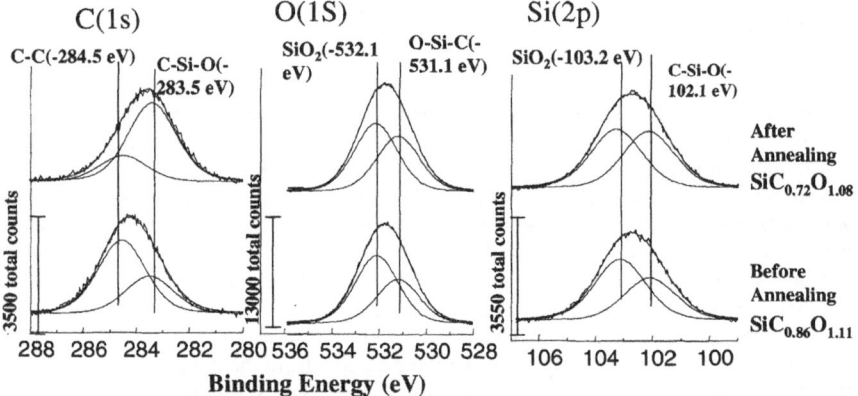

Figure 1: Core level XPS spectra for the OSG film before (bottom) and after (top) annealing at 500 K for 30 min. in UHV.

This particular type of OSG is known to contain substantial amounts of SiO$_2$, consistent with the XPS data. The main effect of annealing in UHV is to have removed a significant amount of adventitious carbon from the sample, resulting in a post-anneal stoichiometry of SiC$_{0.72}$O$_{1.08}$.

XPS spectra of samples containing carbon and ruthenium are complicated by the overlap between the C(1s) spectrum and the Ru(3d$_{3/2}$) component of the Ru(3d) spectrum, as shown in figure 2. For this reason, Ru coverage was determined from intensity of the Ru(3d$_{5/2}$) feature. Since the relative intensities of the Ru(3d$_{3/2}$)/Ru(3d$_{5/2}$) features are well known[6], the estimation of the C(1s) signal intensity by subtraction of the determined Ru(3d$_{3/2}$) intensity is straightforward, as shown (fig. 2).

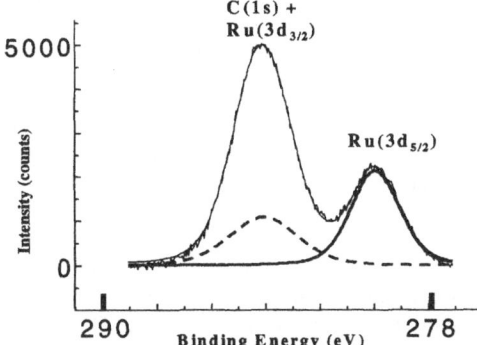

Figure 2: Ru(3d)/C(1s) region, showing the partial overlap of the Ru(3d) and C(1s) core level features. Spectrum corresponds to about 30 minutes Ru deposition (approximately 3 monolayers). Ru(3d$_{3/2}$) component indicated by dashed line was determined from the intensity and binding energy of the Ru(3d$_{5/2}$) component.

The Ru(3d$_{5/2}$) binding energy of 280.1 eV is indicative of Ru in a metallic or zerovalent state[7], and this binding energy was observed independent of Ru coverage, indicating negligible charge transfer between the Ru and the substrate. The variation of Ru(3d) intensity with deposition time on OSG is compared to the behavior on parylene in figure 3 (left), normalized to the substrate C(1s) intensity.

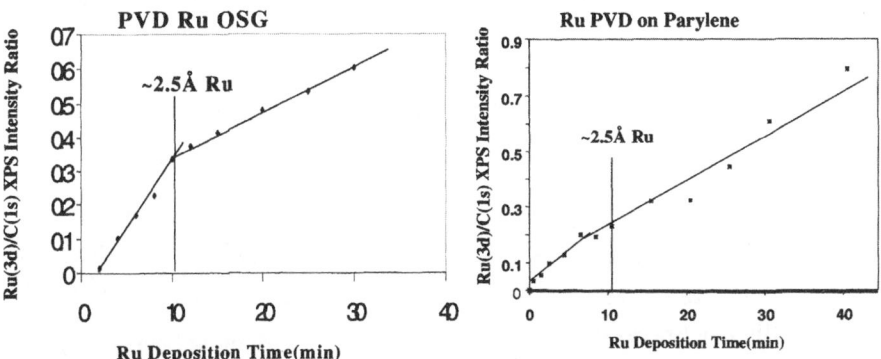

Figure 3: The evolution of the Ru(3d)/C(1s) intensity ratio as function of Ru deposition vs time at 300 K. (Left) Deposition of Ru on OSG. The film thickness at 10 minutes corresponds to 1 monolayer. (Right) Deposition of Ru on Parylene.

The data in figure three (left) indicate a linear increase in Ru intensity on OSG up to 10 minutes deposition time, followed by a change in slope. Such behavior is indicative of initial conformal (2D) growth in the first layer, followed by a transition to 3D growth with the formation of the second layer— SK growth.[8] Although such behavior—when observed for Cu deposition— generally correlates with significant chemical interaction and charge exchange between the metal overlayer and dielectric substrate[4, 9, 10], no evidence for Ru oxidation was observed in the XPS spectra.

In contrast to Ru/OSG, similar data for Ru deposition on parylene shows different behavior (fig. 3, right). There is no sharp change in slope, as for deposition on OSG, and the region where a gradual change is observed occurs well before an average thickness of 1 monolayer is achieved. The data in figure 3 therefore suggest a much less pronounced tendency for initial conformal growth on parylene than on OSG. A comparison of these results in turn suggests a stronger interfacial interaction between Ru and OSG than between Ru and parylene.

In order to examine the suitability of sputter-deposited Ru films for Cu integration, a Ru/OSG sample was prepared for Cu electroless deposition. A sample was prepared by 60 minutes sputter deposition of Ru onto an OSG substrate at 300 K— corresponding to an average thickness of 6 Ru monolayers. The sample was removed from UHV and placed in a formaldehyde-based electroless plating solution (6g/l copper sulfate, 32 g/l ethylenediaminetetraacetic acid (EDTA), 4g/l formaldehyde and sodium hydroxide to adjust the pH of the solution to 11). After 60 minutes, the sample was emersed from the solution, and air-dried. A shiny Cu film was readily visible to the eye. The sample was then subjected to a simple Scotch Tape test in order to provide a rough, qualitative assessment of film substrate adhesion. The Scotch Tape test resulted in the loss of all visible Cu from the sample. XPS spectra of the sample before Cu deposition (but after air exposure) and after Cu deposition and Scotch Tape test, are shown in figure 4.

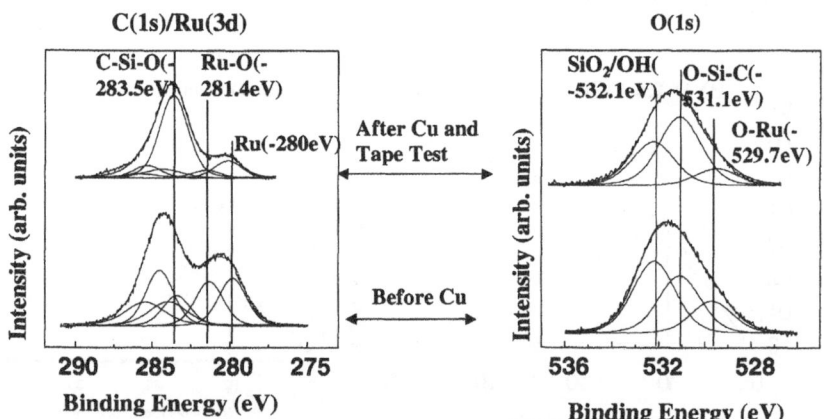

Figure 4. XPS C(1s)/Ru(3d) spectra (left), and O(1s) spectra (right).

The XPS spectra taken after air exposure but before Cu electroless deposition (bottom traces) clearly show the presence of oxidized Ru, as evidenced by a broadening of the Ru($3d_{5/2}$) feature (fig. 4, bottom, left) which is well fit by a component at 281.4 eV—characteristic of oxidized Ru [6, 7]. The presence of oxidized Ru is corroborated by the presence of a component near 530 eV in the O(1s) spectrum[7] (fig. 4, bottom, right). After the Scotch tape test, the Ru($3d_{5/2}$) intensity is reduced by 73% (fig. 4, top, left). This is accompanied by a corresponding loss in intensity of the O(1s) component near 530 eV (fig. 4, top, right). Cu(2p) spectra (not shown) indicated the presence of only trace amounts of Cu after the Scotch tape test.

DISCUSSION

The XPS data shown here indicate that sputter deposition of Ru on OSG at 300 K results in the growth of a conformal interface (SK growth). In contrast, Ru deposition on parylene shows less tendency towards SK behavior, indicative of a weaker interfacial interaction. During deposition on OSG, there is no evidence for Ru oxidation, which would be evidenced by a broadening of the Ru($3d_{5/2}$) component to higher binding energies. Such a broadening is only observed after exposure of the sample to air (fig. 4). Possible XPS evidence for the formation of a Ru carbide interfacial layer is obscured by the overlap of Ru (3d) and C(1s) features in the region of 283-281 eV (fig. 2)—the region in which metal carbide features are often observed.[6] The absence of interfacial carbide formation can be deduced from the fact that the Ru(3d)/C(1s) intensity ratio (fig. 3, left) increases linearly with Ru coverage up to the completion of the first monolayer. Similar Ru/Si data (not shown) similarly strongly suggests an absence of interfacial silicide formation. The weak chemical interaction between Ru and the substrate is consistent with the results of the Scotch tape test on a Cu(electroless)/Ru/OSG sample indicating failure mainly at the Ru/OSG interface.

The lack of interfacial chemical reaction between sputter-deposited Ru and OSG or parylene is in contrast to what is observed for Ta. Ta sputter deposited onto a Si-O-C substrate, where formation of interfacial Ta oxide and Ta carbide is observed[2]. Similarly, Ta deposited on parylene results in the formation of interfacial carbide.[11]. This would suggest that the use of Ru on OSG or parylene sealed dielectric materials should be accompanied by the use of a Ta "glue" layer between the Ru and the dielectric in order to increase interfacial adhesion. Such a scheme, of course, replaces the deposition of a Cu seed on Ta with the deposition of a Ru layer, which can then be used for direct electroless or electrodeposition of Cu. An alternative approach, which would avoid the necessity for a glue layer, would be to hydroxylate the substrate prior to Ru deposition. Such an approach has been shown to enhance the chemical interaction of Cu with a variety of substrates[9, 10]. Other seminoble metals may exhibit similar effects.[12]

CONCLUSIONS

XPS has been used to characterize the chemical interactions of sputter-deposited Ru with OSG and parylene at 300 K. The results indicate the lack of interfacial oxide, carbide or silicide formation on OSG, and the lack of interfacial carbide formation on parylene. Ru exhibits SK growth (conformal growth of the first layer) on OSG, but this tendency is much less clear-cut on parylene. The formation of an electroless Cu layer on a Ru/OSG sample, followed by a Scotch Tape test and subsequent XPS analysis, revealed failure at the Ru/OSG interface, consistent with the weak interfacial chemical interactions observed by XPS. The results indicate that the use of Ru as a diffusion barrier for Cu/low-k integration requires either the use of a "glue" layer (e.g., Ta) between the Ru and dielectric substrate, or possibly some other chemical modification, such as hydroxylation of the substrate.

ACKNOWLEDGMENTS

The authors gratefully acknowledge support for this work from the Semiconductor Research Corporation through the Center for Advanced Interconnect Science and Technology, and from the Robert Welch Foundation (grant no. B-1356.). The authors also thank Prof. Toh-Ming Lu for providing the parylene samples, and thank Texas Instruments for providing the OSG samples.

REFERENCES

[1]   O. Chyan, T. N. Arunagirl, T. Ponnuswamy, *J. Electrochemical Society* **150** (2003) C347.
[2]   J. Tong, D. Martini, N. Magtoto, J. Kelber, *J. Vacuum Science and Technology* **B21** (2003) 293.
[3]   K. Shepherd, J. Kelber, *Applied Surface Science* **151** (1999) 287.
[4]   X. Zhao, M. Leavy, N. P. Magtoto, J. A. Kelber, *Appl. Phys. Lett.* **79** (2001) 3479.
[5]   T.-M. Lu, Rensselaer Polytechnic Institute, private commuciation.
[6]   J. F. Moulder, W. F. Stickle, P. E. Sobol, K. D. Bomben, Handbook of X-ray Photoelectron Spectroscopy, Physical Electronics, Eden Prairie, Minnesota, 1995.
[7]   H. Madhavaram, H. Idriss, S. Wendt, Y. D. Kim, M. Knapp, H. Over, J. Asman, E. Loffler, M. Muhler, *Journal of Catalysis* **202** (2001) 296.
[8]   S. Argile, G. E. Rhead, *Surf. Sci. Repts.* **10** (1989) 277.
[9]   C. Niu, N. P. Magtoto, J. A. Kelber, D. R. Jennison, A. Bogicevic, *Surface Science* **465** (2000) 163.
[10]   M. Pritchett, N. Magtoto, J. Tong, X. Zhao, J. A. Kelber, *Thin Solid Films* **440** (2003) 100.
[11]   G. R. Yang, T.-M. Lu, *Appl. Phys. Lett.* **72** (1998) 1846.
[12]   M. Baumer, H.-J. Freund, *Progress in Surface Science* **61** (1999) 127.

Mat. Res. Soc. Symp. Proc. Vol. 812 © 2004 Materials Research Society   F2.6

In-situ XPS Study of ALD Ta(N) Barrier Formation on Organosilicate Dielectric Surface

Junjun Liu, Junjing Bao, Michael Scharnberg[1], and Paul S. Ho
Laboratory for Interconnect and Packaging, Microelectronic Research Center, the University of
Texas at Austin, Austin, TX 78712-1063, USA
[1]Lehrstuhl für Materuakverbunde, Technische Fakultät der Christian-Albrechts-Universität
zu Kiel, Kaiserstr.2, D-24143 Kiel, Germany

ABSTRACT

Beams of nitrogen and hydrogen radicals were investigated as surface pre-treatment and
process enhancement techniques for atomic layer deposition (ALD) of tantalum nitride barrier
layer on a dense organosilicate (OSG) low k film. *In-situ* x-ray photoelectron spectroscopy
(XPS) studies of the evolution of the low k surface chemistry revealed an initial transient growth
region controlled mainly by the substrate surface chemistry. Pre-treatment of the low k surface
with radical beams, particularly with nitrogen radicals, was found to enhance significantly the
chemisorption of the $TaCl_5$ precursor on the OSG surfaces. The enhancement was attributed to
the dissociation of the weakly bonded methyl groups from the low k surface followed by
nitridation with the nitrogen radicals. In the subsequent linear growth region, atomic hydrogen
species was able to reduce the chlorine content under appropriate temperature and with sufficient
purge. The role of the atomic hydrogen in this process enhancement is discussed.

INTRODUCTION

Atomic layer deposition (ALD) of ultra-thin barrier layers is a key process for
implementation of Cu/low k interconnects. For ALD, the initial chemical reactions at the
substrate surface are important in controlling the barrier uniformity and morphology. This is
particularly important for low k dielectrics generally characterized by weak surface bonds which
have to be properly activated for barrier formation. Plasma pre-treatments with O_2, N_2 and H_2
have been applied to various low k dielectrics for ALD of nitride barrier layers.[1,2,3] O_2 and N_2
plasma or downstream RIE treatments on CVD organosilicates were generally successful in
enhancing chemisorption by activation of the low k surfaces with low energy ions and radicals,
but only in some special cases, the underlying low k film retained its dielectric property. These
studies have generated interests in applying plasma induced radicals for surface modification, but
the effects on barrier formation on low k surfaces are not well understood, particularly regarding
the surface bonds and the chemical reaction for initial barrier formation. Moreover, as porous
OSG is being investigated as a candidate ultra low k for integration in sub-90nm technology
nodes, pore sealing has become a critical issue. Various plasma pretreatments have been
attempted on mesoporous OSG films without much success.[4] An *in-situ* chemical analysis of the
initial barrier formation process would provide information for better understanding of the issues
in pore sealing.

In this study, *in-situ* XPS was used to study the initial formation of Ta-based barriers by
ALD on dense and porous OSG low k dielectrics. To reduce beam damage, nitrogen radical and
atomic hydrogen beams were used to assist barrier formation. The *in-situ* XPS enabled us to
examine in detail the surface modification by N and H radical beams and their impacts on
chemical reactions on the low k surface and the subsequent ALD nitride growth. Our results

demonstrated that the radical beams served as effective reducing agents in initiating and enhancing the precursor chemisorption. This led to an improvement of the barrier film quality and an increase of the deposition rate at a temperature compatible with low k dielectrics (< 400°C).

EXPERIMENTAL DETAILS

All the experiments were carried out in an ultra-high vacuum (UHV) system consisting of an ALD process chamber and an XPS system where interface formation can be examined *in-situ* by XPS during barrier layer deposition. Sample can be transferred among the load lock, the ALD process chamber and the XPS analysis chamber[5] without breaking the vacuum. A mini atom/ion hybrid source was mounted onto the ALD process chamber, providing *in-situ* versatile pre-treatment and process enhancing capabilities. Atomic species were generated by cracking gas molecules with ECR induced plasma. In the atomic beam mode, only neutral species can pass an aperture to reach the process chamber. The atom/ion source was used to generate atomic hydrogen and/or nitrogen radicals in this study by cracking either H_2 or NH_3.

A dense OSG low k film was selected for this study. A metal halide ALD process, $TaCl_5$ + $NH_3 \rightarrow Ta(N) + HCl$, was chosen for its chemical simplicity. The substrate temperature maintained at 390°C. $TaCl_5$ and NH_3 were introduced sequentially, separated by argon purges. The $TaCl_5$ pulse and NH_3 pulse were 2.5 seconds and 4 seconds respectively. NH_3 in molecular form or nitrogen and hydrogen radicals obtained by leaking NH_3 into the hybrid source were used for surface modification. To further reduce the residual chlorine content, experiments were also performed with atomic hydrogen being applied prior to the introduction of ammonia.

The surface modification by nitrogen and hydrogen radical beams was examined by *in-situ* XPS before and after the beam treatments. The evolution of the surface chemistry during ALD processes was monitored by removing samples from the growth chamber to the analysis chamber for *in-situ* XPS analysis at any point of interest during the growths. Growth rates were evaluated based on the attenuation of the photoelectron signal intensity of the substrate, Si2p.

RESULTS AND DISCUSSIONS

The effect of radical beam treatment on the initial growth of tantalum nitride was examined by XPS and the observed evolution of N1s spectra is presented in figure 1. The low k surface had been carefully cleaned with acetone and deionised water[6] followed by outgassing at 350°C for at least 4 hours at low 10^{-7} torr. As indicated by curve 1a, there was pre-existing nitrogen in the original film with a binding energy at -399.0eV. After the film was exposed to 30 ALD cycles, only a small Ta4p peak developed at -403.5eV as a nitride peak appeared at -397.7 eV (curve 1c). With the low k films pretreated with atomic hydrogen and nitrogen beams, its surface was observed to have incorporated a larger nitrogen concentration (curve 1b). When the pretreated film was exposed to 30 identical ALD cycles, the tantalum signal was almost tripled (compare curves 1c and 1d). At this point, the ratio between Ta and Si signals increased from 0.11 to 0.25, indicating that a significant amount of tantalum nitride was already formed.

The surface modification by the radical beams was quantified by spectral analysis of the XPS peak intensities. The changes of the surface chemical compositions are tabulated in table 1. Atomic hydrogen treatment to the dense OSG surface did not cause much surface modification except that the surface chemical composition was modified to a level very close to the bulk

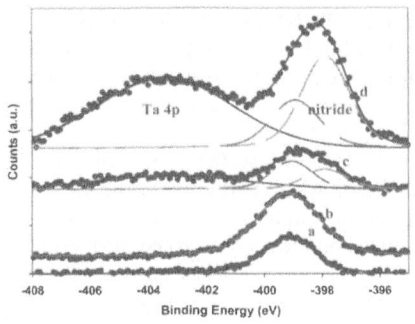

Figure 1: Effects of atomic N pretreatment on ALD Ta(N) growth on the dense OSG low k. N1s spectra.
a) of clean surface;
b) after pretreatments;
c) 30 ALD cycles on untreated film; and
d) 30 ALD cycles on pretreated film.

Table 1: Evolution of Surface Chemical Composition

	Before beam treatments		After H treatment		After N and H treatments ("Surface")
	"Surface"	"Bulk"	"Surface"	"Bulk"	
O	42.7%	40.7%	44.1%	43.7%	44.1%
C	23.6%	25.0%	21.2%	21.0%	16.0%
Si	30.2%	30.7%	30.6%	31.1%	31.6%
N	3.4%	3.6%	4.0%	4.1%	7.0%

Note: The "surface" angle is 30° from the sample surface and the "bulk" angle is normal to the sample surface.

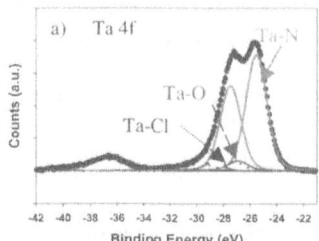 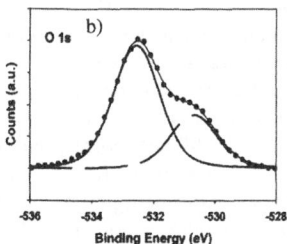

Figure 2: Chemical bonds in the film as deposited. a) Ta4f and b) O1s

chemical composition. The reduction of carbon content could be simply due to the removal of surface contaminants. When atomic nitrogen and hydrogen were both applied by cracking NH_3, the surface carbon concentration was further reduced to appreciably below the bulk carbon concentration, suggesting a depletion of the methyl groups from the surface region. Nitrogen enrichment following the beam treatments was observed as expected since the process was designed to alter the interfacial chemistry by providing more nitrogen species to react with the incoming $TaCl_5$ precursor to promote nitride formation. Enhanced nitride formation was clearly confirmed as depicted in curve 1d.

Deconvolution of the Ta4f spectra in the final film indicated that Ta atoms were primarily bonded with N as shown in fig.2a. The binding energy of Ta-N is located at -25.5 eV, much higher than -22.4 eV for the metallic TaN, suggesting that the Ta(N) formed is in a dielectric phase. N/Ta atomic ratio was found to be around 1.6 in the film. It has a good agreement with the ratio in the known dielectric phase of Ta(N), Ta_3N_5. During the growth, no chemical shifts were observed in C1s and Si2p spectra; however, indicating the chemisorption should be through the formation of Ta-O and/or Ta-N (fig. 1 and 2b).

A complete ALD tantalum nitride growth on the pretreated low k surface is shown in fig. 3. It is evident that there existed an initial transient growth region where part or all of the incoming precursor molecules adsorbed directly onto the low k dielectric surface to form one saturated "monolayer" coverage.[7] Subsequent reaction cycles occurred on the newly formed nitride surface increasing the rate of ALD formation of tantalum nitride. Linear growth behaviour was observed afterwards as a result of a constant surface chemistry in the subsequent reaction cycles.

The decay of the substrate Si2p signal can be used to estimate the thickness of the overlayer and the growth rate in the linear growth region. Using a simple exponential decay model,[8] assuming a "complete" surface coverage at saturation and taking 2.0 nm as the effective attenuation length for Si2p photoelectrons in a partially oxidised tantalum nitride overlayer,[9] the growth rate was estimated to be 0.011nm per cycle. This is close to the growth rates previously reported on ALD tantalum nitride processes on SiO_2,[10,11] suggesting a much improved ALD nucleation on the low k surface resulting from the atomic beam treatments.

Since XPS can not analyse hydrogen, the surface modification induced by atomic hydrogen was not well characterized. While further examination of the role of atomic hydrogen in activating the low k surfaces is required, the apparent lack of reactions between atomic hydrogen species and the surface groups on the dense OSG film may be due to that Si-O bonds (~ 800 kJ/mol) are much stronger than Si-H (~ 300 kJ/mol)[12] and Si-C (~320 kJ/mol)[13] bonds. Modification of the OSG low k surface by atomic nitrogen is better understood. A possible mechanism is that the atomic nitrogen species replaced the methyl groups on the surface since the formation of Si-N bond (~ 470 kJ/mol)[12] is energetically preferred over that of Si-CH_3. Si-N and Si-O bonds can serve as good adsorptive sites for $TaCl_5$ precursor. As a result, the density of surface adsorptive sites was increased, and so was the growth rate.

The effect of the atomic hydrogen in reducing chlorine residue, as well as the step-by-step nature of ALD reactions, was well represented by the evolution of Cl2p spectra in fig. 4. The peaks located around -199.8eV correspond to chlorine in a Ta-Cl bonding environment. After exposure to the $TaCl_5$ precursor (fig. 4a), the sample surface adsorbed a thin layer of the precursor molecules. After being treated with atomic hydrogen and then exposed to NH_3 reagent, the Cl2p peak was completely gone (fig. 4b), indicating a complete reaction between the precursor, the atomic hydrogen and the ammonia. The possibility of the chlorine reduction as a result of reactions with hydrogen molecules or even with any residual oxygen molecules can be ruled out by monitoring the Cl2p evolution following two different treatments of a sample

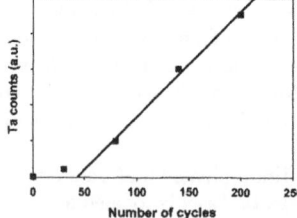

Figure 3: Growth behavior of ALD Ta(N) on the pretreated dense OSG low k.

overexposed to $TaCl_5$ precursor (fig. 4c). The chlorine content was almost halved (fig. 4d) when the sample was treated with atomic hydrogen for only 10 minutes, whereas the chlorine signal was still clearly visible (fig. 4e) after additional 4 hour anneal at 390°C in an H_2 ambient.

The residual Cl content and its chemical states in the ALD Ta(N) films indicated to what extent the precursor molecules reacted with NH_3. It depends much upon the substrate temperature and purging conditions. At our process conditions at 390°C and with long purge (12 seconds Ar purge after $TaCl_5$ pulse), the films as deposited still contain considerable amount of chlorine, about 7 at.%. Further deconvolution of Cl2p spectra revealed the presence of Cl primarily in Ta-Cl bonds, with only a small portion in H-Cl bonds. The reduction of Cl content with atomic hydrogen depends upon the dose of treatment. With the insertion of an H pulse prior to the NH_3 pulse, Cl content was reduced to a level beyond detection (< 1 at. %) on one occasion, as shown in fig. 5.

The observed process enhancement by atomic species can be attributed to their reactivity. In the following reactions of $TaCl_5 + H \rightarrow Ta + HCl$ and $TaCl_5 + N + H \rightarrow Ta(N) + HCl$, the reaction barrier between atomic species and $TaCl_5$ precursors are considerably lower than that between NH_3 gas molecules and $TaCl_5$, as suggested by the high dissociation energies of NH_3 and H_2. As the reactions became more complete, there would be not much partially reacted NHx left in the film to block the adsorptive sites for the incoming $TaCl_5$ precursor.[14]

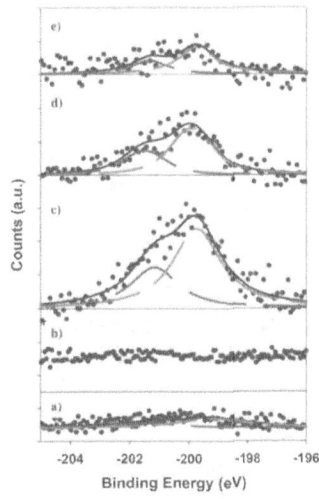

Figure 4: Evolution of Cl2p.
a) after exposure to $TaCl_5$ and purge with Ar;
b) after reaction with atomic H and with NH_3;
c) after additional exposure to $TaCl_5$;
d) after 10 min treatment with atomic H; and
e) after additional 4 hour 390°C anneal in H_2.

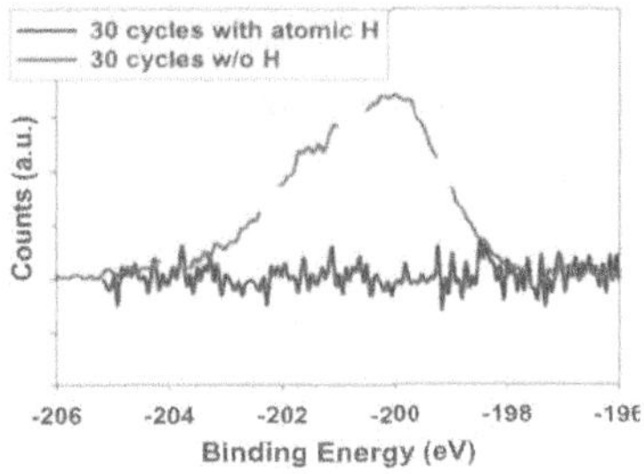

Figure 5: Reduction of Cl content with atomic H.

In summary, the evolution of the surface chemistry during ALD of Ta(N) on the dense OSG low k surface was examined using *in-situ* XPS analysis. Radical beam pre-treatment of the low k surface, particularly with nitrogen radicals, was found to enhance significantly the chemisorption of the TaCl$_5$ precursor. The enhancement was attributed to the dissociation of the weakly bonded methyl groups from the OSG surface followed by surface nitrogen enrichment with the nitrogen radicals. Atomic hydrogen was found to considerably reduce the chlorine impurity content in the Ta(N) film. The improvement of the growth rate observed in this study illustrates the criticality of the surface coverage of the substrate film by the deposited film at saturation for a successful ALD process. It affects the growth rate of the linear growth behaviour and the film uniformity and topography.

ACKNOWLEDGEMENT

The authors gratefully acknowledge ATMI for providing the TaCl$_5$ ampoule. J.L. would like to thank Richard Grooms from MKS instruments, Drs. E. Todd Ryan, Andrew J. Mckerrow, Jae Gab Lee, Ennis Ogawa, Michael Kiene and Yangming Sun for very helpful discussions and suggestions.

REFERENCES:

1. Y. Zhou, G. Xu, T. Scherban, J. Leu, G. Kloster and C. Wu, Char. and Metro. for ULSI Technology, 455-61 (2003)
2. W. Besling, A. Satta, J. Schuhmacher, T. Abell, V. Sutcliffe, A. Hoyas, G. Beyer, D. Gravesteijn, and K. Maex, Proceedings of the IEEE 2002 IITC, 288-91 (2002)
3. A. Satta, M. Baklanov, O. Richard, A. Vantomme, H. Bebder, T. Conard, G. Beyer, K. Maex, W.M. Li, K.E. Elers, and S. Haukka, Micro. Eng., **60**, 59 (2002)
4. E.T. Ryan, M. Freeman, L. Svedberg, J.J. Lee, T. Guenther, J. Connor, K. Yu, J. Sun and D. Gidley, Mater., Tech. and Rel. for Adv. Inter. and Low-k Diel., Mater. Res. Soc. Vol. **766,** p89-94 (2003)
5. P. Abramowitz, M. Kiene and P.S. Ho, Appl. Phys. Let., vol.**74**, no.22, 3293-531 (1999)
6. M.P. Seah and S.J. Spencer, J. Vac. Sci. Technol., **A 21**(2), 345-352 (2003)
7. H. Kim, J. Vac. Sci. Technol., B 21(6), p2231-2261 (2003)
8. D. Briggs and M.P. Seah, Practical Surface Analysis, Vol. 1 (1990)
9. C.J. Powell and A. Jablonski, NIST Standard Reference Database 82, ver. 1.0 (2001)
10. H. Kim, C.Cabral, Jr., C. Lavoie, and S.M. Rossnagel, J. of Vac. Sci. and Tech., **B 20**(4), 1321-1326 (2002)
11. M. Ritala, P. Kalsi, D. Riihelä, K. Kukli, M. Leskelä, and J. Jokinen, Chem. of Mater., **11**, 1712-1718 (1999)
12. D.R. Lide, CRC Handbook of Chemistry and Physics, 81st Edition (2000)
13. Http://chemviz.ncsa.uiuc.edu/content/doc-resources-bond.html
14. M. Laskelä and M. Ritala, Thin Solid Films, **409**, 138-146 (2002)

Mat. Res. Soc. Symp. Proc. Vol. 812 © 2004 Materials Research Society

Nucleation and Growth Dependence of ALD WNC on Substrate Surface Condition

Thomas Abell[1], Jörg Schuhmacher[2], Youssef Travaly[2], Karen Maex[2,3]
[1]Intel affiliate at IMEC, Kapeldreef 75, 3001 Leuven, Belgium (thomas.j.abell@intel.com)
[2]IMEC, Kapeldreef 75, 3001 Leuven, Belgium
[3]K.U. Leuven, Belgium

ABSTRACT

ALD WNC nucleation and growth was observed strongly affected by different substrate materials and surface chemistries. Nucleation was inhibited on most pristine low k surfaces, which is attributed to low concentrations of chemisorption sites (Si-OH). Plasma treatments were used to alter the surface chemistry to improve nucleation. Surface closure and surface roughness of the WNC layer were found to strongly correlate with starting surface condition. Resistivities of the resulting films were also found dependent on starting surface treatment. But the relationship between W content of the films, surface treatment and resistivity was not fully comprehended.

INTRODUCTION

Atomic layer deposition (ALD) is an attractive method to deposit ultrathin Cu diffusion barrier films for semiconductor interconnect applications due to the high conformality of deposition. Reduction of barrier thickness while maintaining adequate barrier properties is becoming increasingly important as feature sizes are reduced as documented in the ITRS roadmap [1]. The ALD process relies upon chemisorption of precursor molecules onto the surface of the substrate for nucleation [2]. Thus, ALD processes are inherently sensitive to the starting surface chemistry of the substrate. Surface chemistries can be significantly altered by several methods including wet chemistry (e.g. HF) and irradiation (e.g. electron beam, UV, or plasma). The nucleation and growth of ALD films are not completely understood with the island growth model as one of several [3]. The island growth model supposes nucleation of the film at chemisorption sites with subsequent growth on the nucleated island. A continuous film is formed when islands grow together and merge.

Integration of ALD with porous low k dielectric materials can be problematic for several reasons. The first is due to the heterogenous nature of the surface chemistry at a molecular level. Most Si-based low k dielectrics incorporate $Si\text{-}CH_3$ bonds that have low polarity, induce microporosity (< 2 nm pore diameters) and result in a hydrophobic surface. These $Si\text{-}CH_3$ terminating bonds are present along with other surface bonds such as Si-O-Si, Si-OH, $Si\text{-}NH_2$ and other contaminant species. It is possible that the number of reactive chemisorption sites for ALD precursors may be the limiting factor determining nucleation density. This would occur in situations where the reaction site concentration is lower than the steric hindrance limit of the precursor molecules on the surface (e.g. maximum packing density of adsorbed molecules).

Another difficulty in integrating ALD with porous low k dielectrics is the ability of the gaseous ALD precursor molecules to penetrate into the porous structure that can result in unwanted internal deposition degrading the electrical properties of the dielectric. It has been shown that precursors penetrate and deposit deeply into mesoporous (>2nm pore diameters) films [4]. Plasma treatments of sufficient power have been shown to seal the surface of microporous films while mesoporous films have not been sealed in this way [5].

This work focused on the WNC ALD process and attempted to assess the sensitivity of the precursor chemistry to the starting surface condition for several material types from thermal oxide to ultra low k dielectric films. An investigation into different surface treatment conditions to modulate the nucleation and growth was also conducted.

EXPERIMENTAL

Blanket films of various dielectrics were deposited onto (100) silicon by CVD, PECVD and spin-on techniques. All dielectrics were Si-based oxides with varying degrees of structure (silicate vs. silsesquioxane), organic incorporation (CH_3) and porosity. Thermally grown oxide was used as a reference and was compared to dense PECVD SiO_2, microporous PECVD SiCOH films and spin-coated mesoporous materials. Ellipsometric porosimetry (EP) was used to characterize the porosity of the low k materials. The films were also characterized by AFM for surface roughness and by static TOFSIMS for initial surface chemistry.

The blanket dielectric films were then exposed to varying plasma conditions in three different plasma chambers: LAM Exelan, LAM Versys and Mattson Aspen Highland. Low pressure conditions in the 10 mTorr range were employed in the Versys and Aspen chambers while the Exelan was operated in the 120 mTorr range for ignition. Power conditions ranged from 100 to 1200 Watts with varying degrees of bias or dual frequency power. Chemistries of Ar, He, N_2 were employed in the plasma to reduce chemical attack of the Si-CH_3 groups. The resulting surfaces were probed for roughness and chemistry with AFM and static TOFSIMS. EP was used to assess sealing to the toluene probe molecule.

ALD WNC films [6] were deposited on the substrates using an ASM Pulsar 2000 reactor involving sequential pulses of TEB, WF_6 and NH_3 at 300 C. Various ALD cycles were used ranging from 20 to 120 for investigation of growth curves and surface closure. These were measured using RBS, XRF and EP. 4 point probe measurements were made for sheet resistance and thicknesses were measured with XRR and TEM. XRR and RBS were used to assess penetration into porous films.

RESULTS AND DISCUSSION

Table I illustrates the range of materials that were considered in this study as substrates for

Table I. Properties of the various materials investigated as WNC substrates

Material	Type	k value	Porosity, % by EP	Pore Diam, nm	Porous Surface, EP
TOX	silicate	3.9	0	0	No
PE-OX	silicate	4.3	0	0	No
CVD1	silicate	3.0	7	<2	Yes
CVD2	silicate	2.7	8	<2	Yes
CVD3	silicate	2.7	8	<2	Yes
CVD4	silicate	2.4	16	<2	Yes
Spinon1	MSQ	2.2	38	<2 + 3.1	Yes
Spinon2	silicate	2.2	35	<2 + 3.1	Yes
PE-SiC	carbide	>4	0	0	No

the ALD WNC process. Thermal oxide, PECVD oxide and PECVD SiC were used as dense reference substrates. The CVD and spin-on materials were used to represent the different classes of Si-based low k materials produced by different methods, tools and precursors. Table I also lists the porosity and pore diameters as measured by EP.

Surface TOFSIMS analysis of CVD2 is shown in Figure 1 and indicates that the pristine material possesses few Si-OH surface groups and a significant quantity of Si-CH$_3$ groups. Figure 1 also reveals the increase in surface Si-OH groups and decrease in surface Si-CH$_3$ groups that can be obtained by increasing levels of plasma exposure. It is notable that the 100/25 W He plasma exposure in the Mattson tool clearly modifies the surface chemistry but the film still appears porous to toluene during EP test. Increasing the power level to 500/25W He in the Mattson appears to accomplish similar surface modification as an N$_2$ plasma ignition in the LAM Exelan chamber. However, an Ar plasma ignition in the same Exelan chamber appears to significantly increase the surface Si-OH groups. Elevated amine groups are also seen after N$_2$ plasma treatment along with organic and halogen contamination for all plasmas.

XRR profiling of plasma treated surfaces (not shown) reveals a significant increase in electron density at the surface similar to SiO$_2$ values. Previous work has shown carbon depletion from the surface region after plasma treatment as measured by EFTEM and TOFSIMS profiling. It is believed that this surface is a dense SiO$_2$-like layer formed from the reconstruction of the Si-O network. This layer is also impermeable to toluene as measured by EP. The plasma treatments also appear to smooth the surface of the low k films and reduce the roughness of WNC layers that are deposited. Figure 2 illustrates the affect of plasma with atomic force microscopy (AFM) measurements.

The number of cycles required for ALD to form a continuous layer is referred to as surface closure. Several methods involving bombardment of the surface to detect substrate signals have been used (e.g. TOFSIMS, ERD). In this work we used the penetration of toluene molecules though a discontinous WNC layer into the porous underlying low k dielectric to detect surface closure. EP measurements of WNC layers on porous low k substrates revealed striking

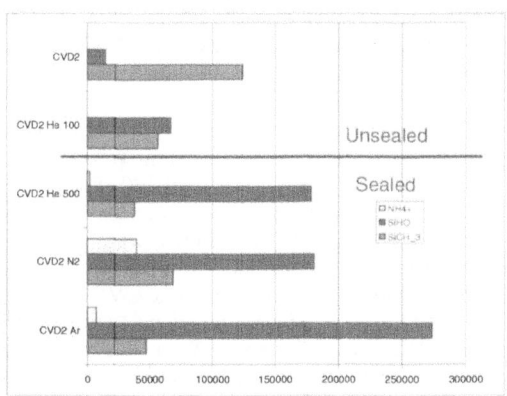

Figure 1. TOFSIMS analysis of pristine and plasma treated CVD2 indicating reduction of Si-CH$_3$ groups and increase of surface Si-OH groups with treatment. Sealing threshold is indicated between 100/25 and 500/25W He plasma powers.

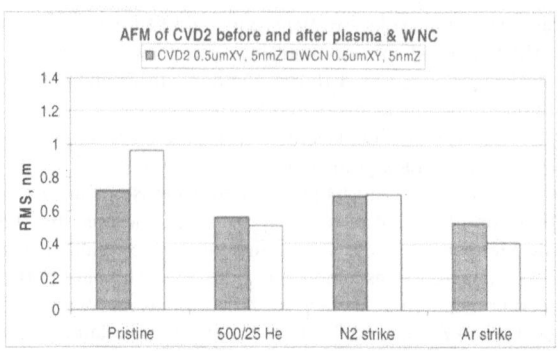

Figure 2. AFM of CVD2 showing reduced surface roughness after plasma treatment and reduced roughness of subsequent WNC films.

Table II. EP sealing of WNC films on porous low k substrates to assess WNC surface closure.

	Treatment	Sealed after Treatment?	50c WNC	65c WNC	85c WNC	105c WNC	120c WNC
CVD1	Pristine	No	No	No	No	No	Yes
CVD2	Pristine	No	No	Yes	Yes	Yes	Yes
CVD2	100/25 He	No	No				
CVD3	Pristine	No	No	No	No	No	No
CVD3	100/25 He	No					Yes
CVD4	Pristine	No					No
CVD4	100/25 He	No					Yes
Spinon1	Pristine	No	No	No	No	No	Yes
Spinon2	Pristine	No					No
Spinon2	Ar strike	No					No

differences in surface growth and closure. Table II depicts the differences. WNC on pristine CVD3 and CVD4 will not form a sealed layer with 120 cycles (~9 nm on dense oxide) indicating very poor nucleation characteristics. However, a low power 100/25W He plasma, that still leaves the low k surface porous to toluene, significantly improved the WNC surface closure on both CVD3 and CVD4.

XRR analysis of the WNC films deposited on the different substrate surfaces was conducted. The x-ray penetration angle into the top surface of the different WNC films was nearly identical for all substrates. This implies that the top surface density of 120 cycle WNC films are similar regardless of substrate. However, this does not reveal detail concerning the bottom of the WNC at the substrate interface. XRR and RBS analysis (not shown) were used to assess penetration of W precursors (and their subsequent deposition) inside porous low k films. Both techniques indicate WNC penetration into the porous structure of pristine CVD2 and CVD4. Plasma treatment that seals the low k's to toluene was found to prevent WNC penetration.

XRF analysis was done on WNC films to determine W content as a function of ALD cycle and surface treatment. Figures 3a and 3b depict the differences in growth on two pristine CVD substrates and the improvements in growth obtained by plasma treatment. The linear growth on CVD1 matches the growth seen on dense PECVD oxide. However, PE-OX has higher W content at 3.1E16 for the same 120 cycles. CVD3 demonstrates clear non-linear growth. Sheet resistance, Rs, is plotted versus XRF W counts in Figure 4 for different plasma treatments on all CVD low k

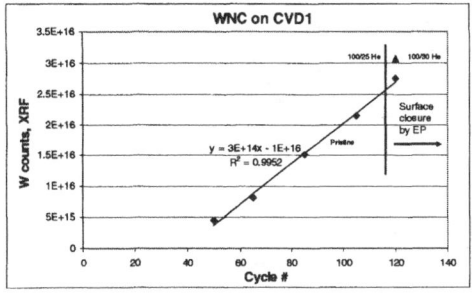
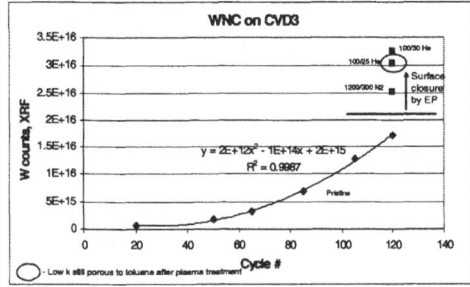

Figure 3. XRF W counts per ALD cycle showing: a)linear growth on pristine CVD1, b)non-linear growth on pristine CVD3. W counts are higher after plasma treatment on both low k's

films and PECVD oxide. The expected relationship between W content and sheet resistance is not evident. In most cases the sheet resistance values fall to 600 to 700 Ohm/sqr after plasma treatment. However, the lower W content seen for 1200/300W N_2 plasma treatment does not seem to strongly affect the sheet resistance. The mechanism for this behavior is not well understood. Groupings based on the type of plasma used are noted. Resistivity values were calculated from the sheet resistance and XRR thickness measurements. Figure 5 shows that higher power plasma treatments appear to reduce the bulk resistivity of the films. This may be caused by the difference between surface-affected growth and growth on WNC. This might account for the difference between the 500~600 uOhm.cm values for thin films (~9 nm) and 350~400 uOhm.cm values reported in the literature for thicker films (~25 nm). It is believed that the microstructure of the films at the nucleation interface are dependent on the surface modifications induced by plasma treatment and that these layers are significant in ultra thin films.

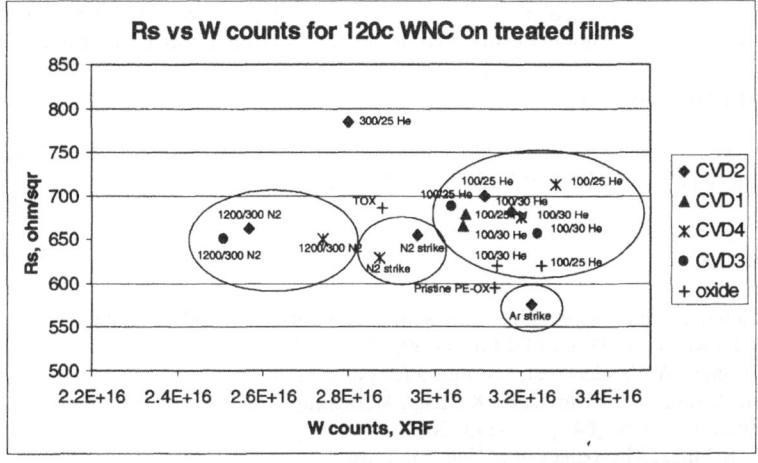

Figure 4. Rs vs. XRF W counts for plasma treated CVD low k substrates and dense oxides. Similar Rs with differing W content as a function of plasma treatment is not well understood.

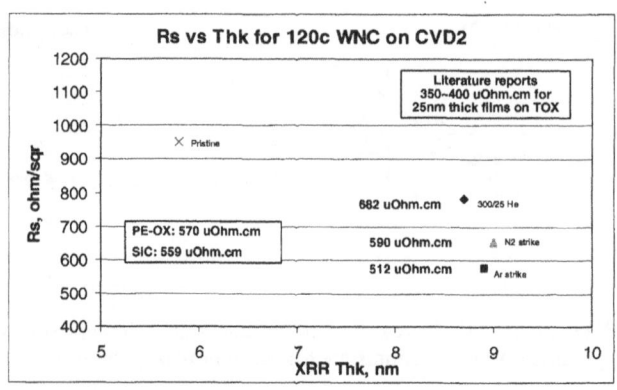

Figure 5. Resistivity of WNC films deposited on plasma treated CVD2 reduces with increasing plasma damage of the substrate.

CONCLUSIONS

ALD WNC nucleation and growth was observed to be strongly affected by different substrate materials and surface chemistries. Nucleation was found to be enhanced on plasma treated low k surfaces, which is attributed to increased chemisorption sites for precursors. Surface closure and surface roughness of the WNC layer was found to strongly correlate with starting surface. Plasma treatments were also shown to seal the surface of some low k materials to penetration of precursor molecules into the films. Resitivity of WNC films was seen to vary with different surface treatment. It is believed that the structure of the WNC ALD films was strongly affected by the initial nucleation site density. However, the relationship between W content, surface treatment and resistivity is not understood but is clearly related to plasma power and chemistry.

ACKNOWLEDGEMENTS

The authors would like to thank Danielle Vanhaeren for AFM measurements, Bert Brijs for RBS measurements and Thierry Conard for TOFSIMS measurements.

REFERENCES

1. International Technology Roadmap for Semiconductors, p75, 2002 Update
2. M.Leskela, M.Ritala, Thin Solid Films **409**, 138 (2002)
3. R.L.Puurunen, W.Vandervorst, submitted for publication
4. G.Beyer, A.Satta, J.Schuhmacher, K.Maex, W.Besling, O.Kilpela, H.Sprey, G.Tempel, Microelectronic Engr. (64) p233-245, 2002
5. T.Abell, K.Maex, Microelectronic Engr., in press.
6. W.M.Li, K.Elers, J.Kostamo, S.Kaipo, H.Huotari, M.Soininen, P.J.Soininen, M.Tuominen, S.Hauka, S.Smith, W.Besling, Proceedings of IITC 2002, (IEEE NY), 2002.

Evaluation of PECVD deposited Boron Nitride as Copper Diffusion Barrier on Porous Low-k Materials

J. Liu[1, 2], W. D. Wang[1], L. Wang[1], and D. Z. Chi[1,]*, and K. P. Loh[2, #]

[1] Institute of Material Research and Engineering, 3 Research Link, Singapore 117602

[2] Department of Chemistry, National University of Singapore, 3 Science Drive 3, Singapore 117543

* e-mail: dz-chi@imre.a-star.edu.sg

Abstract

Ultra low dielectric constant (k) material is needed as the inter-metal dielectrics to reduce RC delay when device dimension is scaled to sub-100nm. Porous dielectric films have been considered as good candidates for the application as inter-metal dielectrics due to their ultra low-k properties. Identifying proper dielectric copper diffusion barrier on the porous low-k films is critical for the low-k/Cu damascene fabrication process. In this study, we have evaluated the compatibility of plasma-deposited amorphous Boron Nitride film as a dielectrics copper diffusion barrier on a MSQ-based porous low-k LKD5109 film (from JSR). Both microwave plasma enhanced CVD (2.45 GHz) and radio-frequency plasma enhanced CVD (13.56 MHz) were applied for the BN deposition in order to evaluate the compatibility of the two plasma processes with the porous film. Growth parameters were optimized to minimize the boron diffusion and carbon depletion in the porous low-k films, which were found to have deleterious effects on the dielectric properties of the low-k films. FTIR and micro-Raman were employed for analyzing the changes in chemical structure of the low-k films after BN growth. Capacitance-voltage measurement was used to characterize the dielectric constants of BN film on Si and the BN-deposited porous low-k film. SIMS characterization was carried out to evaluate the performance of the BN film against copper diffusion.

Introduction

Ultra low dielectric constant (k) material is needed as the inter-metal dielectrics to reduce RC delay when device dimension is scaled to sub-100 nm. Porous dielectric films have been considered as good candidates for the application as inter-metal dielectrics (IMD) due to their ultra low-k properties. Identifying suitable dielectric copper diffusion barrier on the porous low-k films is critical for the successful integration of the porous low-k films into the low-k/Cu damascene interconnect scheme. SiN and a-SiC have been two most extensively studied dielectrics as dielectric diffusion barriers or etch stop layers in the back-end process [1-4]. However, the k values of these films are about 4.0 or even higher and therefore the use of these materials as dielectric barriers will certainly increase the effective k value of dielectric stack when integrated with porous low-k material. As known to have a low dielectric constant (≈2.2), boron nitride (BN) or boron carbon nitride (BCN) thin films have been studied previously on silicon substrates for their potential applications as inter-metal dielectrics [5, 6], but not for dielectric barriers on the porous low-k films. Although low-dielectric-constant BN can act as low-k inter-metal dielectrics, there is a limitation of lowering the k value of BN. Spin-on porous dielectrics have the great extendibility to lower the k value to 2.0 and below by changing the porosity of the films and the extendibility is probably the primary consideration in material selection. Therefore, it is very important to integrate low-dielectric-constant BN as Cu diffusion barrier with porous dielectric films to achieve a low effective k value for future technology nodes.

In this study, we have evaluated the compatibility of plasma-deposited amorphous BN film as copper diffusion barrier on a MSQ-based porous low-k LKD5109 film. Both microwave plasma CVD (MWPECVD) and radio-frequency plasma CVD (RFPECVD) were applied for the BN deposition in order to evaluate the compatibility of the two plasma processes with the porous MSQ low-k film. Growth parameters were optimized to minimize the boron diffusion and carbon depletion in the porous low-k films, which were found to have deleterious effects on the dielectric properties of the low-k films. FTIR and micro-Raman were employed for analyzing the changes in chemical structure of the low-k films after BN growth. Capacitance-voltage measurement was used to characterize the dielectric constants of BN film on Si substrate and the BN-deposited porous low-k film. A k value of about 2.3 was determined for the amorphous BN film while an effective $k_{eff} \approx 2.4$ obtained on the RF plasma CVD BN-integrated porous low k film. SIMS characterization results showed that microwave plasma CVD BN film can act as an effective Cu diffusion barrier. However, a severe damage to the underlying porous low-k films were also observed when microwave plasma CVD was used for BN deposition.

Experimental

A MSQ-based porous low-k film, LKD-5109 from JSR, was used as staring material in this study. The relative dielectric constant k of this film is about 2.2. The meso-pores in the film were determined to have an average pore-size of 2.8 nm from postronium annihilation lifetime spectroscopy (PALS) and the meso-pores were found to be interconnected throughout whole film thickness (about 400 nm).

Both microwave plasma CVD (at 2.45 GHz) and radio-frequency plasma CVD (at 13.56 MHz) were applied for the BN deposition. To determine deposition process temperature window, the LKD films were firstly exposed to 1 hour thermal stress (N_2 ambient) at different temperatures. FTIR and C-V measurement were employed to characterize the low-k films. It was concluded that LKD films cannot withstand 1 hour thermal process at temperatures higher than 500°C.

BN films were grown on LKD5109 and on Si (100, p type) with borozine and N_2 as precursor gases, that were carried by H_2, using microwave enhanced CVD. The deposition process was carried out at different temperatures: 250°C, 300°C, 400°C and 500°C. Microwave plasma power was about 400 w and chamber pressure was 20Torr. The flow rates of borozine, H_2 and N_2 were 5.0sccm, 20sccm and 40sccm, respectively. Radio-frequency plasma enhanced CVD was also used to deposit the BN films on the porous low-k films (with a substrate temperature of 400°C). RF plasma power was around 300w, frequency was 13.56 MHz, and chamber pressure was 2.0×10^{-4} Torr.

After BN deposition, the films were characterized by FTIR and micro-Raman spectroscopy. Using a shadow mask, Au electrodes (200nm) of $1 mm^2$ in area were deposited on BN surface by electron beam deposition to form a metal/insulator/semiconductor structure of Au/BN/LKD/Si. Capacitance-voltage (C-V) measurement was used to evaluate the intrinsic k value of BN on Si substrate and the effective dielectric constant of the BN/low-k stack. SIMS was used to monitor the boron diffusion and the carbon depletion into/in the low-k film after BN deposition. To assess the performance of the BN film as Cu diffusion barrier, 150nm Cu was deposited on BN layers (which were deposited on the low-k films) by PVD sputtering and the samples were annealed at 400°C for 1hour in N_2 ambient. Tof-SIMS was employed to estimate the barrier performance of BN against copper diffusion.

Results and Discussion

It was found that the LKD films are stable under thermal stress at a temperature below 500°C while BN films deposited on low-k films at 250°C and 300°C appeared to be not stable (considerable moisture up-take was observed). Therefore, we focused our study on the BN films deposited at 400°C.

1. Structural and compositional characterizations

Figure 1(a) shows the Raman spectra obtained from as-received LKD film and the LKD film with a thin 60 nm BN layer deposited with microwave plasma enhanced CVD. The absorption peak at $1370cm^{-1}$ in the spectrum of the BN/LKD sample is a typical Raman signal associated with B-N stretching mode [7] and therefore the observation of the peak confirms the formation of a thin BN layer on the porous LKD film. The Raman spectrum of as-received LKD film features two strong absorption peaks at $2916cm^{-1}$ and $2976cm^{-1}$ which are attributed to C-H stretching vibration in CH_3. The absence of these two peaks in the LKD film with BN coating indicates that almost complete depletion of CH_3 groups in the bulk of the LKD film occurred after MWPECVD deposition of the BN film. For Raman characterization of RFPECVD deposited BN, there were no peaks observed due to the extremely high baseline intensity (which is thought due to strong photoluminescence signals from the RFPECVD-BN layer).

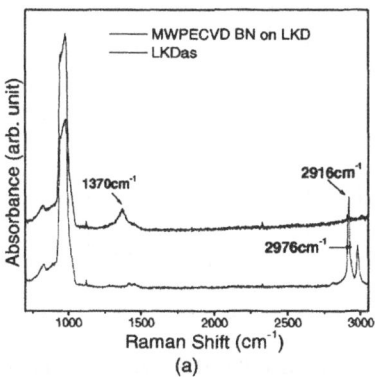 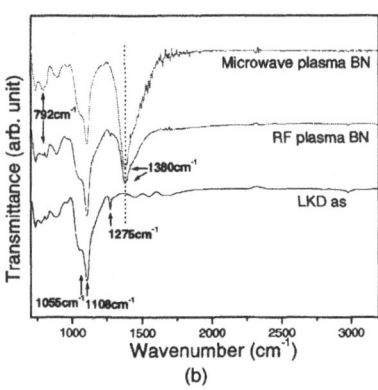

Figure 1. (a) Raman spectrum of as-received LKD film and the low-k film after microwave CVD BN deposited on it at 400°C, (b) FTIR spectrum of as-received LKD, microwave plasma CVD deposited BN on low-*k* film and RF plasma deposited BN on low-k material at 400°C.

A strong absorption peak associated the B-N stretching mode was also observed at $1380cm^{-1}$ in FTIR spectra of BN/LKD structures (for both LKD films coated with MW and RF plasma deposited BN layers), along with a weaker peak at $792cm^{-1}$ that is related to B-N-B bending mode [8]. While the peak at $1275cm^{-1}$ is attributed to Si-CH_3 stretching in the MSQ-based low-k film[9], the peaks at 1108 and 1055 cm^{-1} are assigned to large angle Si-O-Si bonds in a cage structure and the stretching of small angle Si-O-Si bonds in a network structure [10,11]. Comparing the three FTIR spectra, the $1275cm^{-1}$ peak almost disappeared after BN deposition using MWPECVD, while it still can be seen in the LKD film with RFPECVD-deposited BN layer, indicating that the deposition of BN layer using RFPECVD caused less damage to the underlying LKD film as compared to MWPECVD.

Figures 2 (a) and 2(b) compare the SIMS depth profiles of BN/LKD samples prepared with the two plasma processes, i.e., MWPECVD and RFPECVD. It can be seen from these figures that almost a complete carbon depletion occurred in the LKD film when MWPECVD was used for BN deposition, in agreement with Raman and FTIR results, while no significant carbon depletion was observed for the LKD film where BN layer was deposited using RFPECVD (almost the same carbon signal intensity was detected in the as-received LKD film; result is not shown). It is believed that the much higher flux of energetic ions and chemically active radicals in microwave plasma, as compared to RF plasma, are held responsible for the complete alteration of the LKD film after BN deposition. It is also important to note that significant boron diffusion from BN into underlying LKD occurred after BN deposition for both MWPECVD and RFCVD (but a substantially lower B diffusion for RFPECVD, probably due to lower implant flux in RF plasma). This serious boron diffusion is thought due to the small mass of B atom (means larger implant depth for boron ions) and the presence of highly interconnected meso-porous network in the LKD film.

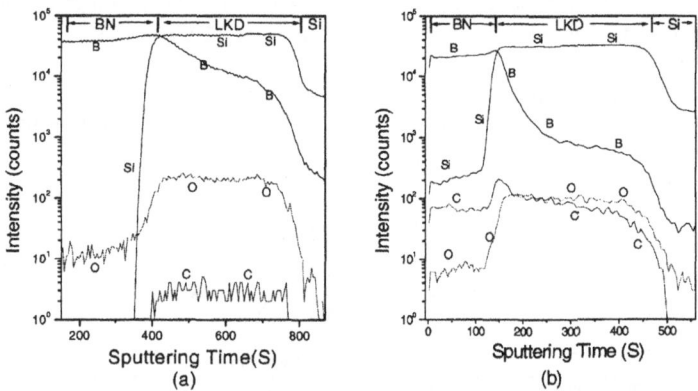

Figure 2. (a) SIMS depth profile of MWPECVD deposited BN on LKD film; (b) SIMS depth profile of RFPECVD deposited BN on LKD film.

2. Electrical Measurement

C-V measurements were carried out on BN films deposited on Si as well as on the LKD films. Figure 3 shows a typical C-V curve obtained on a BN film deposited using MWPECVD, while figures 4 compares the intrinsic k value of the MWPECVD deposited BN film and the effective k values (k_{eff}) of BN (MWPECVD)/LKD and BN (RFPECVD)/LKD stacks.

It can be noted that, though the MWPECVD-BN has a low k value of about 2.3, the BN(MWPECVD)/LKD stack shows a much higher k_{eff} of 5.0, highlighting the integration damage caused to the LKD film during the MWPECVD-BN deposition (due to the severe C depletion and B diffusion). Clearly, optimization of deposition parameters is needed to reduce this degradation of porous low-k film if BN layers are to be used as dielectric diffusion barriers. The observation of lower k_{eff} of 2.4 in the RFPECVD- BN/LKD stack, on the other hand, implies that further optimization of deposition process for minimizing such integration damage is possible.

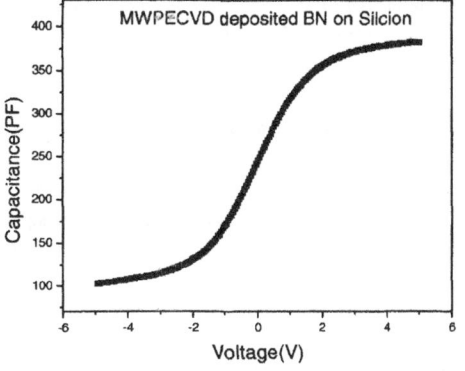

Figure 3. Typical C-V curve of MPCVD-BN on Si substrate.

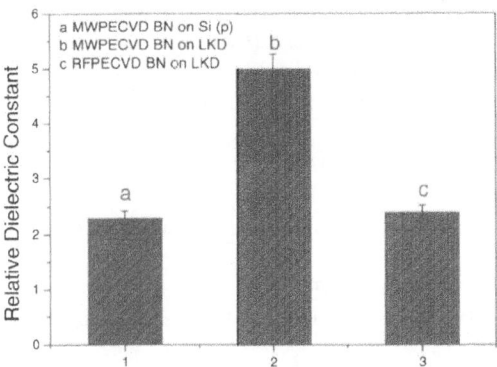

Figure 4. k values of a, b. MWPECVD deposited BN on Si and LKD film, c. RFPECVD deposited BN on LKD film.

3, Barrier Performance against Cu diffusion

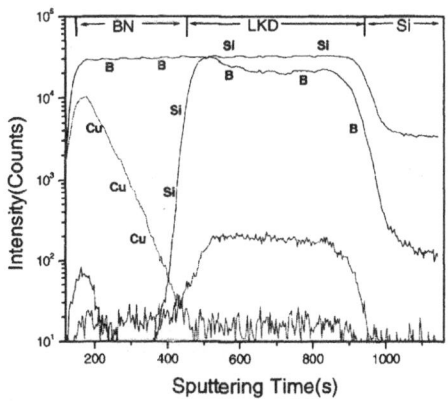

Figure 5. SIMS depth profile of Cu/MWPECVD-BN/LKD/Si structure annealing at 400°C in N_2 ambient for 1 hour.

Figure 5 presents the depth profile of the Cu/MWPECVD-BN/LKD/Si structure during one hour annealing at 400°C. It can be observed that the Cu signal intensity decreases rapidly in the BN layer with a Cu signal intensity almost at noise level (about 10) at the BN/LKD interface, indicating that the MWPECVD-BN layer has successfully blocked the Cu from diffusing into the porous low-k material.

Conclusion

In this work, microwave plasma enhanced CVD was used to prepare low dielectric constant BN films, which were demonstrated to be effective dielectric diffusion barriers against Cu diffusion on porous low-k films. However, severe carbon depletion and boron diffusion were observed after the integration of MWPECVD-BN and porous low-k films. Based on optimized process parameters, the degradation of porous low-k films was obviously reduced after the deposition of RFPECVD-BN film, which is believed to be an effective dielectric diffusion barrier against Cu diffusion.

References:
1. H. Sato, A. Izumi, A. Masuda, H. Matsumura, *Thin Solid Films*, 395, 280-283, (2001)
2. F. Lanckmans, W. D. Gray, B. Brijs and K. Maex, *Microelectron. Eng.*, 55, 329-335, (2001)
3. S. G. Lee, Y. J. Kim, S. P. Lee, H. S. Oh, S. J. Lee, M. Kim, I. G. Kim, J. H. Kim, H. J. Shin, J. G. Hong, H. D. Lee and H. K. Kang, *Jpn. J. Appl. Phys.*, Vol 40 2663-2668 (2001)
4. Y. W. Koh and K. P. Loh, L. Rong, A. T. S. Wee, L. Huang and J. Sudijono, *J. Appl. Phys.*, Vol 93, *No. 2, Jan (2003)*
5. T. Sugino, Y. Etou and T. Tai, *Appl. Phys. Lett.*, Vol. 80, No. 4,649 (2002)
6. T. Sugino, T. Tai, Y. Etou, *Diamond & Relat. Mater.*, 10, 1375-1379, (2001)
7. S. Stockel, K. Weise, D. Dietrich, T. Thamm, M. Braun, R. Cremer, G. Marx, *Thin Solid Films*, 420-421 (2002) 465-471
8. T. Sugino, H. Hieda, *Diamond & Relat. Mater.*, 9 (2000) 1233-1237
9. T. R. Crompton, *The Chemistry of Organic Silicon Compounds*, edited by S. Patai and Z. Rappoport (Wiley, New York, 1989), pp. 416-421.
10. M. G. Albrecht and C. Blanchette, *J. Electrochem. Soc.*, 145, 4019 (1998)
11. M. J. Loboda, C. M. Grove, and R. F. Schneider, *J. Electrochem. Soc.*, 145, 2861 (1998)

Mat. Res. Soc. Symp. Proc. Vol. 812 © 2004 Materials Research Society   F3.10

Structural and functional characterization of W-Si-N sputtered thin films for copper metallizations

Alberto Vomiero[1], Stefano Frabboni[2], Enrico Boscolo Marchi[1,3], Alberto Quaranta[1,4], Gianantonio Della Mea[1,4], Gino Mariotto[3], Laura Felisari[2]
[1] INFN – Laboratori Nazionali di Legnaro, Viale dell'Università 2, I-35020 Legnaro (PD), Italy.
[2] Physics Department, University of Modena and Reggio Emilia
Via Campi 213/a, I-41100 Modena, Italy.
[3] Physics Department, University of Trento, Via Sommarive 14, I-38050 Povo (TN), Italy.
[4] Materials Engineering and Industrial Technologies Department, University of Trento
Via Mesiano 77, I-38050 Trento, Italy

ABSTRACT

Ternary W-Si-N thin films have been reactively sputter-deposited from a W_5Si_3 target at different nitrogen partial pressures. The composition has been determined by 2.2 MeV $^4He^+$ beam, the structure by x-ray diffraction and transmission electron microscope, the chemical bonds by Fourier transform – infrared spectroscopy and the surface morphology by scanning electron microscopy. Electrical resistivity was measured by four point probe technique on the as grown films. The film as-deposited is amorphous with the Si/W ratio increasing from about 0.1 up to 0.55 with the nitrogen content going from 0 to 60 at%. The heat treatments up to 980 °C induce a loss of nitrogen in the nitrogen rich samples. Segregation of metallic tungsten occurs in the sample with low nitrogen content ($W_{58}Si_{21}N_{21}$). Samples with high nitrogen content preserve the amorphous structure, despite of the precipitation of a more ordered phase inferred by FT-IR absorbance spectrum of the layer treated at highest temperature. The surface morphology depends upon the nitrogen content; the loss of nitrogen induces the formation of blistering and in the most nitrogen rich sample the formation of holes. Electrical resistivity preliminary results on the as grown layers range between 500 and 4750 $\mu\Omega$cm passing from the lowest to the highest N concentration.

INTRODUCTION

Amorphous and conductive ternary thin films have properties both scientifically interesting and practically useful. Depending upon composition, annealing temperature and conditions they can remain structurally amorphous or nanostructured, quasi amorphous, up to temperatures as high as 900 °C. The concept of frustration is at the basis of the stability of the amorphous phase. The properties of the films are exploited for various applications as ultra-hard coatings, masks for x-ray lithography, micromachining elements, jet-ink cartridge, optical switching. However the major field of application is in microelectronic devices as diffusion barriers in copper based interconnections [1,2]. Among the various ternary possible compounds, Tm-Si-N compounds (Tm=transition metal), Ti,Ta,W have been considered because of their compatibility with the devices processes. Moreover the amorphous structure and chemical inertness of these compounds give the low diffusivities and the high chemical stability required for a diffusion barrier.

In this work we report on experimental study on W-Si-N alloy, which is a less studied system with respect to Ti-Si-N and Ta-Si-N [3,4]. An interesting difference exists between W-Si-N and the previously cited Ti-Si-N and Ta-Si-N: W does not form stable compound with N [5], although W_2N, WN, and WN_2 have been obtained by sputtering [6] or CVD deposition [7]

Reactively (RF)-sputtered thin layers are characterized from a compositional and structural point of view as a function of the sputtering conditions. Atomic density, microstructure and atomic bonding are investigated in the as grown layers. Heat treatments are performed in order to study the structural and compositional evolution of the layers under annealing. The annealed samples are characterized in the same way the as grown layers are, to obtain information about chemical inertness and stability of the W-Si-N system, eventual formation of crystalline phases and its possible application as diffusion barrier.

EXPERIMENTAL

W-Si-N films were deposited using a 2" source in RF mode (13.56 MHz) from a W_5Si_3 target, via reactive magnetron sputtering in Ar/N_2 gas mixtures. The composition of the Ar/N_2 reactive gas was controlled, in order to obtain films with different N content. During sputtering the total pressure was maintained at 0.4 Pa. Film thickness (~200 nm) was measured using a stylus profilometer. Annealing in vacuum was performed for 1.5 h at 600-750-900-980°C. Stoichiometric composition was obtained by Rutherford backscattering spectrometry (RBS) using a 2.2 MeV $^4He^+$ beam. Film density was calculated by combining the results of RBS analysis and thickness measurements. Surface morphology was investigated using scanning electron microscopy (SEM). Fourier transform-infrared spectroscopy (FT-IR) with a Jasco 660-plus spectrometer, operated in vacuum, was applied, in order to unveil the formation and evolution of chemical bonds. The microstructure was studied by transmission electron microscopy (TEM), acquiring also the signal of the selected area electron diffraction (SAED) pattern. Plan view samples have been prepared by lapping and low energy ion Ar^+ milling until perforation. TEM analysis have been performed on a Jeol JEM 2010 microscope (point resolution 0.19 nm). Images and SAED patterns have been recorded on a slow scan charge coupled device (CCD) camera. Four point probe measurements are performed on samples deposited on oxidized silicon to determine the resistivity of the as grown layers.

RESULTS AND DISCUSSION

Composition and density

Nitrogen between 0% and 60% atomic was obtained by varying N flux in plasma atmosphere. An intense phenomenon of preferential Si re-sputtering was unveiled in the samples with low N incorporation. Si/W atomic ratio as a function of N concentration is reported in fig. 1. The growing N content inhibits Si re-sputtering and Si/W ratio tends to the target composition. The investigation of such a phenomenon is deepened elsewhere [8,9]. Film density exhibits a growing trend with the growing N, in the range $5.3x10^{22}$ at/cc – $7.5x10^{22}$ at/cc. A similar trend is also reported for the Ti-Si-N system [10]. A series of four samples from low to high N was selected for annealing. The samples, labeled from A to D have the following composition: $W_{58}Si_{21}N_{21}$ (A), $W_{41}Si_{17}N_{42}$ (B), $W_{35}Si_{16}N_{49}$ (C), $W_{30}Si_{14}N_{56}$ (D) as measured by RBS.

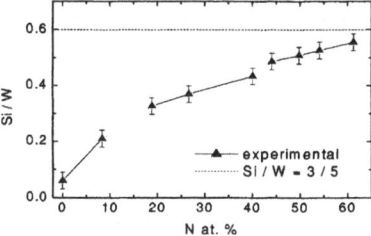

Fig. 1. Si/W ratio vs. N atomic concentration that highlights Si re-sputtering for low N content. Target composition Si/W=3/5 is also reported (dashed line).

From a compositional point of view, the main result of annealing is N loss in the N rich samples. No W or Si loss was found post annealing in all the layers. Fig. 2 presents nitrogen concentration for different heating temperatures in the A-D series.

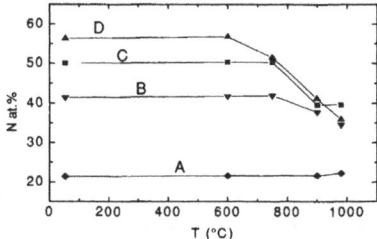

Fig. 2. N atomic concentration as a function of the annealing temperature in the A-D series.

Annealing at 600 °C induced no detectable difference in comparison to the as-grown layers for all the samples. At higher temperatures, the higher N concentration, the more intense is N depletion. All B, C and D layers exhibited N loss, not A. This evidence could be explained in the hypothesis that the Si-N bond is preferentially formed with respect to the W-N bond. In the A sample Si can saturate all N atoms with a strong bond (melting point of silicon nitride at T~1900 °C [11]), which can resist to annealing. For all the other compositions, in which is present N unsaturated from Si, part of N remains only weakly bonded and can be removed during heat treatments. This interpretation is based on the consideration that only weak bonds can be formed between N and W (melting point at T~600 °C [11]), which can be easily destroyed during annealing.

Chemical bonding

FT-IR spectroscopy indicates the formation of Si-N bonds. Absorption spectra of the as deposited films are shown in fig. 3 (a). We observe a broad peak centered at about 880 cm^{-1} characteristic of Si-N bonds [12] in all the layers. The broad conformation of the peak suggests the presence of a structure similar to an amorphous Si_3N_4 network in all the as deposited layers.

The increased intensity of the peak with the increasing N concentration indicates a growing number of Si-N bonds to be formed, as expected from the compositional characterization.

Annealing at 600 °C resulted in no meaningful differences with respect to the as grown films irrespective to N content. It is in agreement with the lack of compositional variations at 600 °C. With the increasing temperatures a new peak grows up at about 802 cm^{-1}, in the high-N containing samples B, C and D (in fig. 3(b) FT-IR spectra relative to sample C). The narrow peak suggests the segregation of a more ordered phase within the previously cited amorphous matrix.

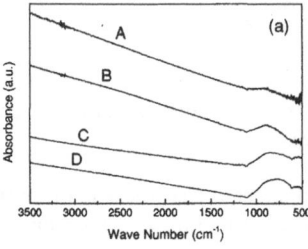
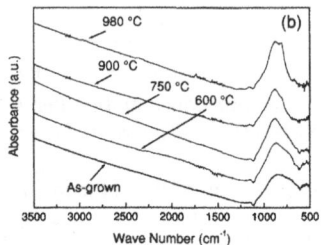

Fig. 3. FT-IR spectra in the A-D series for the as-grown layers (a) and for the C sample annealed at various temperatures (b).

Surface morphology

SEM investigation on the untreated layers highlights an extremely smooth surface for all the compositions. A strong blistering process was detected in the D layer (high N) after annealing. At 600 °C blisters are not exploded, while at higher temperature only exploded bubbles are detected.

In the C sample surface roughening and blisters appear only at 900 and 980 °C in concomitance with an intense N loss. In the two low-N containing layers A and B, instead, blisters are totally absent also at 980 °C, as can be seen in fig. 5, which collects SEM images of the A-D series annealed at 980 °C.

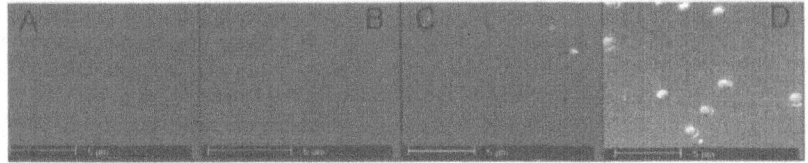

Fig. 5. SEM micrographs of the A-D series after annealing at 980 °C.

Blistering is strictly dependent on N abundance and seems to be connected with N depletion. The previous observations suggest blisters formation to be related with the high mobility of the weakly bonded N atoms in the supersaturated layers: at low temperature blisters can form only at high N concentration, but the internal pressure of the single blister is not sufficient to make bubble exploded. With the increasing temperature, enhanced N mobility allows blisters formation also for a lower N content, and the internal pressure can cause blister explosion, as documented by SEM images.

Crystalline structure

Two selected samples, $W_{58}Si_{21}N_{21}$ (A) and $W_{35}Si_{16}N_{49}$ (C), have been studied by means of TEM and SAED, both as grown and annealed at high temperature (900°C) for 1.5 h. Fig.s 6 (a) and 7 (a) present plan view images of as grown layers, SAED patterns are reported in the inset; both films are characterized by amorphous structures.

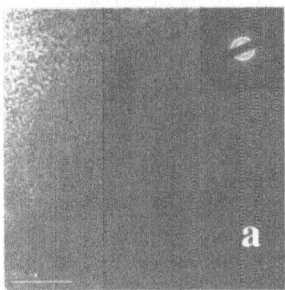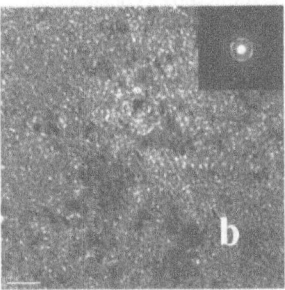

Fig. 6. Plan view image and selected area diffraction of sample $W_{53}Si_{20}N_{27}$ as grown (a) and annealed at 900 °C (b).

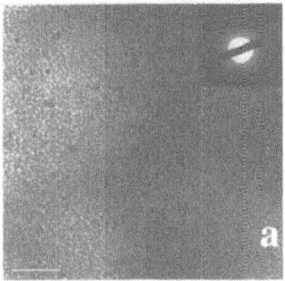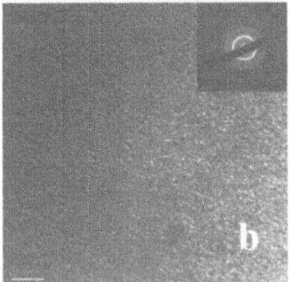

Fig. 7. Plan view image and selected area diffraction of sample $W_{35}Si_{16}N_{49}$ as grown (a) and annealed at 900 °C (b).

Sample with lower N concentration (A) shows the presence of polycrystalline diffraction rings whose lattice spacing correspond to Tungsten (bcc, lattice constant a=0.316 nm), thus suggesting the precipitation of the metallic phase after thermal treatment (fig. 6 (b)). Samples with higher N concentration (C) still preserve their amorphous structure, as indicated by fig. 7 (b), after annealing at high temperature.

Resistivity

Preliminary results on the as grown layers of the A-D series indicate resistivity values depending on N content and ranging between 500 (sample A) and 4750 (sample D) $\mu\Omega$cm. A complete electrical characterization of the films before and after annealing is reported elsewhere [13].

CONCLUSIONS

Ternary W-Si-N thin films of different compositions have been deposited via RF-magnetron sputtering from a single W_5Si_3 target. An intense Si preferential re-sputtering was unveiled for inert sputtering, inhibited by N addition in plasma gas. Strong Si-N bonds were formed with the increasing N concentration. An amorphous structure was found in all the as-grown layers, irrespective of N content. Annealing promoted the loss of N atoms un-bonded to Si or weakly bonded to W, and the precipitation of a bcc crystalline metallic phase of W in the layers with low N content. This behavior is different from the formation of mixed Ti (or Ta) and Si nitrides in Ti (or Ta)-Si-N ternary systems. A blistering process accompanied N loss, probably due the growth of bubbles of N, which lead to the formation of holes on the surface of the N-rich samples.

ACKNOLEDGMENTS

The authors would like to thank Prof. G. Ottaviani for discussions and critical review of the manuscript, Dr. R. Tonini, M. Butturi and G. Das for the acquisition of several experimental results. Research supported by the MIUR project.

REFERENCES

1. R. Rosenberg, D.C. Edelstein, C.-K, Hu, and K.P Rodbell, *Annu. Rev. Mater. Sci.* **30**, 229-262 (2000).
2. A.E. Kaloyeros and E. Eisenbraun, *Annu. Rev. Mater. Sci.* **30**, 363-385 (2000).
3. X. Sun, E. Kolawa, S. Im, C. Garland, and M.-A. Nicolet, *Appl. Phys.* **A65**, 43 (1997).
4. C. Linder, A. Dommann, G. Staufert, and M.-A. Nicolet, *Sens. Act.* **A61**, 387 (1997).
5. A.M. Dutron, E. Blanquet, V. Ghetta, R. Madar, C. Bernard, *J. Phys. IV* **5**, 1141-1148 (1995).
6. J. Reid, E. Kolawa, C.M. Garland, M.A. Nicolet, F. Cardone, D. Gupta and J. P. Ruiz, *J. Appl. Phys.* **79**, 1109 (1996).
7. J. G. Fleming, E. Roherty-Osmun, P. M. Martin Smith, J.S. Custer, Y.D. Kim, T. Kacsich, M.A. Nicolet and C.J. Galewski, *Thin Solid Films* **320**, 10-14 (1998).
8. C. Louro, A. Cavaleiro, and F. Montemor, *Surf. Coat. Technol.* **142-144**, 964 (2001).
9. Vomiero *et al.* to be submitted.
10. X. Sun, J.S. Reid, E. Kolawa and M.-A. Nicolet, *J. Appl. Phys.* **81(2)** 656 (1997).
11. Handbook of Chemistry and Physics 79th edition (1998-99), CRC Press, Inc.
12. N. Wada, S. A. Solin, J. Wong, and S. Prochazka, *J. Non Cryst. Solids* **43**, 7 (1981).
13. A. Vomiero *et al.* E-MRS Spring Meeting 2004, Strasbourg (FR), submitted.

Mat. Res. Soc. Symp. Proc. Vol. 812 © 2004 Materials Research Society

BARRIER LAYER MORPHOLOGICAL STABILITY AND ADHESION TO POROUS LOW-κ DIELECTRICS

R. Saxena [a], W. Cho [a], O. Rodriguez [a], W. N. Gill [a] and J. L. Plawsky [a]
[a] Chemical Engineering Department, Rennselaer Polytechnic Institute, Troy, NY 12180

ABSTRACT

Two particularly important reliability issues facing the integration of low- κ dielectric films are the fracture energy of the barrier-dielectric interface and the barrier layer integrity during processing. We have noticed that the compressive stresses in the barrier layers on low- κ dielectrics lead to spontaneous delamination and formation of telephone-cord like morphologies. These morphologies allow the measurement of fracture energy and are advantageous over artificially contrived features to yield realistic debonding parameters. The fracture energy of common barrier films, TaN and Ta, was determined using this method for varying porosity nanoporous silica and MSQ. Detailed characterization of the telephone cord morphology using a combination of Optical Microscopy, SEM and Profilometry was done. The fracture energy for Ta on different low-κ dielectrics was evaluated using a 1-D model for straight buckles. The kinetic coefficient of buckling was also evaluated.

1. INTRODUCTION

The increased speed in semiconductor devices requires copper and low-κ (<2-2.5) ILD solutions to minimize RC delay. Spin-on silica xerogels obtained by hydrolysis and condensation of tetraethylorthosilicate (TEOS) or its derivates are one class of low- κ porous materials that are being studied extensively [1]. Recent reviews by Plawsky et al. and Maex et al. provide a detailed classification of the different low- κ materials being studied [2,3]. The integrity of contiguous metallization layers deposited on these low-κ dielectrics is of prime importance to the long term reliability of the stack.

Barrier and metal films are usually deposited under high compressive stress and the films tend to relieve that stress. There are many different stress relief mechanisms observed in thin films. One of the most common mechanisms involves the debonding of the film from the substrate in the form of blisters. The buckling deformations can be either straight, or form periodic buckling patterns known as the telephone cord morphology [4].

Several authors have modeled the telephone-cord morphology; however there exists a paucity on available experimental data on the topic. Gioia and Ortiz (1997) have a good summary of available experimental data [4]. More recently, Moon et al. (2002) have observed and characterized these features on diamond-like carbon films on glass substrates [5]. They showed that telephone cords exist for a limited set of conditions, residing within the margin between complete adherence and complete delamination. This limited set of conditions, i.e. improper adhesion and high compressive stress, allow us to observe telephone cord morphologies in barrier films on low-k dielectrics. Volinsky (2003) discusses the only other observations of telephone cords on low-k dielectrics that the authors are aware of [6]. The objectives of this work are thus; to determine conditions under which barrier films fail and result in telephone-cords, and to, completely characterize the telephone-cord morphology to extract the fracture energy of the barrier/dielectric interface.

2. EXPERIMENTS

2.1 Procedure

Varying porosity mesoporous silica (xerogels) substrates were fabricated on silicon wafers with a native oxide layer, using a multi-step procedure described in a previous publication [7]. The porosity of these substrates was controlled by spinning sols containing a relatively non-volatile solvent (ethylene glycol) in an open, ambient environment. The initial TEOS/glycol volume ratio controlled the final porosity of the films. The thickness and refractive index of the films were measured by a variable angle spectroscopic ellipsometer (VASE). The pore size distribution and shape were determined using Ellipsoporosimetry [8]. Details of this technique and the different distributions for our films can be found in a prior publication [9].

Thin Ta and TaN films with varying thicknesses were sputter deposited on the ILD/Si stack using a CVC® DC Magnetron Sputterer. The thicknesses of the barrier thin films were measured using Rutherford Backscattering Spectroscopy (RBS). Microscopy observations of the telephone cord morphology were made using an optical microscope and a JEOL FESEM. Simultaneous observations were also made using a Dektak® Profilometer. It should be noted here that the force on the profilometer stylus was adjusted to 0.05 mg. This was necessary as any higher force caused the thin film to indent under the stylus load.

2.2 Results

Deposited barrier films developed telephone cords beyond a certain critical thickness. To obtain the critical thickness, many different thicknesses of Ta were deposited on different substrates. The substrates were left for a day before taking observations. Beyond this critical thickness, the stress energy in the film exceeds the interfacial fracture energy leading to debonding. Due to the bi-axial stress in the film, the film develops a wavy pattern [6]. The cords start from the wafer edge and propagate (Figure 1, a). The tip of the cord is parabolic and then develops into a fully developed wavy pattern as the tip propagates. In addition to telephone cord, straight sided blisters are also observed in these films, although they are not as ubiquitous as the telephone cords (Figure 1 b and c). To fully characterize the cord morphology, we need the height of the buckle (δ), the width of the buckle (b) and the wavelength of the cord (λ). These are obtained by profilometry and a representative scan is shown in Figure 2.

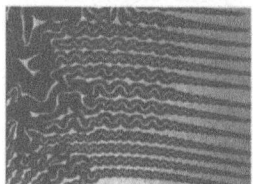  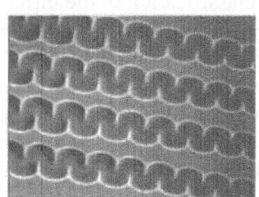

Origin of t-cords from edge (Mag. 5x)   Straight buckles at tip of cords (Mag. 20x)   Fully developed t-cords (Mag. 20x)

Figure 1: Telephone cords (t-cords) for Ta on MSQ of 35 % porosity.

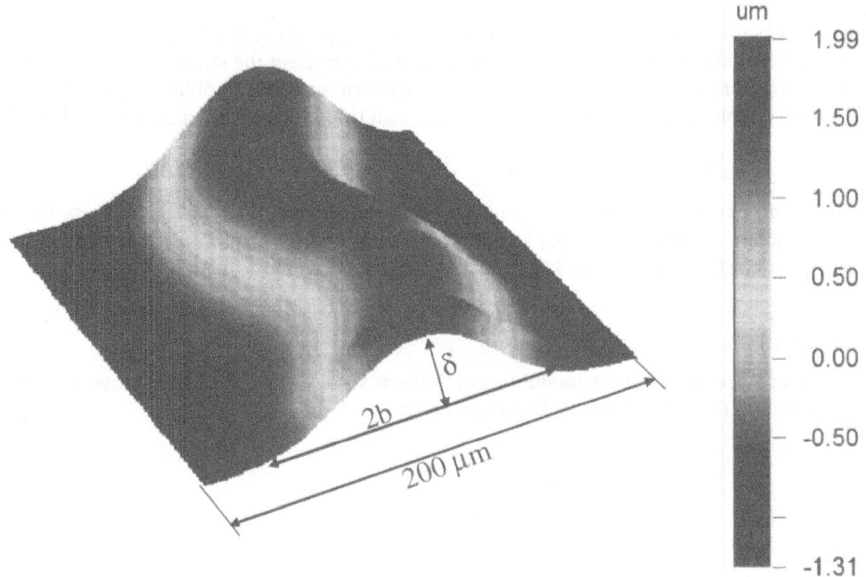

Figure 2: Characterization of the telephone cord features using Profilometry

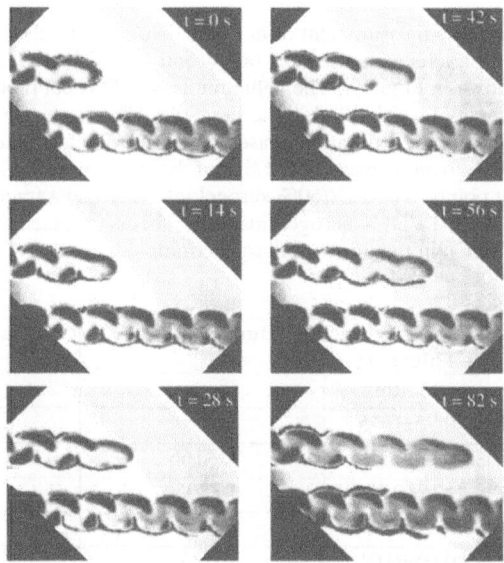

Figure 3: Kinetics of buckling and tip velocity calculations for 0.9 μm Ta on 58 % xerogel. Video is captured using an optical microscope at 20 X magnification.

For films beyond the critical thickness, there exists an additional driving force, G-G_c, which allows the calculation of tip velocities. Here, G and G_c represent the strain energy in the film and the critical fracture energy respectively. A digital camera was used to obtain a video of the propagating telephone cords (Figure 3) and a constant tip velocity was calculated as 4.5×10^{-4} m/s.

3. DISCUSSION

For a thin film of critical thickness, h_c, subject to a equi-biaxial compressive stress, σ, the stress relief mechanism results in a straight sided buckle with the film displacement, δ, normal to the surface and a buckle width of $2b$. The critical stress for buckling is given by [10]:

$$\sigma_c = \frac{\pi^2 E}{12(1-v)}\left(\frac{h_c}{b}\right)^2 \qquad \text{..................} 1$$

where, E and v are the Young's modulus and Poisson's ratio of the film, respectively. The residual stress in the film after buckling is given by:

$$\sigma_r = \sigma_c\left[\frac{3}{4}\left(\frac{\delta}{h_c}\right)^2 + 1\right] \qquad \text{..................} 2$$

Using this 1-D model, the interfacial fracture energy is given by:

$$G_c = \left(\frac{1-v^2}{2E}h_c\right)(\sigma_r - \sigma_c)(\sigma_r + 3\sigma_c) \qquad \text{..................} 3$$

The above model requires the measurement of h_c, b and δ to measure the interfacial fracture energy. These parameters are listed in the table below and the calculated fracture energies for the Ta/ILD interface are shown in Figure 4. The value for TaN/ILD is not plotted as this is a different material and cannot be compared in the same plot. The fracture energy of the Ta/ILD interface decreases exponentially with an increase in porosity. We attribute this loss in adhesion on increasing porosity due to an increased pore size of the films resulting in rougher surfaces and hence fewer attachment points. The TaN/60% xerogel interface had a fracture energy of 2.8 J/m^2 compared to 1.8 J/m^2 for the Ta/58 % xerogel interface. This is expected as TaN is used as an adhesion promoter for Ta/Cu in the present microelectronics stack.

Sample (Porosity, %)	Critical Film Thickness	Buckle Height	Half Width
	(h_c, μm)	(δ, μm)	(b, μm)
Ta/Xerogel: 32	0.75±0.15	6	50
Ta/MSQ: 35	0.6±0.075	2.4166	27.6
Ta/Xerogel: 48	0.75±0.075	4.38034	53
Ta/Xerogel: 58	0.3±0.05	1.08524	13.5
TaN/Xerogel: 60	0.05±0.01	1.25	7.5

The kinetic coefficient of buckling, B, can be calculated from the tip velocity, V_n, using an approach similar to the one shown by Gioia and Ortiz (1998) [11]. The kinetic coefficient of buckling is given by: $V_n = (1/B) (G-G_c)$, where $(G-G_c)$ represents the driving force for crack propagation. From our data for the 58 % porous xerogel and Ta interface, a value of 4.06×10^3 Js/m is determined for the kinetic coefficient of buckling.

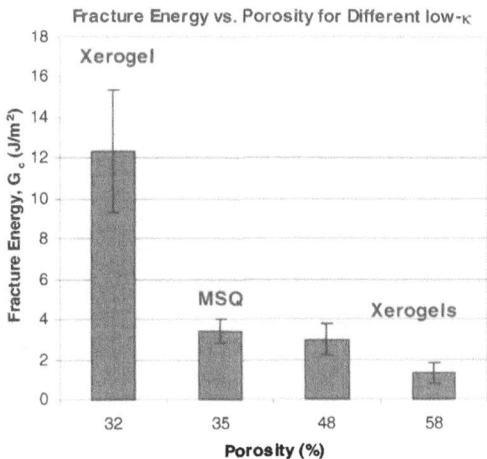

Figure 4: Fracture energy for Ta/ILD for varying porosity ILDs.

4. CONCLUSIONS

a. We have experimentally observed telephone cords as a stress relieve mechanism for barrier films on different low-k dielectrics. The telephone cord morphology was characterized using a combination of SEM and Profilometry measurements.

b. Fracture energy was calculated for different dielectrics using a 1-D model for straight buckles and shows an exponential decrease with increasing substrate porosity. The fracture energy calculations also need to be further verified with industry standard tests like the four-point bending technique.

c. The kinetic coefficient of buckling was calculated for one porosity xerogel and are the first known observations of kinetics of telephone cord buckling.

References

1. S. V. Nitta, V. Pisupatti, A. Jain, P. C. Wayner Jr., W. N. Gill and J. L. Plawsky, *J. Vac. Sci. Technol. B*, 17 (1999) 205.

2. J. L. Plawsky, A. Jain, S. Rogojevic and W. N. Gill, "Nanoporous Dielectric Films: Fundamental Property Relations and Microelectronics Applications", Chapter 4 in Interlayer

Dielectrics for Semiconductor Technologies, S. Murarka, Eisenberg, Sinha (Editors), Elsevier Inc. (2003).

3. K. Maex, M. R. Baklanov, D. Shamiryan, F. Iacopi, S. H. Brongersma and Z. S. Yanovitskaya, *J. Appl. Phys.*, 93(11) (2003) 8793.

4. G. Gioia and M. Ortiz, *Adv. Appl. Mech.*, 33 (1997) 119.

5. M. W. Moon, H. M. Jensen, J. W. Hutchinson, K. H. Oh and A. G. Evans, *J. Mech. Phys. Solids*, 50 (2002) 2355.

6. A. A. Volinsky, *Proc. Mat. Res. Soc.*, W10.7 (2003) 1.

7. A. Jain , S. Rogojevic, S. Ponoth , N. Agarwal , I. Matthew , W.N. Gill , P. Persans , M. Tomozawa , J.L. Plawsky, E. Simonyi, *Thin Solid Films*, 398 –399 (2001) 513.

8. M. R. Baklanov and K. P. Mogilnikov, *Microelectronic Engineering*, 64 (2002) 335.

9. R. Saxena, O. Rodriguez, W. Cho, M. R. Baklanov, K. P. Moglinikov, W. N. Gill and J. L. Plawsky, *J. Non-Crystalline Solids*, Accepted Jan (2004).

10. N. R. Moody et al. *Acta Mater*, 46(2) (1998) 585.

11. G. Gioia and M. Ortiz, *Acta Mater*, 46(1) (1998) 169.

Acknowledgments
This work was supported by the Center for Advanced Interconnect Systems Technology (CAIST), Semiconductor Research Corporation (SRC) contract no. 2002-MC-995, and the New York State Office of Science, Technology and Academic Research (NYSTAR).

Mat. Res. Soc. Symp. Proc. Vol. 812 © 2004 Materials Research Society   F3.13

Atomic Layer Deposition of Tantalum Nitride on Organosilicate and Organic Polymer-based Low Dielectric Constant Materials

Oscar van der Straten, Yu Zhu, Jonathan Rullan, Katarzyna Topol, Kathleen Dunn, and Alain Kaloyeros
UAlbany Institute for Materials & School of NanoSciences and NanoEngineering, University at Albany – SUNY, Albany, NY 12203

ABSTRACT

A previously developed metal-organic atomic layer deposition (ALD) tantalum nitride (TaN_x) process was employed to investigate the growth of TaN_x liners on low dielectric constant (low-k) materials for liner applications in advanced Cu/low-k interconnect metallization schemes. ALD of TaN_x was performed at a substrate temperature of 250°C by alternately exposing low-k materials to tertbutylimido-tris(diethylamido)tantalum (TBTDET) and ammonia (NH_3), separated by argon purge steps. The dependence of TaN_x film thickness on the number of ALD cycles performed on both organosilicate and organic polymer-based low-k materials was determined and compared to baseline growth characteristics of ALD TaN_x on SiO_2. In order to assess the effect of the deposition of TaN_x on surface roughness, atomic force microscopy (AFM) measurements were carried out prior to and after the deposition of TaN_x on the low-k materials. The stability of the interface between TaN_x and the low-k materials after thermal annealing at 350°C for 30 minutes was studied by examining interfacial roughness profiles using cross-sectional imaging in a high-resolution transmission electron microscope (HR-TEM). The wetting and adhesion properties of Cu/low-k were quantified using a solid-state wetting experimental methodology after integration of ALD TaN_x liners with Cu and low-k dielectrics.

INTRODUCTION

Atomic layer deposition (ALD) has been identified as one of the most promising deposition technologies for next-generation interconnect liner-seed applications in back-end-of-line (BEOL) interconnect metallization. While the liner requirements for current technology nodes, i.e. for the 90 nm DRAM half pitch generation – hp90, as denoted in the 2003 edition of the International Technology Roadmap for Semiconductors [1], and possibly for the 65 nm – hp65 generation as well, may be met using existing, sputtering-based liner deposition, ALD is expected to become a viable if not essential alternative in subsequent generations. ALD liner research has been actively pursued in recent years, with an emphasis on developing processes for the deposition of typical liner materials such as refractory metal-based compounds containing Ti, Ta, and/or W [2]. In essence, the source chemistries that had been employed in the development of chemical vapor deposition (CVD) processes for liners over the years are now used to attempt processing in the ALD regime which precludes any source decomposition and allows film growth only by adsorption and "clean" reactions. Hence, the number of commercially available source chemistries suitable for ALD is limited, as the permissible process temperature range may not exceed the temperature at which pyrolysis occurs. Compatibility with existing integrated circuitry (IC) manufacturing presents an additional constraint, since metal-organic chemistries are preferred over the use of corrosive halides.

In current Cu-based interconnect integration schemes, the traditionally employed interlevel dielectric (SiO$_2$) has already been replaced in certain levels by materials with a lower dielectric constant (low-k dielectrics) to reduce contributions of capacitance to the overall RC delay. A relatively large number of candidate low-k materials has recently been investigated, applied by either spin-on deposition methods or by CVD, and the majority of these materials can be classified as based on either organosilicate materials, or organic polymers. In the case of organosilicate-based low-k, Satta et al. found that the growth of TiN on SiOC:H low-k dielectrics is poor, while O$_2$ plasma treatment of the low-k surface prior to liner deposition was shown to promote the TiN nucleation on SiOC:H [3]. Beyer et al. also concluded that the nucleation of barriers/liners by ALD on dense organosilicate-based low-k dielectrics can be achieved after pre-treating the dielectric surface [4]. For organic polymer-based low-k, the nucleation of ALD TiN from a metal-organic chemistry (TDMAT and NH$_3$) directly on the organic polymer SiLK^{TM} was poor, while the nucleation of ALD Al$_2$O$_3$ from TMA and H$_2$O on SiLK^{TM} was facile. An ALD Al$_2$O$_3$ adhesion layer was in this case shown to improve the subsequent nucleation of ALD TiN as liner on SiLK^{TM} [5]. Pretreatment by O$_2$ and N$_2$ plasma of ALD WCN, deposited using WF$_6$, triethylborane [(C$_2$H$_5$)$_3$B] and NH$_3$, was shown to enhance WCN liner nucleation: a continuous liner was obtained at smaller liner thickness in the case of plasma-treated polymer low-k [6].

Research on the integration of liners with porous low-k dielectrics is underway as well, presenting a new set of challenges. In the case of ALD liners particularly, these challenges pertain to the superior conformality of ALD which results in the deposition of liner materials inside the pores of porous low-k. Means to effectively avoid such liner penetration are still to be developed and solutions may be expected to depend on the specific low-k and liner materials in question [4,7].

In this study, the nucleation of ALD TaN$_x$ liners on dense low-k materials was investigated, employing a metal-organic ALD TaN$_x$ process previously optimized on SiO$_2$. Several analytical techniques were used to characterize and compare the nucleation of TaN$_x$ on organosilicate and organic polymer-based low-k dielectrics.

EXPERIMENTAL DETAILS

The metal-organic atomic layer deposition process utilized to deposit TaN$_x$ films has been previously reported for growth on SiO$_2$ and Si [8]. The Ta-source used was tertbutylimido-tris(diethylamido)tantalum (TBTDET), liquid at room temperature, while ammonia (NH$_3$) was employed as co-reactant in a thermal ALD process at 250°C. ALD TaN$_x$ was deposited in a commercially available flow-type ALD reactor by alternately pulsing TBTDET and NH$_3$, separated by Ar purge steps. The organosilicate-based low-k dielectric CORAL^{TM} (k = 2.7) and the organic polymer-based low-k dielectric SiLK^{TM} (k = 2.65) were chosen as substrates. Rutherford backscattering spectrometry (RBS) was used to characterize the dependence of TaN$_x$ film thickness on the number of ALD cycles performed, both on organosilicate and on organic polymer-based low-k, in order to evaluate the growth characteristics of ALD TaN$_x$ on these materials and for comparison to the case of ALD TaN$_x$ on SiO$_2$. To assess the effect of the deposition of TaN$_x$ on surface roughness, atomic force microscopy (AFM) measurements were carried out to characterize the surface morphology prior to and after the deposition of TaN$_x$ on CORAL^{TM} and SiLK^{TM}. The stability of the interface between TaN$_x$ and the low-k materials after annealing TaN$_x$/low-k structures at 350°C for 30 minutes in a forming gas (95% Ar/5% H$_2$)

ambient was studied by examining interfacial roughness profiles, obtained by cross-sectional imaging of focused ion beam (FIB)-prepared samples in a high-resolution transmission electron microscope (HRTEM). Finally, a solid-state wetting experimental methodology involving a high-temperature (600°C) anneal for an extended period of time (48 h.) was employed to quantify the wetting properties of Cu/low-k after integration with ALD TaN_x liners. Wetting characteristics were analyzed using scanning electron microscopy (SEM) at 75° tilt to image the Cu equilibrium crystal shapes (ECSs) formed and to deduce the average contact angle between Cu ECSs and the ALD TaN_x/low-k substrate.

RESULTS

The nucleation of ALD TaN_x on the organosilicate-based low-k was studied by exposing CORAL™ to a varied number of ALD cycles consisting of alternately supplied TBTDET and NH_3, separated by Ar purge steps. The resulting films were analyzed by RBS to compare the Ta peak width for 50, 100, and 200 cycles ALD TaN_x, respectively.

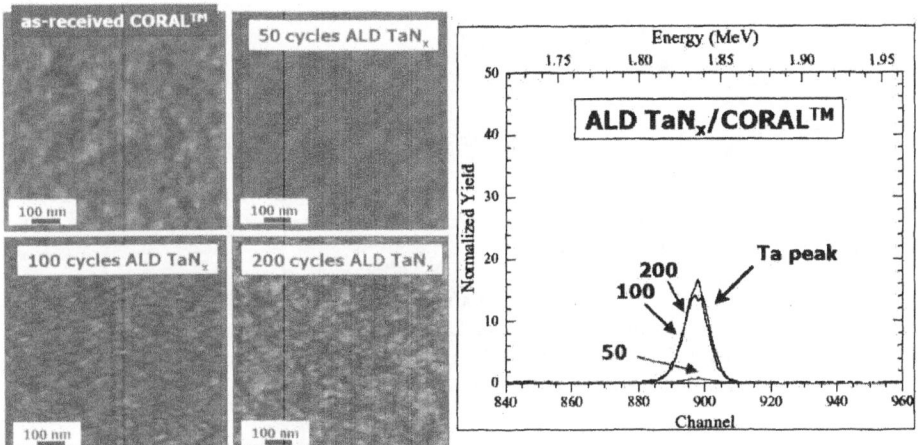

Figure 1. AFM imaging of surface morphology (left), and RBS analysis (right) of 50, 100, and 200 cycles ALD TaN_x on CORAL™, respectively, illustrating the non-linear nucleation rate of TaN_x.

From the AFM analysis of ALD TaN_x on CORAL™ (figure 1), the RMS surface roughness initially remained comparatively constant at approximately 0.70 nm, near the value for as-received CORAL™ (0.65 nm), and increased to 1.1 nm at 100 ALD TaN_x cycles or more. This feature was also reflected in the RBS data which implied the presence of a very low density of Ta atoms on the CORAL™ surface in the case of 50 cycles TaN_x, while exhibiting a significantly higher density for 100 and 200 cycles TaN_x. However, the negligible difference in the Ta peak width between 100 and 200 cycles TaN_x seemed to indicate that the nucleation of ALD TaN_x on CORAL™ was problematic, even after 100 cycles: the similar film thickness observed for both samples may have been the result of a significantly different number of active sites on the starting surface. On SiLK™ (figure 2), the growth characteristics of ALD TaN_x were found to be different: after 50 cycles, the RMS surface roughness has already significantly increased from the value for as-received SiLK™ (0.38 nm) to approximately 0.80 nm. For the case of 100 and 200 cycles TaN$_x$, the surface roughness remained relatively constant at 0.80 nm, while from RBS analysis a nearly linear nucleation rate of TaN$_x$ on SiLK™ was observed, similar to ALD TaN$_x$ on SiO$_2$.

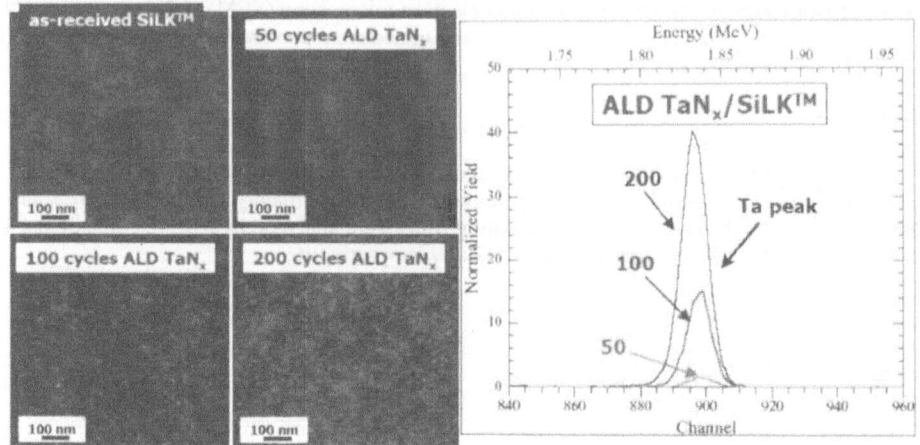

Figure 2. AFM imaging of surface morphology (left) and RBS analysis (right) of 50, 100, and 200 cycles ALD TaN$_x$ on SiLK™, respectively, illustrating the linear nucleation rate of TaN$_x$, similar to the case of ALD TaN$_x$ on SiO$_2$.

The stability of the TaN$_x$/low-k interface after applying thermal stress was studied by exposing both low-k materials to 200 cycles ALD TaN$_x$ and annealing the resulting stacks at 350°C for 30 minutes in forming gas (95% Ar/5% H$_2$). FIB-prepared cross-sectional TEM specimens of the annealed stacks were then analyzed (figure 3).

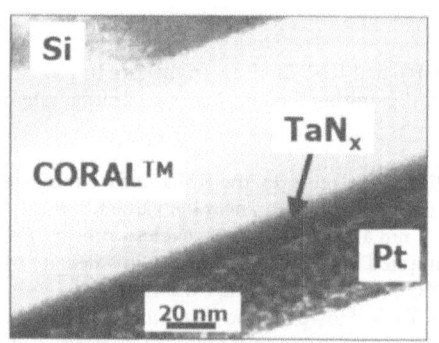 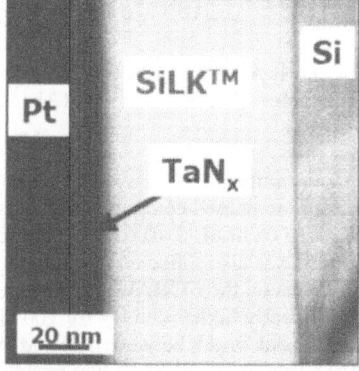

Figure 3. TEM micrographs of annealed TaN$_x$/CORAL™ (left) and TaN$_x$/SiLK™ (right).

A comparison of the micrographs in figure 3 confirms the different TaN_x film thickness on CORAL™ and on SiLK™ after 200 ALD cycles. This is consistent with previously observed RBS data, assuming that no significant film densification has occurred during thermal annealing. In addition, the TaN_x on CORAL™ appears to contain pinholes while the TaN_x film on SiLK™ seems to be continuous: this again may be a result of poor TaN_x nucleation on CORAL™ as opposed to SiLK™. As CORAL™ essentially consists of a matrix based on SiO_2 in which methyl groups have been selectively substituted for oxygen, the CORAL™ surface may be expected to exhibit both –OH (hydroxyl) and –CH termination. Nucleation of ALD TaN_x on SiO_2 (–OH terminated) has not revealed significant problems, and it is therefore suggested that the presence of –CH groups on the CORAL™ surface may inhibit ALD TaN_x nucleation and thus contribute to the relatively large number of ALD cycles required for obtaining a continuous TaN_x film on CORAL™. In the case of SiLK™, an aromatic hydrocarbon polymer network consisting predominantly of phenyl groups, the nucleation mechanism is likely to be more complicated as no reactive surface species are expected to be present. Potentially, the diffusion of reactants (NH_3, and TBTDET) into the SiLK™ polymer, and subsequent formation of a sufficient number of TaN_x clusters could initiate growth, as suggested for ALD TiN on SiLK™ using TDMAT and NH_3 [5].

A solid-state wetting methodology [9] was employed to investigate and quantify the wetting properties of Cu/low-k when integrated with an ALD TaN_x liner. This experimental method involves high-temperature, long-term annealing (600°C for 48 h.) of ultrathin Cu films (20 nm film thickness) in an inert atmosphere (forming gas, 95% Ar/5% H_2). During this annealing process, the polycrystalline Cu film de-wets from the underlying substrate, and separates into islands which subsequently attain the equilibrium crystal shape (ECS) in order to minimize surface energy with respect to their volume.

Figure 4. SEM micrographs (obtained at 75° tilt) for the determination of Cu ECS contact angle in solid-state wetting experiments of Cu on TaN_x/CORAL™ (left), and on TaN_x/SiLK™ (right).

The micrographs in figure 4 clearly show the presence of a large number of Cu equilibrium-shape crystallites. The contact angle θ between Cu ECS and substrate at thermal equilibrium provides a means for quantification of the wetting of Cu on a substrate: a small value of θ indicates superior wetting, while a large value of θ is a sign of poor wetting. The analysis of a statistically significant number of Cu crystallites on TaN_x/CORAL™ and TaN_x/SiLK™ revealed an average contact angle of approximately 80° in the case of CORAL™, and a slightly lower value of approximately 70° for SiLK™. As can be observed in figure 4, Cu appears to have de-wetted less extensively in the case of SiLK™ (the total contact area between Cu and substrate seems larger than for CORAL™), and the average contact angle values suggest improved wetting for SiLK™.

CONCLUSIONS

Employing a metal-organic ALD TaN$_x$ process, the nucleation of TaN$_x$ on both organosilicate-based and organic polymer-based low-k dielectric materials was studied. While the nucleation of TaN$_x$ on SiLK™ was similar to TaN$_x$ on SiO$_2$, the nucleation on CORAL™ appeared to be more problematic: more ALD cycles were required to initiate growth. Finally, solid-state wetting experiments showed that, when integrated with an ALD TaN$_x$ liner, SiLK™ performs better than CORAL™ in terms of minimizing Cu de-wetting.

ACKNOWLEDGEMENTS

The research presented in this study was supported by the Semiconductor Research Corporation under the CAIST program. The authors gratefully acknowledge Dr. Hyungjun Kim (IBM T.J. Watson Research Center) for providing low-k dielectric substrates.

REFERENCES

1. The International Technology Roadmap for Semiconductors 2003 Edition (Semiconductor Industry Association, 2003).
2. H. Kim, J. Vac. Sci. Technol. B 21 (6), 2231-2261 (2003).
3. A. Satta, M. Baklanov, O. Richard, A. Vantomme, H. Bender, T. Conard, K. Maex, W.M. Li, K.-E. Elers, S. Haukka, Microelec. Eng. 60, 59-69 (2002).
4. G. Beyer, A. Satta, J. Schuhmacher, K. Maex, W. Besling, O. Kilpela, H. Sprey, G. Tempel, Microelec. Eng. 64, 233-245 (2002).
5. J.W. Elam, C.A. Wilson, M. Schuisky, Z.A. Sechrist, S.M. George, J. Vac. Sci. Technol. B 21 (3), 1099-1107 (2003).
6. A.M. Hoyas, J. Schuhmacher, D. Shamiryan, J. Waeterloos, W. Besling, J.P. Celis, K. Maex, J. Appl. Phys. 95 (1), 381-388 (2004).
7. E.T. Ryan, M. Freeman, L. Svedberg, J.J. Lee, T. Guenther, J. Connor, K. Yu, J. Sun, D.W. Gidley, in *Materials, Technology and Reliability for Advanced Interconnects and Low-k Dielectrics—2003*, edited by A.J. McKerrow, J. Leu, O. Kraft, and T. Kikkawa (Mater. Res. Soc. Proc. **766**, Pittsburgh, PA, 2003), E10.8.1-E10.8.6.
8. O. van der Straten, Y. Zhu, K. Dunn, E.T. Eisenbraun, A.E. Kaloyeros, J. Mater. Res. **19** (2), 447-453 (2004).
9. O. van der Straten, Y. Zhu, K. Dunn, A. Kaloyeros, in *Materials, Technology and Reliability for Advanced Interconnects and Low-k Dielectrics—2003*, edited by A.J. McKerrow, J. Leu, O. Kraft, and T. Kikkawa (Mater. Res. Soc. Proc. **766**, Pittsburgh, PA, 2003), E10.3.1-E10.3.6.

Mat. Res. Soc. Symp. Proc. Vol. 812 © 2004 Materials Research Society

EFFECT OF ANNEALING ON THE STRUCTURAL, MECHANICAL AND TRIBOLOGICAL PROPERTIES OF ELECTROPLATED Cu THIN FILMS

P. Shukla, A. K. Sikder, P.B. Zantye, Ashok Kumar, and M. Sanganaria
Nanomaterials and Nanomanufacturing Research Center
Department of Mechanical Engineering
University of South Florida, 4202 East Fowler Avenue, Tampa, FL 33620
Novellus System, Inc., San Jose, CA 95134

ABSTRACT

The increasing demand for faster and more reliable integrated circuits (ICs) has promoted the integration of Copper-based metallization. Electroplated Cu films demonstrate a microstructural transition at room temperature, known as self annealing. In this paper we intend to investigate the annealing behavior of electroplated Cu films grown on a seed Cu layer on top of the barrier layers over a single crystal silicon substrate. All the samples were undergone through a multistep annealing process. Grazing incident x-ray diffraction pattern shows stronger x-ray reflections from Cu (111) and (220) planes but weaker reflections from (200), (311) and (222) planes in all the electroplated Cu samples. Transmission electron microscopy was performed on the cross section of the samples and the diffraction pattern showed the crystalline behavior of both seed layer and electroplated Cu. Nanoindentation was performed on all the samples using the continuous stiffness measurement (CSM) technique and it was found that the elastic modulus varies from 110 to 130 GPa while the hardness varies from 1 to 1.6 GPa depending on the annealing conditions. The tribological properties of all the copper films were also measured using the Bench Top CMP tester. Subsequently, Nanoindentation was performed on the samples after polishing the top surface in order to investigate the work hardening and an increase in hardness and modulus was observed. Finite Element Modeling was performed in order to investigate the stress behavior during nanoindentation.

INTRODUCTION

The interconnect RC-delay (resistance-capacitance) has become dominant in determining overall device performance. The RC delay can be reduced by lowering the capacitance using low permittivity (low-k) materials as interlayer dielectrics (ILD) and by lowering the resistance using interconnect material with lower resistivity. Electroplated (EP) Cu has rapidly been adopted in deep submicron ULSI technology due to its lower resistivity and thus resulting in improved electromigration reliability [1–3].

Electroplated Cu films undergo a structural transition even at room temperature, which involves grain growth and texture changes [4,5] called as self-annealing process, which leads to a dramatic drop of the resistivity over time. The self-annealing for EP Cu films has become a constraint for reliability and reproducibility of Cu interconnect process, because changes of grain size and hardness have influence on electromigration and CMP process [6, 7].

For most reliability tests, knowledge of the thin film constitutive mechanical behavior is required. Mechanical properties of thin films often differ from those of the bulk materials. Hence there is a need for evaluation of mechanical properties at nanoscale using state-of-the-art evaluation techniques like Nanoindentation [8, 9]. Here we have investigated the mechanical properties of Si/SiO$_2$/TaN/SeedCu/EPCu stack annealed at different stages with different temperatures. Grazing incidence x-ray diffraction (GIXRD) has been employed to investigate the preferred orientation of crystal growth due to different types of annealing.   Evolution of microstructure with annealing is investigated using atomic force microscopy (AFM). Post-CMP evaluation of mechanical properties of Cu films is also performed in order to investigate the phenomenon of work hardening [10]. Transmission electron microscopy was performed on the cross section of the samples. Finite Element Modeling was performed in order to investigate the stress behavior during nanoindentation. The objective of this study is to explore the effect of post–electroplating treatment on the mechanical and microstructural properties of Cu films.

EXPERIMENTAL

1000 nm thick EP copper films were deposited on a Si/SiO$_2$/TaN/Cu (seed layer) stack. The diffusion barrier layer, TaN (30 nm thick), was deposited by sputtering technique. The Cu seed layer was then deposited onto the barrier layer also using a sputtering method. These wafers were then electroplated in a plating bath to grow a 1000 nm thick copper film. Microstructural information of the annealed Cu films was studied using GIXRD. Experiment was performed using the Philips X'Pert Pro XRD system with 0.5° grazing incidence. An X-ray Cu mirror module with a mask of 10 mm and a divergence slit of 1/32° was used as incident beam optics; whereas diffracted beam optics comprised of a parallel plate collimator (0.27°) with a solar slit. The mechanical properties of the electroplated copper films were evaluated using nanoindenter with a Berkovich tip (XP, MTS System). The continuous stiffness modulation (CSM) technique was used, wherein the contact stiffness was measured continuously as a function of displacement under the load. The hardness and modulus values were calculated in the range of 10% of the film thickness in order to avoid the substrate or underlying hard layer effects. Details of the measurement technique have been discussed elsewhere [11].

The surface and grain boundary characteristics of the electroplated copper films were studied using AFM (Dimension 3100, Digital Instruments). The tribological properties of all the electroplated copper films were also measured using the Bench Top CMP tester. Nanoindentation was performed on the post-CMP samples in order to investigate the work hardening phenomena. Details of the CMP tester have been discussed elsewhere [12].Transmission electron microscopy was performed on the cross section of the samples and also Finite element modeling was performed in order to investigate the stress behavior during nanoindentation using NanoSP1 software from MTS Nano Innovation center, Oak Ridge, TN.

RESULTS AND DISCUSSIONS
Grazing Incidence XRD

Grazing incidence XRD was performed in order to study the structural information of the electroplated Cu films with reduced effect from the bottom layer (TaN) and substrate. GIXRD spectra of different Cu films are shown in Fig. 1. It can be seen from Fig. 1 that all the films exhibit X-ray reflection from Cu (111), (200), (220), (311) and (222) planes. For all the films, strongest x-ray reflections are visible from Cu (111) planes. The intensity of reflections from Cu (220) is second highest. This indicates that crystallization occurs preferentially in (111) and (220) directions. In order to investigate the evolution of crystallites orientations with different types of annealing, percentage relative intensities of all the diffraction peaks (with respect to (111) peak) are tabulated in Table 1. It can be seen from Table 1 that the evolution of (220) and (200) orientation is less pronounced in the samples aged for 7 days in the atmosphere and then annealed in two stages (Sample # 2, 3 and 4). A similar trend could be seen also for the diffraction peaks from (311) and (222) planes. Ageing in air may have helped the crystallites to grow preferably along (111) directions, which is referred to as self-annealing [4]. Subsequent annealing may not be able to recrystallize effectively in other orientations for these samples.

From the XRD analysis, samples 4 and 7 appear to have a similar crystal structure. This is due to the fact that samples 4 and 7 were annealed in one step at 250°C. It can also be seen that evolution of crystallites slightly higher for the samples 3 and 6 than samples 2 and 5. It is prudent to mention that samples 2 and 5 were annealed first at 150°C and then 250°C whereas sample 3 and 6 were annealed first at 200°C and then 250°C. This clearly indicates that electroplated Cu samples may retain the effect of pretreatments received in various stages while they are pretreated in multistep.

Table 1. Comparison of the relative intensity of XRD peaks of different Cu films

Position [°2Th.]	(hkl)	Relative Intensity					
		Samples #					
		2	3	4	5	6	7
43.5004	111	100.00	100.00	100.00	100.00	100.00	100.00
50.7087	200	10.88	11.87	12.47	19.15	19.33	18.28
74.2985	220	23.73	19.67	19.82	39.99	35.68	32.18
89.9930	311	7.53	8.25	9.38	16.27	13.87	9.81
95.2574	222	3.34	4.35	2.76	6.67	5.56	3.64

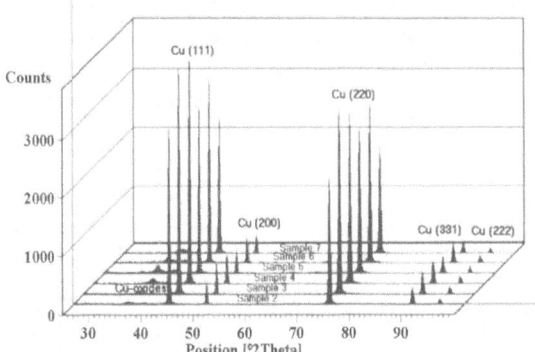

Figure 1. Low angle X-ray diffraction pattern of Cu films annealed at different stages

Nanoindentation

Indentation was performed on all the Cu films annealed at different temperatures. The typical load-displacement curves for all the copper films are shown in Fig. 2(a). It can be seen from the Fig. 2(a) that the loading-unloading curves for all the samples are nearly similar, which may be attributed to the similar crystalline nature of Cu films. The slopes of the unloading curves are higher than those of the loading curves. This is due to the high plastic deformation of these films of metallic nature.

The values of Young's modulus and hardness obtained using Oliver and Pharr's method are summarized in Table 2. It is also observed that Young's modulus and the hardness are higher for the samples where they were annealed in one stage at 250°C. The values were found to increase with increasing the annealing temperature of the first step. A continuous stiffness measurement technique allows measuring the modulus and hardness throughout the indenter displacement. Variations of hardness and modulus with indenter displacements are shown in Fig. 2 (b) and (c), respectively. It can be seen from these figures that hardness and modulus values are greater at the very beginning of the indenter displacement and then decreases towards a plateau region before it starts increasing again with deeper indentation. Initial scattered data with high values may be due to the uncertainty of finding the surface by tip, whereas higher values at the end are due to the effect of the underlying harder barrier layer and substrate.

Aging of the samples in air may allow the growth of oxide layer on the electroplated Cu. In addition, the Cu layer undergoes self-annealing during aging. These may be the reasons for the lower effect of annealing for the samples. XRD results also indicate the same. Annealing at 250°C in one stage seems to have a higher effect on the mechanical properties than the annealing of the samples in two stages. XRD results show the dominance of (111) orientation in the films annealed in one stage (Sample 4 and 7). Higher hardness values for samples 4 and 7 may be explained by an abundance of the (111) oriented crystallites.

Table 2. Results obtained from the Nanoindentation experiments for the copper samples

Sample #	Oxide (nm)	PVD Barrier (nm)	PVD Cu (nm)	EP Cu (nm)	Ageing	1st Anneal	2nd Anneal	Young's Modulus (GPa)	Hardness (GPa)
1	1000	30	150	No	No	No	No	-	-
2	1000	30	150	500	7days in air	150°C, 60min	250°C, 60min	134.85±7.51	1.29±0.14
3	1000	30	150	500	7days in air	200°C, 60min	250°C, 60min	135.26±7.88	1.32±0.15
4	1000	30	150	500	7days in air	250°C, 60min	No	140.49±6.51	1.36±0.14
5	1000	30	150	500	No	150°C, 60min	250°C, 60min	140.70±6.45	1.36±0.13
6	1000	30	150	500	No	200°C, 60min	250°C, 60min	141.35±6.23	1.39±0.14
7	1000	30	150	500	No	250°C, 60min	No	142.13±5.12	1.42±0.10

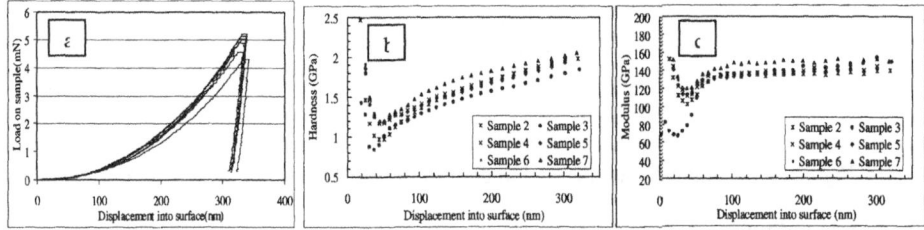

Figure 2 (a) Typical load-displacement curves for all Copper samples,
(b) Hardness and (c) Young's modulus vs. displacement curves.

Surface Characterization by AFM

The surface characteristics of the annealed Cu films are very sensitive to the annealing temperature and time. The surface view of AFM images of all the samples are shown in Fig. 3. Fig. 3 reveals that all the samples except sample 1 have polycrystalline surface morphology. Sample 1 does not show a surface feature since it represents only the thin seed Cu layer. Table 3 represents the surface average roughness (R_a) data estimated from the roughness analysis in AFM. The R_a is the mean of the absolute value of the height of the surface below or above the center. It can be seen from Table 3 that the lowest R_a value is recorded for the sample 1. In general, the surface of a polycrystalline thin film that consists of more (111) oriented crystallites is rougher than the surface that consists of (110) and (100) oriented crystallites. The XRD results confirm the presence of higher (111) oriented crystallites present in sample 4 and 7. A higher value of R_a for sample 4 than sample 7 may be caused by an AFM measurement artifact, which can be seen in form of spikes on the surface view of sample 4 (Fig. 3).

Table 3. The Roughness values of all the samples estimated from AFM roughness analysis.

Sample #	1	2	3	4	5	6	7
(R_a) (nm)	1.03	7.20	6.82	6.83	7.01	7.88	1.03

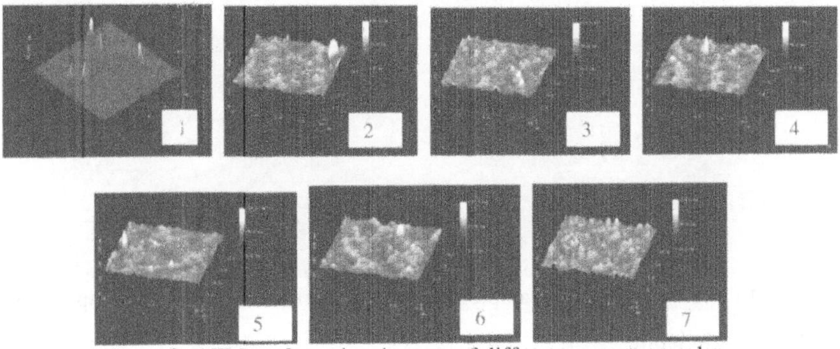

Figure 3. AFM surface view images of different copper samples.

Effect of CMP on Cu Films

The friction behavior of all the copper films was investigated by creating a CMP environment using the Bench Top CMP tester. Polishing was performed at 2 PSI and 150 RPM using a polyurethane pad (IC1000/SubaIV) and Cu slurry containing alumina abrasives. During polishing both coefficients of friction (COF) and acoustic emission (AE) were monitored *in situ*. Average values of COF and AE signal are shown in Table 4. COF and AE values are much lower for the sample 1, which has a very smooth surface than others.

Nanoindentation was performed on the samples after polishing the top surface in order to investigate the mechanical effect of CMP process on the Cu films. The values of Young's modulus and hardness values collected on the post-CMP samples are shown in Table 4. It is observed that Young's modulus and the hardness increase after the CMP process. The differences shown in these figures may be attributed to the work-hardening phenomenon potentially occurring due to the high shear force applied on the Cu films during CMP.

Table 4. Tribological results of copper samples and the results of nanoindentation after CMP.

Sample #	PSI	RPM	Total Polishing Time (sec)	AE (V)	COF (Total)	Young's Modulus (GPa)	Hardness (GPa)
1	2	150	72	0.139	0.369	--	--
2	2	150	72	0.1726	0.3996	139.15±3.99	1.30±0.12
3	2	150	72	0.3483	0.4451	139.55±4.68	1.32±0.13
4	2	150	72	0.3969	0.6151	144.04±5.34	1.35±0.14
5	2	150	72	0.335	0.5949	147.47 ±5.12	1.39±0.10
6	2	150	72	0.1489	0.5826	147.21 ±4.78	1.40±0.14
7	2	150	72	0.3465	0.5627	148.57 ±5.79	1.43±0.10

Transmission Electron Microscopy (TEM)

Transmission electron microscopy was performed on the cross section of the copper samples. Figure 4 shows the TEM picture which clearly shows the oxide layer, then the barrier layer (TaN) and then the Cu seed layer. The diffraction pattern in Figure 5 shows the crystalline behavior of both seed layer and electroplated Cu. It can be seen that the copper planes are randomly oriented showing the planes Cu <000>, Cu <111>, Cu <200>, Cu <220>. This TEM picture confirms the result obtained from XRD analysis.

 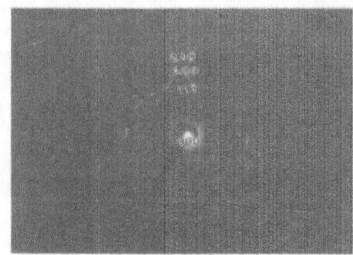

Figure 4 TEM Picture of a Copper Sample   Figure 5 Diffraction Pattern of a Copper Sample.

Finite Element Modeling

Finite Element Modeling is used to investigate the stress behavior of thin films on substrates during nanoindentation using the software package of NanoSP1. The input parameters, which we provide for the substrate, are the elastic modulus, the yield strength, Poisson's ratio, and work hardening modulus of the material. For the film also we all these input parameters and also the film thickness and the indentation depth. It also needs the indenter tip description and also the area function of the tip. After providing these parameters, the software then automatically models the indentation process and we get the output the values of hardness is also obtained. The values of hardness obtained from the FEM exactly matches with the hardness obtained from the nanoindentation method. Figure 6 represents the Von Mises stress contour plot for copper sample # 2.

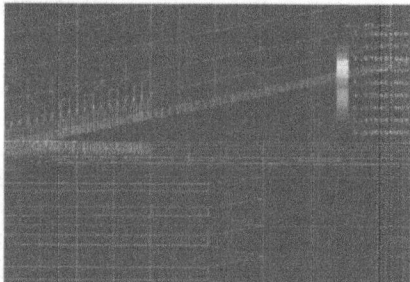

Figure 6. Von Mises Stress Contour Plot for the copper sample 2

CONCLUSIONS

The effect of multistep annealing treatment on electroplated Cu films is investigated using GIXRD, nanoindentation, and AFM techniques. Also, the friction behavior of Cu films during CMP and the effect of polishing on mechanical properties are investigated. GIXRD pattern shows that all the Cu films have good crystallinity with stronger x-ray reflections from

Cu (111) and (220) planes but weaker reflections from (200), (311) and (222) planes. The Cu samples aged in air have lower hardness and modulus values. Indentation experiments reveal that Cu films underwent elastic deformation during indentation and that films are uniform. The hardness and modulus values are larger for the set of the films not aged in air. In addition, the hardness and modulus values increase with the increase in temperature of the first-step annealing. Samples annealed in one stage (sample 4 and 7) show better mechanical properties, which may be due to higher (111) oriented crystallites. The Cu seed layer is very smooth and has lower COF and AE values during CMP. No distinctive trend is found for the COF and AE values for annealed Cu samples. Improved mechanical properties on post-CMP samples may indicate the work hardening of the Cu surface due to the high shear force applied on the sample surface during CMP. The surface roughness of annealed Cu samples increased with the increase in temperature of the first-step annealing. Transmission electron microscopy was performed on the cross section of the samples and the diffraction pattern showed the crystalline behavior of both seed layer and electroplated Cu. Finite Element Modeling results gave the information about the Von Mises stress of Cu samples on the substrate. Annealing at higher temperature may reveal further insight of the microstructural evolution of electroplated Cu films, which is the key in successful implementation of Cu as an interconnect material.

ACKNOWLEDGEMENT

The authors would like to acknowledge NSF CAREER Grant # 9983535 (Ashok Kumar) and NSF GOALI Grant # DMI 0218141 for supporting this research. Part of this research was also supported by USF-Agere High-Tech Corridor Grant # 21-12-142LO. We would also like to thank Novellus Systems for providing samples.

REFERENCES

1. J. Tao, N. W. Cheung, and C. Hu, IEEE Electron Device Lett., **14**, 249-51, (1993).
2. H. K. Kang, I. Asano, C. Ryu, and S. S. Wong, Proc. of IEEE VMIC Conf., 223-229, (1993).
3. K. P. Rodbell, E. G. Colgan, and C. K. Hu, Proc. Of Advanced Metallization for Devices and Circuits – Science, Technology and Manufacturability, 59-70 (1994).
4. S. Lagrange, S.H. Brongersma, M. Judelewicz,, A. Saerens, I. Vervoort, E. Richard, R. Palmans and K. Maex, Microelectronic Engineering **50**, 449-457 (2000).
5. J. M. Harper, C. Cabral, Jr., P.C. Andricacos, L. Gignac, I.C. Noyan, K. P. Rodbell and C.K. Hu, J. Appl. Phys. **86**, 2516-2525 (1999).
6. W.H. The, L.T. Koh, S.M. Chen, J. Xie, C.Y. Li and P.D. Foo, Microelectronics Journal **32**, 579-585 (2001).
7. J. Proost, T. Hirato, T. Furuhara, K. Maex and J.-P. Celis, J. Appl. Phys. **87**, 2792-2802 (2000).
8. T.-H. Fang and W.-J. Chang, Microelectronic Engineering **65**, 231-238 (2003).
9. S.H. Kang, Y.S. Obeng, M.A. Decker, M. Oh, S.M. Merchant, S.K. Karthikeyan, C.S. Seet and A.S. Oates, J. Electron. Mater. **30**, 1506-1512 (2001).
10. Masatoshi Kuroda, Daigo Setoyama, Masayoshi Uno, Shinsuke Yamanaka, Journal of Alloys and Compounds **368** 211–214, (2004).
11. A. K. Sikder, I.M. Irfan, Ashok Kumar and J.M. Anthony, J. Elec. Mater. **30**, 1527-31 (2001).
12. A. K. Sikder, Frank Giglio, John Wood, Ashok Kumar and J.M. Anthony, J. Elec. Mater. **30**, 1522 (2001).

Mat. Res. Soc. Symp. Proc. Vol. 812 © 2004 Materials Research Society   F3.18

Thermal Conductivity of Carbon Nanotube Composite Films

Quoc Ngo[1,2], Brett A. Cruden[2], Alan M. Cassell[2], Megan D. Walker[2], Qi Ye[2], Jessica E. Koehne[2], M. Meyyappan[2], Jun Li[2*], and Cary Y. Yang[1]
[1]Center for Nanostructures, Santa Clara University, 500 El Camino Real
Santa Clara, CA 95050, USA
[2]Center for Nanotechnology, NASA Ames Research Center
Moffett Field, CA, 94035, USA

ABSTRACT

State-of-the-art ICs for microprocessors routinely dissipate power densities on the order of 50 W/cm^2. This large power is due to the localized heating of ICs operating at high frequencies, and must be managed for future high-frequency microelectronic applications. Our approach involves finding new and efficient thermally conductive materials. Exploiting carbon nanotube (CNT) films and composites for their superior axial thermal conductance properties has the potential for such an application requiring efficient heat transfer. In this work, we present thermal contact resistance measurement results for CNT and CNT-Cu composite films. It is shown that Cu-filled CNT arrays enhance thermal conductance when compared to as-grown CNT arrays. Furthermore, the CNT-Cu composite material provides a mechanically robust alternative to current IC packaging technology.

INTRODUCTION

As progress continues in the ultra-large-scale-integration (ULSI) of integrated circuits, microelectronic components including transistors, and more prominently interconnects, have become increasingly more dense and compact. The consequence of increased component density manifests itself in the form of locally high power consumption. An alarming rise in power density with respect to each advancing technology generation has been observed in mainstream microprocessor technologies [1]. The need for addressing this problem is imperative for next-generation IC packaging technology. One potential solution is to find new packaging materials, such as CNTs, that exhibit high thermal conductivity along the axial direction. For a discrete multiwalled nanotube (MWNT), the thermal conductivity is expected to surpass 3000 Wm^{-1}K^{-1} along the tube axis [2]. Through the use of DC-biased, plasma-enhanced chemical vapor deposition (PECVD), as demonstrated in [3], we can fabricate vertically aligned MWNT arrays on silicon wafers of ~500μm thickness and demonstrate their possible application as a heat-sink device, conducting large amounts of heat away from a localized area, such as in critical "hot spots" in ICs. This work focuses on demonstrating that CNT and CNT-Cu composite films are mechanically robust, efficient thermal conductors. Cu serves as a filler material to improve the mechanical stability of the CNT array and to serve as a lateral heat spreader. Our data shows that the structures fabricated in this study are completely reusable, unlike eutectic bonding techniques currently used in packaging technology, which is particularly important for instrument cooling in space applications.

---

* corresponding author: jli@mail.arc.nasa.gov

EXPERIMENT

Description of apparatus

An apparatus consisting of two copper blocks, four resistive cartridge heaters (not shown) embedded in the upper block, and a cooling bath was used to measure the thermal resistance of a given material (figure 1). The upper copper block is surrounded by insulation to minimize heat loss to the ambient, with the exception of the one square inch section designed to contact the material to be measured. The clamping pressure on the sample is controlled by pneumatically manipulating the upper block. Heat is delivered to the system by applying a constant power to the resistive heaters. The steady state temperature difference ($\Delta T = T_B - T_C$) between the two blocks (and consequently, the sample) was measured. From this data, the thermal resistance of the sample is calculated, as shown in equation (1), where Q is the total power in (in W), A is the sample area, C_L is the constant heat transfer coefficient and T_B, T_C, and T_{amb} represent the temperature of the upper block, the chilled lower block (20°C), and the ambient environment, respectively. C_L is used to estimate the heat loss to the ambient in this measurement configuration and is determined by placing a thick insulator between the two blocks and measuring the steady state ΔT at a variety of constant powers. This analysis yields a constant heat transfer coefficient of 0.0939 W/K, representing the heat loss (in W) per Kelvin to the ambient, which is factored into the final determination of the measured thermal resistance

$$R = \frac{A(T_B - T_C)}{Q - C_L(T_B - T_{amb})} \qquad (1)$$

It should be noted, however, that the dominant thermal resistance mechanism in this measurement configuration is that of the contact interfaces between the sample and the copper blocks. To minimize this contact resistance, two steps were taken: 1) polishing both copper blocks to reduce the effect of surface roughness and 2) making use of a high thermally conductive, conformal material, Microfaze A6 (AOS Thermal Compounds, LLC) to reduce contact resistance on the backside of a silicon wafer, the substrate on which the investigated films were fabricated.

Sample preparation

Carbon nanotubes were synthesized using the procedure and reactor conditions detailed in [3]. A layer of chromium (2000Å) was used as an adhesion layer for the thin layer of nickel catalyst used for the MWNT array growth. The resulting as-grown tubes are shown in figure 2. Using scanning electron microscope (SEM) data, we estimate the length of the tubes to be ~7.5μm. Following nanotube synthesis, copper filling between individual MWNTs (hereafter known as nanotube trenches) was accomplished through electrodeposition. In a three-electrode

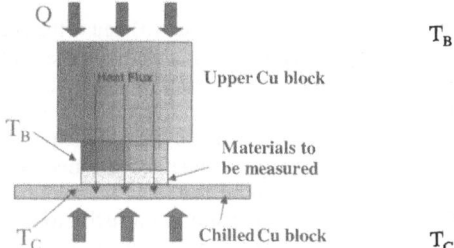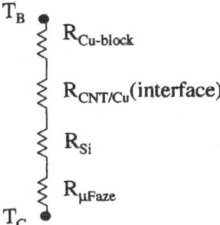

Figure 1. Apparatus used for thermal resistance measurement and equivalent thermal resistance model. T_B is the temperature of the block, measured using an embedded thermocouple. T_C is the temperature of the chilled lower copper block (20°C).

setup with the MWNT sample as the working electrode, a Saturated Calomel Electrode (SCE) as the reference electrode, and a one square inch platinum foil as the counter electrode (CE), set in parallel with the MWNT sample. Both the Cr substrate and MWNTs serve as electrodes during the electrodeposition. Various additives were added in the solution to achieve optimum gap filling into the high-aspect-ratio forest-like MWNT arrays. The recipe of the electrolyte solution used in this study is based on the methodology reported for deep-trench filling of Cu interconnects for damascene processes [4]. Typically, the Cu was deposited at –0.20 to –0.30 V (vs. SCE) at a deposition rate of ~430nm/min. The resulting CNT-Cu composite material has ~30% submicron voids as shown in figure 3.

RESULTS AND DISCUSSION

To summarize the structure used, figure 1 includes the equivalent thermal resistance model for the CNT-Cu composite sample. The resistance of the CNT-Cu composite can be obtained by de-embedding the thermal resistance contribution of the copper block ($R_{Cu\text{-}block}$), silicon wafer (R_{Si}), and the Microfaze material ($R_{\mu Faze}$). The thermal resistance of the copper block, $R_{Cu\text{-}block}$, must be taken into account due to the placement of the thermocouple (approximately 1.3 inches from the copper block surface). $R_{Cu\text{-}block}$ for this configuration is determined as 0.83 cm^{2}K/W based on bulk copper properties. To summarize, we can determine the resistance of the CNT-Cu composite film by equation (2).

$$R_{CNT\text{-}Cu} = R_{total} - R_{Cu\text{-}block} - R_{Si} - R_{\mu Faze} \qquad (2)$$

$R_{\mu Faze}$ is determined using two control measurements. The first measurement involves measuring the thermal resistance of a piece of silicon with Microfaze on the backside of the wafer, resulting in $R_{control,1} = R_{Cu\text{-}block} + R_{block\text{-}Si} + R_{\mu Faze}$, where $R_{block\text{-}Si}$ is the interface resistance between the copper block and silicon wafer. The second resistance measurement involves a piece of double-sided polished silicon, resulting in $R_{control,2} = 2R_{block\text{-}Si} + R_{Si}$. Assuming both Si-Cu interfaces in the second control measurement are similar, we can divide this value in half and use the simple relation in equation 3.

$$R_{\mu Faze} = R_{control,1} - R_{Cu\text{-}block} - 0.5(R_{control,2} - R_{Si}) \qquad (3)$$

The intrinsic silicon contribution to the thermal resistance in equations (2) and (3) can effectively be neglected. For the 500μm thick silicon wafer used in this study, the intrinsic silicon thermal resistance can be calculated as 0.034 cm^{2}K/W, which proves to be two orders of magnitude less than the final measured values of the CNT-Cu sample. One caveat to this analysis is in regards to the thermal resistance of Microfaze with respect to the amount of power applied to the upper block. The thermal resistance of the first control sample decreases approximately exponentially for increasing powers (hence different temperature gradients), but can be corrected for in the final analysis as will be demonstrated. The double-sided silicon sample shows no such power dependence, exhibiting a constant resistance of 11.10 cm^{2}K/W, resulting in 5.55 cm^{2}K/W per silicon interface. Subtracting the silicon resistance, which is constant with respect to power, we can also determine $R_{\mu Faze}$ at different powers. The power dependence of the Microfaze is captured in figure 4(a).

Now that the power dependence of the Microfaze material is quantified, we proceed with the analysis of the CNT/Si/Microfaze and CNT-Cu/Si/Microfaze stacks. From the previous discussion, we can expect these samples also to exhibit the same power dependence, which indeed is the case and is clearly seen in figure 4(b). Combining the power dependence with the measurements in figure 4(b), we summarize the values of measured thermal resistance in table I.

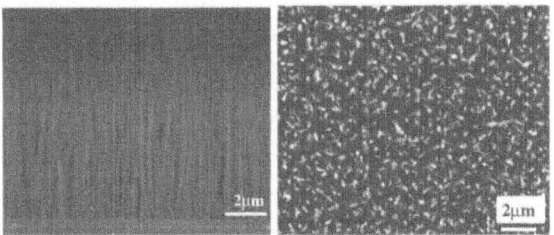

Figure 2. SEM micrographs (cross-section and top-down) of as-grown MWNT film. Length of nanotubes is estimated as ~7.5μm.

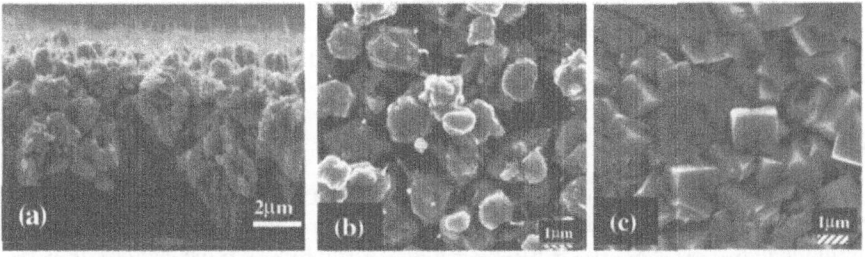

Figure 3. SEM micrographs (a) cross-section (b) top-down of CNT-Cu composite film after filling nanotube trenches by copper electrodeposition using recipe in [4]. (c) shows a top-down view of copper electrodeposition with low additive concentration.

All measurements were performed at similar clamp pressures, ~6.8 psi. Errors causing the standard deviation in the measurements can be attributed to two main factors: 1) variations in contact area due to varying CNT length distribution (see fig. 2) and 2) variations in measurement of total power, ΔT, and ambient temperature loss. However, even at the upper bounds of the measured thermal resistance values for the CNT-Cu composite films, this worst-case scenario represents values that are on the order of the thermal budgets for a variety of commercial microprocessor systems [1].

It is noted that the Cu deposited in the MWNT array used in this study was not a solid film. Instead, it forms a porous film with ~70% Cu and CNTs and ~30% voids. This increases the mechanical strength so that the sample can be repeatedly measured under different clamping pressures. In addition, it leaves space such that the composite film can be deformed to make maximal contact with the hot surface. However, studies conducted on the buckling force of discrete MWNTs [5-7] demonstrate the tremendous amount of force that these structures can withstand. Based on this analysis, we can speculate that nanotubes will not buckle under the force applied in this study, roughly two orders of magnitude less than the calculated CNT buckling force. This speculation is confirmed by SEM characterization before and after the thermal resistance measurement, seen in figure 5, showing no effect on the CNT-Cu composite after compressive stress. Measurements conducted with higher clamp pressures are underway to confirm the buckling force and optimize thermal conductance. The exposed CNTs serve as the thermal interface material, while Cu is only filled at the bottom portion of the structure to serve as both a mechanical anchor and lateral heat spreader.

CONCLUSION

This work demonstrates the fundamental usefulness of CNTs and CNT-Cu composite films as efficient heat conductors. Our study confirms that these novel films can accomplish effective heat conduction by increasing contact area. In addition, it has been shown that CNTs provide the added benefit of high mechanical stability and reusability. Further characterization and investigation are in progress to integrate such a process into packaging process flows.

ACKNOWLEDGEMENTS

MDW is an undergraduate intern at NASA Ames Research Center from the University of Oklahoma, Tulsa. BAC and AMC are also with the University Affiliated Research Center at University of California, Santa Cruz. QY is with ELORET. NASA contract number NAS2-03144 supports both UARC and ELORET.

Table I. Thermal Resistance measurement summary

Material	Thermal Resistance ($cm^2 K/W$) ± STDEV
CNT film	2.42 ± 0.33
CNT-Cu composite film (#1)	1.05 ± 0.22
CNT-Cu composite film (#2)	0.96 ± 0.13
Bare double-sided silicon	11.10 ± 0.65

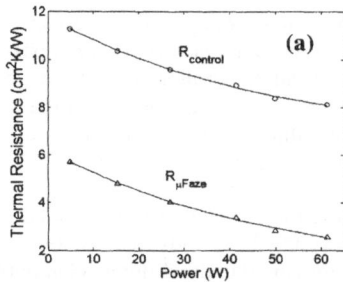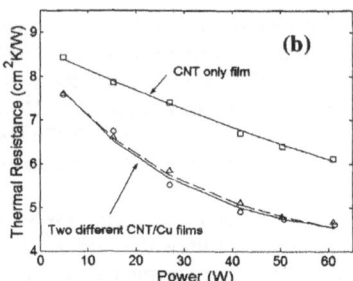

Figure 4. (a) Resistance versus power characteristics for first control sample and Microfaze. (b) Resistance versus power characteristics for CNT and CNT-Cu samples (before de-embedding Microfaze and upper copper block). The symbols represent measured data; the lines represent exponential fits to the data.

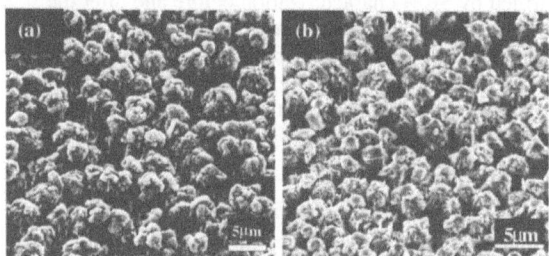

Figure 5. SEM micrographs (a) before and (b) after compressive thermal resistance measurement. The morphology of both the CNTs and Cu are similar in both cases.

REFERENCES

1. R. Viswanath, V. Wakharkar, A. Watwe, and V. Lebonheur, "Thermal Performance Challenges from Silicon to Systems", *Intel Technology Journal*, Q3 (2000).
2. P. Kim, L. Shi, A. Majumdar, and P.L. McEuen, "Thermal Transport Measurements of Individual Multiwalled Nanotubes", *Phys. Rev. Lett.*, **87**, p. 215502-1 (2001).
3. B.A. Cruden, A.M. Cassell, Q. Ye, and M. Meyyappan, "Reactor Design Considerations in the Hot Filament/Direct Current Plasma Synthesis of Carbon Nanofibers", *J. Appl. Phys*, **94**, 4070 (2003).
4. K. Kondo, N. Yamakawa, Z. Tanaka, and K. Hayashi, "Copper Damascene Electrodeposition and Additives", *J. of Electroanalytical Chemistry*, **559**, 137 (2003).
5. H. Dai, J.H. Hafner, A.G. Rinzler, D.T. Colbert, and R.E. Smalley, "Nanotubes as Nanoprobes in Scanning Probe Microscopy", *Nature*, **384**, 147 (1996).
6. H. Dai, N. Franklin, and J. Han, "Exploiting the Properties of Carbon Nanotubes for Nanolithography", *Appl. Phys. Lett.*, **73**, 1508 (1998).
7. J. Li, A.M. Cassell, and H. Dai, "Carbon Nanotubes as AFM Tips: Measuring DNA Molecules at the Liquid/Solid Interface", *Surf. and Interf. Analysis*, **28**, 8 (1999)

Mat. Res. Soc. Symp. Proc. Vol. 812 © 2004 Materials Research Society   F3.19

Silver Patterning by Reactive Ion Beam Etching for Microelectronics Application

L. Gao, J. Gstoettner, R. Emling, P. Wang, W. Hansch, D. Schmitt-Landsiedel
Institute for Technical Electronics, Technical University Munich
D-80333, Munich, Germany

ABSTRACT

Dry etching of silver for the metallization in microelectronics is investigated. Etching is performed using an electron-cyclotron-resonance reactive-ion-beam-etching system (ECR-RIBE) in an Ar/CF$_4$ or Ar/CF$_4$/O$_2$ mixture. The etch characteristics are strongly affected by ion energy (beam voltage and microwave energy); the O$_2$ concentration in the reactive mixture has only a small effect. An anisotropic, smooth etch profile and clean surface are obtained. Focused ion beam (FIB) and atomic force microscopy (AFM) have been used to study the etched profile and the roughness, respectively.

INTRODUCTION

As devices continuously shrink in ultra large-scale integration (ULSI), the RC delay of the interconnection system becomes one of the most critical limitations on IC performance [1]. Because of its high electrical conductivity and its good electromigration properties, copper replaces aluminium in some microelectronic applications [2, 3]. Silver, with a lower bulk resistivity than that of aluminium and copper, is one of the candidates for future interconnect metallization [4-8]. It has been shown that a sputtered Ag metallization is possibly more suitable for ULSI than sputtered Cu due to a much higher electrical conductivity in features sizes below 100 nm [4]. Furthermore, it was measured that sputtered Ag can reach the same electromigration resistance as CVD or PVD deposited copper [5,8]. To evaluate the potential application of silver metallization, one of the important aspects is the pattern transfer by etch processing. Several investigations on Ag etching have been reported. The etching speed of wet etching using in a Fe(III)-nitride-H$_2$O or in a NH$_4$OH/H$_2$O$_2$-H$_2$O can be very fast. However, it is very difficult to control and easily causes underetching [8]. A further work has been done with a hybrid dry-wet etching technique [7], which leads to a better result. However, it is complicated and time consuming. In another dry etching report a conventional technique of reactive ion etching (RIE) is established in an oxygen plasma [6]. The drawback here is that a low etch rate is obtained.

To achieve anisotropic etching, physical sputtering along with radical reaction is essential. A plasma generation method using an ECR has been applied for etching and deposition [9,10]. In contrast to conventional RIE [11] the ion energy and flux can be controlled independently in the RIBE system [12] and yields better ion directionality due to its high mean-free path and high ion-to-radical density ratio by a divergent magnetic field method. In order to optimise the etch rate and obtain smooth etch morphology in the RIBE process, it is important to know the effects of process variables such as beam voltage and current, and etching gas concentration.

EXPERIMENTAL DETAILS

In our investigation, we use an Ar, CF$_4$ and O$_2$ based ECR plasma in RIBE (ECR-RIBE), which is shown in Figure 1 (Technics Plasma GmbH RIB Etch 160 instrument).

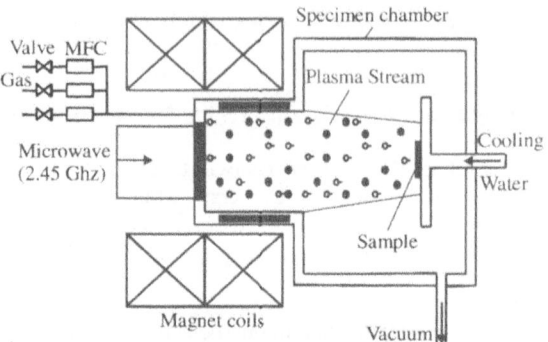

Figure 1. ECR-RIBE etching apparatus schematic diagram.

The ECR source is excited by a 2.45 GHz microwave power supply. A broad beam Kaufman-type ion source delivers a controlled beam of reactive gas to the surface for accelerating ions toward the substrate. The magnet coils are arranged around the periphery of the plasma chamber to achieve a microwave ECR condition (magnetic flux density, 1800 G) that enables the plasma to effectively absorb the microwave energy at low gas pressures of 10^{-3} to 10^{-5} mbar.

Silver films were sputter-deposited in an Alcatel RF-magnetron system, using a 99.99% pure target and pure argon process gas. The base pressure of the system was 2×10^{-7} mbar. The silver films were deposited directly on Si or on SiO_2/Si (300 nm SiO_2 was thermally grown in a furnace with the temperature of 1000^0C). Deposition is carried out at an RF power of 50 W and a processing pressure of 7×10^{-3} mbar [5]. For a part of the samples, thin films of titanium nitride (TiN) are used as one of the most popular diffusion barrier layers for metallization applications. In addition, the Ag/TiN/SiO$_2$/Si-stack structure is also investigated using CF_4 in the same facilities. The samples are covered with a photoresist and patterned using lithography.

The standard etch conditions used in the study were room temperature, 350 V or 500 V beam voltage, 100 mA beam current, 1089 G ECR magnet, 180 W microwave power, 1.2×10^{-4} mbar pressure, -50 V Acceleration voltage, 0.99 A neutralization current and Ar/CF$_4$ (10 sccm/10 sccm). It is shown in the following where some of these parameters are varied systematically that these standard conditions represent the optimal process. Etch depths are determined from profilometer (Dektak3ST, Veeco Instruments) measurements. The step of thickness is measured at six different places and the average is taken as the step height. Surface morphology was examined with a FIB system and atomic force microscopy (AFM).

RESULTS AND DISCUSSION

RIBE etch rate

The effect of beam voltage on the etch rates of silver were examined varying the beam voltage from 250 V to 600 V. The results are shown in figure 2. It can be seen from the figure that the etch rate increases monotonically with increased ion energy. By changing the beam voltage from 250 V to 600 V, there is an increase in etch rate from 3.3 nm/min to 112.6 nm/min. This is expected since an increase in beam voltage helps to increase the sputter desorption of the etch products, leaving behind a clean surface exposed to the reactive ion species. The etching process is controlled by physical sputtering at higher beam currents. By increasing the beam current, plasma density also increases mainly due to higher concentrations of reactive species and higher ion flux which increases the bond breaking efficiency and the sputter adsorption component of the etching mechanism.

The dependence of etch depth rate on microwave power is shown in figure 3 for the CF_4/Ar discharge at a pressure of 1.2×10^{-4} mbar with a beam voltage of 500 V. As power increases, the plasma density and associated concentration of reactive species (F atoms) should also increase, producing higher etch rates. This is a fairly typical result for ECR discharge and is ascribed to the increased ion current and active species at higher powers. It is interesting to note that an etching delay is observed in the onset of etching between 10 to about 30 seconds, with the amount of the time delay decreasing as power increases. It is caused most certainly by the presence of a surface oxide which must be removed before etching of the silver can commence.

Etching characteristics of silver with an $Ar/CF_4/O_2$ mixture are shown in figure 4. The CF_4/O_2 mixture is investigated because this mixture is widely used in the dry etching process in semiconductor manufacturing. Fluorine atoms are created in a CF_4 plasma. When oxygen is added to a CF_4 plasma, it reacts with carbon species to form molecules such as CO_2, limiting the recombination of F atoms to CF_4 and setting them free for the etching process. At the same time, silver can also be oxidized in an oxygen plasma producing silver oxide. In the range of the O_2 flow rate between 0 to 8 with a total flow of 50 (50 here is 10 sccm), no etch residues were observed. By the changing the CF_4/O_2 flow rate from 50/0 to 42/8 (50 here is 10 sccm) with a beam voltage of 500 V, the etching rate is nearly constant between 51~53 nm/min. The etch rate of resist, however, increases with the increase of O_2 flow rate, so the etch selectivity of silver to the resist decreases under conditions without etch residues. Therefore, in using CF_4/O_2 mixture, it is difficult to ensure the etch selectivity in high aspect ratio structures with hard photoresist as a mask. At the same time, with O_2 as additional discharge gas, it is difficult to remove the photoresist after higher power RIBE etching. On the other hand, slight deposition of SiO_2 may occur when using an ECR Plasma with CF_4 containing O_2.

We also have studied the effect of substrates tilted with a certain degree to the plasma stream. In this paper, we have done the experiments with the inclination angle of 0^0, 10^0 and 50^0 with a beam voltage of 500 V. Additionally, the substrate is rotated during RIBE etching. Figure 5 shows the angle dependence of the etch rate of silver. With the rotation of the substrate, the same etching characteristics were obtained, except that the etch rate is somewhat lower at 50^0.

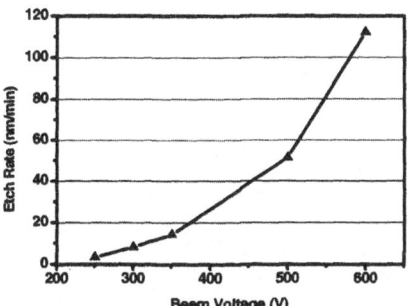

Figure 2. Ag etch rate as a function of beam voltage in CF_4/Ar discharge.

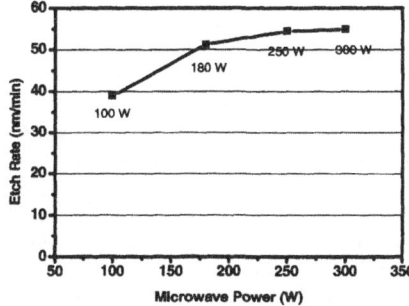

Figure 3. Ag etch rate as a function of microwave power in CF_4/Ar discharge with 500 V beam voltage.

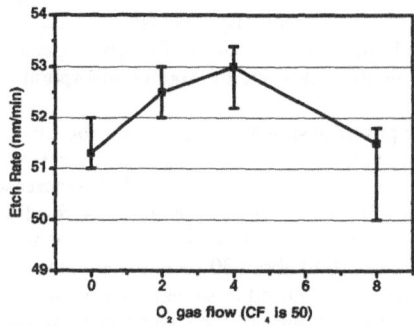

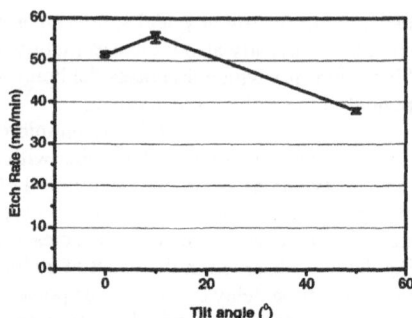

Figure 4. Ag etch rate as a function of gas mixture with 500 V beam voltage.

Figure 5. Ag etch rate as a function of tilt angle with 500 V beam voltage.

Etch morphology and etch profile

Figure 6 shows micrographs (Electron Beam-SEM, Focused Ion Beam-SEM) of silver etched at the standard etch conditions given in the experiment section 2.

Figure 6. FIB-SEM micrographs of silver etched at 350 V beam voltage
(a) FIB-micrograph (primary beam: Ga^+-ions (50keV, 5pA); image formation by detection of secondary electrons; angle of incidence: 45 deg) shows the surface of the patterned silver lines; (b) FIB-micrograph (50keV Ga^+-ions, 5pA, SE-detection, angle of incidence: 45 deg) gives a closer view to one of the silver lines and shows its grain structure; (c) SEM-micrograph (primary electron beam at 5kV (angle of incidence: 30 deg), SE-detection) shows the cross-section of a silver line); (d) FIB-micrograph (Ga^+-ions, 5pA, SE-detection, angle of incidence: 45 deg) depicts the cross-section of a silver line and shows its grain structure.

188

Figure 6(a) shows the surface of the patterned silver lines. The FIB figure 6(b) here gives a closer view to one of the silver lines and shows its grain structure. In this image as peculiarity of focused ion beam imaging the grain structure of the metal lines is revealed: The difference in brightness of the the various single crystal grains is caused by changes in the secondary electron yield. This quantity is influenced by ion channelling of the primary Ga^+-ions into the grains. As grains show different orientations the ion channelling effect and therefore the secondary electron yield varies. [13] The cross-section of a silver line is shown in figure 6(c). Figure 6(d) gives the cross-section of a silver line and shows its grain structure with FIB micrograph. The surface is extremely clear and sidewalls are straight. The slight ripple in the side wall of the sample is thought to be caused by pattern transfer from the photolithography. Therefore, we can assume that the vertical corrugations seen in the figure are transferred from roughness on the initial photoresist sidewalls.

Thin films of TiN, one of the most popular diffusion barrier layers for metalization applications, and the integration with silver as Ag/TiN stack layers, were also processed using CF_4 in the same RIBE equipment. The etch rate of TiN is 5.12 nm/min at 500 V beam voltage, slower than that of Ag, which means a high selectivity of Ag etching to TiN diffusion barrier. In the RIBE process, physical etching due to ion bombardment leads to a roughness which depends on the etching conditions. Different ion bombardment conditions leads to different surface roughness. Figure 7 gives the result for samples of $Ag/SiO_2/Si$ multilayer structures.

The roughness after etching at a pressure of 1.2×10^{-4} mbar, beam voltage of 500 V and Ar/CF_4 or $Ar/CF_4/O_2$ is quantitatively measured by AFM. The root-mean-square (RMS) roughness at 500 V beam voltage with only CF_4 reactive gas is 14 nm less than that of 500 V beam voltage with reactive gas CF_4 and O_2 (9:1), and also the surface after etching is clearer when etching was performed without oxygen. Similar results were reported in [6] for RIE silver etching, where rough films were obtained with oxygen discharge.

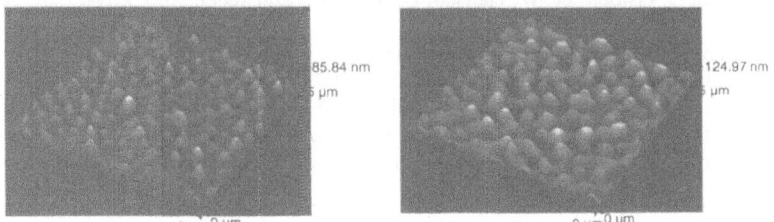

Figure 7. AFM images of the surface of RIBE etched samples($Ag/SiO_2/Si$): at 500 V beam voltage with Ar/CF_4 (left) and at 500V beam voltage with $Ar/CF_4/O_2$ (right).

CONCLUSIONS

Silver, a promising metallization material for future microelectronics applications, was patterned in a ECR reactive ion beam etching system in a Ar/CF_4 or $Ar/CF_4/O_2$ gas mixture. The influences of beam voltage and current, microwave power, O_2 concentration and tilt angle on etch rate were investigated. The etch characteristics are strongly affected by ion energy (beam voltage and microwave energy). A high etch rate, anisotropic and smooth etch profiles are obtained. The O_2 concentration in the reactive mixture has only a small effect on the etch rate but causes slightly increased roughness of the surface. The present ECR-RIBE system proves to be an appropriate tool for silver line patterning in silver metallization. Further work needs to be done using other mask materials and performing electrical measurements on the patterned Ag structures.

ACKNOWLEDGMENTS

This work was supported by German Research Foundation under grant No. FOR 395.

REFERENCES

1. International Technology Roadmap for Semiconductors, Semiconductor Industry Association, (2003).
2. S. Venkatesan, et al, Techn. Digest Int. Electron Devices Meeting, **31**, 769 (1997).
3. D. Edelstein, et al, Techn. Digest Int. Electron Devices Meeting, **31**, 773 (1997).
4. R. Manepalli, F. Stepniak, S.A. Bidstrup-Allen, P. Kohl, IEEE Trans. Adv. Pack. **22**, 4 (1999).
5. M. Hauder, J. Gstoettner, W. Hansch, D. Schmitt-Landsiedel, Sensors and Actuators A **99**, 137 (2002).
6. Y. Zeng, L. Chen, Y.L. Zou, P.A. Nguyen, J.D. Hansen and T.L. Alford, Mater. Chem. Phys., **66**, 77 (2000).
7. M. Hauder, J. Gstoettner, W. Hansch, D. Schmitt-Landsiedel, Microelectro. Eng. **60**, 51 (2002).
8. M. Hauder, J. Gstoettner, W. Hansch, D. Schmitt-Landsiedel, App. Phys. Lett. **78**, 838 (2001).
9. Y.B. Hahn, J.W. Lee, et al, J. Vac. Sci. Technol. B **17**, 366 (1999).
10. A.K. Dutta, J. Vac. Sci. Technol. B **13**, 1456 (1995).
11. A. Bertz, T. Werner, Applied Surface Science **91**, 147 (1995).
12. S.J. Pearton, J.W. Lee, et al, J. Vac. Sci. Technol. B **14**, 118 (1996).
13. D. L. Barr, L. R. Harriot, W. L. Brown, J. Vac. Sci. Technol. B **10**, 3120 (1992).

Mat. Res. Soc. Symp. Proc. Vol. 812 © 2004 Materials Research Society

Free-standing line patterns of nanocrystalline electrodeposits

Karen Pantleon[1], Henrik Myhre Jensen[2*], Marcel A.J. Somers[1]

The Technical University of Denmark, DK – 2800 Kgs. Lyngby, Denmark
[1]Department of Manufacturing Engineering and Management (IPL)
[2]Department of Mechanical Engineering (MEK)
(*now: Aalborg University, Department of Building Technology and Structural Engineering)

ABSTRACT

Free-standing Cu- and Ni-line patterns with varying line dimensions in the range of a few micrometers were manufactured by means of electrodeposition through a lithographically prepared mask. XRD studies revealed for Cu-lines a pronounced influence of the pattern geometry on crystallographic texture and peak broadening. Information on the strain distribution within individual Cu-lines was obtained from FEM.

INTRODUCTION

Electrochemical deposition has been successfully implemented in various microfabrication technologies [1-3]. For instance, plating through a mask results in free-standing patterns, which can either directly be used as metallic patterns in micro-systems or act as mould for the specially developed injection moulding process. The two most important materials in this field are Cu, which has become the dominating material for interconnects in integrated circuits in microelectronics, and Ni, a promising material to realize movable structures for micro-electro-mechanical applications. Successful miniaturization of the feature sizes in microelectronics, sensor and actuator technologies requires a thorough understanding of the correlation between manufacturing process and resulting properties of patterned electrodeposits.
The present paper reports about free-standing line patterns of electrodeposited Cu- and Ni-films. X-ray diffraction (XRD) studies were carried out in dependence on the geometry of the line patterns, in order to quantify crystallographic texture and to investigate the shape of diffraction peaks as a measure of lattice imperfections. Supplementary to XRD, finite element modeling (FEM) of the strain distribution within individual Cu-lines was performed.

EXPERIMENT AND SIMULATION

Electrochemical deposition

In order to ensure electrical conductivity during the deposition, a polycrystalline Au-layer (thickness 200 nm) was deposited on a glass wafer. UV-lithography (photo-resist thickness 6.5 μm) was carried out in clean room atmosphere in order to obtain a three-dimensional template, which was subsequently filled with Cu and Ni, respectively, by electrochemical deposition. Deposition parameters are summarized in Table 1. During deposition, a current thief was placed around the wafer in order to improve the current density distribution over the surface area.

Table 1: Parameters used for electrochemical deposition of Cu- and Ni-deposits ([a] PEG - polyethylene glycol, molar mass: 3400 g/mol, [b] MPSA - 3-mercapto-1-propanesulfonate).

	Cu-electrodeposits	Ni-electrodeposits
electrolyte (air agitation)	acidic Cu-bath: 0.56 mol/L $CuSO_4 \cdot 5H_2O$ + 1.4 mol/L H_2SO_4 + $1.1 \cdot 10^{-3}$ mol/L Cl^- + $8.8 \cdot 10^{-5}$ mol/L PEG^a + $1.5 \cdot 10^{-3}$ mol/L $MPSA^b$	Ni-sulphamate bath: 1.9 mol/L $Ni(NH_2SO_3)_2 \cdot 4H_2O$
temperature	room temperature	40 °C
average current density on wafer	400 A/m^2	50 A/m^2
substrate	PVD-Au-layer	PVD-Au-layer

The patterns consisted of parallel lines with varying line widths (W), line lengths (L) and interline distances (D) in the range of a few micrometers. Figure 1 shows the pattern geometry. For comparison with the line patterns continuous non-patterned films were deposited with identical deposition parameters (both for Cu- and Ni-deposits). The thickness of the electrodeposits amounted to about 5 µm.

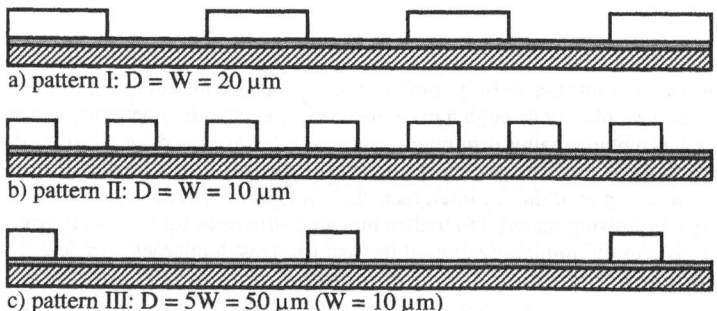

a) pattern I: D = W = 20 µm

b) pattern II: D = W = 10 µm

c) pattern III: D = 5W = 50 µm (W = 10 µm)

Figure 1: Dimensions of Cu- and Ni-line patterns. Here only some of the lines are shown; in reality, each of the patterns contains several hundreds of parallel lines: I) 675 lines, II) 1350 lines, III) 580 lines.

X-ray diffraction

X-ray measurements were performed with a powder diffractometer D8 Discover (BRUKER AXS), which is equipped with 1/4-circle Euler cradle. Quantitative texture analysis was performed by calculating the orientation distribution function (ODF) from experimental pole figures. With grazing incidence diffraction applying small incidence angles of 5°, the broadening of XRD-peaks obtained from (220) diffracting lattice planes of Cu and Ni, respectively, was investigated.

Finite Element Modeling

Finite element modeling was carried out for Cu-lines with an aspect ratio of W/d = 4 (d is film thickness), which corresponds to pattern I. A two-dimensional model was applied. The model bases on the following assumptions: i) plane strain condition, e.g. lines are considered infinitely long, ii) isotropic elastic properties of non-textured Cu-lines, iii) stiff (glass) substrate and iv) a constant tensile strain is imposed as boundary condition at the interface. The total thickness of the film is divided into 8 sublayers with equal thickness at various depths z.

RESULTS AND DISCUSSION

X-ray diffraction

For the Ni-deposits, ideal fibre textures consisted of multiple fibre axes: a <100> of moderate strength, a weaker <311> and an even weaker <111>, see Table 1. Results show almost no dependence of crystallographic texture on the geometry of Ni-line patterns and furthermore, no differences between Ni-patterns and corresponding continuous Ni-deposits were observed.

Table 2: Texture components and orientation densities of Ni-electrodeposits determined from local maxima in the calculated ODF. (*existing, but the local maxima can not be determined precisely because of overlap with broad <100>)

Ni on Au	Component (fibre axis ‖ surface normal)		
	<100>	<311>	<111>
Ni-pattern I	3.6	2.7	1.8
Ni-pattern II	3.6	2.6	2.0
Ni-pattern III	3.8	≈2.5*	0.9
Ni-no pattern	4.3	≈2.5*	1.9

XRD-peak profiles measured for the Ni-(220) diffracting lattice planes were identical for the various line patterns and the continuous film. Peak profiles are presented in Fig. 2; it is seen that no differences between the individual curves can be resolved.

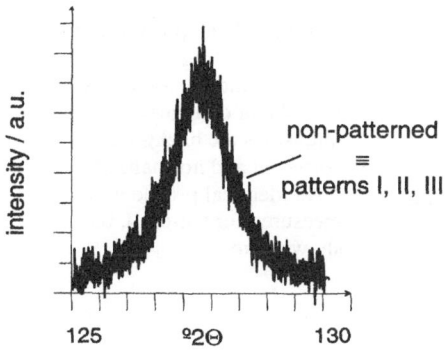

Figure 2: Peak profiles of Ni-(220) diffracting lattice planes for various pattern geometries. Peaks obtained for patterns I, II, III and non-patterned show identical profiles, i.e. they can not be distinguished here. For the sake of comparison, intensities are background corrected and normalized.

In contrast to Ni-electrodeposits, a strong dependence of crystallographic texture and XRD-peak broadening was observed for electrodeposited Cu-patterns, see Table 3. Patterns I and II, i.e. patterns with line dimensions of D = W, exhibited a very strong <111> and a weaker <511>. In contrast, for pattern III with increased distance between the lines, i.e. D = 5W, a relatively weak texture was observed consisting of multiple fibre axis <100>, <221>, <111>, <511> and <211>. For the continuous Cu-film, the texture consisted of a <100> as a major component, a <111> and a weak <211>.

Table 3: Texture components and orientation densities of Cu-electrodeposits determined from local maxima in the calculated ODF. ✔ The relatively broad <100> overlaps with the <511>, which can therefore not be quantified accurately.

Cu on Au	Component (fibre axis ‖ surface normal)				
	<111>	<100>	<511>	<221>	<211>
Cu-pattern I	17.2	-	3.8	-	-
Cu-pattern II	15.0	-	2.4	-	-
Cu-pattern III	2.3	2.9	✔	2.4	1.7
Cu-no pattern	2.6	5.8	-	1.3	-

Similarly to the observed dependence of crystallographic texture on the geometry of the Cu-line patterns, also XRD-peak profiles measured for the Cu-(220) diffracting lattice planes depend on the pattern dimensions, see Fig. 3. For patterns with line widths equal to the interline distances (D = W, pattern I and II) a relatively broad peak was measured indicating a high amount of lattice imperfections. They result either from a small size of coherently diffracting domains (crystallites) and/or a high amount of lattice distortion (microstrain); a separation of both effects was not carried out here. For pattern III containing Cu-lines with increased distances between neighboring lines (D = 5W) a very narrow peak, which even allows resolving the $K\alpha_2$-component, was measured. Similarly, such a small peak broadening was also observed for the non-patterned Cu-deposit.

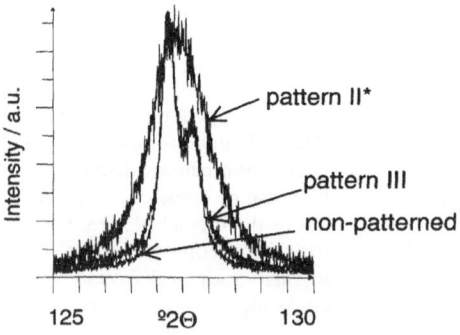

Figure 3: Peak profiles of Cu-(220) diffracting lattice planes for various pattern geometries. For the sake of comparison, intensities are background corrected and normalized. * An identical profile was measured for pattern I, but is not shown here.

Finite Element Modeling

Fig. 4 shows the strain field within an individual Cu-line as calculated by means of FEM in dependence on the distance to the surface for ψ-values up to 90° (the angle ψ defines the angle between the surface normal and direction vector n within the film).

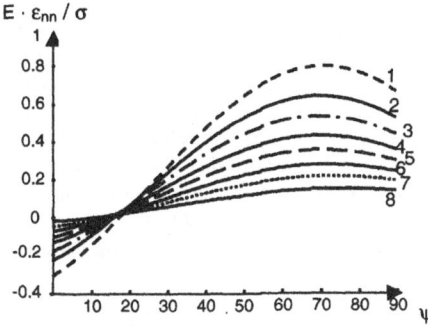

Figure 4: Finite element modeling of the depth dependence of strain (averaged over the line width) as a function of the inclination angle ψ. The different curves refer to different depth (8 – sublayer at surface, 1 – sublayer at interface).

While in Fig. 4 the strain is averaged over the width of a Cu-line, the actual distribution of strain within one Cu-line is seen in Fig. 5 for two selected ψ-angles ($\psi = 0°$, $\psi = 37.5°$). FEM calculations indicate fairly inhomogeneous strain distributions; both in depth (cf. Fig. 4) as well as over the Cu-line width (cf. Fig. 5).

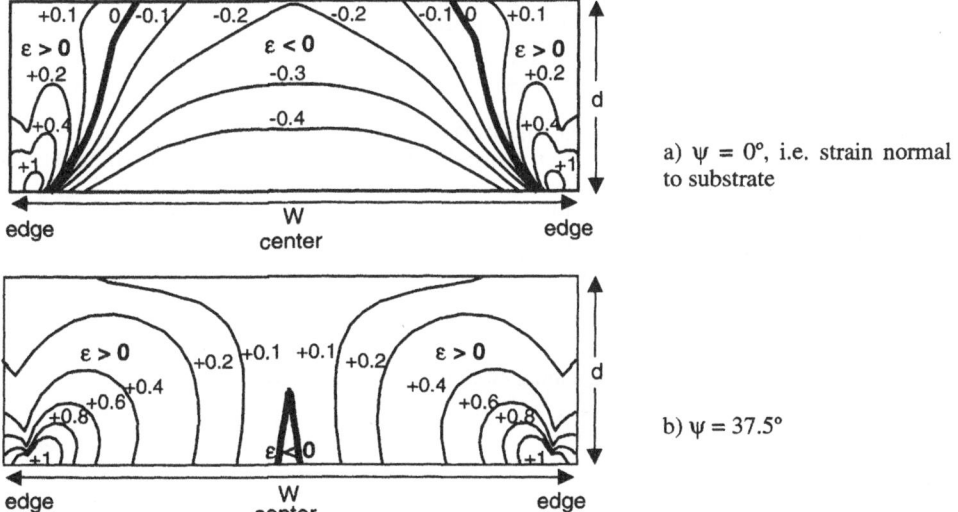

a) $\psi = 0°$, i.e. strain normal to substrate

b) $\psi = 37.5°$

Figure 5: Distribution of strain over the width of a Cu-line with aspect ratio W/d = 4.

CONCLUSIONS

A pattern dependence of microstructure and related properties seems to be a characteristic feature only for the electrodeposited Cu-lines, whereas for Ni-lines no dependence on the line dimensions and furthermore, no differences to non-patterned Ni-films were observed. Self-annealing effects of Cu-electrodeposits (see for example [4]) are expected to play an important role for the observed pattern dependence of both the crystallographic texture and the peak broadening. The onset of self-annealing is evident from the observed minor texture components <511> and <221>, since they represent recrystallization twins of <111> and <100> oriented grains, respectively [5]. For those Cu-deposits, which have predominantly a <100> fibre texture (<100> is known as typical recrystallization texture of Cu [4,5]), simultaneously a very small peak broadening was measured. Both results indicate a further progress of self-annealing for Cu-lines with D = 5W compared to Cu-lines with D = W. Alternative interpretations are presently considered, for example a pattern dependence of the growth rates during electrodeposition and mechanical interactions between the strain fields of neighboring lines in the patterns [6].

It is shown that the strain distribution within free-standing electrodeposited lines can be gathered from FEM. FEM performed on single lines can supplement XRD studies, which always average over many hundreds of identical lines.

ACKNOWLEDGEMENT

The authors gratefully acknowledge the Danish Technical Research Council (STVF) for financial support (grant 2020-00-0016) and the Department of Micro- and Nanotechnology at DTU for carrying out the process of photo-lithography.

REFERENCES

[1]   M. Datta, D. Landolt, Electrochimica Acta **45**, 2535 (2000).
[2]   P.C. Andricacos, C. Uzoh, J.O. Dukovic, J. Horkans, H. Deligianni, IBM J. Res. Develop. **42**, 567 (1998).
[3]   T.E. Buchheit, A.A. LaVan, J.R. Michael, T.R. Christenson, S.D. Leith, Metallurgical and Materials Transactions **33A**, 539 (2002).
[4]   I.V. Tomov, D.S. Stoychev, I.B. Vitanova, Journal of Applied Electrochemistry **15**, 887 (1985).
[5]   K. Pantleon, J.A.D. Jensen, M.A.J. Somers, Journal of The Electrochemical Society **151**, C45 (2004).
[6]   K. Pantleon, M.A.J. Somers, submitted to Acta materialia (2003).

Mat. Res. Soc. Symp. Proc. Vol. 812 © 2004 Materials Research Society   F3.21

Pulsed MOCVD of Cu Seed Layer using a (Hfac)Cu(3,3-Dimethyl-1-Butene) Source Plus H₂ Reactant

Jaebum Park, Heejung Yang, and Jaegab Lee

School of Advanced Materials Engineering, Kookmin University, Seoul 136-702, Korea
E-mail: lgab@kookmin.ac.kr

Abstract

Pulsed metalorganic chemical vapor deposition (MOCVD) of conformal copper seed layers, for the electrodeposition Cu films, has been achieved by an alternating supply of a Cu(I) source and H_2 reactant at the deposition temperatures of 50 − 100°C. The Cu thickness increased proportionally to the number of cycle, and the growth rate was in the range of 3.5 to 8.2 Å /cycle, showing the ability to control the nano-scale thickness. As-deposited films show highly smooth surfaces even for more than 100nm. In addition, about a 90% step coverage was obtained inside trenches, with an aspect ratio greater than 30:1. H_2, introduced as a reactant gas, can play an active role in achieving highly conformal coatings, with increased grain sizes.

Introduction

Al-based metallization has been rapidly replaced with Cu as the dimension of microelectronic devices has been scaled down to below 100-nm.[1-7] Cu electroplating is the most common technology employed for void-free filling of via and hole structures with high growth rates in the damascene structure.[8,9] However, this method requires the conformal deposition of a thin Cu seed layer prior to the electroplating of a thicker Cu layer. The conventional physical vapor deposition encounters several problems, such as poor conformal coverage, especially on the sidewalls and the overhang at the opening of vias.[2,3,10] Therefore, current research appears to focus on developing an alterative deposition technique for the seed layer, which must be continuous, smooth, conformal, and thin enough for a critical dimension of 100-nm and below.[11] Furthermore, nano-scale thickness control is needed for the design of the interface properties between the Cu seed layer and the barrier, which significantly affects reliability and adhesion property.[12] Recently, in order to control the nano-scale deposition, a copper atomic layer deposition (ALD) process has been studied using an organometallic Cu(II) source and H_2 reducing agents, showing various degree of success in producing highly conformal layers of Cu thin films with atomic thickness control.[13-18] Compared with ALD Cu using Cu(II), pulsed MOCVD by a Cu(I) source provides a wide range of growth rate, depending on the deposition temperature, and the amount of Cu precursor introduced into the chamber.[19] Moreover, the additional reactants pulsing can influence the deposition characteristics such as the surface mobility of the Cu adatoms and the additional reaction of the adsorbed Cu precursor on the Cu surface, and thus resulting in the enhanced film quality and conformal coatings.

In this study, the pulsed MOCVD of a Cu seed layer, using a Cu(I) source and H_2 reactant, have been investigated. H_2 was introduced, followed by Ar purging, to promote the decomposition of the (hfac)Cu(DMB) adsorbed onto the substrate, thus improving the quality of the Cu layer. Moreover, the use of H_2 pulse may provide the clean Cu surface, and thus allowing the enhanced surface mobility of Cu adatoms, and the increased grain-size and improved conformality of Cu films over high aspect ratio contacts and vias.

Experimental

MOCVD TiN (-30nm) was deposited using tetrakis-dimethyl-amino titanium (TDMAT) as a precursor on either a 4 in. (100), p-type Si wafer, or patterned pieces on it, at 300 °C, followed by pulsed MOCVD Cu deposition by alternating the supply of the (hexafluoroacetylacetonate) $Cu^{(I)}$ (3,3-dimetyl-1-butene), [(hfac)Cu(DMB)] and H_2 reactant. The MOCVD TiN system consists of a cold-wall type reactor in which a 4 in. Si wafer is placed on a resistive heater and the TiN precursor introduced through a shower plate. The pulsed MOCVD system consists of a cold-wall low-pressure CVD chamber (4 cm in diameter and 2 cm high) in which the substrate is radiatively heated with a 500W halogen lamp through the bottom quartz window and the in-situ reflectivity of the growing Cu film is measured using an He-Ne laser beam ($\lambda = 632.8$ nm). The substrate is a piece of Si wafer (1.5x1.5 cm^2), on which a thin MOCVD TiN layer was deposited. The experimental variables included the pulse time, temperature, and pressure. A typical pulsed deposition cycle consisted of a 3 sec (hfac)Cu(DMB) cycle, then a 6 sec Ar purge, followed by 3 sec H_2 reactant pulse, and finally, a 6 sec Ar purge. The Cu source was heated at 30 °C, with the gas delivery line maintained at 40 °C. The experimental conditions for the deposition of the TiN and Cu films are listed in Table 1.

Table 1. Deposition conditions of MOCVD TiN and pulsed MOCVD Cu.

Conditions	MOCVD TiN	Pulsed MOCVD Cu
Precursor	TDMAT	(hfac) Cu (DMB)
Deposition Temperature (°C)	300	50 — 250
Carrier gas flow rate (Ar, sccm)	200	10
Carrier gas flow rate (H_2, sccm)	·	13
Working pressure (torr)	1	0.4
Bubbler Temperature (°C)	45	30
Duration Time	25 sec	50 — 200 cycles

The resistivity and Cu film thickness were measured by a four point probe and Rutherford backscattering spectroscopy (RBS), respectively. The conformal coatings of the Cu films, over high aspect ratio trenches, were identified by field emission scanning electron microscopy (FESEM). The surface roughness of the films was measured by atomic force microscopy (AFM).

Results and Discussion

The typical deposition cycle employed in this experiment was determined by examing the effects of the variables such as Cu and H_2 pulse time, substrate temperature on the growth rate and resisitivity.

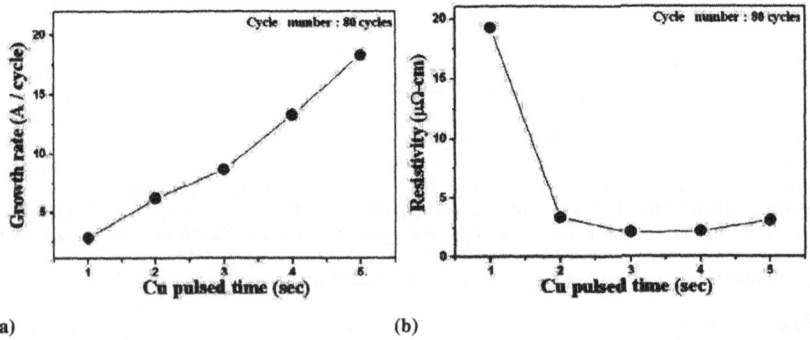

(a)   (b)

Figure 1. The effects of Cu pulse time on (a) the growth rates and (b) resistivies of Cu deposited on 30-nm thick MOCVD TiN film at 100°C.

Fig. 1 shows the effects of Cu pulse time on the growth rates of Cu deposited on 30-nm thick MOCVD TiN coated Si piece, and the corresponding resistivities. The deposition condition for pulsed MOCVD Cu is as follows; the substrate temperature is 100 °C, the process pressure is 0.4 torr, Ar and H_2 pulse time are 6 sec and 3 sec, respectively. As the Cu pulse time increases from 1 to 5sec , the growth rate linearly increases from 2.8 to 18.3 Å / cycle and the resistivity drops rapidly from 19.2 to 3.1 $\mu\Omega$-cm and then slowly decreases to the minimum value of 2.1 $\mu\Omega$-cm. The dependence of Cu growth rate and resistivity on H_2 pulse time also has been investigated, as shown in Fig, 2.

As the H_2 pulse time increases, the growth rate continues to decrease and then tends to remain constant. In addition, the resistivity also decreases with increasing H_2 pulse time.

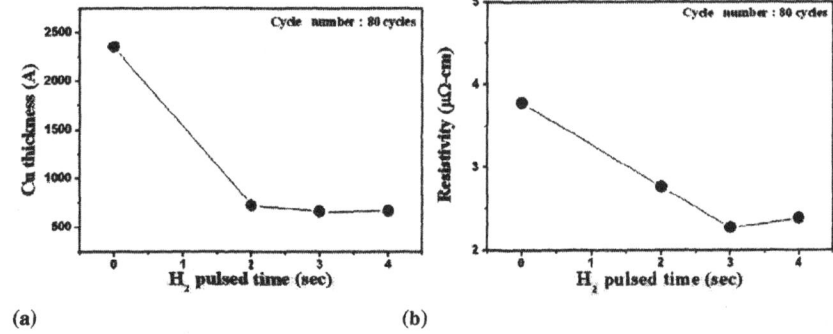

(a)   (b)

Figure 2. The effects of H_2 pulse time on (a) the growth rates and (b) resistivies of Cu deposited on 30-nm thick MOCVD TiN film at 100°C.

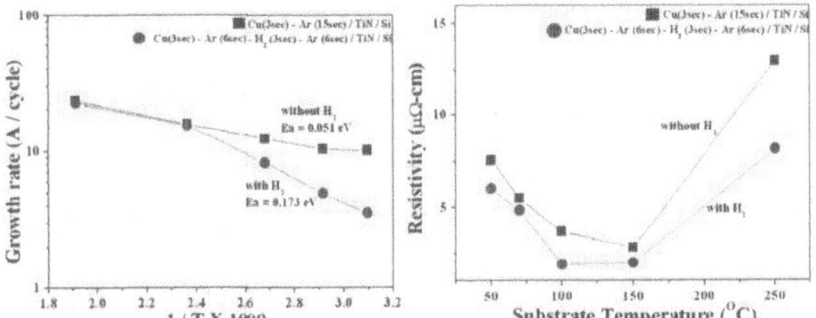

Figure 3. The temperature dependence of the growth rate of pulsed MOCVD Cu films with H_2 (or without H_2).

Figure 4. The H_2 effects on resistivity of pulsed MOCVD Cu films at at various temperatures.

Pulsed MOCVD of Cu, at 50 − 100 °C, for the cycle of 3 sec (hfac)Cu(DMB) pulse following by 15 sec Ar purging, produced a growth rate in the range of 10 to 11.2 Å/cycle, and shown as a weak function of the temperature(Fig. 3). It is also noted that the 3 sec H_2 pulse added to the deposition cycle tends to decrease the Cu growth rate in the temperature range of 50 to 150°C. Also, the Arrhenius plot shows the higher activation energy of 0.18 eV with H_2 pulse obtained in the low temperature regime, possibly indicating the added hydrogen could affect the adsorption of the (hfac)Cu(DMB) precursor. In addition, the H_2 cooling effects on the substrate temperature in the halogen lamp heating system and the atomic hydrogen reacting with the incompletely decomposed Cu precursor could be contributors to the decreased growth rate of Cu at the low temperature.

The dependence of resistivity on the substrate temperature and H_2 pulsing is shown in Fig. 4. The added H_2 pulse effectively reduces the resistivity of the Cu film to 2.0 $\mu\Omega$-cm. Fig. 5 shows the Cu thickness linearly increased, with increasing the number of cycles during the pulsed MOCVD of Cu at 100 °C and revealing the growth rate of about 9.3 Å/cycle. Further, the RMS (root mean square) of the pulsed MOCVD Cu films exhibited the significantly low value of about 28 Å for 140 nm-thick Cu film, which was comparable with that of a sputtered Cu film. This might be a result of 2-dimensional growth occurring in H_2-pulsed MOCVD Cu.

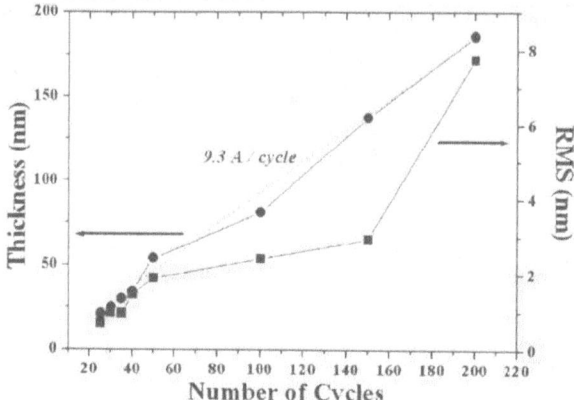

Figure 5. Thickness of pulsed MOCVD Cu films as a function of the number of deposition cycles on a 30-nm thick TiN at 100°C , and the corresponding rms roughness.

The H_2 pulse, following Ar purging, can dissociatively adsorb onto the Cu surface, promoting the dissociation of the Cu(I)-hfac, via the reaction below: [20, 21]

$$2[Cu(I)\text{-}hfac]_{(ad)} + [H]_{(ad)} \rightarrow 2Cu_{(s)} + H(hfac)_{2(g)} \qquad [1]$$

The H adsorbed onto the Cu surface react with a Cu(I)-hfac surface intermediate, which is difficult with dissociation via the bimolecular disproportionation reaction at temperatures below about 100 °C, and producing contamination free Cu metal.

Fig. 6 shows the surface morphology of about 100 nm-thick Cu films deposited at 100°C with and without H_2 pulse, revealing the difference in grain size and surface roughness. MOCVD Cu films of the similar thickness were grown at 100°C and the pressure of 0.4 torr with (or without) H_2, and the surface morphology of the films was compared. The results show the added H_2 tends to increase the grain size and improving the surface morphology. In addition, the comparison of the pulsed Cu and MOCVD Cu films reveals the significantly increased grain size in the pulsed Cu films, indicating the pulsed MOCVD facilitates a lateral grain growth, and thus resulting in larger grains. Furthermore, the Cu films deposited with H_2 pulse exhibit much larger grain sizes of about 580 nm, compared with those of about 250 nm without H_2 pulse. Moreover, the much smoother surface can be seen in H_2 pulsed Cu films. This may indicate that the adsorbed hydrogen clean the Cu surface via the reaction of the adsorbed H and the surface intermediates, which allows the fast path for the surface diffusion of the Cu adatoms and promoting the two-dimensional growth and larger grain growth.

Fig. 7 shows a FESEM image of a 30nm thick Cu film, grown with H_2, on a 30nm-thick MOCVD TiN coated trench substrate. The bottom of the MOCVD TiN coated trench was

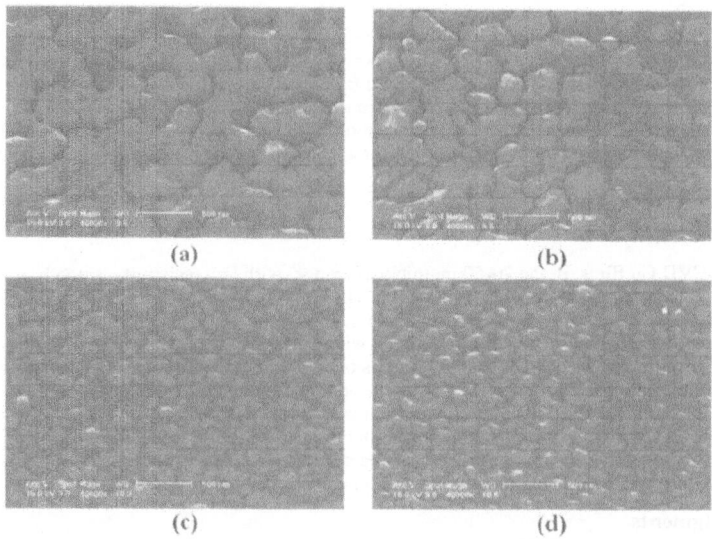

(a)

(b)

(c)

(d)

Figure 6. Plan-view SEM of about 100 nm-thick Cu films deposited at 100°C by using (a) pulsed MOCVD with H_2 pulse, (b) pulsed MOCVD without H_2 pulse, (c) MOCVD with H_2, and (d) MOCVD without H_2.

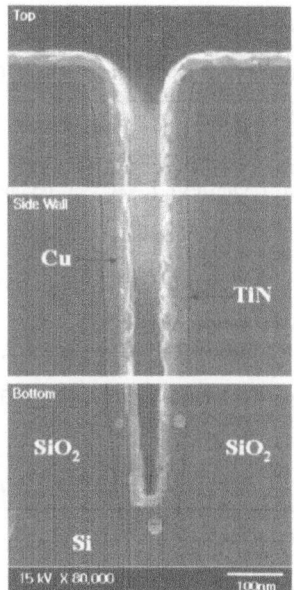

Figure 7. Conformal coverage of a Cu film, on a trench pattern (A/R 22: 1), covered with a TiN film, by pulsed MOCVD Cu and H_2.

about 85nm wide and 2.5μm-deep, which was approximately equal to an aspect ratio of 30. The Cu film, deposited over the extremely high aspect ratio trench, revealed a very smooth surface, with about a 90 % step coverage. In contrast, the Cu film deposited without H_2 showed about a 54 % step coverage over the MOCVD TiN coated trenches. This increased step coverage with H_2 pulse may be attributed to the enhanced surface mobility of Cu adatoms on the clean Cu surface as a result of the removal of the adsorbed (hfac)Cu(DMB) intermediates by the atomic hydrogen cleaning.

Conclusions

Pulsed MOCVD Cu films, in the 20-50nm thickness range, with low resistivity, smooth, a highly conformal coating and good adhesion to the TiN, were obtained. The addition of H_2 can help the dissociation of the Cu(I)-hfac intermediate on the Cu surface, thus decreasing the impurity content, and achieving the low resistivity, with excellent adhesion. In addition, this allows for highly conformal depositions of Cu films over a high aspect ratio of 30:1, possibly resulting from the enhanced surface diffusion of the adatoms. The linear relationship of film thickness to the number of deposition cycles, which is the typical advantage of ALD, was accomplished. But, the growth rate is higher than that of the ALD process using a Cu(II) precursor . Consequently, the present process can be considered as one of the solutions for the deposition of very thin seed layer at high aspect ratio contacts and vias.

Acknowledgments

The authors gratefully acknowledge the financial support through the Center for Nanostructured Materials Technology, by the Korean Ministry of Science.

References

1. D. Edelstein, et al, 1997 IEEE Int. Electron Devices Meet. Digest, p. 773(1997)
2. N. Awaya and Y. Arita, J. Elctron. Mater. 21, 959 (1992).
3. A. Jain, T. Kodas, R. Jairath, and M. J. Hampden-Smith, J. Vac. Sci. Technol. B 11,2107 (1993).
4. J. Lin and M. Chen, Jpn. J. Appl. Phys., Part 1 38, 4863 (1999).
5. S. P. Murarka and S. Hymes, Solid State Mater. Sci. 20, 87 (1995).
6. Y. J. Park, V. K. Andleigh, and C. V. Thompson, J. Appl. Phys. 85, 3546 (1999).
7. C. Whitman, M. M. Moslehi, A. Paranjpe, L. Velo, and T. Omstead, J. Vac. Sci. Technol. A 17, 1893 (1999).
8. V. M. Dubin, et al, Proc. of the 1998 Advanced Metallization Conference for ULSI Applications, p. 405, 1998.
9. P. C. Andricacos, C. Uzoh, J. Dukovic, J. Horkans, and H. Deligianni, IBM J. Res. Dev. 42, 567 (1998).
10. A. F. Burnett and J. M. Chech, J. Vac. Sci. Technol. A 11, 2970 (1993).
11. W. H. Lee, Y. K. Ko, I. J. Byun, B. S. Seo, J. G. Lee, P. J. Reucroft, J. U. Lee, and J. Y. Lee, J. Vac. Sci. Technol. A. 19(6), 2974 (2001).
12. C-K. Hu, L. Gignac, S. G. Malhotra, and R. Rosenberg, Appl. Phys. Lett. 78, 904 (2001).
13. M. Juppo, M. Ritala, M. Leskela, J. Vac. Sci. Technol. A. 15 2330 (1997).
14. P. Martensson and J. O. Carlsson, *Chem. Vap. Deposition*, 3. 45 (1997).
15. P. Martensson and J. O. Carlsson, J. Electrochem. Soc., 145, 2926 (1998).
16. M. Juppo, M. Vehkamaki, M. Ritala, M. Leskela, J. Vac. Sci. Technol. A 16 2845 (1998).
17. R. Solanki and B. Pathangey, Elctrochem. Solid-State Lett., 3, 479 (2000).
18. B. S. Lim, A. Rahtu, and R. G. Gordon, Nature Materials Vol. 2, 749 (2003).
19. K. Kim and K. Yong, Electrochem. Solid-State Lett., 6, 106 (2003).
20. S. L. Cohen, M. Liehr, and S. Kasi, Appl. Phys. Lett., 60. 1585 (1992).
21. S.W.Rhee, S.W.Kang, and S.H. Han, Electrochem, Solid-State Lett.,3, 135 (2000).

Mat. Res. Soc. Symp. Proc. Vol. 812 © 2004 Materials Research Society F3.23

Preparation and Characterization of Copper Film on Plastic Substrate by ECR-MOCVD Coupled with a DC Bias

Bup Ju Jeon and Joong Kee Lee
Eco-Nano Research Center, Korea Institute of Science and Technology, P.O.Box 131,
Cheongyang Seoul 130-650, Korea

ABSTRACT

Copper films were prepared at room temperature under a $Cu(hfac)_2$-Ar-H_2 atmosphere in order to obtain metallized plastics by using ECR-MOCVD (Electron Cyclotron Resonance Metal Organic Chemical Vapor Deposition) coupled with a DC bias system. Structural analysis of the films by ECR showed that fine copper grains were embedded in an amorphous plastic matrix with good adherence. Considering the AES result of the film prepared by ECR-CVD, we construe that the copper film is chemically bonded with a plastic substrate. The delamination force determined by the nano-scratch tester® showed in the range from 10 to 20mN.

INTRODUCTION

The advantages of both metal and plastic films have been incorporated into metallized film, makes application in many industrial fields, for example, microelectronic packing, flexible printed circuits, plastic solar cell, gas barrier and electromagnetic interference (EMI) shielding. The deposition of a metal film on plastic, particularly copper, is of considerable interest due to its low resistivity and resistance to electromigration. The employed conventional methods for the deposition of copper film on plastics are electrodeposition and evaporation (thermal and e-beam). Electrolytic and electroless process require extensive substrate preparation before deposition be carried out. Moreover, the use of hazardous chemicals and precious metal catalysts are necessary for these processes to obtain satisfactory conductor characteristics. Evaporation is technically difficult to get good selectivity and characteristics due to poor adhesion, substrate deformation and an aging effect.

The adhesion of copper film to the plastic substrate is the most important and basic property in any application above mentioned. Thus, chemical vapor deposition processes for the preparation of copper films are of considerable significance since continuous deposition of large areas can be easily achieved with good adherence. However, the chemical vapor deposition is typically carried out at high substrate temperatures. Therefore it is not suitable for polymeric and other temperature sensitive substrates. Operating temperatures in the range of 150 to 200°C have to be used to prepared copper thin films from the organometallic precursors. The deposition of thin copper film on plastic substrates by the MOCVD (Metallic Organic Chemical Vapor Deposition) method at room temperature has never been tried.

Recently, we found that MOCVD is possible at room temperature when periodic negative voltage is applied under ECR (Electron Cyclotron Resonance) plasma system. The periodic negative voltage induces ions and radicals to have nucleation reaction on the surface of the substrate. The high efficiency in exciting the reactants in ECR plasma coupled with periodic negative voltage allows the deposition of films at room temperature. In the present study structural and chemical analyses of the copper films were carried out by comparing those of magnetron sputtering, and their delamination force were determined as a function of H_2/Ar mole ratio and DC voltage.

EXPERIMENTAL

An Astex-1000 ECR plasma generator, a CVD reactor and a high vacuum system were employed to carry out the experiments. It consisted of two chambers, the plasma chamber and the deposition chamber. The gases are introduced through two separate inlets; hydrogen and argon molecules are introduced into the upper plasma chamber, while organometallic precursor with carrier gas is fed into the lower deposition chamber. The DC bias is connected to the grid near the substrate, in order to induce metal ion and radicals on the surface of non-conductive polymer substrate. The copper films were prepared at room temperature by ECR-MOCVD under the following conditions: working pressure of 25 mTorr, bubbler pressure of 200 Torr, H_2/Ar ratio of 0-0.33, total flow rate of 100sccm, negative DC bias voltage of 700-1100V, upper and lower magnetic current of 120 and 170A, respectively, and microwave power of 700 W. Cu(hfac)$_2$ (hfac: 1,1,1,5,5,5,-hexafluoro-2,4-pentandione) with purity 99.9% and hydrogen were used as a source of copper and reactive gas, respectively. Polyethylene terephthalate(PET) sheet with 5x5cm was used as substrate. Scratch testing of the deposited Cu/C:H films were measured in accordance with Nano scratch tester(NTS, CSM Instruments).

In order to compare the characteristics of ECR-CVD with those of physical vapor deposition, a magnetron sputtering system was employed. The films were deposited in a cylinder type reactor(D400mm×H300mm) with an rf planar magnetron facing down with the substrate placed 10cm from the target. A copper target is used as the metal source and Ar is fed as carrier gas. The copper target was placed on the cathode and connected to an 13.56 MHz radio frequency with matching network and power source of max. 1kW. Thin films were prepared by changing the rf power of sputter from 500W after maintaining flow 30sccm of argon gas with a mass flow controller. A constant working pressure of 20mTorr was maintained throughout

RESULTS AND DISUSSION

Figure 1(a) shows that typical variation of Cu, C and O Auger signals versus sputtering time for the specimen prepared by ECR-CVD. In the specimen of copper film from ECR-CVD, AES depth profile shows the formation of homogeneous copper layer with average composition about 80% at Cu and 20%. Oxygen exists only on the surface, which shows the oxidation at the film surface. Fluorine was not found within the film, but fluorine is a component of Cu(hafc)$_2$, copper precursor. Those results clearly show that the byproduct incorporation in the film is negligible. By observing the chemical shift of the Auger carbon peak at the beginning of the profiles of Figure 1(b), we can construe that some of carbon converted to carbide form due to chemical bonding and organic carbons network was formed in the deposited film. On the other hand, the copper film prepared by a magnetron sputtering shows pure copper content with clear interfacial layer on the plastic substrate as be shown in Figure 2.

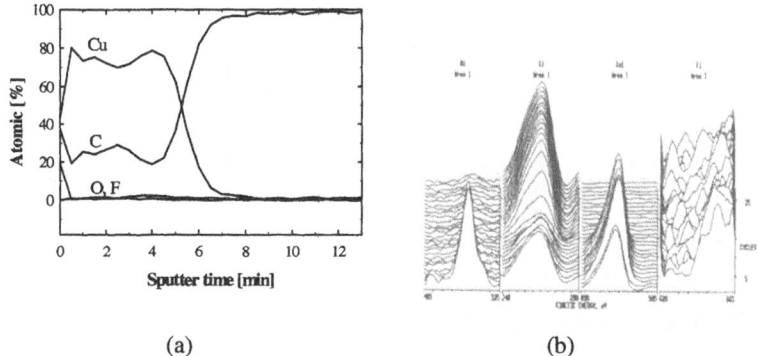

(a)   (b)

Figure 1. AES (a) concentration depth profile with etching time and (b) O1s, C1s and Cu1s peak through depth profile of the Cu layer structure prepared by ECR-MOCVD.

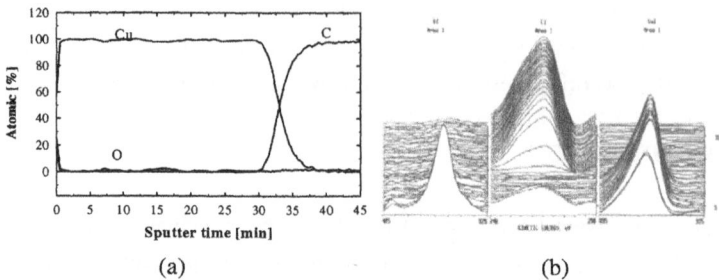

(a)   (b)

Figure 2. AES (a) concentration depth profile with etching time and (b) O1s, C1s and Cu1s
peak through depth profile of the Cu layer structure prepared by magnetron sputtering.

Cross-sectional TEM micrograph was employed to investigate the interface morphologies of
samples prepared by ECR-CVD and magnetron sputtering, respectively. As shown in Figure 3(a),
the copper film prepared by ECR-CVD has very little void between columns and has tightly
packed columns. Also, the copper atoms be embedded into the polymer just below the surface
during the growth process due to the high ion energy. By contrast, much void with a rather
smooth interface between the film and plastic substrate was observed for sample prepared by a
magnetron sputtering (Fig.3(b)). The film prepared by the magnetron sputtering looks like a
columnar shape.

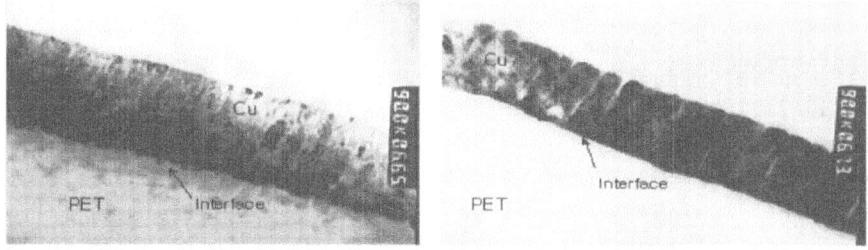

Figure 3. Cross-sectional TEM images at Cu/PET interface of deposited Cu film by (a) ECR-
MOCVD and (b) magnetron sputtering.

Table 1 shows the test conditions for nano-scratch tester of copper films. For scratches with
larger scratch widths, this is the most reliable method to detect coating damage. This technique is
able to differentiate between cohesive failure within the coating and adhesive failure at the
interface of the coating-substrate system. The scratch test method consists of the generation of
scratches with a spherical stylus(tip radiu 2 μm), which is drawn at a constant speed across the
coating substrate to be tested, under progressive loading at a fixed rate.

Table 1. Test conditions for nano-scratch tester

Load type	Progressive
Initial load	0.1 mN
Final type	50 mN
Loading rate	2994 mN/min
Scanning load	0.1 mN
Speed	355.28 mm/min
Scratch length	6 mm
Indenter	Rockwell C diamond, tip radius 2 μm
Cantilever	ST-008

Figure 4 shows the depth curves (penetration depth Pd and residual depth Rd) and force curves (normal force Fn and tangential force Ft) plotted against nominal force (Fn) and scratch length. The penetration depth corresponds to the depth measured during the scratch minus the depth measured in the presacn. The residual depth is the difference between the depth measured during postscan and prescan. For progressive loading, the critical load (Lc) is defined as the smallest load at which a recognizable failure occurs. As can be seen in Figure 4(a), the determined critical load was 12mN. Figure 4(b) shows the image of scratch for the copper film prepared by ECR-MOCVD.

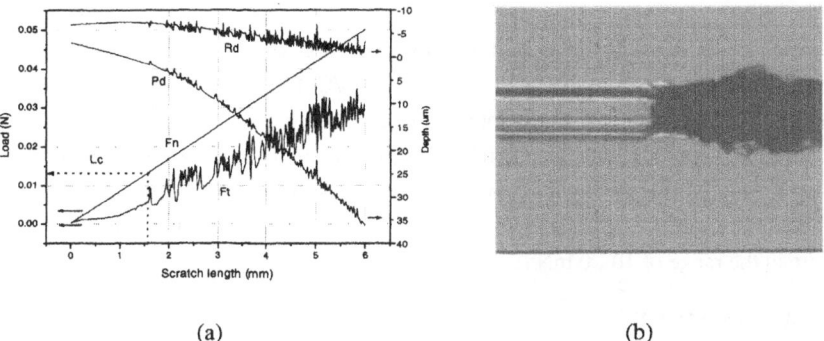

(a)   (b)

Figure 4. Nano-scratch test result of (a) the adhesive load and (b) image of scratch pattern for the film prepared by ECR-MOCVD. (Pd : penetration depth, Rd : residual depth, Fn : normal force, Ft : tangential force, Lc : Critical load)

Figure 5 shows delamination force from critical load of copper films deposited on PET by ECR-CVD as function of H_2/Ar ratio and DC bias. The difference in delamination for the copper films was insignificant with H_2/Ar ratio and DC bias, and the values are in the range of 10-20mN. However, as shown in Figure 5(b), the highest delamination force was observed at 0.25 H_2/Ar ratio.

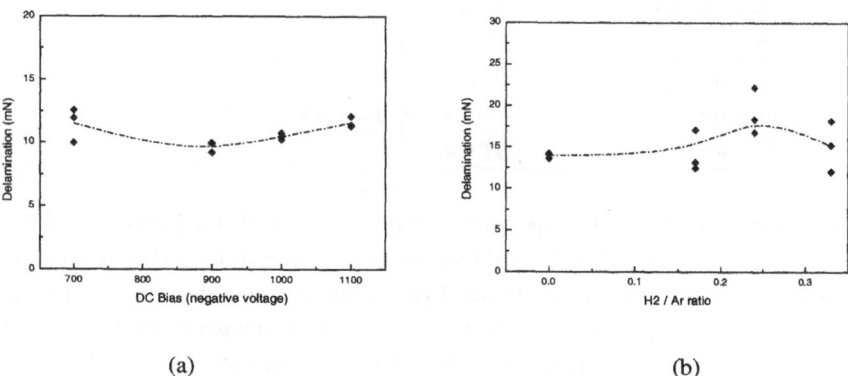

(a)   (b)

Figure 5. The delaminate variation of copper films deposited on PET as a function of (a) (-)DC bias voltage and (b) H_2/Ar mole ratio.

CONCLUSION

Combination of ECR plasma with high-density ionization and a periodic negative voltage of optimum frequency makes it possible to deposit Cu/C films on the polymer substrates by MOCVD at room temperature. It was found through structural and chemical analyses of the Cu films that fine copper grains were embedded in an amorphous plastic matrix with good adherence possibly from chemical boning with a plastic substrate. The adhesion of copper layer on PET was not changed significantly by DC bias voltage and H_2/Ar ratio, and the delamination values are in the range of 10-20 mN .

ACKNOWLEDGMENT

Research was supported by DUTP (Dual Use Technology Project) under contract number 01-IT-MP-04.

Mat. Res. Soc. Symp. Proc. Vol. 812 © 2004 Materials Research Society   F3.24

Thermal Stability and Electrical Properties of Ag(Al) Metallization

Hyunchul C. Kim[1], N. David Theodore[2], James W. Mayer[1], and Terry L. Alford[1]

[1]Department of Chemical and Materials Engineering, Arizona State University, Tempe, AZ 85287, U.S.A.
[2]Digital DNA™ Labs., Motorola Inc., 2100 E. Elliot Rd. MD-EL622, Tempe, AZ, 85284, U.S.A.

ABSTRACT

The thermal stability and electrical resistivity of Ag(Al) alloy thin films on SiO_2 are investigated and compared to pure Ag thin films by performing various analyses: Rutherford backscattering spectrometry (RBS), X-ray diffractometry (XRD), transmission electron microscopy (TEM), and four-point probe. The susceptibility to agglomeration of Ag on SiO_2 layer is a drawback of Ag metallization. Ag(Al) thin films show good thermal stability on SiO_2 layer without any diffusion barrier. The films are stable up to 600 °C for 1 hour in vacuum. Electrical resistivity of as-deposited Ag (5 at % Al) thin film is slightly higher than that of pure Ag thin film. However, the resistivity of Ag(Al) samples annealed at high temperatures (up to 600 °C for 1 hour in vacuum) remains constant due to the improvement of thermal stability (large reduction of agglomeration). This finding can impact metallization for thin film transistors (TFT) for displays, including flexible displays, and high-speed electronics due to lower resistivity value compared to Cu thin film.

INTRODUCTION

Silver has been considered as a potential interconnect material for ultralarge-scale integration (ULSI) technology due to its lower bulk electrical resistivity (1.57 μΩ-cm at room temperature) when compared with other interconnect materials (Al 2.7 μΩ-cm and Cu 1.7 μΩ-cm)[1-3] since lower resistivity interconnects can reduce RC delays and power consumption. Also, silver has been attractive for display technologies due to low absorption in the visible ranges and low electrical resistivity[4,5].

However, it has been reported that Ag film on most substrates, including SiO_2, is reshaped and disrupted at high temperature. Eventually, the electrical resistance of the Ag film is increased such that the Ag's function of a conductor is destroyed[1]. This phenomenon is called agglomeration of polycrystalline films. Agglomeration of thin films is a mass transport process that occurs at high temperatures in order to reduce the total system energy[6,7]. This phenomenon is initiated by grain boundary grooving and is followed by void, hillock, and island formation[8]. Changes in surface morphology of thin films increase electrical resistivity of thin films by enhancing scattering of conduction electrons through the films[1].

Therefore, the enhancement of thermal stability of Ag films by minimizing agglomeration has been considered as a reliability issue for Ag metallization. Several techniques have been suggested for the improvement of thermal stability of Ag film, including plasma treatment of low-k dielectrics[3] and the use of diffusion barriers[9]. Ag can be a good candidate of future interconnects if the drawback of weak thermal stability resulting from thin film agglomeration at high temperature, is removed without any additional processes or diffusion barrier.

In this study, the use of Ag thin film containing a small amount of Al is investigated as a method to improve the thermal stability of pure Ag on SiO_2 at high temperatures, as analyzed by various techniques.

EXPERIMENT

Pure Ag and Ag(Al) alloy thin films were deposited on SiO_2 substrates using electron-beam evaporation. Base pressure and operation pressure were 5×10^{-7} and 4×10^{-6} Torr, respectively. In order to fabricate the Ag(Al) thin film, Ag(Al) alloy targets for electron-beam evaporation were prepared by mixing pure Ag slug with pure Al slug targets (90 atomic % Ag - 10 atomic % Al and 95 atomic % Ag - 5 atomic % Al) and melted completely by electron beam in the evaporator. Ag(Al) thin films were deposited onto substrates with the same thickness as the Ag film to avoid thickness dependence on thin film agglomeration[1]. Thermal anneals in vacuum ($\sim 5 \times 10^{-8}$ Torr) were performed for 1 hour for investigation of the thermal stability of Ag and Ag(Al) thin films on SiO_2 substrates. The thermal stability of these thin films was characterized by Rutherford backscattering spectrometry (RBS), X-ray diffractometry (XRD), and in-line four-point probe analysis. In addition, transmission electron microscopy (TEM) was used to evaluate agglomeration, and to observe microstructure changes in the films as influenced by Al alloying effects.

RESULT & DISCUSSION

Figure 1 shows the RBS spectra for Ag(5 at. % Al) films, as-deposited and annealed for 1 hour in vacuum on oxidized silicon. The anneal temperature ranged from 400 °C to 600 °C.

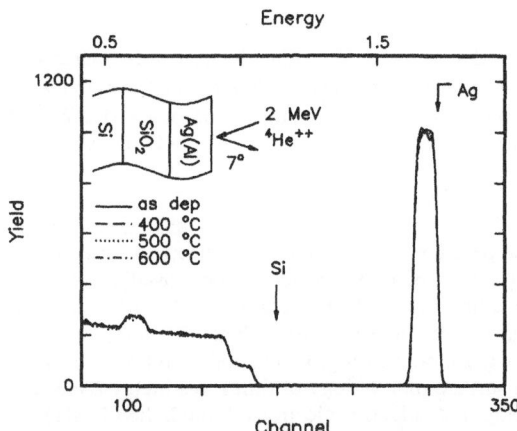

Figure 1. RBS spectra of as-deposited and annealed Ag (5 at. % Al) thin films on SiO_2 substrate. Vacuum anneals were performed for 1 hour at 400 °C, 500 °C, and 600 °C.

From figure 1, the intensity and width of the Ag peak from as-deposited and annealed Ag (5 at. % Al) thin films are the same (unchanged). Also, the step in the spectrum arising from the SiO_2 layer is stable up to 600 °C. These suggest that no significant interdiffusion occurs between Ag and SiO_2 layer and there is no significant change in surface morphology caused by agglomeration at high temperature. Al was not detected in the RBS data, as shown in figure 1, due to the very small amount of Al in the Ag film and low detection limits of the RBS analysis. However, the height of the Ag peak in figure 1 has decreased slightly compared to the pure Ag peak obtained from RUMP[10] simulation. This means that the density of the Ag(Al) films are slightly different from that of the pure Ag film. The thickness of these films is approximately 95 nm and its value is same with pure Ag thickness.

The RBS spectra from as-deposited and annealed pure Ag thin films are shown in figure 2. The anneal processes were performed with identical conditions as with Ag(Al) thin film. It is found that film morphology was changed from the sample annealed at 400 °C for 1 hour. A typical shape of RBS spectrum for agglomerated thin film is that the intensity of Ag peak is decreased due to the reduction of number of counts for backscattered ions from the agglomerated Ag thin film and a sloping back edge of Ag peak produced by hillocks formation[9]. Also, a Si peak is found at the surface of the sample due to exposure of SiO_2 substrate resulting from formation of voids. From the RBS data, Ag films on SiO_2 substrate showed poor thermal stability when compared with Ag(Al) alloy film on SiO_2.

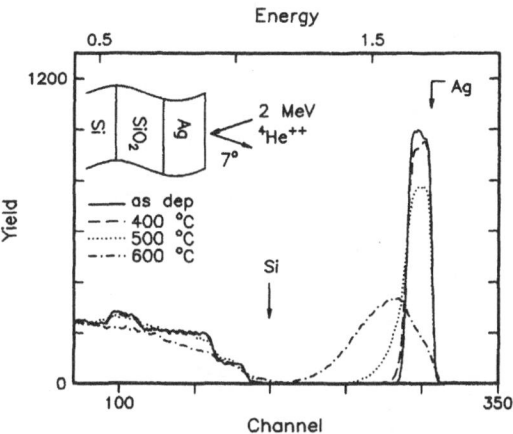

Figure 2. RBS spectra of as-deposited and annealed Ag thin films on SiO_2 substrate. Vacuum anneals were performed for 1 hour at 400 °C, 500 °C and 600 °C.

Figure 3 shows XRD spectra obtained from as-deposited and annealed Ag (5 at. % Al) and pure Ag film. A glancing angle (1° tilt) technique was used to collect more information from the thin film. All peaks present in figure 3 correspond to Ag and no phase changes or changes in lattice constant are found in both Ag(Al) alloy and pure Ag film on SiO$_2$ up to 600 °C.

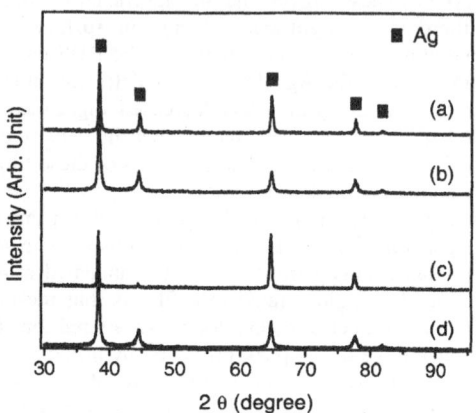

Figure 3. XRD patterns of as-deposited and annealed Ag (5 at. % Al) and Ag thin films on SiO$_2$: (a) Ag(Al), 600 °C; (b) Ag(Al), as-dep; (c) Ag, 600 °C; (d) Ag, as-dep.

For Ag(Al) alloy films, the peak intensity is increased and peak width is decreased when annealed at 600 °C. This sugggests that the grain size of the Ag(Al) film is enlarged and crystallinity is enhanced when annealed at high temperature. For pure Ag film, film texture is found to have changed after annealing at 600 °C. It is found that agglomerated Ag film has more preferred <111> texture perpendicular to the substrate than stable Ag(Al) alloy film on amorphous SiO$_2$ substrate[11].

Figure 4 presents the microstructures of Ag (10 at. % Al) and pure Ag film and comparison of microstructure changes after annealing at 600 °C in vacuum for 1 hour.

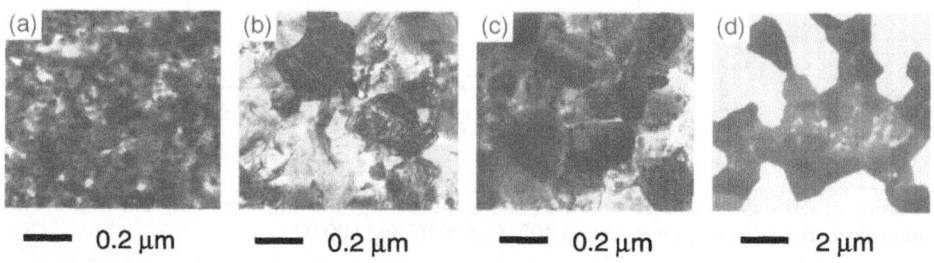

Figure 4. Transmission electron micrographs of Ag (10 at. % Al) and Ag thin films on SiO$_2$ substrate, as-deposited and annealed at different temperatures for 1 hour in vacuum: (a) Ag(Al), as-dep; (b) Ag(Al), 600 °C; (c) Ag, as-dep; (d) Ag, 600 °C.

When we observe the microstructure of as-deposited pure Ag(Al) films and pure Ag film (figure 4 (a) and (c)), it is found that the grain size of Ag(Al) is smaller than that of pure Ag film due to the alloying effect of Al. After annealing the films at 600 °C, the Ag film is agglomerated and disrupted with bigger grains. The Ag(Al) film shows increased grain size, which is similar to the grain size of as-deposited pure Ag film and stable microstructure.

The resistivity changes of pure Ag, Ag(5 at. % Al), and Ag(10 at. % Al) as a function of different anneal temperatures are shown in figure 5. The resistivity of pure Ag thin film is the lowest compared to the other films. Ag(10 at. % Al) film has the highest resistivity of the as-deposited Ag and Ag(Al) films due to enhancement of impurity scattering of conduction electrons in Ag(Al) alloy film. Due to the increase in crystallinity and grain size for annealed Ag(Al) alloy film, as confirmed by XRD and TEM analyses, the resistivity of the annealed Ag(Al) thin films has decreased compared to the as-deposited Ag(Al) film. Ag(5 at. % Al) and Ag(10 at. % Al) thin films show constant resistivity after annealing at 400 °C. This means that the Ag(Al) on SiO_2 is a thermally stable system without significant morphology change as seen by RBS, XRD, and TEM analyses. For pure Ag thin films, the resistivity of Ag film annealed at 400 °C has decreased slightly due to the enhanced crystallization and grain growth although agglomeration is initiated as shown in figure 2. The results shown in this work are evidence of the improved thermal stability of Ag(Al) thin films on SiO_2.

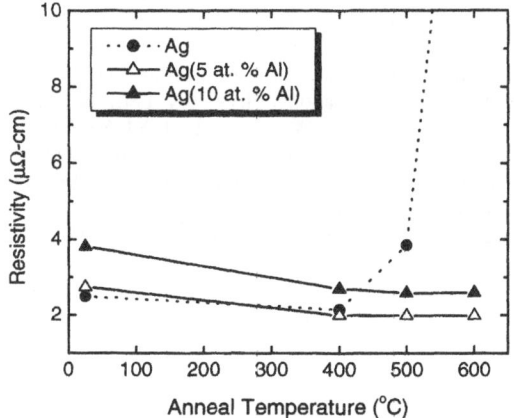

Figure 5. Electrical resistivity of Ag, Ag(5 at. % Al), and Ag(10 at. % Al) thin films on SiO_2 substrate annealed at various temperatures in vacuum for 1 hour.

CONCLUSIONS

We have improved the thermal stability of Ag thin films on SiO_2 substrate by adding small amounts of Al to pure Ag film. Ag(Al) alloy thin films studied in this work have resistivity values that are comparable with pure Ag thin film for as-deposited samples. The alloy films maintain lower resistivity than Ag thin film, when annealed at 400 °C on SiO_2. In addition, the Ag(Al) alloy films are thermally stable up to 600 °C without surface treatment of the substrate or use of other diffusion barriers. This finding can impact metallization processes for thin film transistors (TFT), or organic light emitting diodes (OLED) for flexible electronics and flat panel displays as well as high speed electronic devices.

ACKNOWLEDGMENTS

This work was partially supported by a grant from the NSF (DMR-0308127, L. Hess). The authors also acknowledge partial support from the ASU Ira A. Fulton School of Engineering (Dr. Paul Johnson).

REFERENCES

1. H. C. Kim, T. L. Alford, and D. R. Allee, Appl. Phys. Lett. **81**, 4287 (2002).
2. P. Nguyen, Y. Zeng, and T. L. Alford, J. Vac. Sci. Technol. B **19**, 158 (2001).
3. Kaustubh. S. Gadre and T. L. Alford, J. Vac. Sci. Technol. B **18**, 2814 (2000).
4. J. –H. Lee, S. –H. Lee, K. –L. Yoo, N. –Y. Kim, and C. K. Hwangbo, Surf. Coat. Technol. **158-159**, 477 (2002).
5. M. Fahland, P. Karlsson, and C. Charton, Thin Solid Films **392**, 334 (2001).
6. R. Seemann, S. Herminghaus, and K. Jacobs, Phys. Rev. Lett. **86**, 5534 (2001).
7. C. Pennetta, L. Reggiani, and G. Trefán, Phys. Rev. Lett. **84**, 5006 (2001).
8. T. P. Nolan, R. Sinclair, and R. Beyers, J. Appl. Phys. **71**, 720 (1992).
9. T. L. Alford, Lingui Chen, and Kaustubh S. Gadre, Thin Solid Films **429**, 248 (2003).
10. L. R. Doolittle, Nucl. Instrum. Methods, Phys. Res. **B 9**, 344 (1985).
11. H. C. Kim, N. D. Theodore, and T. L. Alford, in press J. Appl. Phys. (2004).

Mat. Res. Soc. Symp. Proc. Vol. 812 © 2004 Materials Research Society   F3.25

Morphology of Ti₃₇Al₆₃ Thin-Films Deposited by Magnetron Sputtering

N. David Theodore[1], Hyunchul C. Kim[2], Kaustubh S. Gadre[3], James W. Mayer[2], and Terry L. Alford[2]

[1]DigitalDNA™ Labs., Motorola Inc., 2100 E. Elliot Rd., MD-EL622, Tempe, AZ 85284, U.S.A.
[2]Department of Chemical and Materials Engineering, Arizona State University,
Tempe, AZ 85287, U.S.A.
[3]Intel Corp., RA3-402, 2501 NW 229ᵗʰ St., Hillsboro, OR 97124, U.S.A.

ABSTRACT

TiAl based thin-films possess high oxidation-resistance and high melting points, making them possible candidates for application in electronics. The behavior of the films upon exposure to various temperatures is of interest for such application. In the present study, $Ti_{37}Al_{63}$ thin films were deposited onto SiO_2 substrates using RF magnetron sputtering from a compound target. Anneals were performed in vacuum at temperatures ranging from 400 °C to 700 °C. The phases and microstructural behavior of the films were evaluated as a function of annealing. Microstructural behavior was correlated with resistivity changes in the films. The behavior of Ti-Al films as potential under-layers for silver metallization was also evaluated. The Ti-Al was observed to enhance the thermal stability of pure Al thin-films. The results are relevant for potential application of the films to electronics.

INTRODUCTION

Titanium aluminide thin films have high melting points, good high-temperature strengths, light weights, and good oxidation resistances [1,2] making them attractive for structural coatings. Some of these properties can make the layers relevant for electronics as well. Sputter-deposited titanium aluminide films have been studied for mechanical applications at high temperatures [3-6]. However, there have been very few studies for application in electronics. Silver has potential for application as an interconnect metal, due to its low resistivity. However Ag tends to agglomerate upon annealing. If Ag is to be used for interconnects, its thermal stability needs to be improved.

In the present work, we study the structural, electrical, and Ag-underlayer properties of sputter-deposited $Ti_{37}Al_{63}$ thin films on SiO_2/Si substrates, to evaluate the potential usefulness of the films for electronics applications.

EXPERIMENTAL DETAILS

RF magnetron sputtering was used to deposit titanium aluminide thin films onto SiO_2 /Si substrates. A compound target was used, with a composition of 40 at. % Ti with 60 at. % Al. The depositions used 350 W power with 5 mTorr Ar gas at room temperature. The thin films were annealed in a vacuum furnace for 1 hour, at temperatures ranging from 400 °C to 700 °C. Base pressure was approximately 5×10^{-9} Torr. The resulting films had a composition of $Ti_{37}Al_{63}$ as measured by Rutherford backscattering spectrometry (RBS).

Transmission electron microscopy (TEM) was performed using a JEOL 200 CX instrument, operating at a voltage of 200 KV. The microstructures of the as-deposited and annealed titanium aluminide films were characterized. Crystallinity and phase formation in the films were evaluated using TEM and selected-area electron diffraction. Stability of the titanium aluminide films was evaluated after high temperature annealing. Four-point probe analysis was used to measure electrical resistivities of the films as-deposited and after anneals at various temperatures. Silver was electron-beam evaporated onto Ti-Al films. The wafers were then annealed at 400 °C and 600 °C for 1 hour each in vacuum (~ 5 x 10^{-8} Torr). The morphology of the Ag was then evaluated using optical microscopy, and was compared with the morphology of Ag on SiO$_2$ annealed under the same conditions.

RESULTS AND DISCUSSION

A plan-view TEM micrograph is presented in Fig. 1, together with a representative electron-diffraction pattern obtained from the as-deposited Ti$_{37}$Al$_{63}$ film. The contrast in the micrograph, and the diffuse ring in the diffraction pattern, indicate that the as-deposited film is amorphous in structure. In Fig. 1c, indexed ring positions are presented for TiAl and TiAl$_2$. The position of the diffuse ring in the diffraction pattern obtained from the as-deposited Ti$_{37}$Al$_{63}$ matches the positions of the (111) ring of TiAl and the (116) ring of TiAl$_2$. This matching can arise from short-range order present within the amorphous film, even though long-range order is absent.

Figure 2 presents a plan-view TEM micrograph and a representative electron-diffraction pattern obtained from the Ti$_{37}$Al$_{63}$ thin film after annealing at 500 °C for 1 hour in vacuum. The film annealed at 500 °C has crystallized with large numbers of small grains. Continuous rings are seen in the electron-diffraction pattern shown in Fig. 2b due to random in-plane orientation of these small grains. The rings are sharper and more numerous than the diffuse ring from the amorphous as-deposited film. In Fig. 2c, indexed ring positions are presented for TiAl and TiAl$_2$. Matching is observed between the diffraction rings obtained from the annealed film, and the (111), (200), (202), (311) rings of TiAl, and the (116), (020), (220), (1 1 18) rings of TiAl$_2$.

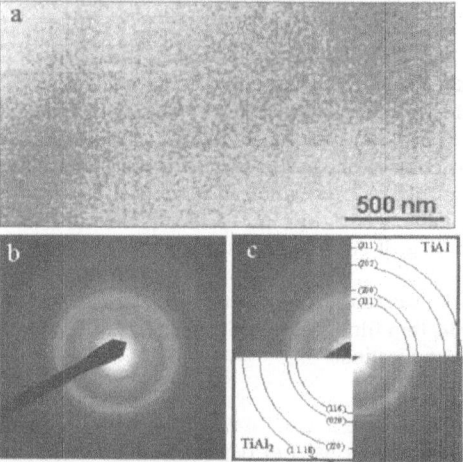

Fig. 1 Plan-view (a) transmission electron micrograph, and (b) electron diffraction pattern obtained from as-deposited Ti$_{37}$Al$_{63}$ thin film; (c) diffraction pattern from 'b' with indexed ring positions from TiAl and TiAl$_2$.

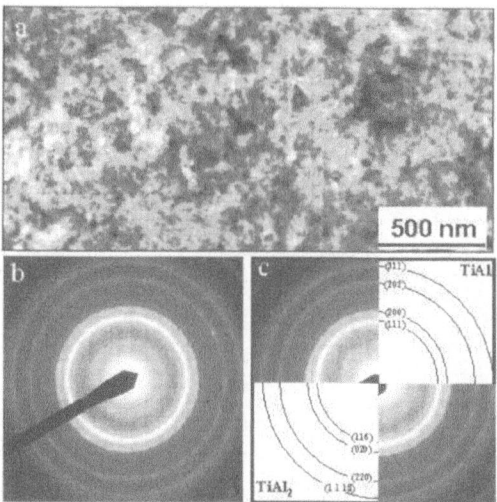

Fig. 2 Plan-view (a) transmission electron micrograph, and (b) electron diffraction pattern obtained from Ti$_{37}$Al$_{63}$ thin films, annealed at 500 °C, in vacuum for 1 hour; (c) diffraction pattern from 'b' with indexed rings from TiAl and TiAl$_2$.

A plan-view TEM micrograph is presented in Fig. 3, together with a representative electron-diffraction pattern obtained from the Ti$_{37}$Al$_{63}$ thin film after annealing at 700 °C for 1 hour in vacuum. Annealing at 700 °C results in growth of the grains compared to the 500 °C annealed sample. The electron-diffraction rings are not as continuous as in the diffraction pattern from the film annealed at 500 °C. Increased grain-sizes can cause such discontinuity in diffraction patterns due to fewer grains being present in the selected-area used to form the pattern. In Fig. 3c, indexed ring positions are presented for TiAl and TiAl$_2$. Matching is observed between the diffraction rings obtained from the annealed film, and the (111), (200), (202), (311) rings of TiAl, and the (116), (020), (220), (1 1 18) rings of TiAl$_2$. In addition, occasional diffraction spots match the (011) and (017) rings of TiAl$_2$.

Based on the Ti-Al phase diagram, for Ti$_{37}$Al$_{63}$ thin films (with their composition confirmed by RBS analysis), TiAl$_2$ and a small amount of γ-TiAl are possible equilibrium phases that could be present [7]. TEM diffraction patterns do indicate the presence of multiple phases in the annealed Ti$_{37}$Al$_{63}$ thin films.

The as-deposited Ti$_{37}$Al$_{63}$ thin film shows non-crystalline amorphous structure. Transformation to the crystalline structure occurs when the Ti$_{37}$Al$_{63}$ films are annealed above 400 °C for 1 hour in vacuum. The electron-diffraction rings match both TiAl$_2$ (JCPDS: 47-1177) and TiAl (JCPDS: 05-0678) phases. The similarity of the JCPDS powder diffraction data, and the width of the electron-diffraction rings, makes the phases difficult to resolve. The phases observed by TEM are stable up to 700 °C.

When TiAl thin films are deposited by sputtering, an amorphous phase is produced and crystallized phases form when the films are exposed to heat treatment [3-6]. Thermal annealing with appropriate temperature and time can cause the amorphous phase to crystallize because the amorphous phase is thermodynamically a metastable state [6].

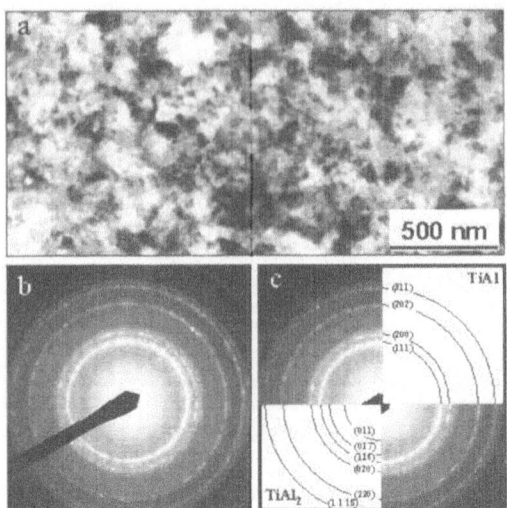

Fig. 3 Plan-view (a) transmission electron micrograph, and (b) electron diffraction pattern obtained from $Ti_{37}Al_{63}$ thin films, annealed at 700 °C, in vacuum for 1 hour; (c) diffraction pattern from 'b' with indexed rings from TiAl and $TiAl_2$.

Wang et al. [5] reported that crystallization of 50 at. % Ti - 50 at. % Al thin films can occur when the films are deposited at substrate temperature > 573 K, with 10^{-9} m/s deposition rate. Calculated mean diffusion distances were also reported. Insufficient atomic mobility, at lower temperatures, is found to prevent crystallization of the titanium aluminide. Our TEM diffraction data obtained from the as-deposited $Ti_{37}Al_{63}$ thin films on SiO_2 are consistent with the above explanation and with other studies in the literature [3-6].

Transformation from amorphous to crystalline phases occurs at 400 °C, when the $Ti_{37}Al_{63}$ thin films are annealed for 1 hour. In an atomistic view, self-diffusion of Ti or Al atoms in the titanium aluminides plays a significant role in the crystallization process. Long-range ordering in the titanium aluminide is influenced by the motion of atoms and point defects. It has been shown by Herzig et al. [8,9] that the effective activation-energies for both Ti and Al lattice self-diffusion in Al-rich bulk titanium aluminide are lower than corresponding values for Ti-rich titanium aluminide and $Ti_{50}Al_{50}$. Similar behavior was seen for the effective activation-energy for grain-boundary diffusion in titanium aluminide [10]. The lower temperature observed, in the present study, for crystallization of $Ti_{37}Al_{63}$ thin films when compared with $Ti_{53}Al_{47}$ thin films [11], is consistent with a lower activation energy for self-diffusion in Al-rich titanium aluminide.

Figure 4 presents electrical resistivity values measured from the $Ti_{37}Al_{63}$ thin films annealed at different temperatures, for 1 hour, in vacuum. The resistivity of the as-deposited films is ~220 µΩ-cm. Resistivity values are seen to drop as the films are annealed. For the film annealed at 700 °C, the resistivity has dropped to ~110 µΩ-cm.

Figure 4 shows that the resistivities of $Ti_{37}Al_{63}$ thin films depend on anneal temperature. So, thermal annealing at high temperatures ~ 700 °C can be a useful method for reduction of the resistivity of as-deposited titanium aluminide thin films. The resistivity, even after 700 °C annealing, is not as low as in the case of pure metals such as Al.

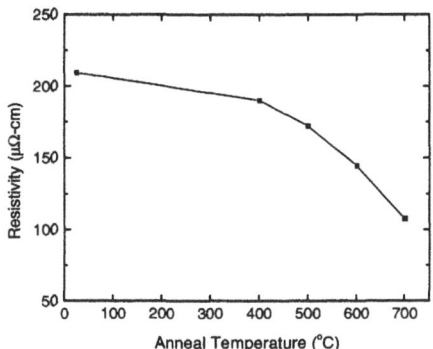

Fig. 4 Electrical resistivity of $Ti_{37}Al_{63}$ thin films annealed at different temperatures for 1 hour in vacuum.

As mentioned above, TEM diffraction data from the $Ti_{37}Al_{63}$ thin films show that the films, as deposited on SiO_2 substrates, are amorphous. Charge carriers, such as electrons, are highly scattered by amorphous materials due to their disordered structure. The increased scattering results in high resistivity values when compared to crystalline materials. This is consistent with the mean free path for conduction electrons being on the order of the short-range order in the material [12].

When resistivity is high, ≥ 100 $\mu\Omega$-cm, electrical conduction is controlled by atomic-scale defects such as anti-site defects (Ti_{Al} & Al_{Ti}) and vacancies (V_{Ti} & V_{Al}), rather than by macroscopic defects such as grain boundaries [13,14]. The drop in electrical resistivity, as the $Ti_{37}Al_{63}$ film transforms from amorphous to crystalline phase, can be explained as arising from a reduction in scattering of conduction electrons by atomic-scale defects. At higher anneal temperatures, point defects are annihilated, resulting in decreased densities of such defects. This results in reduced scattering of conduction electrons, leading to a reduction in electrical resistivity.

Based on the TEM results, it appears that the degree of crystallinity increases as the sample is annealed. The grain size of the film annealed at 700 °C has increased slightly when compared with the 500 °C annealed film. However, the similarity of grain sizes at these two temperatures suggests that grain-size effects or grain-boundary effects are not the main reason for decreasing resistivity as anneal temperature is increased.

The atomic structure of the $Ti_{37}Al_{63}$ thin film moves to a more ordered state, and more atomic-scale defects are annihilated, upon annealing at high temperature. A decrease in electrical resistivity then arises due to reduction of scattering between conduction electrons and atomic-scale defects.

Figures 5a-c present optical micrographs obtained from Ag thin-films on SiO_2, as deposited, and annealed at 400 °C, and 600 °C respectively. The as-deposited Ag appears smooth and featureless, whereas isolated voids are visible in the 400 °C annealed Ag, and the 600 °C Ag is almost entirely agglomerated. Optical micrographs of Ag thin films on Ti-Al, as deposited, and annealed at 400 °C, and 600 °C (Figs. 5d-f) show improved thermal stability of the Ag. The as-deposited and 400 °C annealed Ag thin-films appear featureless, whereas isolated voids are visible in the 600 °C annealed film. The improved thermal stability of the Ag is likely due to lower interfacial energy between Ag and Ti-Al compared to the Ag-SiO_2 interfacial energy.

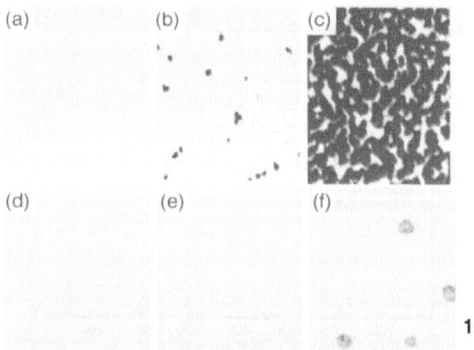

Fig. 5 Optical micrographs of Ag thin films on SiO_2, (a) as deposited, (b) annealed at 400 °C, (c) annealed at 600 °C. Optical micrographs of Ag thin films on Ti-Al (d) as deposited, (e) annealed at 400 °C, (c) annealed at 600 °C. All anneals were performed for 1 hour in vacuum.

CONCLUSIONS

Titanium aluminide thin films of $Ti_{37}Al_{63}$ composition were deposited by sputtering using a compound target. The thin films on SiO_2 showed good thermal stability at high temperature (700 °C), and reasonable electrical resistivity when appropriate anneal conditions were used. The resistivity is however higher than that of metals such as Al. The behavior of Ti-Al films as potential under-layers for silver metallization was evaluated. The Ti-Al was observed to enhance the thermal stability of pure Al thin-films. These conclusions are supported by data from transmission electron microscopy, four-point probe, and optical microscopy analyses.

REFERENCES

1. J. Zhang, B. V. Reddy, S. C. Deevi, Scripta Mater. **45**, 645 (2001).
2. D. Veeraraghavan, U. Pilchowski, B. Natarajan, V. K. Vasudevan, Acta Mater. **46**, 405 (1998).
3. T. Viera, B. Trindade, A.S. Ramos, J.V. Fernandes, M.F. Vieira, Mat. Sci. Eng. **A329-331**, 147 (2002).
4. N. Senkov, M.D. Uchic, Mat. Sci. Eng. **A340**, 216 (2003).
5. Z. Wang, G. Shao, P. Tsakiropoulos, F. Wang, Mat. Sci. Eng. **A329-331**, 141 (2002).
6. R. Banerjee, S. Swaminathan, J.M.K. Wiezorek, R. Wheeler, H.L. Fraser, Metall. Mater. Trans. **A 27**, 2047 (1996).
7. C. Lei, Q. Xu, Y.Q. Sun, Mat. Sci. Eng. **A313**, 227 (2001).
8. C. Herzig, T. Przeorski, Y. Mishin, Intermetallics **7**, 389 (1999).
9. C. Herzig, M. Friesel, D. Derdau, S.V. Divinski, Intermetallics **7**, 1141 (1999).
10. C. Herzig, T. Wilger, T. Przeorski, F. Hisker, S. Divinski, Intermetallics **9**, 431 (2001).
11. H.C. Kim, N.D. Theodore, K.S. Gadre, J.W. Mayer, T.L. Alford, *in press*, Thin Solid Films, (2004).
12. C. Kittel, *Introduction to Solid State Physics*, 7th ed. (Willey, New York, 1996), p. 530.
13. J. H. Mooij, Phys. Stat. Sol (**a**) **17**, 521 (1973).
14. Y. Shirai, K. Masaki, T. Inoue, S. R. Nishitani, M. Yamaguchi, Intermetallics **3**, 381 (1995).

Mat. Res. Soc. Symp. Proc. Vol. 812 © 2004 Materials Research Society   F5.10

Reliability of Dielectric Barrier Films in Copper Damascene Applications

Albert S. Lee, Annamalai Lakshmanan, Nagarajan Rajagopalan, Zhenjiang Cui, Maggie Le,
Li Qun Xia, Bok Heon Kim, and Hichem M'Saad
Dielectric Systems & Modules Product Business Group
Applied Materials Inc., Santa Clara, CA 95054, U.S. A.

ABSTRACT

The film properties of two PECVD deposited dielectric copper barrier films have been optimized to improve BEOL device reliability in terms of electromigration. Two critical aspects that affect electromigration are the dielectric barrier film hermeticity and adhesion to copper. We use a method to quantify the barrier film hermeticity and have optimized the hermeticity of the BLOκ™ low-κ dielectric barrier film to be similar to that of silicon nitride. By using FT-IR we find that the film porosity has a much stronger effect than the film stoichiometry on hermeticity. In addition, the interfaces between Damascene Nitride™ with copper, as well as BLOκ with copper have been engineered to improve the interfacial adhesion energy to >10 J/m^2 for both Damascene Nitride and BLOκ.

INTRODUCTION

With the advent of copper dual damascene, dielectric copper barriers are becoming the cornerstone for back end of line device reliability. This paper will address two concepts for dielectric thin film reliability: hermeticity of the bulk barrier film and adhesion between the barrier and copper layers. It is known that the interface between copper and the dielectric barrier is the fastest diffusion path for copper electromigration, and that improved adhesion of the barrier film to copper reduces copper electromigration [1,2]. In this paper, we show that adhesion is improved by interfacial engineering of the barrier film. We introduce the concept of hermeticity, which is the film's ability to prevent moisture from penetrating into the underlying layer. As we transition to lower-κ barriers, the barrier film density decreases and porosity increases. Hence, the ability of the barrier to stop impurities or moisture from reaching the underlying copper becomes a challenge. A non-hermetic film can allow copper oxide formation at the barrier-copper interface, leading to adhesion loss and poor electromigration performance. Thus, the hermeticity of the dielectric barrier becomes an important criterion for barrier integrity. In this work we have optimized the hermeticity of BLOκ™ [3] low-κ copper barrier film. In addition, we have examined methods to modify the interface between copper and barrier films in such a way as to dramatically improve the adhesion of the BLOκ and Damascene Nitride Si$_x$N$_y$ [4,5] films to copper.

EXPERIMENTAL DETAILS

We deposited both BLOκ and Damascene Nitride in a PECVD Producer® SE Twin Chamber™. Films were deposited on 300mm substrates. The processing temperature was

350°C to 400°C, and gas pressures ranging from 2-8 Torr were maintained in the chamber during deposition. RF powers ranging from 300-1500W were used to deposit the film. Damascene Nitride was deposited using silane (SiH_4) as the silicon precursor and ammonia (NH_3) as the nitrogen source. BLOκ, which is essentially a nitrogen-doped silicon carbide, uses trimethylsilane (TMS) as the silicon and carbon source and ammonia as the nitrogen source. The film dielectric constant and leakage were measured using an SSM mercury probe. To characterize the copper barrier properties, the barrier films were deposited onto blanket copper and annealed for three hours at 400°C. After the anneal, Secondary Ion Mass Spectroscopy (SIMS) was used to determine the copper diffusion depth into Damascene Nitride and BLOκ.

We used a method to quantify the film's hermeticity, where 500Å of the dielectric barrier is deposited on top of a 1μm TEOS oxide film with tensile stress, and the stack is subjected to 85°C and 85% humidity for 17 hours. A TEOS oxide with tensile stress readily absorbs moisture, upon which the film stress becomes compressive. A stress change in the barrier/oxide stack after 85°C and 85% humidity indicates moisture uptake by the underlying oxide film, and a stress change of >20 MPa indicates a non-hermetic barrier film.

We used the four-point bend technique [6,7] to quantify the adhesion of BLOκ and Damascene Nitride to copper. The blanket copper substrates were prepared by depositing TEOS oxide onto bare silicon substrates, PVD metal barrier and copper seed deposition onto the oxide, electrochemically plating (ECP) copper onto the barrier/seed, and finally chemical mechanical polishing (CMP) the ECP copper. After deposition of the dielectric barrier onto copper, pieces from the same substrate are bonded together face to face using an epoxy, and are then cut into sample pieces for the four-point bend test. Figure 1 shows the sample stack and four-point bend geometry. The interfacial adhesion energy is obtained from the plateau of the load versus

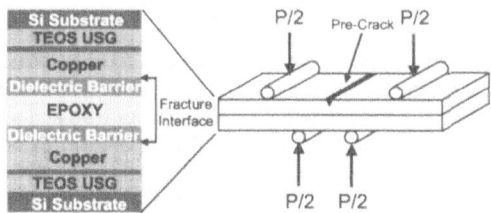

Figure 1. Four-point bend test scheme for dielectric barrier adhesion to copper.

	Damascene Nitride	BLOk I
Compressive Stress (MPa)	100-200	200-300
Hardness (GPa)	18.6	10.9
Cu Diffusion Depth, 3h 400°C anneal (Å)	<250	<250
Relative Dielectric Constant	6.8	4.9
Leakage @ 1MV/cm (A/cm2)	$4x10^{-10}$	$2x10^{-10}$
Breakdown Field (MV/cm)	9.5	6

Table 1. Properties of Damascene Nitride and BLOκ

displacement curve. For each dielectric barrier deposition condition, three to five sample pieces were measured and averaged to obtain the interfacial fracture energy.

DISCUSSION

Table 1 summarizes the basic film properties of Damascene Nitride and BLOκ. Both films exhibit low leakage and high breakdown fields. Figure 2 shows SIMS Cu diffusion depth profiles for Damascene Nitride and BLOκ after annealing. We define the copper diffusion depth as the distance at which the Cu concentration decreases by four orders of magnitude. As seen in Figure 2, the copper diffusion depths into Damascene Nitride and BLOκ are <250Å, indicating that both films are excellent copper barriers.

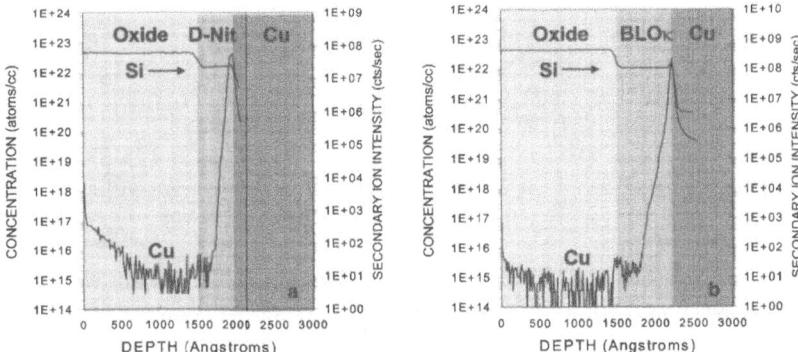

Figure 2. SIMS Cu diffusion depth profiles for (a) Damascene Nitride and (b) BLOκ after 3h 400°C N_2 anneal.

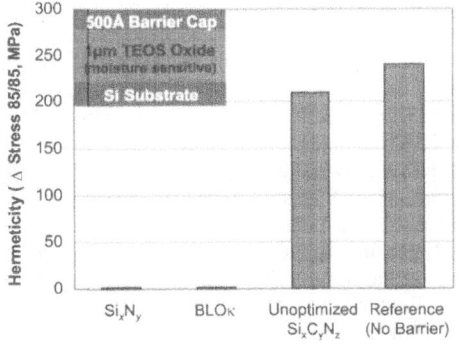

Figure 3. Hermeticity of barrier films, expressed as stress change after 17hrs in 85°C/85%. A smaller Δ stress indicates a more hermetic film.

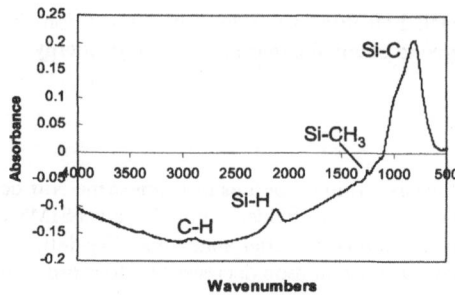

Figure 4. Typical FT-IR spectrum of BLOκ

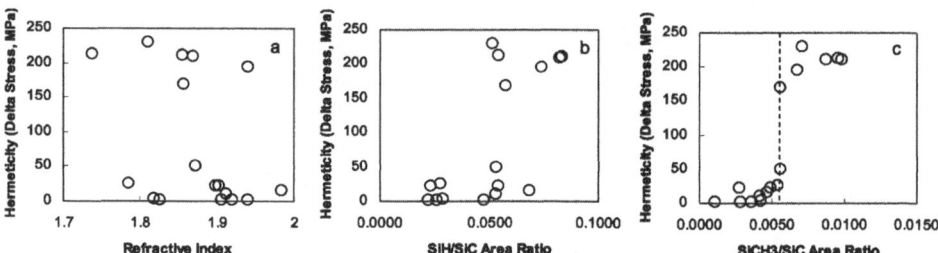

Figure 5. Correlations between BLOκ hermeticity and (a) refractive index, (b) SiH/SiC FT-IR peak area ratio, and (c) SiCH₃/SiC FT-IR peak area ratio. A smaller Δ stress indicates a more hermetic film.

We adjusted the BLOκ bulk film properties by varying the process deposition conditions to optimize the hermeticity. Figure 3 compares the hermeticity of four films: 1) the optimized BLOκ, 2) Damascene Nitride, 3) an unoptimized $Si_xC_yN_z$ film and 4) a reference with no barrier cap. As seen in the figure, the optimized BLOκ allows a stress change of <10MPa in the underlying oxide layer, and the hermeticity of BLOκ is comparable to that of silicon nitride.

We use FT-IR and refractive index measurements to try to understand the film properties which influence the hermeticity of the film. Figure 4 shows a typical FT-IR spectrum of BLOκ. We correlate the Si-H/Si-C and Si-CH₃/Si-C FT-IR peak area ratios with hermeticity to understand how Si-H and Si-CH₃ bonding in the film contribute to the film hermeticity. As seen in Figure 5a, there is no correlation between hermeticity and the film refractive index, indicating that hermeticity is not a function of the film stoichiometry. Figure 5b shows no correlation between Si-H bonding and hermeticity. In contrast, Figure 5c shows a strong correlation between the amount of Si-CH₃ bonding in the film and hermeticity, where increasing the amount of Si-CH₃ bonding causes the film to become less hermetic. Because CH₃ is a bulky functional group, Si-CH₃ bonding in the film tends to make the film more porous. Decreasing the amount of Si-CH₃ bonding makes the film less porous and more dense, causing the film to be more hermetic. Through this FT-IR study we conclude that porosity and not stoichiometry is the dominant factor which affects the film hermeticity.

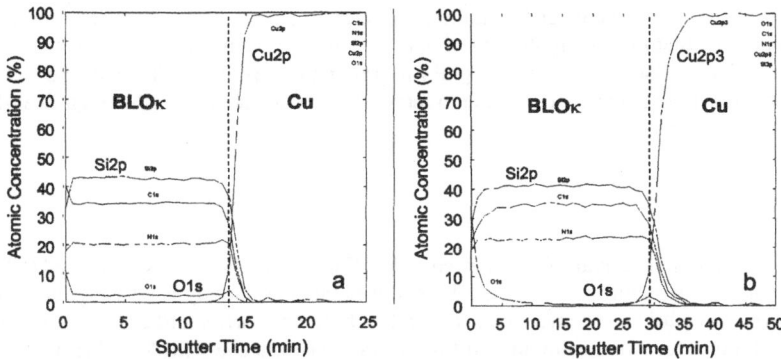

Figure 6. XPS depth profiles of BLOκ deposited on top of copper (a) without annealing and (b) after 24h annealing at 275°C in an N_2/O_2 atmosphere.

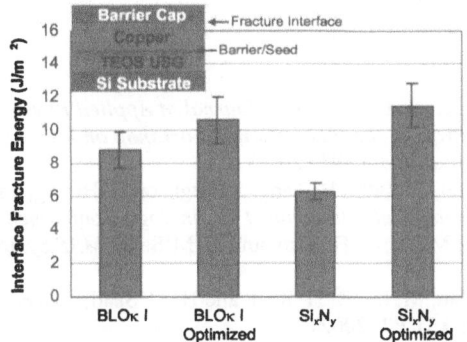

Figure 7. Barrier film to copper adhesion strengths measured using the 4-point bend method.

In addition, we have examined the ability of BLOκ to prevent oxygen from diffusing to the BLOκ / Cu interface to form copper oxide. We deposit 350Å of BLOκ onto copper and then anneal at 275°C for 24h in an N_2/O_2 atmosphere. As seen in the XPS depth profiles in Figure 6, we see that no additional copper oxide forms at the BLOκ / Cu interface even after 24h annealing, and that in addition to being hermetic, the optimized BLOκ film is able to prevent O_2 from oxidizing the copper interface.

We engineered the interface between the dielectric barrier and copper to promote adhesion of the two materials. During deposition of Damascene Nitride and BLOκ, an in-situ pre-treatment step is used to remove copper oxide prior to depositing the barrier film. We optimized the transition between the pre-treatment and deposition to improve the barrier film adhesion to copper. Figure 7 shows adhesion strengths of BLOκ and Damascene Nitride to copper. As seen in the figure, by modifying the interface between the dielectric barrier and copper, we were able to significantly improve the adhesion strengths to >10 J/m^2 for both films. Dixit and Padhi, as well as Lane, et al. have shown that the activation energy for electromigration failure, Q, can be directly correlated with the barrier film adhesion energy to copper [1,2]. Electromigration experiments show that increasing the adhesion energy from 3.5 J/m^2 to 10 J/m^2 results in an increase in Q from 0.66eV to ~1.04eV [2]. By improving the BLOκ and Damascene Nitride adhesion to copper we believe we are able to increase Q, which should lead to significantly improved electromigration performance.

CONCLUSIONS

By using a method to quantify hermeticity, the BLOκ hermeticity was optimized to be similar to that of silicon nitride. We use FT-IR to find that hermeticity is a strong function of the film porosity. In addition, the interface between the dielectric barrier and copper was engineered to increase the interfacial adhesion to >10 J/m^2 for both Damascene Nitride and BLOκ. By optimizing both the bulk film properties (hermeticity) and the interface with copper (adhesion), we have demonstrated barrier film performance that minimizes reliability issues related to copper electromigration.

REFERENCES

[1] M. W. Lane, E. G. Liniger, and J. R. Lloyd, *Journal of Applied Physics* 93, 1417 (2003).
[2] G. Dixit and D. Padhi, *Proc. of the International Workshop on Physics of Semiconductor Devices (IWPSD), Chennai, India* 393 (2003).
[3] P. Xu, K. Huang, A. Patel, S. Rathi, B. Tang, J. Ferguson, J. Huang, and C. Ngai, *Proceedings of the International Interconnect Technology Conference 1999* 109 (1999).
[4] A. Lee, N. Rajagopalan, M. Le, B. H. Kim, and H. M'Saad, *MRS Symp. Proc. Vol. 715* A10.7.1 (2002).
[5] A. S. Lee, N. Rajagopalan, M. Le, B. H. Kim, and H. M'Saad, *Journal of The Electrochemical Society* **151** F7 (2002).
[6] Q. Ma, H. Fujimoto, P. Flynn, V. Jain, F. Adibi-Rizi, and R. H. Dauskardt, *MRS Symp. Proc. Vol. 391* 91 (1995).
[7] M. Jenkins, J. Snodgrass, R. H. Dauskardt, and J. C. Bravman, *MRS Symp. Proc. Vol. 629* FF5.12.1 (2000).

Mat. Res. Soc. Symp. Proc. Vol. 812 © 2004 Materials Research Society   F6.8

Effect of Surface Chemistry on the diffusion of Copper in nanoporous dielectrics

Oscar Rodriguez, Woojin Cho, Ravi Saxena, Ravi Achanta, William N. Gill and Joel L Plawsky
Department of Chemical and Biological Engineering
Rensselaer Polytechnic Institute, Troy NY 12180 USA

ABSTRACT

This work is aimed at understanding the nature of the interactions between metal interconnects and nanoporous dielectrics in integrated circuits. Electrical testing of MIS capacitors is used to assess Cu diffusion and charge injection in the dielectric in the presence of an electric field. We have found that surface modification of nanoporous silica reveals the importance of chemically bound or adsorbed water species in the dielectric and how they trigger metal diffusion. We propose that a combination of moisture-related species in the dielectric and interfacial oxygen oxidize Cu. The copper oxide acts as a source for Cu ions available for diffusion. A quantitative analysis of Cu drift in nanoporous dielectrics that shows the importance of surface chemistry is presented and the mechanism of metal diffusion and charge injection in nanoporous dielectrics is discussed.

INTRODUCTION

In an interconnect structure, the dielectric is in contact with other materials, such as a metal (copper), a diffusion barrier (for example Ta or Ta nitride), or a hard mask (Si nitride). It is important to minimize the interactions between these different materials to insure long-term device stability though some minimal interaction is necessary to provide bonding and good adhesion between layers.

The increased speed in semiconductor devices requires copper and low-k ILD solutions to minimize RC delay. Porous materials have been proposed to replace SiO_2 as the interlayer dielectric. Copper is known to be a fast diffuser through SiO_2, several porous low-k dielectrics and Si [1-3]. However, a mechanism of how Cu diffuses has not been established. In this work, we show that surface chemistry plays an important role in stopping Cu diffusion, which may lead to an understanding of Cu drift in nanoporous dielectrics.

EXPERIMENTAL DETAILS

Bias-temperature stressing (BTS) measurements in Metal-Insulator-Silicon (MIS) capacitors were used to study Cu drift in nanoporous dielectrics [4]. Epitaxial n-type silicon wafers with an epi layer resistivity of 3-5 ohm-cm were used. The insulator films used include nanoporous silica (xerogel), Methyl-Silsesquioxane (MSQ) and sintered nanoporous silica.

The nanoporous silica films were spin-coated on Si wafers using the two-solvent sol-gel method described in a prior publication [5]. A surface modi?cation step was done using 4% vol. solution of trimethyl chlorosilane (TMCS) in hexane. Sintering of nanoporous silica was done at 900 C in air for 2 hours. After sintering, the films were treated with TMCS to remove any moisture in the film.

Copper dots of 1 mm in diameter and 300 nm thick were e-beam deposited on top of the dielectric. To complete the MIS structure, the backside of the Si wafer was carefully etched and a ~500 nm thick layer of Al was sputter deposited to make an electrical contact. Finally, the capacitors were annealed at 250 C for 1 hour in a $3\%H_2/Ar$ atmosphere.

Film thickness and porosity (calculated from the refractive index measurement) were determined by variable angle spectroscopic ellipsomety (VASE). The MSQ films used in this study were ~40% porous and 500 nm thick. Nanoporous silica of two porosities were tested: 24%, 754 nm; 38%, 985 nm. Sintered nanoporous silica was 18% and 400 nm thick. FTIR was used to characterize the chemistry of the materials tested.

High Frequency Capacitance-Voltage measurements were used to assess the diffusion of metal in the oxide after different Bias Thermal Stressing (BTS) conditions. All capacitance-voltage (C-V) measurements were made on HP4280A 1 MHz capacitance meter/C-V plotter. A small ac signal of 10 mV rms was superposed on the applied dc bias. The sample was vacuum held on the hot chuck inside an MSI electronics light shield. The measurements were done in a N_2 ambient and the BTS conditions were chosen so that a clear V_{fb} shift in the C-V curve can be observed after a relatively short period of stress-time.

RESULTS

Figure 1 shows the Capacitance-Voltage (CV) measurements of the different MIS capacitors before and after Bias Thermal Stress. A shift to the left in the CV curves after BTS is observed in all the cases, which is an indication of positive charges from Cu being deposited in the dielectric. The CV traces for the Cu/MSQ/Si capacitor show that the minimum capacitance may increase after BTS. This increase in capacitance may be attributed to an accumulation of Cu ions at the Dielectric/Si interface [2].

The Cu+ Ion Concentration can be obtained from the CV traces [6],

$$[Cu^+] = \frac{-\Delta V_{fb} C_{dielectric}}{q}$$

Where, $[Cu^+]$ is the concentration of Cu ions in the dielectric per unit area (of the metal gate in the capacitor); $C_{dielectric}$: Capacitance of dielectric at accumulation, per unit area; q is the elementary charge; ΔV_{fb} is the voltage shift between the "before BTS" and the "after BTS" curves. Figure 2 and 3 show the variation of Cu+ Concentration for the different materials. The rate of injection of Cu+ in all the materials is initially high but then decreases rapidly at longer times. The relative order of resistance for Cu diffusion is as follows: Sintered Nanoporous Silica < Nanoporous Silica < MSQ. This result indicates that although the matrix of all the materials is silica-based, the process of Cu+ injection and diffusion is strongly dependent on the surface chemistry. It is important to note that the sintered nanoporous films were tested at less stringent BTS conditions than the conditions used for as-deposited nanoporous silica. More severe BTS conditions lead to rapid failure (due to faster Cu diffusion) of the capacitor under test.

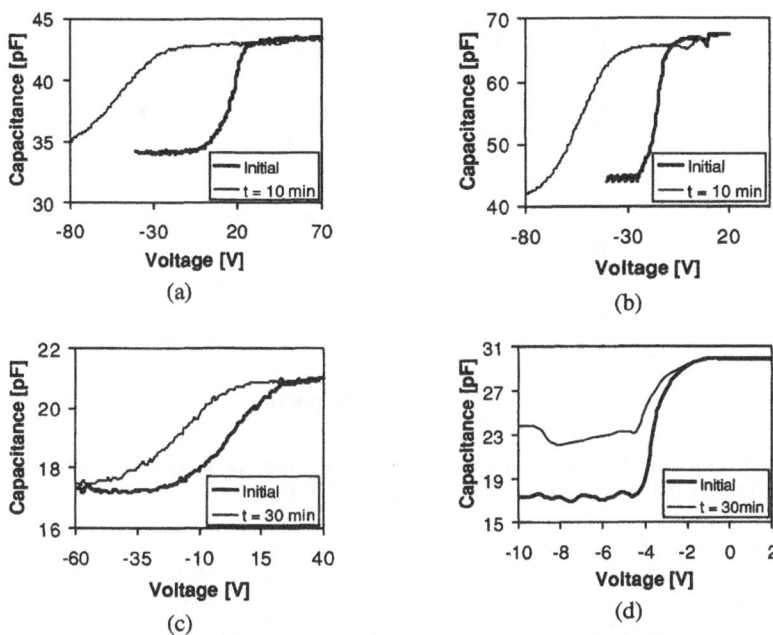

Figure 1. CV Traces showing comparable shifts for (a)Xerogel, 24% 1 MV/cm, 150 C; (b) Sintered Nanoporous Silica, 18% 0.25 MV/cm, 150 C; and (c) Nanoporous Silica, 38%, 0.5 MV/cm, 150 C; (d) MSQ, 40%, 1.5 MV/cm, 200 C.

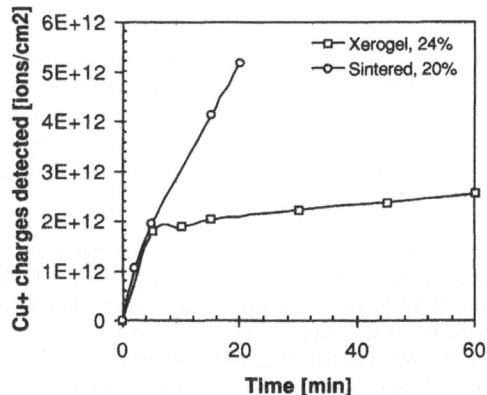

Figure 2. Comparison between charges detected for 20% porous Sintered Nanoporous Silica (0.25 MV/cm; 150 C) and 24% porous As-Deposited Nanoporous Silica (1 MV/cm, 150 C).

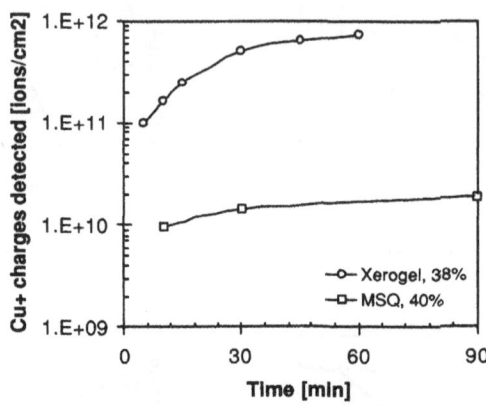

Figure 3. Comparison between Charges detected for 38% porous Nanoporous Silica (0.5 MV/cm, 150 C) and 40% porous MSQ (1.5 MV/cm, 200 C)

Figure 4 shows the FTIR spectra for 40% porous MSQ and 38% Nanoporous Silica. There is no sharp absorption band at ~3750 cm^{-1}, related to O-H stretching vibration of free or single silanol groups or absorption at 3700 cm^{-1} and 3500 cm^{-1} related to O-H stretching of bonded silanol groups and due to adsorbed water, respectively. At ~2900 cm^{-1} we observed a peak related to the asymmetric and symmetric C-H stretching modes of methyl in trimethylsylil groups; at 1260 cm^{-1} we find absorptions due to –CH$_3$ bending, and at 848 cm^{-1} and 866 cm^{-1} due to -CH$_3$ vibrational modes [7].

DISCUSSION

For the diffusion of metallic Cu enhanced by the action of an electric field into SiO$_2$ or other dielectrics, the metal must be ionized first. The ionization occurs by oxidation. The Cu oxide, which results from the ionization, acts as the source of the Cu ions that are available for diffusion. This is similar to the diffusion of silver in glass as discussed by Kapila and Plawsky [8]. The oxidizing species may be interfacial oxygen and hydroxyl species generated from water (in the form of silanol groups) in the dielectric [9].

Two major methods are used to remove –OH groups from silica and silica-based materials: thermal dehydroxylation (sintering) and surface modification [10,11]. By sintering, we are able to reduce the amount of SiOH groups present in nanoporous silica. The mechanism of sintering relies on condensation reactions occurring on the surface at high temperatures of the form 2SiOH = Si-O-Si + OH + H$^+$. We speculate that the combination of moderate temperatures (<300 C) and an external electric field may induce a similar condensation reaction in as-deposited nanoporous silica. Hydroxyl ions that reach the surface may react with copper to form Cu(OH), which is not stable and decomposes to form CuO.

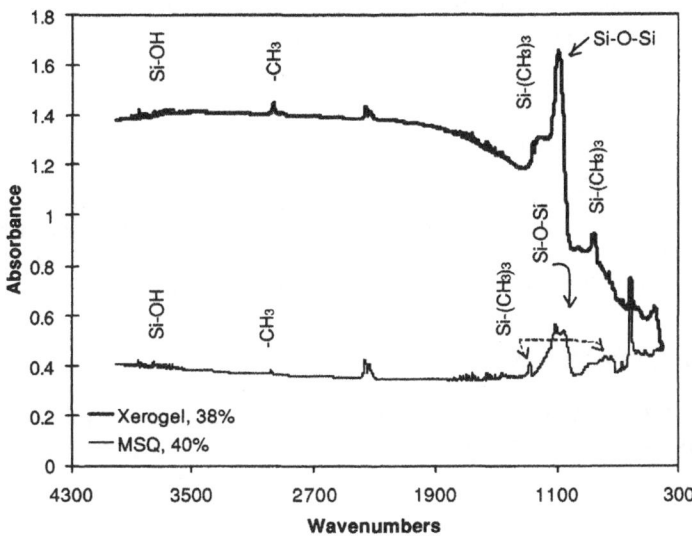

Figure 4. FTIR Spectra of Nanoporous Silica and MSQ indicating the presence of organic groups after fabrication.

Rogojevic *et al* showed that the amount of adsorbed moisture scales with surface area of the material when exposed to an ambient of ~100% relative humidity at room temperature [12]. That is, the more porous films adsorb more moisture. Moisture adsorbs in Si-O-Si and –OH sites. There is a significant reduction of hydroxyl groups due to the surface modification step in the fabrication of nanoporous silica, which replaces the hydrogen in silanol groups with a trimethylsylil group from TMCS; however, the trimethylsylil groups may provide incomplete shielding of the silica surface from contact with moisture (the area occupied by each $–(CH_3)_3$ group is 0.416 nm^2) [13]. The oxygen that is exposed to the metal may decrease with increasing porosity since as the surface area decreases the organic groups may provide better surface coverage. If larger-chain surface modifiers are used, the increasing effect of steric forces and other effects will reduce the efficiency of the removal of –OH groups.

We postulate that the source of Cu+ available for diffusion is a metal oxide formed at the Cu/SiO_2 interface. Cu interacts with the interfacial oxygen and hydroxyls formed from silanol groups to form a non-stochiometric oxide (Cu_3O_2) or CuO. The growth of this oxide is not a thermally activated diffusion reaction at low temperatures, but it is induced by the external electric field. This metal oxide will form very quickly after the high temperature and the electric field are applied. The Metal or the oxidizing species must diffuse through the metal oxide layer for the reaction to continue. Since Cu forms p-type oxides (e.g. Cu_2O) the diffusion of metal ions through this metal-oxide layer towards dielectric occurs in preference to diffusion of larger OH- toward metal. The injection of metal ions leads to a charge build-up in the dielectric and an increase in dielectric leakage [3]. When the density of Cu ions in the dielectric reaches a critical value, a conduction path that links the cathode and anode may form which triggers a breakdown.

CONCLUSION

Electrical characterization of Cu diffusion in MIS capacitors reveals that Cu diffusion is strongly related to surface chemistry of the dielectric. For the diffusion of metallic Cu enhanced by the action of an electric field into nanoporous dielectrics, the metal must be ionized first. The ionization occurs by oxidation. The Cu oxide, which results from the ionization, acts as the source of the Cu ions that are available for diffusion. Hydroxyls formed from Silanol groups and interfacial oxygen due to incomplete shielding by the organic groups may act as the oxidizing species.

ACKNOWLEDGMENTS

This work is funded by the Semiconductor Research Corporation under task SRC 995.009, and New York State through the New York State Science and Technology Foundation, NYSTAR.

REFERENCES

1. J. D. McBrayer, R. M. Swanson, and T. W. Sigmon, *Journal of the Electrochemical Society*, 133, 1242 (1986)
2. Y. Shacham-Diamand, A. Dedhia, D. Hoffstetter and W.G. Oldham. *Journal of The Electrochemical Society*, 140 (8) 2427-2432 (1993)
3. L. S. Loke, J. T. Wetzel, P. H. Townsend, T. Tanabe, R. N. Vrtis, M. P. Zussman, D. Kumar, C. Ryu, and S. Wong, *IEEE Transactions on Electron Devices*, 46, 2178 (1999).
4. E.H. Nicollian and J.R. Brews. *MOS Physics and Technology*. Wiley, New York. (1982) pp 435-440
5. Nitta, S., Jain, A., Gill, W.N., Wayner, Jr., P.C., and Plawsky, J.L. (1999). *J. Applied Physics*, 86, 5870.
6. D.K. Schroder. *Semiconductor Material and Device Characterization*. John Wiley & Sons, New York (1990) p. 261
7. S. Fruehauf, I. Streiter, S.E. Schulz, E. Brendler, C. Himcinschi, M. Friedrich, T. Gessner, D.R.T. Zahn. *Advanced Metallization Conference 2001* 287-294
8. D. Kapila and J.L. Plawsky. *Chemical Engineering Science* 50 (16) 2589-2600 (1995)
9. S. Rogojevic, A. Jain, W.N. Gill, and J.L. Plawsky *Journal of The Electrochemical Society* 149 (9) F122–F130 (2002)
10. C. J. Brinker and G. W. Scherer, *Sol-Gel Science* (Academic Press, New York, 1990)
11. R.K. Iler, *The Chemistry of Silica* (Wiley, New York, 1979)
12. S. Rogojevic, A. Jain, W.N. Gill and J.L. Plawsky. *Electrochemical and Solid-State Letters*, v 5, n 7, July, 2002, p F22-F23
13. L.A. Belyakova, A.M. Varvarin, *Colloids and Surfaces A* **154** (1999) 285–294

Mat. Res. Soc. Symp. Proc. Vol. 812 © 2004 Materials Research Society   F8.1

MEMS Metallization

Christian Lohmann[1], Knut Gottfried[2], Andreas Bertz[1], Danny Reuter[1], Karla Hiller[1], Michael Kuhn[1] and Thomas Gessner[1, 2]
Chemnitz University of Technology (CUT), Center of Microtechnologies[1]
Fraunhofer IZM, Department MDE[2]
Chemnitz, Germany

ABSTRACT

Silicon is the dominating material for the fabrication of MEMS devices, especially in high volume production. However, metals with their typical properties are used to enhance or enable the functionality of MEMS. In contrast to microelectronic technologies, not only the electrical but also the mechanical and optical behavior of metals could be helpful. New requirements in MEMS technologies demand optimized processes in metallization for the fabrication of microstructures.

This paper presents some metallization applications and related technology development in the field of MEMS.

INTRODUCTION

Based on its widespread application in microelectronics technology and the mechanical properties as well, crystalline silicon is the dominant material for MEMS fabrication. Its usage is not limited to act as the base plate (substrate) but as the functional film or component too. The electrical resistivity of crystalline silicon can be varied in a wide range (0.005 up to 1000 Ωcm) offering a large flexibility as MEMS material. However, from an application and technology point of view in certain cases metal films are to be used. For example for rf MEMS a resistivity as low as possible is desired in order to minimize electrical losses. Metals with a resistivity of a few $\mu\Omega$cm meet this requirement. Furthermore light reflection from MEMS components can also be increased by metal coating of surfaces. Several effects for signal transducing offer other application fields: e.g. thermal – electrical, thermal – mechanical (bimorph) and magnetic-electrical transducers. Further fields for metal application are given by technology. The galvanoforming process within the LIGA technology [1] represents a powerful method for shaping materials, even if they are difficult to etch. Electroplating and electroless plating processes can easily be combined with wet etching processes in order to get released structures.

For mass fabrication by using bulk technology [2] nowadays metal is often used as an electrode and contact pad material. In this case the requirements are for instance: good adhesion, low thermal mismatch to adjacent films (substrate) and high long-term stability (relaxation, migration). This is quite comparable to IC fabrication. Surface micromachining by sacrificial layer etching underneath metal films as known from rf devices (switches, variable capacitances) [3] and laminated *post*-CMOS MEMS [4] illustrates much more challenges with respect to the mechanical properties. Thin film stress, yield strength, hardness and Young's modulus are critical issues. Although structure bending can be minimised by stress compensation, stress hysteresis due to relaxation can influence reliability and long-term behaviour deeply. Thus even metal surrounded crystalline silicon core structures like SCREAM devices [5] indicate strong temperature dependence [6]. On the other hand it is shown that metal based devices can offer excellent reliability through continuous research. TI's Digital Mirror Device is obviously one of the most impressive one. Data as published recently [7] indicate that problems like hinge fatigue, hinge memory, temperature and light influence can be solved by material and technology adaptation. Metal based fixtures for the fabrication of highly capacitive isolated anchor structures are proven to be stable enough for its application within certain MEMS [8].

Further challenges for MEMS metallization within special applications are given when the operation temperature is increased up to 400°C. Thus the interconnect system, the connections to the next wiring level as well as the whole packaging concept has to be adapted. The huge number of possible interactions within and between several materials of the functional parts at elevated temperatures are critical. Nevertheless, as shown recently [9] metallization concepts exists and have been verified even at the harsh atmosphere of a car exhaust and a surrounding temperature up to 400°C.

Metal used as optical reflection layer

Micro mirror devices (MMD) are typical MEMS products. They are used for a wide variety of applications like imaging systems or optical switches. The optical and mechanical characteristics of a MMD are dominated by the properties of the movable mirror plate. Two different structures are mainly used for the fabrication, pure metal plates, like the DMD [7], or metal coated plates [10]. The DMD structure includes mirror plates consisting of an aluminium alloy. Therefore, the optical and mechanical properties of the movable structure are limited to those of the alloy. Another method for the fabrication of a MMD is to coat a single crystal silicon element with a thin metal layer. The advantage of this method is the individual optimization of mechanical and optical properties by a separation of the materials. The mechanical properties are defined by the core material of the component (e.g. single crystal silicon) and the optical properties by the deposited reflection layer (e.g. Al, Ag, Au).

A special implementation of metal coated single crystal silicon based micro mirrors is the combination of a micro mirror and a grating directly on top of the moveable plate [11]. This kind of MEMS enables high quality digitising of images without the restriction to a limited colour space of the conventional RGB technique.

This micro mirror is a complex system consisting of an elastically hinged mirror plate actuated by an electrostatic field between plate and fixed electrodes on top of a carrier. It is fabricated by silicon bulk micromachining and uses an active optical area of 5.12 mm x 3.0 mm and a resonance frequency of about 1 kHz. Due to technological reasons the gratings, mechanical scribed aluminium grating as well as electron beam patterned SiO_2 gratings, are fabricated at the beginning of the wafer process. The silicon membranes (40 µm thick) for the mirror plate is wet etched from the backside of the actuator wafer. During this process, the aluminium gratings are protected by a wafer holder. Finally the membranes are patterned into mirror plates and torsion springs by dry etching from the front side.

Due to the stress of thermal oxide the micro mirrors with SiO_2 grating are bended in a convex shape. This can be compensated by a deposition of a surface layer, in that case aluminium, inducing a well defined tensile stress. But the required thickness would change the profile of the grating in an unacceptable way. One possible solution is to deposit a PECVD oxide on the back-side of the micro mirrors. This layer compensated the bending of the mirror caused by the SiO_2 grid on the frontside. The stress of this oxide is negative and can be adjusted by annealing of the wafer.

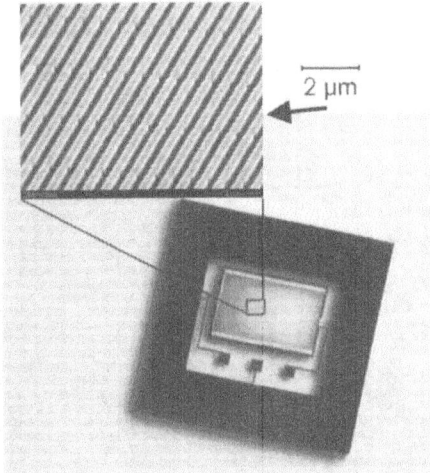

Figure 1. Micro mirror, consisting of a metal coated silicon element, with a grating directly on top of the moveable plate.

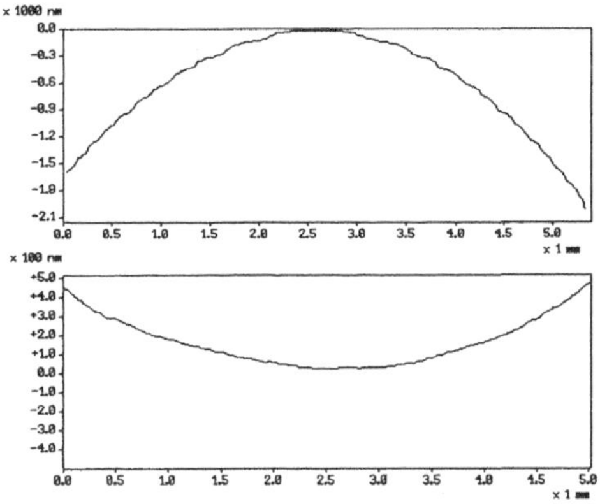

Figure 2. Bending of micro mirror surface before (top) and after compensation (bottom), note the different scale of the diagrams

Figure 2 shows the bending before and after compensation. Thereafter, with exception of the mechanical scribed gratings, an aluminium layer (55 ... 70 nm) is deposited on both sides of the mirrors using sputter masks. This layer serves as a reflector on the mirror front side and as a conducting and stress compensation layer on the mirror back side. Finally the silicon wafer with the mirrors is anodically bonded to a glass wafer which carries the fixed electrodes.

Metal used as electrical functional layer

The SCREAM – Technology (Single Crystal Reactive Ion Etching and Metallization) was developed at the Cornell University [12] for the fabrication of "High Aspect Ratio Microstructures" (HARMS). The silicon structures are formed by a sequence of deep silicon etch, sidewall passivation and release etch. Due to a consequent usage of self aligned processes, only one single mask for the deep silicon etch process is necessary. Finally the microstructure is coated by a metal layer, which enables wire bonding on top of the structure, provides a low resistivity ohmic contact between the bond wire and the MEMS structure and minimizes the electrical losses inside the structure by minimizing the electrical resistance. Figure 3 shows the principle process flow for the fabrication of a SCREAM structure.

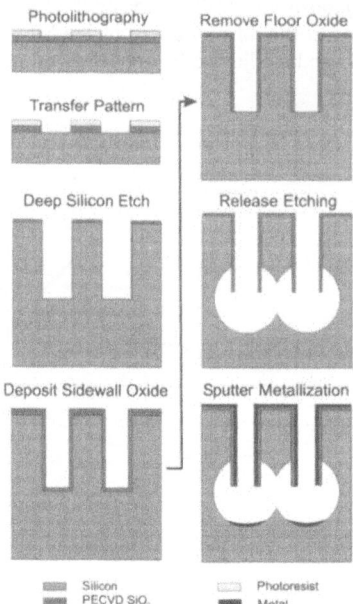

The deposition of metals onto oxide covered, etched sidewalls in deep silicon trenches is one of the biggest challenges in the fabrication of SCREAM devices. The electrical characteristics of the device depends strongly on the quality of the metal layer. The crucial points for the development are the step coverage and its uniformity of the deposited layer. Figure 4 illustrate the result of PVD deposition process and its simulation.

Silicon — PECVD SiO$_2$ — Photoresist — Metal

Figure 3. Process flow for the fabrication of SCREAM structures

However, the multilayer structure of a SCREAM device results in a deformation of the movable structure depending on the mechanical stress of the deposited layer. This deformation is sensitive to temperature changes and therefore critical.

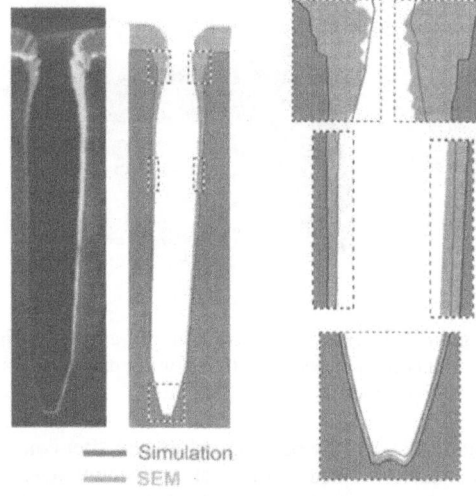

Simulation
SEM

Figure 4. Comparison of PVD deposition process (left) and its simulation result (right)

Metal used as electrical and mechanical functional layer

A novel CMOS–compatible technology for the fabrication of MEMS based on standard single crystal silicon wafers was presented recently [8]. High Aspect Ratio Microstructures are manufactured using a three mask level technology and dry processing throughout. The released micromechanical components consist of monocrystalline silicon without additional thin films after processing. As a result of the novel process flow the structures are surrounded by air gaps and fixed by special anchors (Fig. 5). These Air gaps Insulated Microstructures (AIM) were fabricated and tested with respect to mechanical stability, temperature dependence and electrical behavior.

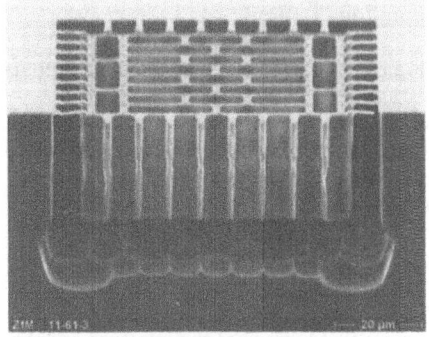

Figure 5. Released test structure with 34 interconnection beams consisting of Al

The mechanical and electrical contact between the seismic mass and the bulk material, represented by the conducting layer, is realized by interconnection beams, which consist of aluminum or a martial stack of a SiO_2-, PE-SiN and aluminum to ensure good mechanical stability and electrical conductivity. Therefore, the electrical and mechanical properties of the beams depend on properties of the used metal. The material stack has to provide tensile stress, which acts as a force on the released structure. For compressive stress an unsteady mechanical system would be obtained and tilting of the seismic mass is possible (Fig. 6). Tensile stress produces forces with opposite directions compared to the compressive stress. The mechanical system is in a steady state and the po-

sition of the seismic mass is fixed. For the characterization of the tensile stress, the temperature dependencies and the long-term behaviour are measured and drawn in Fig. 7 and Fig. 8. The tested material stack of 250 nm SiO_2, 250 nm PE-SiN and 300 nm Al indicates tensile stress over the whole temperature range (25°C to 150°C). After the heating cycle

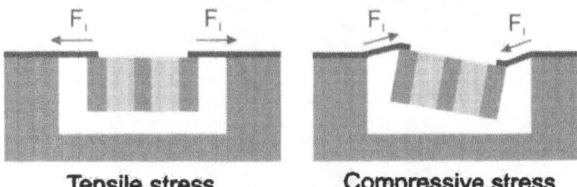

Tensile stress   **Compressive stress**

Figure 6. Results of different stresses depending on the material stack properties

only insignificant changes were detected. The long-term behaviour shows that the tensile stress of the material stack changes only within the first two weeks. After this period of time no changes could be found.

As a result the tensile stress is temperature and long-term stable. Therefore this kind of layer stack can be used for the interconnection beams.

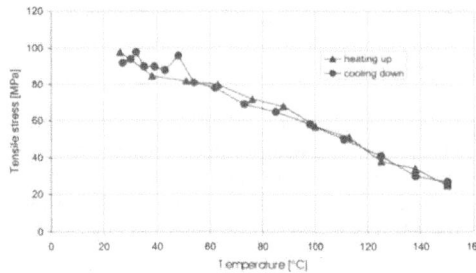

Figure 7. Long-term behaviour of the tensile stress of a 250 nm PE-SiN, 250 nm SiO_2 and 300 nm Al layer stack

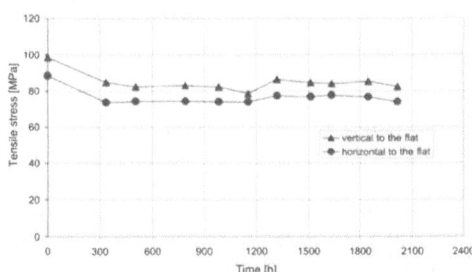

Figure 8. Temperature dependency of the tensile stress of a 250 nm PE-SiN, 250 nm SiO_2 and   300 nm Al layer stack

237

Wafer-level packaging based on low temperature bonding

Low temperature wafer level MEMS-capping technologies are developed to enable the combination of wafer bond techniques and metallized MEMS structures.

One technique to cap microstructures at the wafer level is adhesive wafer bonding. The advantages in comparison to anodic and silicon fusion bonding are low temperatures, independence from surface roughness and material, and no need in high voltages.

The contact pressure, bonding temperature, and layer thickness of the SU8 are critical process parameters. The formation of voids is inspected by infrared light or ultrasonic scan in case of aluminum. On the basis of the developed bonding technique, a metal coated SCREAM-like capacitive in-plane vibration sensor is capped with the new technology (Figure 9). The cavity between the sensor and the top wafer is formed by the SU8. To seal the cavity, the isolation trenches have to be covered by the SU8. Figure 10 shows a SEM image of the interface and the covered trench.

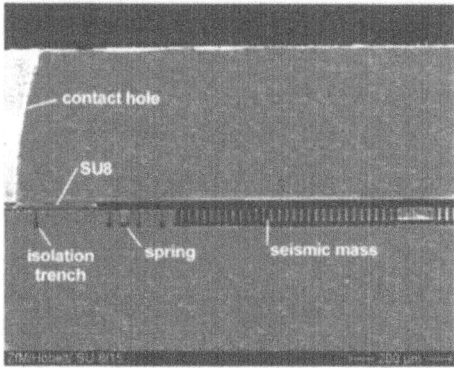 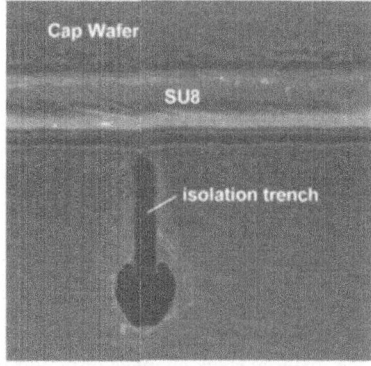

Figure 9. SEM image of a capped SCREAM structure

Figure 10. SEM image of the isolation trench covered by the bonded SU8 layer

After dicing the bonded wafer into single chips, the sensors are electrical interconnected by wire bonding through contact holes in the cap wafer, patterned by DRIE using the aluminium on top of the bondpads as the etch stop layer. The frequency response of the sensor is recorded before and after the capping process. No significant changes in the behaviour of the sensors and thus no influence on the sensor by the bonding process is detected.

In contrast to low temperature adhesion bonding a promising technology for low temperature silicon direct bonding is presented recently [13]. Direct bonding without intermediate layers allows an exact adjustment of gap distances, while the low temperature enable metallic components like electrodes for drive and control. This low temperature bonding is also usable for integrating of mechanics and electronics by bonding the MEMS wafer directly onto an electronic wafer as demonstrated [13].

Metallization for high temperature applications

The physical properties of most of the substrate materials used in MEMS fabrication offer a high capability for an application under extreme conditions like temperatures above 125 °C and up to 1000 °C. The most limiting factors in that case are the device metallization and the packaging concept. One main reason is the huge number of possible interactions within and between the materials at elevated temperatures. Within the metallization of temperature charged MEMS interactions like diffusion, solid state reactions, crack formation and propagation as well as film delamination have to be expected as shown in Figure 11. Moreover, oxidation and

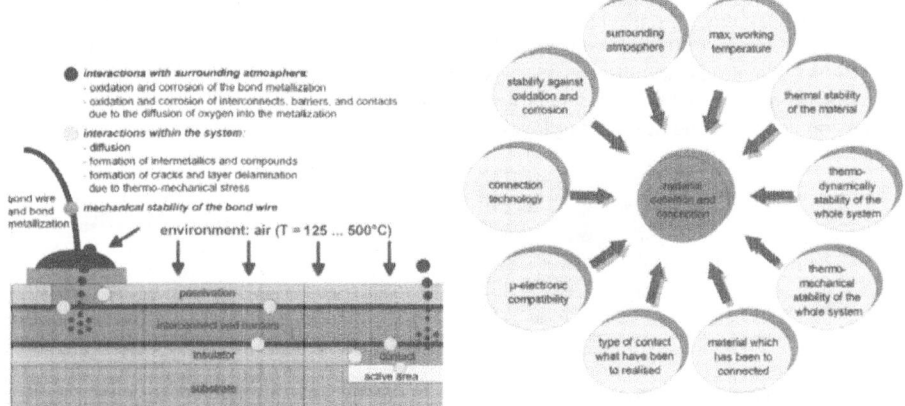

Figure 11. Possible interactions in temperature charged interconnect systems

Figure 12. Factors for choice of materials for
temperature loaded metallization schemes

corrosion problems are arising. Diffusion processes and solid state reactions mainly take place at interfaces of different materials. The formation of cracks as well as the delamination of films are related to thermo-mechanical stress. Reasons for stress are a mismatch in the CTE values as well as changes in the crystalline structure of the materials itself. However, these problems are partly solvable by the choice of appropriate materials and/or introduction of supporting materials like diffusion or reaction barriers. Further serious problems are the oxidation and corrosion of the metal layers. The use of a hermetic package is mostly not possible for MEMS due to their desired function. Moreover, guaranteed hermetic packages for elevated temperatures are a problem in general. Therefore, technological solutions on the chip level are required, which ensure the necessary stability against corrosion and oxidation. Basically, the whole interconnect system needs to be encapsulated by passivation layers from silicon oxide (SiO_2) and/or silicon nitride (Si_3N_4). The limiting factor in this case is thermo-mechanical stress. Most of the passivation materials show compressive stress. This leads to tensile stress in the underlying conductors that enhances the formation of voids. To overcome these problems, an optimisation of the deposition technologies and of the film thickness is necessary. Finally, a passivation can not solve all corrosion related problems. The connections to the next wiring level require an opening of the passivation at certain points and the deposition of a bond metallization. This metallization either must not react with oxygen or the reaction must be self limiting. Furthermore, a diffusion of oxygen through the bond metallization has to be prevented. Figure 12 summarizes once again all factors regarding the material choice of a temperature loaded interconnect and wiring system.

In the following the metallization of a high temperature pressure sensor based on silicon carbide (SiC) is briefly described. The sensor was designed to monitor the pressure inside a car engine during the combustion cycle [14] and should be placed in the (cooled) cylinder head of the engine. Temperatures up to 400 °C have to be expected at this place. Even under that condition a precise and reliable function must be given. The sensor is based on a piezo-resistive work principle because it is widely used for that purpose. A silicon carbide layer (3C-SiC), epitaxially grown on SOI-substrates [15], has been used for the fabrication of the piezo-resistors. Figure 13 shows the complete sensor device itself as well as the chip, Figure 14 shows a schematic structure of the chip. The necessary interconnect system comprises ohmic contacts to the SiC-substrate, interconnects, and the bond metallization. Moreover, a wiring technology is needed, to realise the connection to the next wiring level.

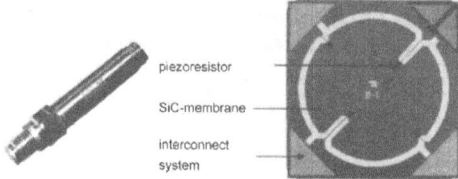

Figure 13. SiC-pressure sensor, complete device and sensor chip with metallization

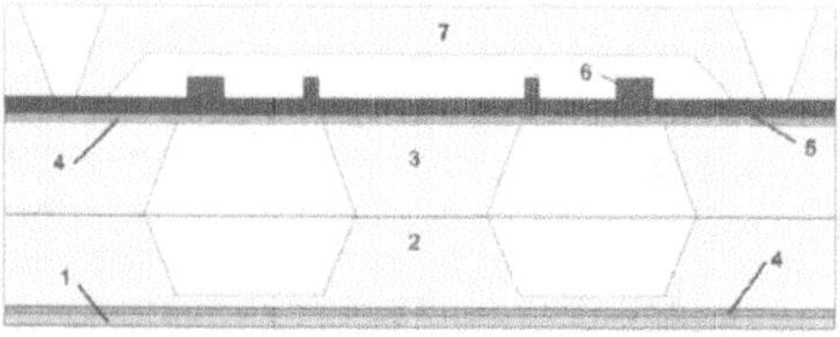

Figure 14. SiC-pressure sensor element, 1...SiC protection membrane, 2...lower membrane SOI-wafer, 3...upper membrane SOI-wafer, 4...Oxide layer, 5...SiC membrane, 6...piezo-resistors, 7...cap wafer

For the ohmic contacts to 3C-SiC WSi_2 and $MoSi_2$ have been chosen. The reasons therefore are their high thermal stability regarding SiC [16][17] and promising results concerning the reachable contact resistivity [18][19]. Both silicides have also been used as interconnects. The connection to the next wiring level have been realized using an aluminium bond metallization together with aluminium heavy wire bonding. The use of heavy wires is based on a simple reason. Commonly used bond wires with diameters far below 100 μm are alloys of aluminium with up to 2 % weight content of silicon. That is necessary to ensure the mechanical stability of the wire. At elevated temperatures the silicon starts to diffuse. The result is drastic loss in mechanical stability [20]. Heavy wires do not need that silicon content, because of the large diameter their own stability is sufficient. Aluminium as bond metallization offers an another advantage. It is well known, that aluminium in connection with oxygen forms an extremely dense oxide (Al_2O_3). Only a few nanometers of this oxide are necessary to prevent a further diffusion of oxygen through the oxide film itself. This leads to a self limiting oxidation process and ensures the protection of the underlying silicide against oxidation and corrosion.

The described metallization has been successfully tested for up to 2000 hours at 400 °C in air ambient. Neither electrical nor other kind of degradations (film delamination, cracks) could be observed. In figure 15 the overall resistance of one SiC resistor connected using the MoSi$_2$-Al metalliaziton is shown. No changes in the resistance are visible during the thermal storage. After the storage the initial room temperature values are reached again.

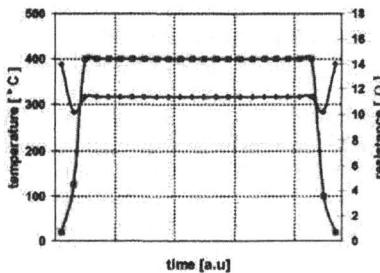

Figure 15. Behaviour of the overall resistance of the MoSi$_2$-Al metallization during a thermal storage for 2000 hours at 400 °C in air

CONCLUSION

In this article, several metallization issues are described with a focus on metal coated single crystal silicon, the dominant material for MEMS fabrication. The optical, mechanical or electrical properties of functional metal layers are utilized to enhance the performance of silicon based MEMS devices significantly.
Also related technology development for MEMS capping or metallization for high temperature applications are presented.

ACKNOWLEDGMENTS

The authors would like to thank the staff of the center of Microtechnologies for assistance in processing the samples and devices.This work was funded by the German Research Foundation (DFG) within the Collaborative Research Centre SFB 379 "Micromechanical Sensor- and Actuators- Arrays", by the "Sächsische Aufbaubank " and by the "Bundesministerium für Bildung und Forschung"

REFERENCES

1.	W. Ehrfeld et al., "LIGA Process: Sensor Construction Techniques via X–Ray Lithography", Technical Digest, IEEE Sensor and Actuator Workshop, pp. 1-4, 1988
2.	H. Kuisma, "Inertial Sensors for Automotive Applications", Transducers '01, 2A1.01, pp. 430-433, 2001
3.	G.M. Rebeiz, "RF MEMS Switches", Transducers '03, 4A1A, pp. 1726-1729, 2003
4.	D.F. Guillou, S. Santhanam, L.R. Carley , "Laminated, Sacrificial –Poly MEMS Technology in Standard CMOS", Sensors and Actuators 85 (2000) pp. 346-355

5. M.T.A. Saif, N.C. Mc Donald, "Planarity of Large MEMS", J. of MEMS, Vol. 5, No.2, June 1996, pp. 79-97
6. T. Gessner, A. Bertz, C. Steiniger, U. Wollmann, "Material and Technology Approaches of Surface Micromachining", Proc. of the Micro Matreials '97, Berlin, Germany, 1997, pp.90-96
7. M.R. Douglass: DMD Reliability, "A MEMS Success Story", Proc. of SPIE Vol. 4980 (2003) pp. 1-11
8. A. Bertz, M. Küchler, R. Knöfler, T. Gessner, "A Novel High Aspect Ratio Technology for MEMS Fabrication Using Standard Silicon Wafers", Sensors and Actuators A 97-98 (2002), pp. 691-701
9. T. Gessner, K. Gottfried, R. Hoffmann, C. Kaufmann, U. Weiss E. Charetdinov, P. Hauptmann, R. Lucklum, B. Zimmermann, U. Dietel, G. Springer, M. Vogel,"Metal oxide gas sensor for high temperature application"; Microsystem Technologies Vol. 6, No. 5, August 2000
10. T. Gessner, S. Kurth, C. Kaufmann, J. Markert, A. Ehrlich, W. Dötzel."Micomirrors and micromirror arrays for scanning applications" Proceedings of SPIE, Vol. 4178 – 35, pp. 338-347, 18-21 September 2000, Santa Clara, CA, USA
11. Kuhn, M.; Flaspöhler, M.; Krönert, S.; Kaufmann, C.; Gessner, T.; Hübler, A.; Frühauf, J, "Microactuators with Diffraction Grating", Micro System Technologies 2003, Poster-session, München, October 7-8, 2003
12. Z.L. Zhang, N.C. MacDonald, "An RIE process for submicron silicon electromechanical structures", J. Micromech. Mircoeng., 2 (1992) pp.31-38.
13. K. Hiller, S. Kurth, N. Neumann, R. Hahn, C. Kaufmann, M. Hanf, S. Heinz, T. Gessner, W. Dötzel, G. Ebest, "Application of low temperature direct bonding in optical devices and integrated systems", Proceedings of MICRO SYTEM Technologies 2003, pp. 102-109
14. J. von Berg, "Piezoresitiver Brennraumdrucksensor auf der Basis neuer Substrat-materialien für den Serieneinsatz im Automotor", Dissertation TU Berlin, UFO Atelier für Gestaltung und Verlag GbR, Allensbach, ISBN 3.930803-88-7, 2000
15. G. Krötz, H. Möller, M. Eickhoff, S. Zappe, R. Ziermann, E. Obermeier, J. Stoemenos, "Hetroepitaxial Growth of 3C-SiC on SOI for sensor applications", Mat. Science and Eng. B61-62, 1999, pp. 516-521
16. F. Goesemann, R. Schmid-Fetzer, "Stability of W as electrical contact to 6H-SiC, phase relations and interface reactions in the ternary system W-Si-C", Materials Science and Engineering, 1995, Volume B34, pp. 224 – 231
17. Costa, Silva, M. J. Kaufman,"Phase Relations in the Mo-Si-C System Relevant to the Processing of MoSi2-SiC Composites", Metallurgical and Materials Transactions A, 1994, Volume 25 A, pp. 5 – 15
18. L. M. Porter, F. Davis, "A critical review of ohmic and rectifying contacts for silicon carbide", Materials Science and Engineering, 1995, Volume B34, pp. 83 – 105
19. N. Lundberg, M. Östling, "CoSi2 ohmic contacts to n-type 6H-SiC, Solid-State Electronics", 1995, Volume 38 No.12, pp. 2023 – 2028
20. J.T. Benoit, R.R. Grzybowski, D.B. Kervin, "Evaluation of aluminium wire bonds for high temperature (200 °C) electronic packaging", Third international high temperature electronics conference, 1996, Trans. vol. 1, pp. III-17 - III-23

Mat. Res. Soc. Symp. Proc. Vol. 812 © 2004 Materials Research Society

"Chlorine-based Reactive Ion Etching Process to Pattern Platinum
for MEMS Applications"

Sung H. Choi, Jon V. Osborn, Brent A. Morgan
The Aerospace Corporation
2350 E. El Segundo Blvd., M2-244
El Segundo, CA 90245

ABSTRACT

We have developed platinum (Pt) deposition and chlorine-based Reactive Ion Etch (RIE) processes that are needed to deposit, pattern and embed electrodes deep within a MEMS process flow. Various combinations of chlorine-based gases were tested to find the optimum gas mixture for RIE. A 1: 0.4 mixture ratio of pure chlorine to argon and 100-150 Watts of RF microwave power were found to be optimum conditions for the RIE of platinum metal. The addition of argon gas to chlorine was found to contribute to the anisotropic etching of platinum, obtaining vertical shaped sidewall patterns. A simple model of the platinum etching mechanism is proposed. Following the plasma enhanced formation of platinum and chlorine ions, volatile products of platinum chlorides were formed and driven away at elevated temperature. As a demonstration of our RIE process, micron-sized platinum patterns with vertical sidewalls were fabricated. Etch chemistry was investigated using ToF SIMS analysis. This etch technique will be useful to device developers intending to use the unique properties of platinum metal as an electrode that is deeply embedded within a MEMS process flow.

INTRODUCTION

Platinum is used in many commercial and industrial applications. As a sensor material platinum is used to detect oxygen, carbon-monoxide, and nitrous-oxides. As a catalyst, platinum is used in fuel cells and automobile emission control systems to facilitate the controlled conversion of hydrogen containing fuels to carbon-dioxide, water and heat. In biomedical applications platinum electrodes are generally inert, do not corrode within the body, and rarely cause allergic reactions. In many of these applications the high melting point and/or corrosion resistant characteristic of platinum as well as its catalytic chemical properties are exploited. These conventional applications of platinum indicate that Microelectromechanical Systems (MEMS) could benefit from the use of this material by exploiting its unique properties.

Particular benefits of using Platinum in MEMS processes are related to its unique material properties. Table 1. Shows several properties of Platinum as it compares to other common semiconductor process metals. For example

Platinum may be used in lower layers of a MEMS device due to its high melting point, such that subsequent high temperature depositions and anneals may be performed without significant migration or chemical modification of the metal. In many electrical contact applications, there is a need to have a low loss metal electrode, deeply embedded in a MEMS process flow. Platinum is a good candidate metal due to its hardness, relatively low resistance and high melting point as well as its unique oxidation resistant chemical property. Furthermore, Platinum may be used as part of MEMS based biological sensors, laboratory on-a-chip devices, and micro-fuel cells due to its unique acid resistant and catalytic properties.

Table 1. Platinum Bulk Properties[1-3]

	Pt	Al	Au	Cu	Ti	W
Melting Point(°C)	1768.4	660.3	1064.2	1084.6	1668	3422
Resistivity (*10^{-8} Ω-m)	9.6	2.4	2.05	1.54	39	4.82
Hardness (mohs)	3.5	2.75	2.5	3.0	6.0	7.5
Thermal Conductivity (W/cm °K)	0.716	2.37	3.17	4.01	0.219	1.74
Coefficient of Linear Expansion (*10^6/°K)	8.8	23.1	14.2	16.5	8.6	4.5

As part of our prior research program to evaluate and develop MEMS technology for use in space systems, we have developed a MEMS process for the prototype of novel MEMS devices, Figure 1. A few of the unique aspects of our in-house process are: I) It is design rule compatible with the Cronos (acquired by MEMSCAP Inc.) MUMPS process [4], II) We allow for the insertion of novel MEMS materials and process steps within the baseline MEMS process flow, and III) It is a non-commercial process used exclusively for research and development and/or one-of-a-kind prototype development efforts of MEMS devices for use in space system applications. This process has been previously used to successfully prototype novel chemical sensors, micro-thrusters, and to develop a novel dry release method [5-9].

In our current research, we have been investigating the use of refractory metals for the purposes of embedding electrostatic actuator electrodes and RF switch contact electrodes deep within a MEMS device process flow. In our current application the high melting point, chemical resistance and hard contact properties of platinum will enable us to develop the embedded metal structures of a novel MEMS RF switch that we have previously patented [10].

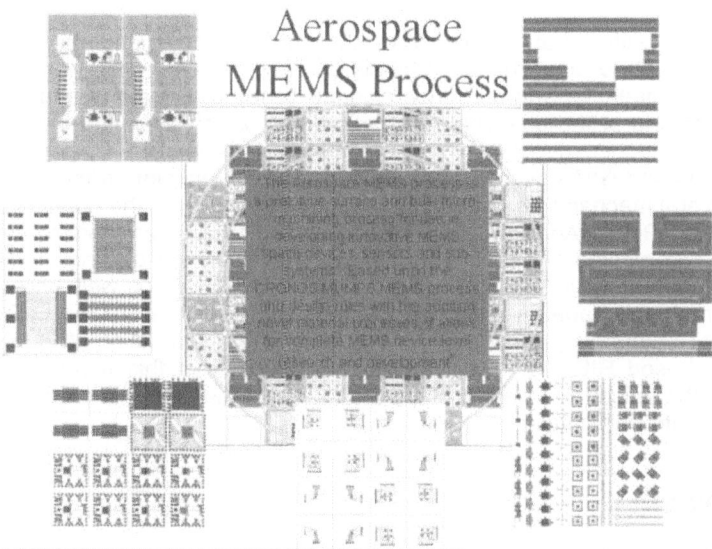

Figure 1. The Aerospace MEMS process is a surface and bulk micromachine process that is Cronos design rule compatible. The process includes thick TEOS oxide layers, embedded TiC hardened surface materials and embedded Pt metallization layers.

In order for us to evaluate the use of platinum metal within our MEMS process we needed to develop the means to pattern platinum thin films after deposition. We have found in the literature and used conventional e-beam evaporation techniques to deposit platinum metal as a thin film [11], but there are very few references to etching and patterning thin Pt films at micron-scales within silicon based MEMS processes [12]. Given the desired micron scale metal trace-widths with vertical sidewall profiles, we determined that dry etching of platinum was a good candidate process. However, platinum does not readily form volatile products and therefore is known to be very difficult to etch in conventional plasma systems. Until this work, most of the attempts to pattern Pt have focused on sputtering, which results in poor selectivity. Our research aims at exploring new chemistries and techniques to develop a dry etching process for patterning platinum, using a more chemically oriented etch process, offering higher selectivity at appreciable etch rates.

Dry etching involves subjecting a photo-lithographically patterned sample to a high density RF plasma of chemically reactant ions under low vapor pressure. With a decade of experience on the microelectronics foundry line, the chlorine based dry Reactive Ion Etch (RIE) method has proven quite efficient at producing finely featured structures of semiconductor materials with smooth sidewalls. RIE etching has been widely used in the semiconductor industry for high fidelity pattern transfer and sidewall etch anisotropy, which is critical as the minimum feature size shrinks. Etching of deep sub-micron CMOS polysilicon gates is frequently performed in a chlorine based RIE plasma and a thorough understanding of the chemical mechanism has been developed there to model and predict etching directionality and feature profile evolution. Chlorine etching of the silicon and GaAs systems has been extensively studied in the past decade, and several mechanisms have been proposed to elucidate the dominant reaction pathways [13-15]. Additionally, chlorine-containing gases are commonly used for the plasma etching of aluminum materials. However, dry etching processes for patterning materials such as gold and platinum are not well known in the literature. In this paper, we introduce a chlorine-based Reactive Ion Etching (RIE) process to patterning Pt materials. We demonstrate fabrication of Pt patterns by chlorine-based RIE. Sidewall profiles that result from this process appear to confirm the anisotropic nature of the etch while at the same time good control of the etch rate is maintained.

EXPERIMENT

An SLR770 RIE system, located at the UCLA Nanofabrication Laboratory, with an ICP plasma source was utilized to develop a chlorine based RIE process for fabricating Pt materials. This RIE system can produce nearly monoenergeric, low energy (500V-100kV) and high flux (0.1-2 mA/cm^2) chlorine ions. Monatomic Cl$^+$ is found to be the primary ion by mass spectrometry measurement. Typically, the chlorine ions have an energy distribution with approximately 10eV full-width at half maximum. Throughout these experiments the ICP power was fixed at 500W and sample bias was 360-370V.

The process experiments included manipulation of three parameters: Cl$_2$ & Ar mixture ratio, process pressure, and RF power. Before RIE processing, 2 μm, μm, and 0.5 μm thick Pt layers were deposited by e-beam evaporation onto both coated silicon substrates and quartz substrates. A thin (10-30 nm) Ti adhesion layer was used to coat the substrates prior to Pt deposition. Next a 1.2 to 1.4 μm photoresist (AZ5214) pattern was spun onto the substrates as the pattern mask material and then subsequently exposed in a standard Suss MA-6 mask aligner to create test structures. Various combinations of chlorine-based gases were tested to find optimum etch conditions. Additionally, various mixture ratios of pure chlorine to argon were also tested to determine the best sidewall etch profiles for this process. Good results were found for pure Cl$_2$ in Ar and these results are presented here. A simplified picture, shown in Figure 2, illustrates the basic RIE arrangement.

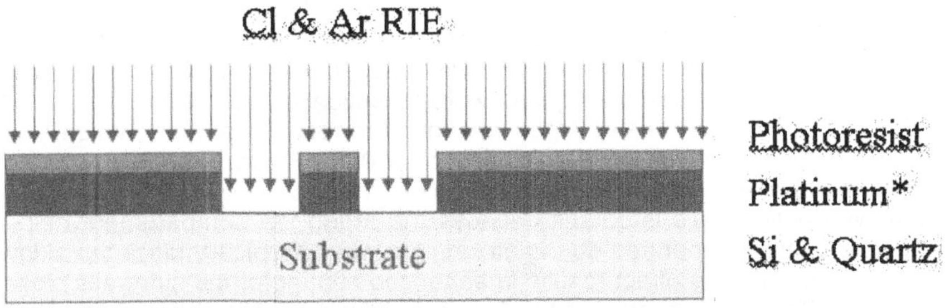

Cl & Ar RIE

Photoresist

Platinum*

Substrate

Si & Quartz

Figure 2. Simplified representation of the Pt RIE process. *Platinum e-beam evaporation depositions were 2 μm, 1μm and 0.5 μm. Platinum depositions included a 10-30nm Ti adhesion layer deposition. Pt features were 1.5 μm wide.

RESULTS & DISCUSSION

Gas Mixture Ratio vs. Etch Profile

A gas mixture composed of pure chlorine to argon in the ratio of 1:0.4 while using 100-150 Watts RF power were found to be optimum conditions of RIE for Pt materials. The addition of argon gas to chlorine is found to contribute to the anisotropic etching, and results in more vertical sidewall profiles. Following the plasma enhanced formation of Pt and chlorine ions, volatile products of Pt-chlorides were formed and driven away at elevated temperature. In our early experiments using this process, with a gas mixture of 1:0.25 pure chlorine to argon, RF power 200W, and Pressure of 50 mTorr, we obtained an etch rate of 500 Å/min. With these etching conditions micron-sized Pt patterns with low aspect ratio sidewalls were fabricated as shown in Figure 3. Aspect Ration here is defined to be the ratio of the height of the sidewall to the extension of the base of the sidewall from the vertical at the top of the sidewall. In our initial experiments this ratio was approximately one-to-one (1:1).

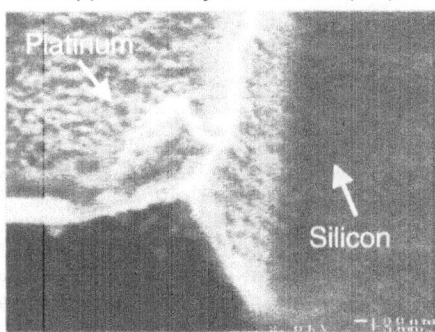

Figure 3. SEM image of 2 μm thick Pt pattern generated by chlorine-RIE process. This sample was etched using a gas mixture ratio of 1:0.25, pure chlorine to argon, with 200W RF power at 50 mTorr.

To achieve better vertical sidewalls profiles of Pt patterns, various gas-mix ratios were investigated during our processing. Based on the general observation that Ar can enhance an etch rate of polymer side-walls close to the etch surface, various mix ratios of Cl_2 to Ar gases were investigated. A mixture ratio of 1:0.4 composed of pure chlorine to argon was found to result in a more vertical shape of the sidewalls than our initial mixture ratio 1:0.25. Figure 4 shows a profile of the sidewalls of Pt patterns resulting from RIE processes with the higher gas mix ratios. We believe this is occurs because the argon ion bombardment helps to remove polymers generated during etching that would typically block the sidewall etching process. This effect results in enhanced etching of the sidewalls close to the etch surface. As can be seen in Figure 4, increasing the percentage of argon from 23% to 29% results in better vertical shape of side-walls. In these revised experiments our aspect ratio is greater than ten-to-one (10:1).

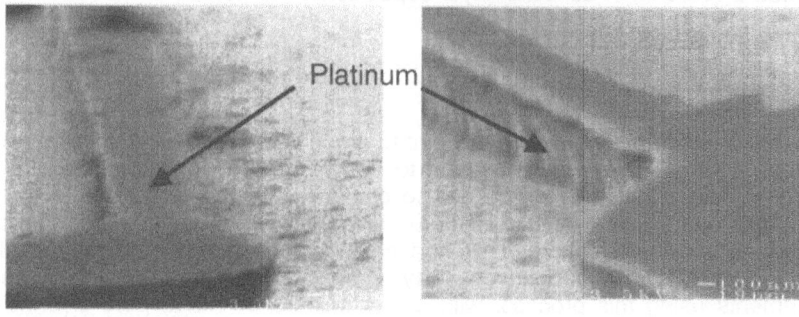

| (a) | (b) |

Figure 4. Scanning Electron Micrograph of 500-1000 nm thick Pt patterns, 1.5 µm wide prior to etch, generated by a chlorine-RIE process with a gas mixture of 1: 0.3 of pure chlorine to argon (a) and 1:0.4 (b). The aspect ratio and profile of vertical sidewalls was improved by increasing the proportion of argon in the gas mixture.

Process Pressure vs. Etch Rate

Extrapolating from the results of other RIE process research efforts, etch rate typically increases with process pressure [16-17]. In our experiments, various process pressures were tested to determine the effect of process pressure on Pt etch rate. Chlorine RIE runs with various process pressures ranging from 50 mTorr to 100 mTorr confirmed that general observation, as shown in Figure 5. Above 80 mTorr, our etch rate increased significantly, up to 900 Å/min, at 100 mTorr. This effect may be due to overheating the substrate at high process pressures as a result of interaction with the high-density plasma.

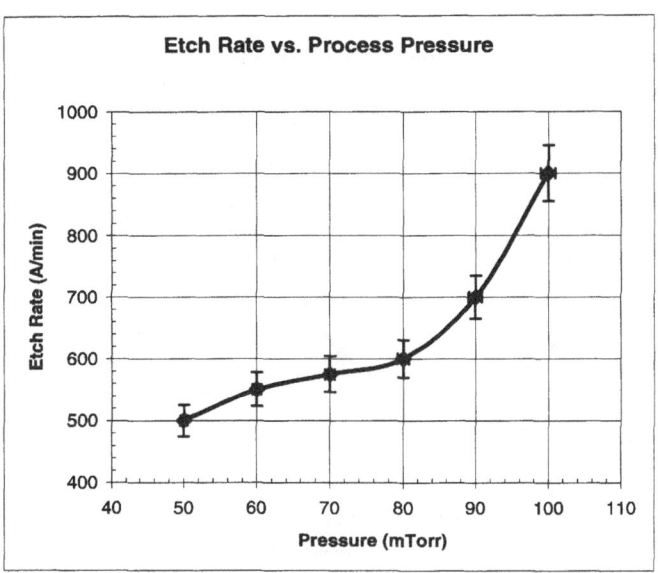

Figure 5. Pt etch rate vs. process pressure based on chlorine-RIE process. Other conditions for RIE process were fixed at 150W RF power and 1:0.25 gas mix ratio of chlorine to argon.

RF Power vs. Etch Rate

It is generally true that etch rate increases with RF power. In our etch experiments this trend is also seen as shown in Figure 6. However, our etch rate is found to taper off as the power increases beyond about 200W. This may be due to saturation of the plasma above a certain level of RF power, perhaps due to RF losses or thermal losses to the chamber. Our RF generator has the capability of delivering up to 600 watts of forward power, but we found that the photoresist (AZ5214) we have used as a mask material starts to char at settings as low as 250 watts, therefore we did not test beyond this level.

Chemistry

We believe that during our chlorine-based RIE processing, following the plasma enhanced formation of Pt and chlorine ions, volatile products of Pt chlorides were formed and driven away at elevated temperature. A simple model of the Pt etch mechanism is proposed below:

$Pt + Cl_2 + Ar_2 \rightarrow PtCl_x\uparrow$ *(Complexes)+ Ar(+Polymers)*↑   *at elevated temperature*

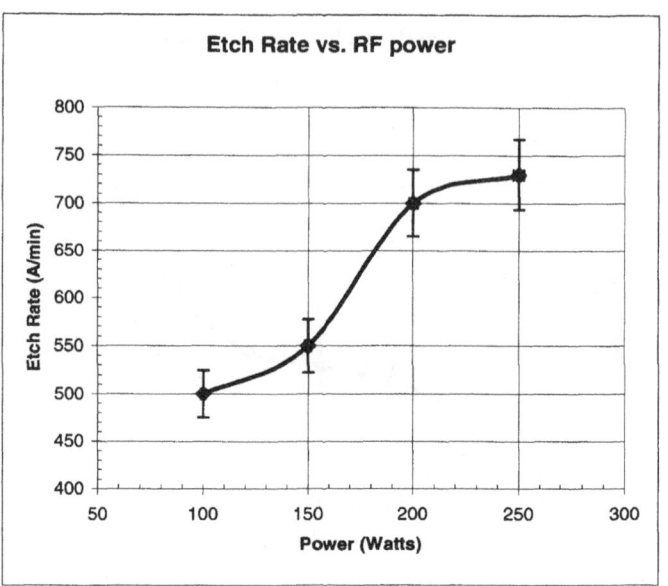

Figure 6. Pt etch rate vs. RF power based on chlorine-RIE process. Other conditions for RIE were fixed at 1:0.25, chlorine to argon gas mix ratio and process pressure of 60 mTorr.

The chemistry behind this etching process for Pt was studied by using static secondary ion mass spectroscopy (SIMS) on Pt field regions after exposure to the chlorine-RIE process. Static SIMS measurements were made with a Physical Electronics Trift2 time of flight (ToF) SIMS. Static SIMS is a measurement technique that allows us to analyze surfaces and the near-surface region for trace contaminants and reaction products of both elemental and molecular types. Our system uses a pulsed Indium primary ion beam to impact the surface of interest. Some of the material from the surface of interest is sputtered during this collision process and some further fraction of this sputtered material is singly ionized. These "secondary ions", are then accelerated away from the sample and travel over a known path length. Their arrival time at a detector allows one to determine their mass. The energetics of the sputtering process are such that large molecular fragments often remain intact after sputtering, allowing compound as well as elemental specificity with this technique.

In our experiments, ToF SIMS analysis was performed on silicon and quartz substrates containing the Pt metal structures etched by our chlorine-based RIE. We used these results to better understand the resulting surface chemistry. Figure 7 shows results of this analysis of (7a) Pt/Si and (7b) Pt/quartz; both after being etched by our chlorine-RIE process. In general, PtClx compounds have a distinctive isotope pattern which facilitates mass spectral identification. For example, the isotopic distribution of $PtCl_2$ ranges from 262 to 272, with a peak at 266.

As can be seen in Figures 7 and 8, $PtCl_2$ (most prominent peak = 266 amu) is among the various $PtCl_x$ complex molecules observed with Static SIMS on both samples. Mass spectral observation of $PtCl_2^+$ is consistent with the following SIMS-induced ionization reaction that takes place on the surface of the sample.

$PtCl_3 \rightarrow PtCl_2^+ + Cl^-$

Likewise, observation of $PtCl_3^+$ is consistent with the following reaction

$PtCl_4 \rightarrow PtCl_3^+ + Cl^-$

Large numbers of Cl^- ions are observed among the negative ions, also consistent with this SIMS-induced ionization reaction. These observations suggest that the etch mechanism proposed earlier is reasonable.

$PtCl_3$ is a stable Pt-chloride with a melting point of 400C. $PtCl_4$ has a melting point only slightly less at 370C. The similarity of melting points suggests that for process temperatures of several hundred C, one would expect to see more or less the same decomposition rate of the two Pt-chlorides. A more stable Pt-chloride, Pt_6Cl_{12}, melts at 581C [3] but no mass spectral peaks were observed that require the presence of that compound for explanation.

While Pt is being etched, the temperature of the substrate is increased close to the melting point locally, resulting in much of the $PtCl_3$ and $PtCl_4$ vaporizing. Process temperatures may not be high enough to drive this reaction to 100% completion, resulting in a small amount detectable by Static SIMS. Yield of $PtCl_2$ is found to be different by about two orders of magnitude between the silicon and quartz substrates. Because quartz has a lower thermal conductivity than Si, it may have been hotter during the RIE process which in turn drove off more of the residual $PtCl_3$ and $PtCl_4$ from the surface.

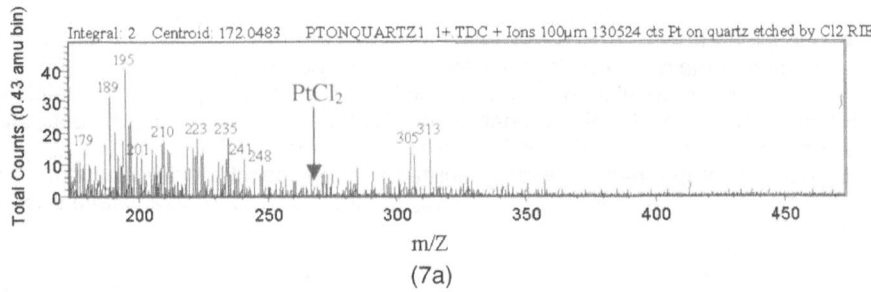

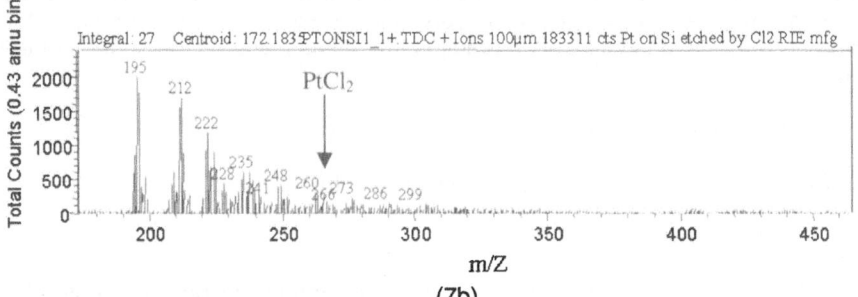

Figure 7. Surface analysis data by TOFSIMS for: (7a) Pt/Si etched by chlorine-based RIE and (7b) Pt/quartz both etched by our chlorine-based RIE process.

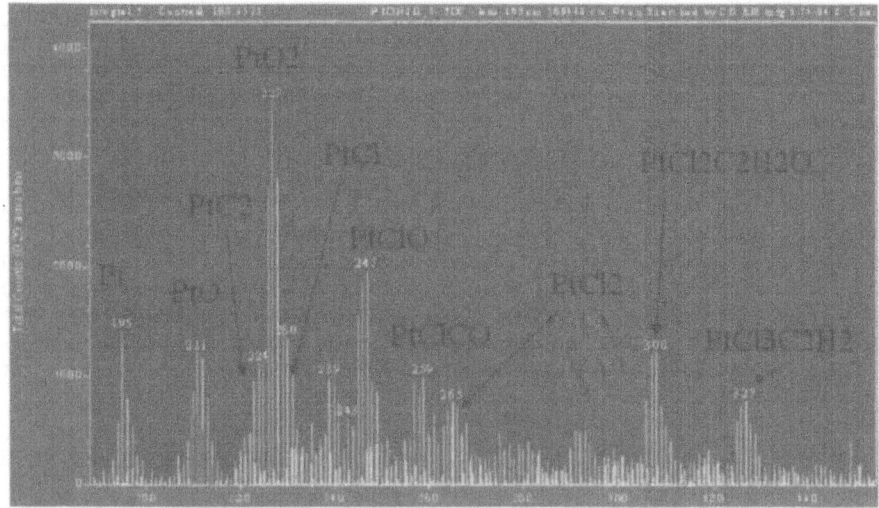

Figure 8. ToF SIMS Analysis of Pt etched patterns on Silicon, showing the relative proportions of the predominantly Pt based molecules created as a result of our chlorine-based Pt RIE process. Labels are associated with the location of previously known spectra for the given molecules.

Conclusions

We have developed a novel MEMS dry etching process for patterning Pt metals on silicon and quartz substrates. The optimum conditions of the RIE process to result in anisotropic vertical sidewalls with high-aspect ratio (greater than 10:1) and good etch rate control were demonstrated as a function of gas mix ratios, RF power and process pressure. We have investigated the resulting RIE chemistry using ToF SIMS and conclude that the resulting surface chemistry analysis is consistent with the formation of several $PtCl_x$ compounds. This process is compatible with conventional MEMS and Microelectronic process technology and will enable the insertion of platinum metal electrodes deep within a MEMS device structure.

Acknowledgements

The authors thank Mr. Steve Franz of the UCLA nanofabrication laboratory for the use of the SLR770 RIE System and Dr. Steve Moss of The Aerospace Corporation for advice and critical review of the material presented here.
This work was supported under The Aerospace Corporation's Mission Oriented Investigation and Experimentation program, funded by the U.S. Air Force Space and Missile Systems Center under Contract No. FA8802-04-C-0001.

References

1. G.V. Samsonov (Ed.) in *Handbook of the Physicochemical Properties of the Elements*, IFI-Plenum, New York, USA, 1968.

2. D.R. Linde (Ed), *Handbook of Chemistry and Physics 75th Edition*, CRC Press, London, UK, 1995.

3. Webelements on-line chemical and material reference.
http://www.webelements.com/webelements/compounds/text/Pt/cl12Pt6-10025657.html

4. MEMSCAP Multi-User MEMS Process (MUMPS) website
http://www.memscap.com/index.html

5. R.C. Cole, R. Robertson, J. Swenson, J.V. Osborn, "MEMS Packaging Techniques Using HF Vapor Release," IMAPS, Reno NV, 2002.

6. Bruce H. Weiller, Peter D. Fuqua and Jon Osborn, "Characterization and Thermal Failure Analysis of a Micro Hot Plate Chemical Sensor," *Proceedings of the Electrochemical Society*, vol. PV 2002-6, 2002, p. 36 - 42.

7. Bruce H. Weiller, Peter D. Fuqua, and Jon V. Osborn, "Fabrication, Characterization, and Thermal Failure Analysis of a Micro Hot Plate Chemical Sensor Substrate" Journal of The Electrochemical Society, 151 (3) H59-H65 (2004)

8. D.H. Lewis Jr., S.W. Janson, R.B. Cohen, and E.K. Antonsson, "Digital Micropropulsion," *Sensors and Actuators*, p. 143-154, **80,** 2000.

9. S.W. Janson, H. Helvajian, and K. Breuer, "MEMS, Microengineering and Aerospace Systems," AIAA paper 99-3802, 30th AIAA Fluid Dynamics Conference, Norfolk, VA, June 1999.

10. Jon V. Osborn, *RF MEMS Switch,* USPTO no. 6,426,687, July 20, 2002.

11. G.S. Lodha, S. Pandiz, A. Gupta, R. V. Nandedker, and K. Yamashita, Applied Physics A62, 29, 1996.

12. F. Ulashia, J. Pettic, and D. Mcvitti, Journal of electrochemical Society, vol. 135, No.6, 1521, 1988.2. S.C. McNevin and G.E. Becker, *J. Vac. Sci. Technol.,* B3, 485 (1985)

13. R.A. Baker, T.M. Mayer and W.C. Pearson, *J. Vac. Sci. Technol.,* B1, 37 (1983)

14. U. Gerlach-Meyer, J.W. Coburn and E. Kay, *Surf. Sci., 103,* 177 (1981)

15. J.W. Coburn and H.F. Winters, *J. Appl. Phys.,* 50, 3189 (1979)

16. Gray, I. Tepermeister, and H.H. Sawin, *J. Vac. Sci. Technol.,* B11, 1243 (1993)

17. Steinbruchel, *Appl. Phys. Lett.,* 55, 1960 (1989)

Mat. Res. Soc. Symp. Proc. Vol. 812 © 2004 Materials Research Society   F8.3

Silver Metallization with Reactively Sputtered TiN Diffusion Barrier Films

L. Gao[1], J. Gstöttner[1], R. Emling[1], Ch. Linsmeier[2], M. Balden[2], A. Wiltner[2], W. Hansch[1],
D. Schmitt-Landsiedel[1]
[1]Institute for Technical Electronics, Technical University Munich,
D-80333, Munich, Germany
[2]Max-Planck-Institut für Plasmaphysik, EURATOM Association,
D-85748, Garching, Germany

ABSTRACT

The physical and electrical properties as well as thermal stability of reactively sputtered titanium nitride (TiN) film serving as a diffusion barrier was studied for silver (Ag) metallization. The thermal stability of Ag/TiN metallizations on Si with 12-nm-thick TiN barriers, as-deposited and after annealing at 300-650^{0}C in N_2/H_2 for 30 min, was investigated with sheet resistance measurement, X-ray diffraction, focused ion beam-scanning electron microscopy, atomic force microscopy and X-ray photoelectron spectroscopy. According to electrical measurement no change of sheet resistance was found after annealing at 600^{0}C, but an abrupt rise appeared at 650^{0}C annealing. There are two causes by which the Ag/TiN/Si structure became degraded. One is agglomeration of the silver layer, and the other is oxidation and diffusion which are also associated problems during thermal annealing.

INTRODUCTION

A high performance interconnection network on a chip is becoming increasingly important for ultra large-scale integration (ULSI) of Si integrated circuits. Continued shrinking of devices has led to a discrepancy between the transistor and interconnect performance [1]. Because of its high electrical conductivity and its good electromigration properties, copper replaces aluminum in some microelectronic applications [2]. However, Cu resistivity increases substantially in sub-100-nm technology due to the size effect. Silver, with the lowest bulk resistivity among the conductor metals and similar or even higher electromigration resistance, is one of the best potential candidates for future interconnect metallization [3-5].

Until now, a large number of barriers against Cu diffusion have been investigated. The usage of Ag interconnects in Si ULSI devices requires also development of barrier layers which prevent Ag diffusion into the Si substrate and SiO_2 insulators at elevated temperatures. Thin aluminum oxynitride ($Al_xO_yN_z$) diffusion barriers have been formed in high temperature by annealing Ag/Al bilayers on oxidized Si substrates in ammonia ambient, in which Ag is deposited by electron-beam evaporation [6]. Diffusion and electromigration behavior was investigated for sputtered Ag lines with different barrier layers of Al_2O_3, Si_3Ni_4 and Ti [5].

Titanium nitride (TiN) barriers, which are thermally stable at high temperature, have been widely investigated to prevent interdiffusion of Cu in ULSI [7,8]. This report discusses process development and characterization of TiN as a diffusion barrier for silver metallizations. The sample structure in this work is Ag (200 nm)/TiN (12 nm)/Si. Finally, a possible failure mechanism of TiN barrier in preventing Ag diffusion is discussed.

EXPERIMENTAL DETAILS

The n-type Si samples were prepared according to the standard RCA wafer cleaning procedure prior to loading into a vacuum chamber. After Ar plasma cleaning and without breaking the vacuum, TiN thin films were deposited by reactively sputtering Ti (pure target) in a mixture of Ar and N_2 at a constant total pressure of $3x10^{-3}$ mbar in an Alcatel RF-magnetron system. Pure Ti or TiN films with five different compositions were deposited by changing the N_2:Ar flow rate ratio to 0, 0.2, 0.5, 0.8 and 1.2 with RF power of 200 W. The base pressure of the system with a liquid nitrogen trap was $2x10^{-7}$ mbar. Without breaking vacuum after the deposition of TiN films, Ag films were sputter-deposited directly on the TiN film using a 99.99% pure silver-target in pure argon process gas. Deposition was carried out at a RF power of 50 W and a processing pressure of $7x10^{-3}$ mbar. To evaluate the barrier capability of the TiN layers, samples of the Ag/TiN/Si were thermally annealed in a N_2/H_2(95%:5%) gas flow furnace for 30 minutes at temperatures ranging from 300 to 650^{0}C. X-ray diffraction (XRD) analysis was used to determine a crystallographic structure of the films, and focused ion beam-scanning electron microscopy (FIB-SEM) as well as atomic force microscopy (AFM) were employed to observe the surface morphology and the change of the surface microstructure. Sheet resistance of the films was measured using a four-point probe. A Dektak profilometer was used to measure the thickness of the deposited thin films. X-ray photoelectron spectroscopy (XPS) was carried out using a commercial system (PHI 5600) with a monochromatic Al Kα source and a hemispherical analyzer, operated at a pass energy of 93.9 eV for survey spectra and 29.35 eV for the spectra used to gain the depth profile. The analysis spot diameter at the sample was 0.8 mm.

RESULTS AND DISCUSSION

Surface morphology of the as-deposited and annealed films

Figure 1 shows surface morphology of the Ag surfaces of Ag/TiN/Si multilayer structures after annealing at various temperatures using FIB-SEM. A smooth and uniform surface of the as-deposited Ag film can be seen, and at low temperature annealing the surface morphology remained stable. Higher temperature annealing of the multilayer films caused an increase in the number and the size of the holes, and also, as expected, the film surface became much rougher. Agglomerated islands were observed clearly after 500^{0}C annealing, (which is also confirmed by a following AFM measurement) and the conducting islands became more and more isolated after higher temperature annealing. This is shown in the enlarged micrograph of figure 1(d). This observation is consistent with the results of the sheet resistance measurement reported in the following paragraph. Since XRD did not indicate any new phase formed upon annealing, the islands were thought to be agglomerations of Ag films.

Figure 2 shows the surface morphology of Ag/TiN/Si samples obtained by AFM analysis over (a) 1 μm x 1μm and (c) 20 μm x 20 μm areas. Figure 2(a) is from an as-deposited film, and figure 2(b) shows the height profile plot of the line marked on the surface image in figure 2(a). After 500^{0}C annealing, the surface morphology and height profile plot of the line on the surface image appear as in figure 2(c) and 2(d). The AFM analysis is measured over a 20 μm x 20μm area because of the grain growth after high temperature annealing. The AFM measurements confirm again that after 500^{0}C annealing the surface morphology gets rougher and also the grain size grows.

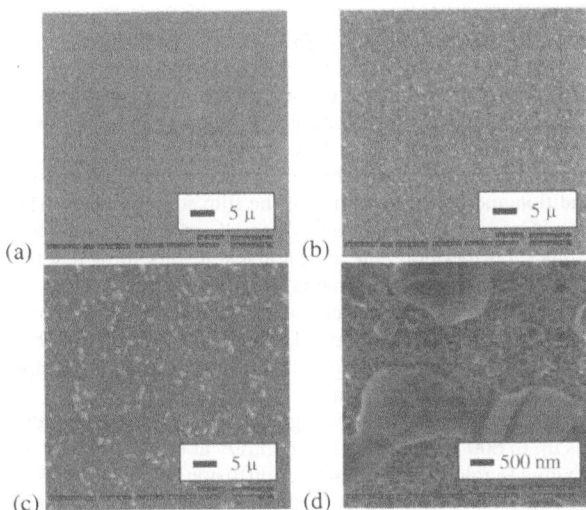

Figure 1. SEM photographs of the surfaces of multilayer structures (Ag/TiN/Si), (a) as-deposited and after annealing at (b) 300^0C, (c) 500^0C and (d) 650^0C.

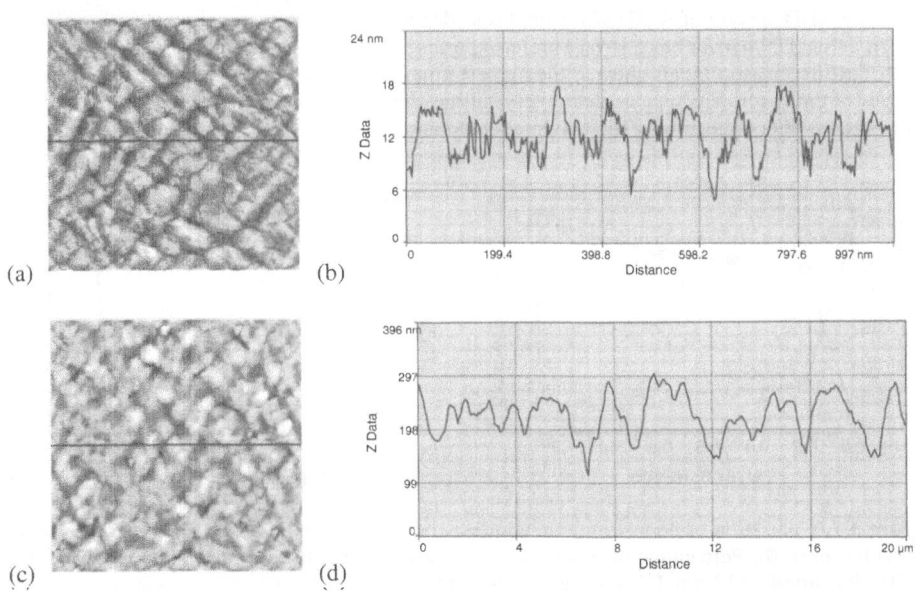

Figure 2. Surface morphology of Ag/TiN/Si substrates: from as-deposited (a) AFM image of the surface and (b) height profile plot of the line marked on the surface image in (a); after 500^0C annealing (c) AFM image of the surface and (d) height profile plot of the line marked on the surface image in (c).

257

Sheet resistance measurement of Ag/TiN/Si

Figure. 3(a) shows the dependence of the TiN thin film deposition rate and resistivity on the Ar/N_2 flow ratio. In the chamber, the nitrogen percentage increased with the nitrogen flow rate in the sputtering gas mixture, with the Ar flow kept at a constant rate of 10 sccm and the chamber pressure maintained at $3x10^{-3}$ mbar. The TiN thin film electrical resistivity depends on the deposition condition and is very sensitive to the film stoichiometry. The electrical resistivity of the pure titanium film initially increased as a small amount of nitrogen was added to the sputtering chamber. Furthermore, the adhesion test results show that there is a good adhesion between Ag and TiN barrier (Ar/N_2 flow ratio: 10 / 5 sccm).

The electrical resistivity of the as deposited Ag (200 nm) thin films on TiN (12 nm) is about 1.93 $\mu\Omega$ cm, similar to previously obtained values for sputtered silver films [4]. Figure 3(b) shows the percentage change of sheet resistance versus annealing temperature for the samples of Ag/TiN/Si with 12 nm TiN barrier film.

From figure 3(b), we can see that the Ag/TiN/Si sheet resistance remains stable for annealing at temperatures up to 600°C and then rises abruptly at 650°C annealing. This is due to the surface morphology change (agglomeration) and out diffusion of silicon after annealing of the samples at different temperatures. But from the micrographs shown in figure 1 it is obvious that the main cause is the agglomeration of the silver film at high annealing temperatures.

To investigate the annealing-induced material changes in these samples, material characteristics of the multilayer structure after annealing were examined with XRD. Figure 4 shows the XRD spectra of Ag/TiN/Si samples as deposited and after annealing at 650°C. The Ag film on a thin TiN barrier has a strong preferred orientation of its crystallites determined by XRD. Texture measurements showed the highest intensity for Ag(111). However, the XRD analysis of the specimens reveals that no crystalline chemical compounds of Ag_xSi_y were formed in the system after annealing at the extreme temperature of 650°C.

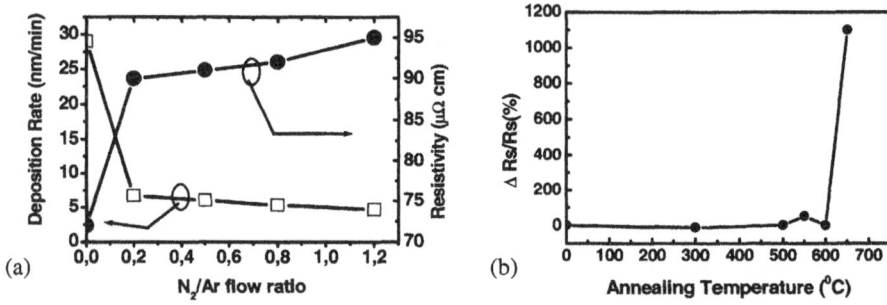

Figure 3. (a) Variation of deposition rate and resistivity as a function of Ar/N_2 flow ratio for TiN films (100 nm); (b) Percentage of sheet resistance variation against annealing temperature for Ag/TiN/Si samples (12 nm TiN covered by 200 nm Ag).

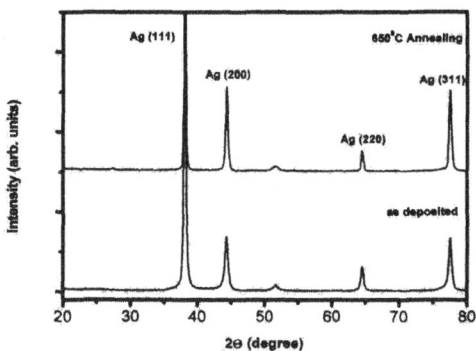

Figure 4. XRD spectrum for the Ag/TiN/Si samples as-deposited and annealed at 650°C.

Failure mechanism of TiN barrier

XPS is a very sensitive technique for the chemical analysis of the first several monolayers of a solid. In combination with a sputtering ion beam, depth profiles can be measured. Here we applied XPS using 3 keV Ar^+ ions to check the purity of the silver layer in the as-deposited film of a Ag/TiN/Si sample and to measure the elemental depth profile of a Ag/TiN/Si sample after 650 °C annealing. The high purity level of Ag of the as-deposited layer was recorded throughout the whole Ag layer. O contamination levels within the Ag layer are <1 at% and C <0.5 at%. Ti and N were present in near-stoichiometric concentrations with slightly larger Ti amounts (Ti/N = 1.0...1.3, varying between spectra taken at several depths, within the TiN layer zone). Annealing of a Ag/TiN/Si sample at 650°C causes out diffusion of Si across the TiN diffusion barrier. Ag was still present at the surface; however it contained 20 at% oxygen. A constant oxygen concentration was found by XPS measurement also at extended depths where a decreasing Ag fraction is replaced by silicon. This is attributed to oxidation during the high temperature annealing. Ti is no more located only at the interface, but is present over a wide depth range at constant concentration. In contrast to the measurements at the as-deposited sample, however, no N is detected after 650 °C annealing. This can be understood in view of the different enthalpies of formation for TiN and TiO_2, respectively. TiO_2 is about 3 times more stable and under the presence of oxygen a conversion of TiN into TiO_2 is plausible [9].

As an additional investigation, the top layer of Ag was removed from the annealed samples using Fe(III)-nitride. A large number of randomly distributed reaction spots were found after 600°C annealing on the TiN/Si surface (figure 5), becoming larger for the higher temperature annealed samples. The surface morphology remained stable however at 500°C annealing. That indicates the TiN barrier film failed to prevent Ag diffusion at the highest annealing temperatures. The etching pits occuring on 600°C annealed samples cause numerous local defects. The failure of the reactively sputtered TiN diffusion barrier is presumably due to diffusion through the barrier layer by way of localized defects.

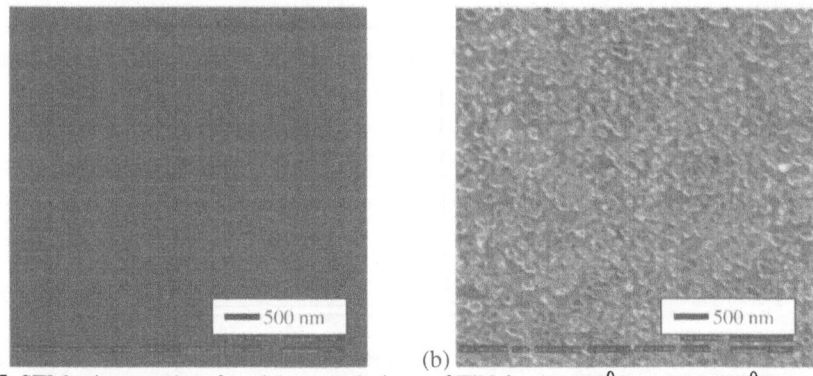

(a)   (b)

Figure 5. SEM micrographs of surface morphology of TiN for (a) 500^0C and (b) 600^0C annealed samples of Ag/TiN(12 nm)/Si after removal of the Ag overlayer.

CONCLUSION

The properties of TiN thin film as a barrier against Ag diffusion in silver interconnect multilayer structure of Ag (200 nm) / TiN (12 nm) / Si have been studied by sheet resistance measurement, XRD, FIB-SEM, XPS and AFM measurement. After annealing at various temperatures in N_2/H_2 (95%/5%) gas flow, we could find that the sheet resistance remains stable until 600^0C. It rises abruptly after 650^0C annealing, and surface agglomeration becomes much larger, but no new chemical phases or compounds are formed. FIB-SEM and XPS revealed that this is due to the surface morphology change after annealing of the samples at high temperatures and the destruction of the TiN film. It can be concluded that TiN thin films are promising as a barrier material for silver metallization.

ACKNOWLEDGMENTS

This work was supported by German Research Foundation under grant No. FOR 395.

REFERENCES
1. International Technology Roadmap for Semiconductors, Semiconductor Industry Association, (2003).
2. D. Edelstein, J. Heidenreich, et al, Techn. Digest Int. Electron Devices Meeting, 31 (1997).
3. R. Manepalli, F. Stepniak, S.A. Bidstrup, P. Kohl, IEEE Trans. Adv. Pack. **22**, 4-8 (1999).
4. M. Hauder, W. Hansch, J. Gstoettner, D. Schmitt-Landsiedel, Appl. Phys. Lett. **78**, 838-840 (2001).
5. M. Hauder, J. Gstoettner, W. Hansch, D. Schmitt-Landsiedel, Sensors and Actuators A **99**, 137-143 (2002).
6. Y. Wang, T.L. Alford, Appl. Phys. Lett. **74**, 52-54 (1999).
7. M. Moriyama, T. Kawazoe, M. Tanaka, et al., Thin Solid Films **416**, 136-144 (2002).
8. J.S. Chen, K.-Y. Lu, Thin Solid Films **396**, 204-208 (2001).
9. I. Barin, Thermochemical Data of Pure Substances, 3rd ed., Vol. 1&2, Weinheim (1995).

Mat. Res. Soc. Symp. Proc. Vol. 812 © 2004 Materials Research Society   F8.4

Electrical Behavior of Nano-Scaled Interconnects

M. Engelhardt, G. Schindler, W. Steinhögl, G. Steinlesberger, M. Traving
Infineon Technologies, Corporate Research, Otto-Hahn-Ring 6, D-81730 Munich, Germany

Abstract

Sub-lithographic copper damascene lines were fabricated to investigate already today the physical phenomena and scaling limits of metallic conductors in the metallization systems of chip generations which are believed to be in production 10 years from now and later. Using standard manufacturing processes and state-of-the-art process tools, including standard lithography tools, narrow copper lines were fabricated at the expense of a relaxed pitch by use of a removable spacer technique. These copper nano interconnects were passivated and subjected to electrical measurements. Our results show that continuous down scaling to increase device performance will result in an unfavorable increase of the electrical resistivity of copper in state-of-the-art metallization schemes. Electrical measurements over a wide range of temperatures down to cryogenic temperatures reveal the limited potential of cooling to reduce resistivity of conductors as lateral dimensions will be shrinked down to the sub-100nm regime. By down scaling of copper diffusion barriers in damascene trenches, barrier functionality was demonstrated after high temperature anneals and excessive bias-temperature stress tests for films meeting or even exceeding end-of-roadmap thickness requirements. An analysis of the temperature dependence of the leakage current measured at very high electric fields applied between neighboring damascene lines suggests the conduction mechanism in the SiO_2 used as intermetal dielectric to be Frenkel-Poole type rather than Schottky emission. Electromigration life times of sub-100nm copper lines embedded in oxide were found to be comparable with those obtained for similar structures fabricated with today's feature sizes.

Introduction

A look at Table 81b in the current International Technology Roadmap for Semiconductors (ITRS, /1/) displaying the MPU interconnect long term requirements reveals that about 75% of the fields are highlighted in red, standing for "manufacturable solutions are not known". These ITRS statements reflect that future interconnect technology is very challenging and gives rise to what people sometimes call the "interconnect crisis". As a consequence, RC delays associated with interconnects will gain increasing importance on the overall performance of future highly integrated circuits /2/. For an early experimental assessment of the interconnect challenges ahead and of these "red brick walls", the scaling behavior of copper damascene metallization was investigated by fabrication of hardware for an electrical characterization.

Experimental

Next generation lithography (NGL) required for pattern definition for the fabrication of far-in-the-future technology generations is still under investigation. Direct writing methods such as e-beam lithography are capable to define such tight feature sizes but suffer from the intrinsic tradeoff of very low throughput. Even in the research arena throughput is required to obtain a substantial number of test wafers with sufficiently extended test structures for destructive tests and analyses which is inevitable for unit process adaptations. To by-pass this lithography bottleneck standard manufacturing lithography was used in combination with a removable spacer technique /3/ to fabricate already today damascene copper lines with the critical dimensions

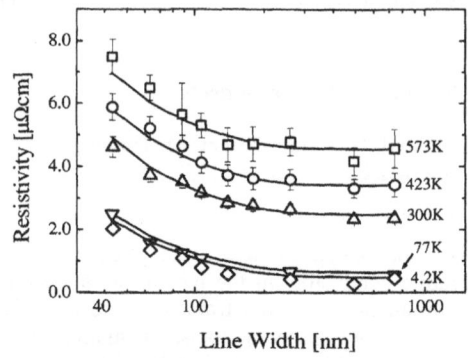

Fig. 1: Dependence of the electrical resistivity of copper on the CD of the damascene trench over a large temperature range. The drawn lines are the result of a model calculation /6/

(CDs) of local interconnects in future semiconductor products. The use of a stepper for lithography provides a sufficient high number of test structures that even an evaluation of the electromigration behavior can be performed in the reliability methodology facilities. The penalty one has to pay in this approach is that the aggressive CDs are obtained at the expense of a relaxed pitch. With this technique a reduction of the CDs given by the exposure wavelength used in lithography of more than an order of magnitude was demonstrated. With the use of yesterday´s i-line lithography the narrowest sub-lithographic damascene structures reliably obtained with this technique were in the 20nm regime /4/; copper damascene interconnects embedded in SiO$_2$ were obtained with feature sizes down to 40 nm /3/ for a comprehensive investigation of the electrical behavior. For all unit processes of the metallization part the state-of-the-art process modules and equipments used for a copper damascene based BEOL of a mature 0.35µm RF technology were utilized /3/. For the experiments on barrier scaling /5/, the thick PVD diffusion barrier bi-layer (10nm TaN + 40nm Ta) used in these RF products served as reference.

Results and Discussion

The dependence of the electrical resistivity of copper on the linewidth was measured over a large range of temperatures (Fig. 1). The strong increase of the resistivity with decreasing CD results from size effects coming into play as the CD approaches the dimension of the mean free path (λ_{Cu}=40nm, /6/) of the charge carriers. The experimental data are in excellent agreement with model calculations /7/ which take into account both the contributions of charge carrier scattering at internal and external surfaces, i. e. grain boundaries and interfaces between conductor and insulator, respectively. The curves obtained for the various temperatures display all the same shape, demonstrating that the size effect contribution to the resistivity originating from charge carrier scattering on the conductor surface, structural defects and/or imperfections is temperature independent; thus, conductor cooling gets less an less effective for future technology nodes as CDs

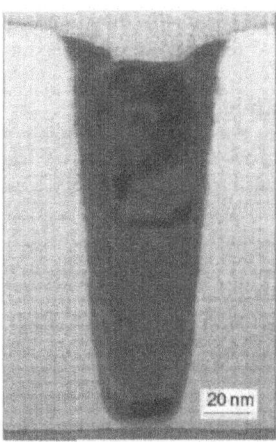

Fig. 2: TEM micrograph of nano interconnect displaying two copper grains; the structure meets the ITRS requirement for barrier thickness (3.5nm) and CD for the 32nm node. The aspect ratio (4:1) has twice the value required

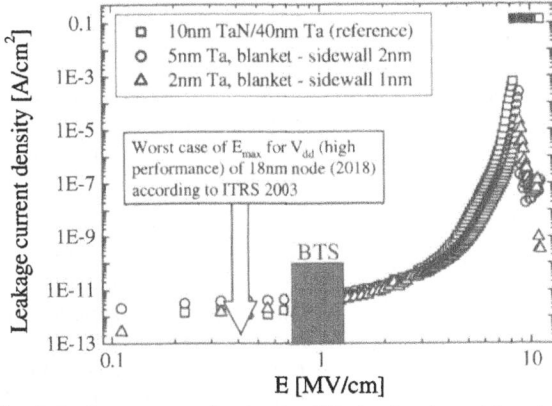

Fig. 3: Leakage current density for diffusion barriers with end-of-roadmap thicknesses; for a comparison the result for the reference barrier is also shown (see text!)

get tighter and tighter. This result shows that the ITRS requirement for conductor resistivity ($\rho = 2.2\mu\Omega\cdot$cm) can be met for the 40nm technology node when cooling down to liquid nitrogen temperatures. Experiments with various thermal treatments to enlarge the average copper grain size /8/ in nano grooves in SiO_2 did not result in the expected grain growth /9/ and an associated decrease of electrical resistivity; the results suggest that the maximum average grain size is limited by the smallest geometric dimension (i.e. by the CD) of the damascene groove structure (Fig. 2). The roadmap requirement for conductor resistivity which is the same for all technology generations to come until the end of the roadmap (2018) is the only ITRS requirement which is supposed to remain a "red brick wall".

The barrier film in future copper metallization schemes beyond the 50nm node is also rated as "manufacturable solutions are not known" by the ITRS. The leakage current measurements (Fig. 3) performed after applying thermal stress in anneals (120min @ 450°C) reveal no difference between the thick reference barrier and Ta barriers with down-scaled thicknesses of

5nm and 2nm as measured on the wafer surface. High resolution TEM micrographs show that the physical thicknesses of these barrier films on the sidewall of the damascene trenches are 1.7nm and sub-1nm, respectively and hence below the ITRS thickness requirement (2nm) for end-of-roadmap products. For a first check, that the deposited ultrathin barrier films are contingent and pinhole-free, the structures were investigated in the SEM after a special sample preparation (Fig. 4). The electrical measurements obtained under worst case assumptions for the electric field strength (0.4MV/cm) in high performance 18nm-node products also show that the leakage current between neighboring metal lines of different barrier thicknesses is at the

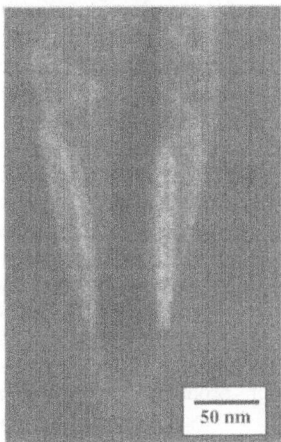

Fig. 4: SEM micrograph of nano-scaled barrier film in a damascene trench after wet chemical removal of copper and an SiO_2 recess etch

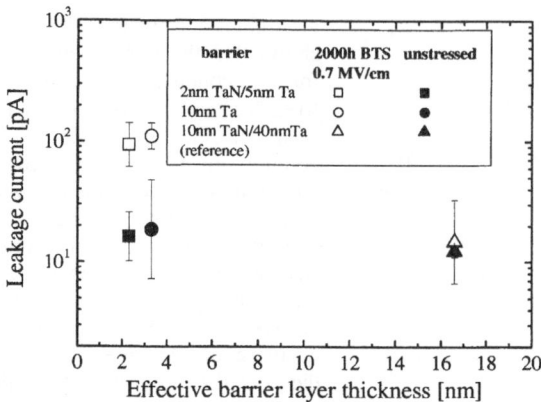

Fig. 5: BTS tests: Leakage currents for various barrier systems

same low level (Fig. 3). From the temperature dependence of the leakage current in the range of electrical field strengths around 1MV/cm can be concluded /5/ that the leakage current transport mechanism in this regime is dominated by a Frenkel-Poole mechanism via traps in the intermetal dielectric (SiO$_2$ with SiN etch stop layers) used.

After applying simultaneously thermal (200°C) and electrical (E=0.75-1.2MV/cm, see Fig. 3) stress in excessive (t=0...2000h) bias-temperature-stress (BTS) tests /5/, no essential difference in the electrical behavior i. e. leakage currents measured at room temperature could be seen between stressed and unstressed samples for all barrier thicknesses studied; compared to the samples with the thick reference barriers, metallizations with end-of-roadmap barrier thicknesses displayed only a marginal increase of leakage current after 2000h of BTS stressing (Fig. 5), if any.

In destructive tests of nano interconnects the critical current density at which the samples failed was found to be substantial higher (up to 100MA/cm^2, /10/) than the values obtained for metallization systems with relaxed CDs.

For an overall assessment of nano interconnect metallization and the integration of the metallization in the insulation system, test line structures were subjected to the very same procedures for electromigration investigations as current products. Using a failure criterion of 20% for the relative resistance increase $\Delta R/R$ (Fig. 6), the mean-time-to-failure (MTF) was determined /10/ at 3 ambient temperatures for 1 current density (4.4MA/cm^2) and for 3 current densities at one ambient temperature (275°C), to obtain the activation energy (E$_a$=0.88eV) and the current density exponent (n=1.93) using Black's equation; based on the company's reliability specifications for today's semiconductor products the resulting lifetime for damascene copper lines (CD=65nm) embedded in SiO$_2$ was found to be comparable (>100a) with the value known for current technologies.

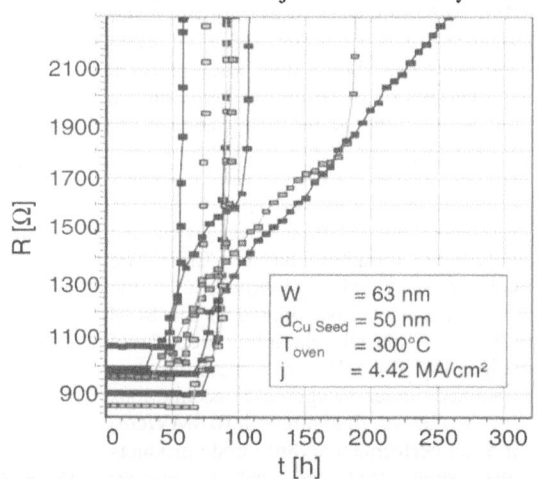

Fig. 6: Drift of electrical resistance of damascene copper lines observed during investigation of the electromigation behavior

Conclusions

For an assessment of future copper damascene metallization, the electrical behavior of conductor lines with the CDs of local interconnects in future technology generations was investigated. The electrical characterization of deep sub-100nm copper lines, obtained with relaxed pitch by use of standard lithography and a removable spacer technique reveals that the ITRS requirement for conductor resistivity in future technology generations will not be met and that even cooling provides only a limited potential to reduce resistivity. Experiments to enlarge grain size by thermal treatments resulted in no decrease of resistivity for such narrow feature sizes since the maximum achievable grain size seems to be limited by the CD of the damascene structure. The electrical characterization of diffusion barrier films with effective physical thicknesses below those required by the ITRS for end-of-roadmap semiconductor products displayed as low leakage current levels as thick barriers in 0.35μm products, serving as reference. Subjecting samples with nano-scaled barrier films to excessive BTS testing resulted in similar leakage current behavior as observed for the reference samples. The electromigration behavior of nano interconnect lines was found to be comparable to that of conductor lines in current technology generations.

Acknowledgements

The authors want to thank W. Hönlein for valuable discussions and the process engineers of the Infineon Munich cleanroom H84 for technical assistance, in particular the members of the metallization group, the reliability methodology group, and the failure analysis group. Thanks are also due to E. Unger for wet chemical sample preparations.

References

/1/ International Technology Roadmap for Semiconductors, 2003 Edition of the ITRS, http://public.itrs.net/

/2/ G. Schindler, W. Steinhögl, G. Steinlesberger, M. Traving, M. Engelhardt, Microelectronic Engineering, Vol. 70, p. 7-12 (2003)

/3/ M. Engelhardt, G. Schindler, K. Mosig, G. Steinlesberger, W. Steinhögl, G. Gebara Advanced Metallization Conference, Proc. Vol. , AMC 2001, p. 11 (2001)

/4/ M. Engelhardt, G. Steinlesberger, I. Jansen, K. Schober, Proc. of the 4th International AVS Conference on Microelectronics and Interfaces, ICMI 2003, p. 174 (2003)

/5/ M. Traving, W. Steinhögl, G. Schindler, G. Steinlesberger, M. Engelhardt Advanced Metallization Conference, Proc. Vol. , AMC 2003, p. 391 (2003)

/6/ T. S. Kuan et al, Mat. Res. Soc. Proc., Vol. 612 MRS, D7.1.1 (2000)

/7/ W. Steinhögl, G. Schindler, G. Steinlesberger, M. Engelhardt, Phys. Rev. B **66**, 075414, (2002)

/8/ G. Steinlesberger, M. Engelhardt, G. Schindler. W.Steinhögl, M. Traving, W. Hönlein, E. Bertagnolli, Mat. Res. Soc. Symp. Proc. Vol. 766, Materials Research Society, E4.2.1 (2003)

/9/ G. Schindler, V. Klandzevski, G. Steinlesberger, W. Steinhögl, M. Traving, M. Engelhardt Advanced Metallization Conference, Proc. Vol. , AMC 2003, p. 213 (2003)

/10/ G. Steinlesberger, A. von Glasow , M. Engelhardt G. Schindler, W. Hönlein, M. Holz, E. Bertagnolli, Proc. of International Interconnect Technoloy Conference, IITC 2002, p. 265 (2002)

Mat. Res. Soc. Symp. Proc. Vol. 812 © 2004 Materials Research Society F8.6

The Influence of Temperature and Concentration on Copper Deposition Kinetics in Supercritical Carbon Dioxide

Yinfeng Zong and James J. Watkins
Department of Chemical Engineering,
University of Massachusetts Amherst,
Amherst, MA 01003

ABSTRACT

The kinetics of copper deposition by the hydrogen-assisted reduction of bis(2,2,7-trimethyloctane-3,5-dionato)copper in supercritical carbon dioxide was studied as a function of temperature and precursor concentration. The growth rate was found to be as high as 31.5 nm/min. Experiments between 220 °C and 270 °C indicated an apparent activation energy of 51.9 kJ/mol. The deposition kinetics were zero order with respect to precursor at 250 °C and 134 bar and precursor concentrations between 0.016 and 0.38 wt.% in CO_2. Zero order kinetics over this large concentration interval likely contributes to the exceptional step coverage obtained from Cu depositions from supercritical fluids.

INTRODUCITON

Copper is the material of choice for interconnect structures in advanced integrated circuits due to its low electrical resistance and superior electromigration resistance. Recently, we demonstrated that high purity copper films could be deposited with exceptional step coverage within high aspect ratio features in a single step by supercritical fluid deposition (SFD)[1]. The advantages of SFD are related to the physicochemical properties of the deposition medium. Supercritical fluids (SCFs) such as carbon dioxide exhibit low viscosity, high diffusivity, zero surface tension and pressure-dependent densities that can equal or exceed those of liquid solvents. In fact, many organometallic compounds, including a wide range of CVD precursors, exhibit significant solubilities in SCFs[2-4]. The solubility of metal precursors in SCFs obviates precursor volatility constraints often encountered in conventional vapor phase deposition and eliminates mass transport limitations to uniform step coverage. Accordingly, SFD is essentially a hybrid technique that combines the advantages of solution-based chemistry with exceptional transport properties typical of the gas phase. Moreover, the technique can be extended to a broad range of metals and metal alloys[5-9].

Since SFD is a new technique, little information is available regarding deposition kinetics or mechanisms. Here we report dependence of film growth rate during depositions by the hydrogen assisted reduction of bis(2,2,7-trimethyloctane-3,5-dionato)copper, [Cu(TMOD)$_2$] in carbon dioxide as a function of temperature and precursor concentration.

MATERIALS AND METHODS

Bis(2,2,7-trimethyloctane-3,5-dionato)copper [Cu(TMOD)$_2$] was obtained from Epichem, Inc. (Allentown, PA) and used as received. The chemical structure of the compound is shown in Figure 1. H$_2$ (Ultra High Purity grade) and CO$_2$ (Coleman grade) were obtained from Merriam Graves Corp. (Charlestown, NH). The deposition experiments were carried out in a 170 ml, custom-designed high-pressure reactor (Figure 2). The reactor consists of opposed stainless steel flanges sealed with an o-ring. The reactor contains a 19 mm diameter aluminum stage heated with three internal cartridge heaters (Omega Engineering Inc.).

Depositions for the kinetic study were carried out in batch mode at low precursor conversion (< 15%). In a typical experiment, a 12 mm × 12 mm TiN coated Si wafer was secured on the stage. A predetermined mass of precursor was loaded into the reactor, the reactor was sealed, flushed with CO$_2$ at low pressure, and heated to 60 °C. Supercritical CO$_2$ was then loaded into the reactor at pressures between 110 and 138 bar. The system was maintained at these conditions for approximately one hour to ensure complete dissolution of the precursor. Three samples of the reactor solution were then collected using high pressure sample loops (Valco Instruments Inc.). Just prior to deposition, a known quantity of hydrogen was loaded into the reactor. The stage was then heated to the desired temperature (between 220 °C and 270 °C). The set point was reached within 40 seconds. The stage was maintained at the deposition temperature for 5 minutes then the power was cut to the heater. The temperature of the stage dropped below 200 °C within ten seconds. Samples of the fluid phase were collected after stage cooling. All reactor samples were analyzed by UV-Vis Spectrometry to obtain precursor concentration before and after the reaction. After the reaction, supercritical CO$_2$ was used to purge the reactor to remove the remaining precursor and any by-product. The effluent was passed through an activated carbon packed bed prior to venting.

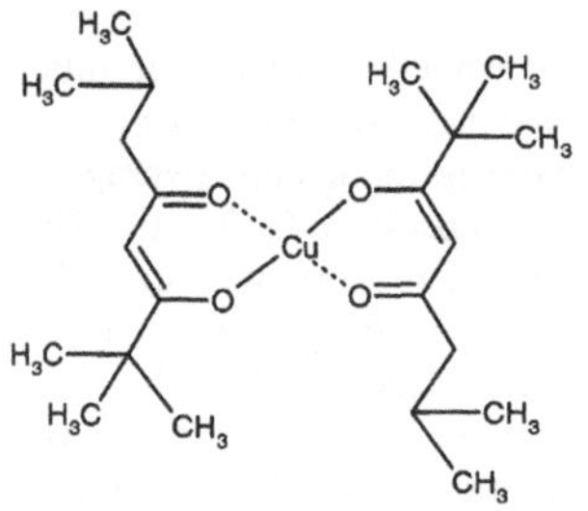

Figure 1. Structure of Bis(2,2,7-trimethyloctane-3,5-dionato)copper(II).

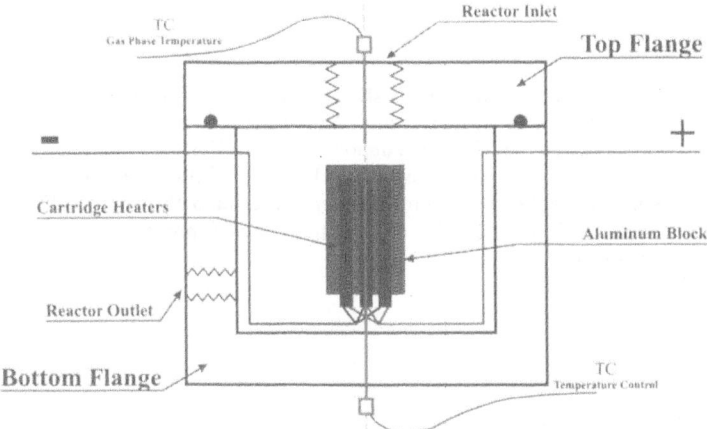

Figure 2. Schematic diagram of the high-pressure stainless steel reactor used for the kinetic study.

Film thickness was measured using a Sloan DEKTAK[3] Surface Profiler. Film growth rate was calculated by dividing the film thickness by the reaction time. Therefore, it should be considered as the time-averaged growth rate. The sheet resistance of the film was measured using a Jandel four point probe. Film composition at the surface and throughout the bulk was studied using X-ray Photoelectron Spectroscopy (XPS) (Quantum 2000 Scanning ESCA Microprobe (Physical Electronics USA)) equipped with an Ar^+ ion-sputtering gun. The instrument used monochromatic Al Kα x-rays (1486.6 eV). The X-ray setting was 15 kV, 25 W with 45° incident angle with 100 μm beam size. The ion gun sputtering setting was 1 kV, 700 nA, with a 1 mm × 1 mm crater dimension. The XPS data were analyzed using Multipak (version 6.1A, Physical Electronics USA).

RESULTS AND DISCUSSION

Film properties

All the copper films deposited were virtually free of contamination. A typical XPS depth profile is shown in Figure 3. The data indicate that oxygen and carbon are only present on the surface of the film and can be attributed to surface contamination and ambient oxidation after the film was exposed to air. The sheet resistance was dependent on film thickness. For thick films, the sheet resistance was typically 2.5 – 3.0 μΩ-cm, close to the bulk value for pure Cu (1.7 μΩ-cm). When the film was only ~50 nm thick, the sheet resistance increased to ~ 6 μΩ-cm. Thickness dependent sheet resistance result is consistent with data for Cu films deposited by other techniques[10].

Temperature dependence

The temperature dependence of film growth rate is shown in Figure 4. Temperatures between 220 and 270 °C were studied in 10 °C increments. An overall activation energy of 51.9 kJ/mol was calculated from the data. For MOCVD Cu deposition using Cu(hfac)$_2$, activation energies between 75 kJ/mol – 80 kJ/mol have been reported[11, 12]. In those MOCVD studies, the surface reaction was assumed to be the rate-controlling step. Reaction pathways for SFD are under study in our lab at present, but it is not at all surprising that the SFD process could have a different rate-controlling mechanism.

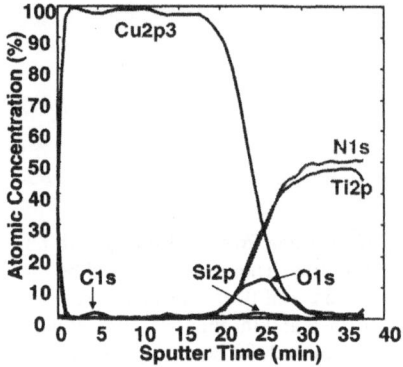

Figure 3. Typical XPS depth profile of a copper film deposited by the hydrogen-assisted reduction of Cu(TMOD)$_2$ in supercritical CO$_2$. (Deposition conditions: 250 °C, 134 bar, 0.12 wt.% precursor and 0.1 wt.% H$_2$.)

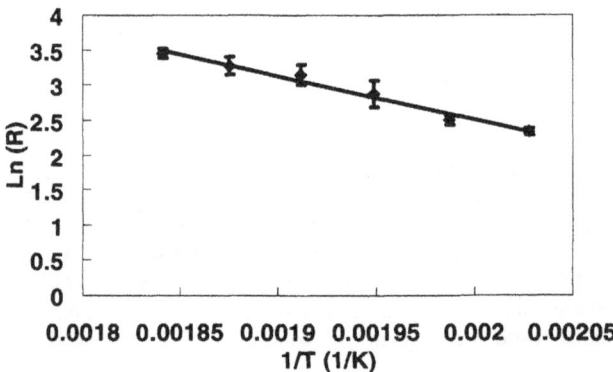

Figure 4. Temperature dependence of Cu film growth rate during H$_2$ –assisted reduction of Cu(TMOD)$_2$ at a precursor concentration of 0.12 wt.% and a H$_2$ concentration of 0.1 wt.%.

Precursor concentration dependence

The effect of precursor concentration on growth rate is shown in Figure 5. The deposition conditions for all the experiments were 250 °C, with 134 bar initial CO_2 loading and 0.1 wt.% H_2. The result shows that between 0.016% to 0.38% wt.% precursor, the growth rate remains constant at ~ 25 nm/min. This zero order concentration dependence suggests two possible rate-limiting steps: surface reaction or by-product desorption. Zero order kinetics favors good step coverage. By depositing at high precursor concentration and maintaining zero order kinetics for the surface reaction, deposition rate can be kept constant even if the precursor concentration evolves along the trenches.

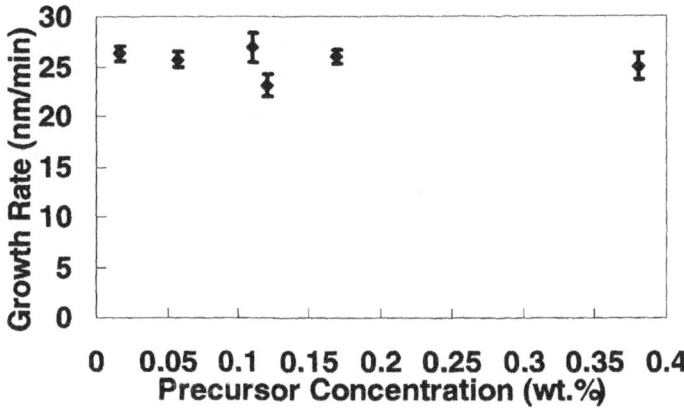

Figure 5. Precursor concentration dependence of the growth rate of Cu SFD at 250 °C, 134 bar and a H_2 concentration of 0.1 wt.% in CO_2.

CONCLUSIONS

The deposition of Cu by the hydrogen assisted reduction of Cu(TMOD)$_2$ in supercritical carbon dioxide at temperatures between 220 and 270 °C exhibits an overall activation energy of 51.9 kJ/mol. The reaction follows zero order kinetics with respect to precursor over broad range of precursor concentration (0.016% to 0.38% wt% in CO_2). Zero order kinetics likely contributes to the exceptional step coverage observed for Cu depositions from supercritical carbon dioxide.

ACKNOWLEDGMENTS

Financial support from NSF (CTS-0245002) and the Camile and Henry Dreyfus Foundation is gratefully acknowledged. Instruments supported by the Materials Research Science and Engineering Center at University of Massachusetts Amherst were used in the study.

REFERENCES

1. J. M. Blackburn, D. P. Long, A. Cabanas, and J. J. Watkins, *Science* **294**, 141-145 (2001).
2. N. G. Smart, T. Carleson, T. Kast, A. A. Clifford, M. D. Burford, and C. M. Wai, *Talanta* **44**, 137-150 (1997).
3. A. F. Lagalante, B. M. Hansen, T. J. Bruno, and R. E. Sievers, *Inorg. Chem*, **34**, 5781-5785 (1995).
4. W. Cross, A. Akgerman, and C. Erkey, *Industrial & Engineering Chemistry Research* **35**, 1765-1770 (1996).
5. J. M. Blackburn, PhD. Dissertation, University of Massachusetts Amherst, 2001.
6. J. M. Blackburn, D. P. Long, and J. J. Watkins, *Chemistry of Materials* **12**, 2625-2631 (2000).
7. D. P. Long, J. M. Blackburn, and J. J. Watkins, *Advanced Materials* **12**, 913-915 (2000).
8. J. J. Watkins, J. M. Blackburn, and T. J. McCarthy, *Chemistry of Materials* **11**, 213-215 (1999).
9. A. Cabanas, D. P. Long, J. J. Watkins, *Chemistry of Materials* (in press).
10. J. W. Lim, K. Mimura, and M. Isshiki, *Applied Surface Science* **217**, 95-99 (2003).
11. D. H. Kim, R. H. Wentorf, and W. N. Gill, *Journal of the Electrochemical Society* **140**, 3267-3272 (1993).
12. Y. D. Chen, A. Reisman, I. Turlik, and D. Temple, *Journal of the Electrochemical Society* **142**, 3903-3911 (1995).

Integration and Reliability

Mat. Res. Soc. Symp. Proc. Vol. 812 © 2004 Materials Research Society

Process-Oriented Stress Modeling and Stress Evolution During Cu/Low-K BEOL Processing

Charlie Jun Zhai, Paul R. Besser, Frank Feustel, Amit Marathe, Richard C. Blish II
Advanced Micro Devices, Inc
1 AMD Place, MS 79, Sunnyvale, CA 94088-3453

ABSTRACT

The damascene fabrication method and the introduction of low-K dielectrics present a host of reliability challenges to Cu interconnects and fundamentally change the mechanical stress state of Cu lines. In order to capture the effect of individual process steps on the stress evolution in the BEoL (Back End of Line), a process-oriented finite element modeling (FEM) approach was developed. In this model, the complete stress history at any step of BEoL can be simulated as a dual damascene Cu structure is fabricated. The inputs to the model include the temperature profile during each process step and materials constants. The modeling results are verified in two ways: through wafer-curvature measurement during multiple film deposition processes and with X-Ray diffraction to measure the mechanical stress state of the Cu interconnect lines fabricated using 0.13um CMOS technology. The Cu line stress evolution is simulated during the process of multi-step processing for a dual damascene Cu/low-K structure. It is shown that the in-plane stress of Cu lines is nearly independent of subsequent processes, while the out-of-plane stress increases considerably with the subsequent process steps.

INTRODUCTION

Conventional stress measurement methods include wafer curvature and X-ray diffraction. Stress measurement, which is limited to simple test structures, has insufficient spatial resolution to determine the detailed stress distribution in a complicated interconnect system. Hence, finite element stress modeling has been extensively used for detailed stress analysis. [1-4] During the last few years, X-ray diffraction measurements [3-6] have been used to validate stress modeling. Traditionally, mechanical stress in interconnects has been modeled under the assumption that the overall structure after final process is in zero stress state at a presumed temperature, i.e. stress-free. Although this method is simple and often adequate to most needs, it does not provide adequate understanding of the stress evolution during the process. [4-6] For example, the fabrication of a simple M1-V1-M2 Cu BEoL structure requires multiple deposition processes including etch, CVD, Cu plating, CMP, etc. Each process step either adds or removes certain materials at a specific process temperature, and the stress state in the material will be affected by each step. Therefore, it is desirable to develop a process-oriented modeling approach, which takes into account of the effect of process changes on the stress evolution in all of the materials. [2] In order to better understand the stress evolution during the process, researchers have developed process-oriented stress modeling techniques. Among these, Cifuentes and Shareef [1] developed a numerical technique using artificial nodes to simulate material interfaces bond or debond to achieve a stress history. Lee and Sauter Mack [2] developed their own in-house code employing a technique similar to the element birth and death capability in ANSYS. They demonstrated through modeling that stresses in metal lines and dielectric film depend not only on the final temperature excursion but also depend on the process sequences and process conditions. They also concluded that intrinsic stress in the dielectrics has little effect on the metal line stress. However, their model is limited to the 2D domain, and their assumption that the stress free temperature is 400C for all materials is not supported by more recent experimental results. [6]

In the present study, a process-oriented modeling technique has been developed and fully implemented in ANSYS Parametric Design Language. The model is fully parametric on the 3D domain. All the process conditions (mostly process temperatures) and feature dimensions are parametric inputs. In order to accommodate the most recent advancement in Cu based BEoL manufacturing technology, the model incorporates both single and dual-damascene configurations.

MODELING

It has been observed that narrow Cu lines exhibit linear elastic behavior while wide Cu lines or films exhibit more plasticity during temperature cycling, due to the difference in constraint conditions [3-4]. For simplicity, this study assumes that linear elasticity prevails for all materials without accounting for microstructure-dependent nonlinear behavior of Cu. The geometrical modeling involves element birth and death technique offered by ANSYS. First, the complete geometric model with final configuration is built, but all the elements are specified as "inactive", namely "dead elements" which do not contribute to solving the linear equations for finite element analysis. Then, depending on the process, elements will be selected for reactivation. Only activated elements are involved in the FEA. At each process, more elements will be reactivated to replicate the material deposition. The activated elements are assigned appropriate thermal-mechanical boundary conditions at each process. It is well known that intrinsic stress will arise from deposition of dielectrics including SiO_2, Si_3N_4 and carbon-doped oxide based low-K dielectrics. In this study, a concept of "Virtual Deposition Temperature" will be introduced to simplify the account of intrinsic stresses. However, Lee and Sauter Mack [2] have shown that metal stress changes very little when the dielectric intrinsic stress ranges from 300 to 500 MPa. Table 1 lists the properties of materials used in the modeling. [3-6]

Table 1. Material Properties (at Room Temperature).

Material	Modulus (GPa)	CTE (ppm/C)
Cu	115	17.7
OXIDE	71.4	0.51
CVD Low-K	6	20
Spin on Low-K	2	66

MODELING CALIBRATION WITH WAFER CURVATURE

The model is initially calibrated with wafer curvature measurements during sequential deposition of dielectric film stacks (no metals) as seven process steps. The dielectric films deposited include TEOS (tetra-ethyl-ortho-silicate), FTEOS (fluorine-doped TEOS), a Carbon-doped oxide low-K dielectric film (Low-K) and a Si-C dielectric etch stop layer. The wafer curvature data was collected at room temperature after each process, and accumulative film stress is computed using Stoney's formula [8]. A simple model was hence developed in accordance to the process-oriented modeling procedure as described previously, by using a technique "Virtual Deposition Temperature." The overall film stress σ is the sum of the thermal stress σ_T induced by CTE mismatch between films and substrate and a temperature-independent intrinsic stress σ_I. The total stress can be written as

$$\sigma = \sigma_T + \sigma_I, \qquad (1)$$

and the thermal stress is defined as

$$\sigma_T = \frac{E}{1-v}\Delta\alpha \cdot \Delta T = \frac{E}{1-v}\Delta\alpha \cdot (T_D - T_R) \qquad (2)$$

where $\Delta\alpha$ is the difference in CTE between film and substrate, T_D is the deposition temperature and T_R is the room temperature. We introduce a virtual deposition temperature T_V, and use it to express the magnitude of the intrinsic stress by varying its magnitude relative to the deposition temperature. The virtual deposition temperature T_V for each specific film material is obtained from wafer curvature measurements and used as the stress free temperature in the modeling afterwards. For ease of manipulation, the virtual deposition temperature is defined as

$$T_V = \frac{\sigma_I}{\dfrac{E}{1-v}\Delta\alpha} + T_D$$

$$\qquad (3)$$

Substituting (2) and (3) into (1) yields an expression for stress.

$$\sigma = \frac{E}{1-v}\Delta\alpha \cdot (T_D - T_R) + \sigma_I = \frac{E}{1-v}\Delta\alpha \cdot (T_V - T_R) \qquad (4)$$

Figure 1 is a plot of the overall film stress at the room temperature after ILD deposition process step. Included in this plot are the stresses from modeling and from wafer curvature measurement. The initial TEOS layer creates a compressive stress state (step 1) that is magnified with the deposition of the Si-C etch-stop layer (step 2). After deposition of a C-doped oxide film (step 3), the stress becomes less compressive (suggesting that the C-doped oxide is under tensile stress); however an oxide cap ILD (step 4) will again make the stress more compressive. The measured stress evolves as shown in steps 4, 5, 6 and 7 with each subsequent deposition of a composite film (etch stop layer + oxide + C-doped oxide + cap). Based on this sequence of ILDs, the overall stress is becoming less compressive as more dielectric layers are deposited. The overall in-plane film stress remains compressive; however, it is noteworthy that the compressive stress decreases with deposition of low-K films that have much higher CTE value than the Si substrate. The correlation between the modeling results and measurements is excellent, as the maximum error between them is less than five percent.

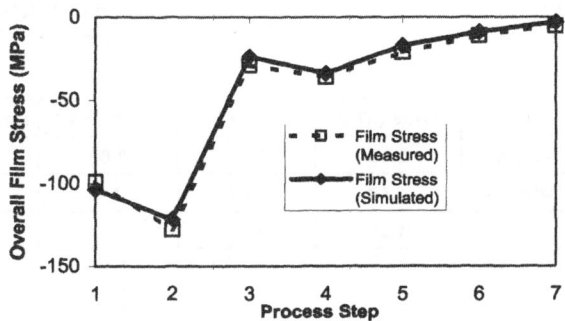

Figure 1. Stress evolution during film deposition.

MODELING CALIBRATION WITH XRD

In addition to calibrating the model by wafer-curvature measurements, the model was calibrated with X-Ray diffraction (XRD) measurements recently published by Besser and Jiang [7]. The structure used in their work is illustrated in Figure 2 and includes the SiN-passivated Cu line structure fabricated in either Cu/Oxide and Cu/Low-K (porous organic spin-on material) dielectrics. The Cu lines were 0.25 μm wide, including a 250 Å Ta barrier. The post-plating anneal temperature was 150°C, and the cap layer was 1000Å SiNx deposited at ~400°C. The model utilizes the interconnect line dimensions and ILD thicknesses detailed in the reference [7] and uses the deposition process conditions as inputs. The Cu was assumed to be in a stress-free state at 250°C for the FEA model.

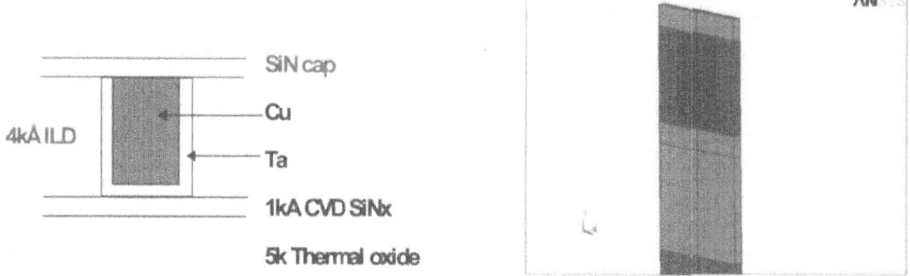

Figure 2. Schematic cross-section (left) of the damascene-fabricated Cu lines used to calibrate the model. The FEA model of passivated Cu line is shown at right.

Table 2 lists the stress variation during unit temperature change (in the unit of MPa/°C) for Cu/Oxide and Cu/Low-K, obtained by XRD and FEA, respectively. Stress variation is defined as the slope of a linear fit to the measurements of mechanical stress as a function of temperature. The results of the FEA model correlate well with the XRD results. The error is less than 15%, except the transverse stress σ_y. The transverse stress predicted by FEA is 50% lower than that measured by XRD. While the agreement between experiment and model is good, the FEA tends to overestimate the stress level parallel to line (longitudinal) and underestimate the stress level transverse to line.

Table 2. Stress variation during unit temperature change for Cu lines fabricated in oxide and low-K ILD, determined from XRD[7] and finite element methods.

	Oxide (XRD)	Oxide (FEA)	Low-K (XRD)	Low-K (FEA)
$-d\sigma_x/dT$	2.48	3.00	2.01	2.56
$-d\sigma_y/dT$	1.46	1.42	1.2	0.81
$-d\sigma_z/dT$	1.26	1.10	0.3	0.28

STRESS EVOLUTION IN CU LINES DURING PROCESSING

The stress evolution in the Cu line is vital to understanding the reliability risk associated with process and material changes. Finite element models were constructed for a hybrid Cu/Low-K technology in which the line level dielectric is carbon-doped CVD low-K material while the via level dielectric is FTEOS. The etch stop layer is Si-C dielectrics. All the material properties are listed in Table 1. Figure 3 (a) shows a dual damascene structure which consists of a M1-V1-M2 layer and a top passivation layer. Figure 3(b) shows a detailed structure of the Cu line/via embodied by the Ta barrier metal in this particular study. The structure spans only half of the pitch size in the Z-direction (out-of-plane) to take advantage of symmetry. A coupled boundary condition was applied to the lateral and back surfaces.

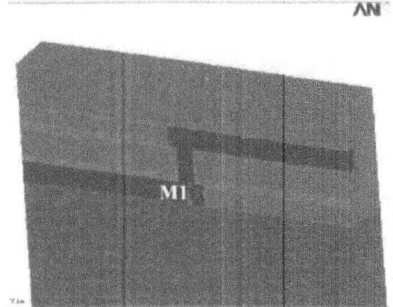

| Figure 3(a). M1-V1-M2 FEA model. | Figure 3(b). Cu/barrier structure (FEA). |

Deposition processes including CVD, electroplating, etch and CMP involve material addition or removal, as simulated by the elemental birth and death in the process-oriented modeling procedure. For the purpose of simplicity, only volume averaged stresses (σ_x, σ_y and σ_z) in M1 at three key stages during the process are reported here. The model was constructed in 3 steps.

Step 1: M1, deposition of M1 and Cap (etch stop) layer, cool down to 25C.
Step 2: M1-V1, deposition of via layer (V1 dielectric and V1 Cu), cool down to 25C.
Step 3: M1-V1-M2, deposition of M2, cool down to 25C.

Figure 4 shows the history of three stress components (σ_x, σ_y and σ_z) vs. process steps for Cu lines fabricated in Low-K. Without considering the nonlinear behavior of Cu, the in-plane stresses (σ_x and σ_y) are nearly independent of subsequent process steps. These in-plane stresses (σ_x and σ_y) slightly decrease at the completion of the M1-V1 process, then slightly increase at the completion of M1-V1-M2. However, the out-of-plane stress σ_z increases considerably with the deposition of V1 and M2 layers. After M2 deposition, σ_z increases by more than 25%. The out-of-plane normal stress σ_z is indicative of the peeling stress between layers, and a continuous increase in σ_z with deposition processes may be a concern for dielectric/metal delamination, especially at the lower-level of interconnects.

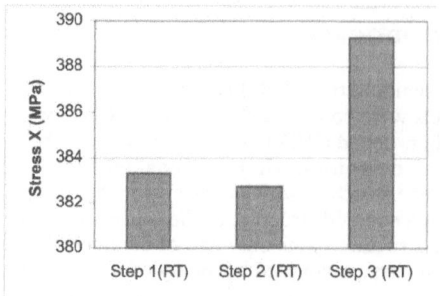
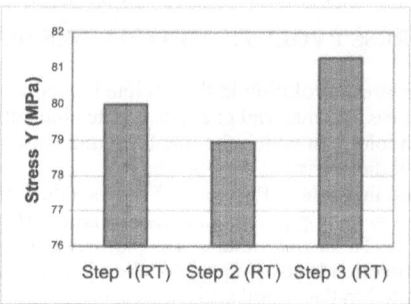
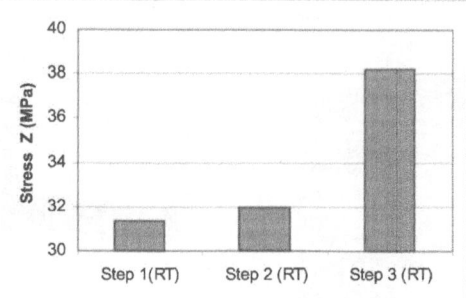

Figure 4. Stress Evolution of σ_x, σ_y and σ_z for a M1 line during subsequent processing.

CONCLUSIONS

A process-oriented finite element analysis technique was developed by using "element birth and death" in ANSYS. The model is calibrated with both wafer-curvature blanket film measurements and X-Ray diffraction (XRD) measurement of metal line stress. The Cu line stress is investigated as a function of process step for a Cu/low-K dual damascene hybrid structure. The in-plane stress of Cu lines is nearly independent of subsequent processes, while the out-of-plane stress increases considerably with the subsequent process steps. The modeling approach can be readily applied to study of a variety of BEoL structures for different Cu/low-K integration technologies.

REFERENCES

[1] A.O. Cifuentes and I.A. Shareef, IEEE Trans. Semiconductor Manufacturing v15, 128 (1992).
[2] J. Lee and A. Sauter Mack, IEEE Trans. Semiconductor Manufacturing v11, 458 (1998).
[3] D. Gan, G. Wang and P.S. Ho, Proceedings of the International Interconnect Technology Conference, 271 (2002).
[4] S-H. Rhee, Y. Du and P.S. Ho, Proceedings of the International Interconnect Technology Conference, 89 (2001).
[5] G. Reimbold, et al., Proceedings of IEEE IEDM 2002, 745 (2002).
[6] P.R. Besser et al., Journal of Electronic Materials 30 (4), 320 (2001).
[7] P.R. Besser and Q.T. Jiang, Accepted for publication in MRS Symp. Proc. 795 (2004).
[8] G. Stoney, The Tension of Metallic Films Deposited by Electrolysis, Proc. Roy. Soc., London, A 82, *1909*

Mat. Res. Soc. Symp. Proc. Vol. 812 © 2004 Materials Research Society   F2.8

Ultra Low-dielectric-constant Materials for 65nm Technology Node and Beyond

Hao Cui, Darren Moore, Richard Carter, Masaichi Eda, Peter Burke
LSI Logic Corporation, 23400 NE Glisan St., M/S AR-220, Gresham, OR 97030, U.S.A.
David Gidley and Huagen Peng
Department of Physics, University of Michigan, Ann Arbor, MI 48109, U.S.A

ABSTRACT

Pore characteristics including pore size distribution, porosity, and pore interconnectivity of PECVD SiCOH inter-layer dielectric (ILD) materials with different dielectric constant (κ) values have been studied. Oxygen plasma damage to SiCOH low-κ films increases dramatically as the κ value decreases. Simulations showed that, compared to the ILD film, the overhead dielectric films have a significant impact on the overall effective κ (κ_{eff})of the BEOL interconnects. Reducing the κ values of these overhead films helps to alleviate the pressure on the κ value requirement of the ILD materials while still meeting the κ_{eff} target. Ultra low-κ (ULK) PECVD hydrogenated silicon carbide (H:SiC) films with a κ of 3.0 have been studied for the etch-stop applications. Studies of the chemical composition and bonding structure suggest that less Si-C networks are formed and more micro-porosity are incorporated in the ULK H:SiC film. The leakage current of the ULK H:SiC film is found to be about 5 times lower than the H:SiC and H:SiCN films with higher κ values. The etch rate of ULK H:SiC film using a standard SiCOH ILD etch chemistry has been found to be negligible. Such an extremely high etch selectivity makes these films very good etch-stop layers.

INTRODUCTION

As ultra-large scale integrated circuit (ULSI) technology advances, low-κ dielectric materials are needed for ULSI interconnect applications to address the increasingly significant interconnect-related issues including interconnect delay, dynamic power consumption and cross-talk noise. Evident from the history of the International Technology Roadmap for Semiconductor (ITRS) projections, the implementation of low-κ materials has been continuously pushed out [1]. For example, the technology node where low-κ materials with a κ_{eff} of about 2.0 were projected to be implemented has been pushed out from the 90 nm node in the 1999 ITRS to the 32 nm node in the 2003 ITRS [1]. Such a delay has been caused by the revelations of substantial process and reliability challenges as researchers and engineers started to evaluate and integrate materials with lower κ values.

The dielectric constant is a physical measure of how easy a material can be polarized in an external electric field. The quantitative relation between the dielectric constant and properties of the atoms and molecules of the material is described by the Clausius-Mossotti equation

$$\frac{\varepsilon_r - 1}{\varepsilon_r + 2} = \frac{1}{3\varepsilon_0} \sum_j N_j \alpha_j \qquad [1]$$

where ε_r is the dielectric constant, N_j and α_j the volume density and polarizability of the jth type of atom or molecule. The symbol κ, which has been adopted by the VLSI industry to represent the dielectric constant, will be used in the discussions of this work. As shown in equation 1, κ can be reduced by modifying the chemical structure of the materials and reducing α, which usually includes electronic, ionic and dipolar components. A more straightforward and common way to reduce κ is reducing N_j, the number of atoms or molecules per unit volume, by introducing free volume (porosity) into the materials. The excellent correlations between κ and the density of various siloxane-based low-κ thin films reported by others suggest that the κ reduction of this type of material is mainly driven by the decrease in film density [2,3]. Ultra low-κ (ULK) ILD materials for future ULSI technology nodes most likely will have a certain amount of porosity. The incorporation of pores into dielectric materials raises significant process integration and reliability issues. Usually, these issues become more pronounced as the porosity increases. The uses of overhead dielectric films ease the difficulties to integrate the porous ILD. However, their impact on the κ_{eff} needs to be evaluated.

EXPERIMENTAL DETAILS

The SiCOH films studied in this work were deposited in capacitvely-coupled PECVD systems using a reaction of organosilane and oxygen gases. Deposition of the ULK H:SiC films was carried out using a organosilane-based chemistry in a capacitively-coupled PECVD system. Positron Annihilation Lifetime Spectroscopy and Ellipsometric Porosimetry were used to investigate the pore characteristics of these films. The films chemical compositions were studied using Rutherford backscattering spectrometry (RBS) and hydrogen forward spectrometry (HFS) measurements using He^{++} ions with an incident energy of 2.275 MeV. The Fourier transform infrared spectroscopy (FTIR) spectra of these films were measured using a BIO-RAD QS-2200 FTIR spectrometer to investigate the changes in chemical bonding structures. A spectroscopic ellipsometer system with a wavelength range from 200 nm to 800 nm was used to measure the film thickness and refractive index (RI). Metal-insulator-semiconductor (MIS) capacitors were used for the capacitance-voltage (C-V) and current-voltage (I-V) tests of these low-κ dielectric films. The κ values were extracted from the C-V curve of MIS capacitors measured at 1MHz.

RESULTS AND DISCUSSIONS

Porous low-κ ILD

A large number of materials have been investigated for low-κ applications. These materials can be deposited using either chemical vapor deposition (CVD) or spin-on coating techniques. From the material type point of view, most low-κ materials can be divided into three categories: inorganic, organic, and hybrid. Carbon-doped silicon oxide thin films deposited using plasma enhanced chemical vapor deposition (PECVD) have thus far gained more favor in the ULSI industry for use as low-κ ILD. Carbon-doped silicon oxide materials are often designated as SiCOH, and are also called organosilicate glass (OSG) or methyl-doped oxide. For the discussions on porous ILD materials, this paper will focus on the PECVD SiCOH thin films.

From the studies of chemical composition and FTIR spectra, the Si-O bonds are found to be the backbone of the carbon-doped silicon oxide thin films [4]. For the porous PECVD SiCOH materials, the pores can be introduced by the disordering of the dense Si-O network and the formation of loose Si-O structures such as the Si-O cage structures. The incorporation of methyl groups also interrupts the dense Si-O network by terminating the Si atoms with CH$_3$ and H. The desorption of large amounts of thermally unstable groups from the films during the deposition is also thought to play an important role in forming the pores in the porous SiCOH thin films. The introduction of pores, micro-pores or meso-pores, into the ILD materials, however, raises serious process integration and reliability issues including poor mechanical strength, susceptibility to plasma damage, poor coverage and integrity of metallic barrier on porous ILD sidewall, penetration of wet chemicals and gases into porous ILD, low thermal conductivity, etc [5-7]. The pores of ILD materials can be divided into two types according to their effective average size: the micro-pores with an effective diameter of less than 2.0 nm and the meso-pores with an effective diameter of 2.0 nm and greater. Compared to the micro-pores, the incorporation of meso-pores reduces the κ value more effectively due to their larger sizes. However, these meso-pores, especially the interconnected ones [8], should be minimized or avoided since they pose some of the most challenging issues to integrate porous low-κ materials. Studying the pore characteristics of porous low-κ materials is of great importance to understand and address some of the key integration and reliability issues of these materials.

Positron Annihilation Lifetime Spectroscopy (PALS) and Ellipsometric Porosimetry (EP) are two very useful analytical techniques to probe the pore characteristics of porous materials [8,9]. Three PECVD SiCOH thin films with κ values of 3.0, 2.4 and 2.0 were studied using PALS and EP to understand the relations between the κ value and pore characteristics. The porosity, pore size distribution and pore interconnectivity of these three films are summarized in table I. As expected, the film porosity increases as the κ decreases. Of more interest and importance are the changes in pore size and pore interconnectivity. Although all three films have isolated micro-pores with an average diameter of 0.7-1.1 nm, the films with κ values of 2.4 and 2.0 start to contain ~ 1.9 nm pores. Interconnected meso-pores with a diameter of 2.9-3.5 nm, which can be considered killer pores for 65 nm technology node and beyond, start to appear when the dielectric constant approaches 2. Such an increase in the porosity, pore size and pore interconnectivity of porous ILD can lead to significant process and reliability challenges.

Table I: The pore characteristics of three PECVD SiCOH films with different κ values

Films	k value	Porosity	Pore size (nm)			Pore Inter-connectivity
			Micro	Meso 1	Meso 2	
SiCOH # 1	3.0	8%	0.7-1.1	none	none	No
SiCOH # 2	2.4	22%	0.6-1.1	1.7-2.0	none	No
SiCOH # 3	2.0	38%	0.7-1.1	1.9	2.9-3.5	Yes

For PECVD and spin-on siloxane-based low-κ ILD films, the film damage by various plasma treatments is an issue that needs attentions [4,6]. Impact of κ value and pore characteristics on the oxygen plasma damage of three PECVD SiCOH materials with κ values of 3.0, 2.4 and 2.0 has been investigated. The changes in κ, FTIR spectrum, film thickness and refractive index of the films after the plasma treatment have been studied. Figure 1 shows the FTIR spectra of the films after a reactive ion etch (RIE) O_2 plasma treatment (300 mT, 550 W, 30 °C) for 40 seconds. The κ values of these films before and after the O_2 plasma treatment are summarized in Fig. 2. The film with a κ of 3.0 is found to be resistant to O_2 plasma and it experienced little change in κ, FTIR spectrum, thickness and RI. The same O_2 plasma treatment, however, caused significant κ increases of the films with lower κ values. The κ increases of these films were accompanied by the dramatic changes in the FTIR spectra where the signal intensity of Si-CH₃ bending and C-H stretching peaks decreased significantly while a wide –OH peak arose at around 3650 cm⁻¹. The degree of plasma damage to the film is found to increase significantly as the film κ value decreases. The oxidative plasma damage to SiCOH films usually involves the Si-CH₃ bond breakings by the oxidative radicals that diffuse into the film, causing significant moisture absorption and film densification. As the porosity of SiCOH films increases, the oxidative radicals diffuse faster in the film, making the films more susceptible to the plasma damage. Innovative plasma ash process or alternative techniques to remove the photo resist may be needed in order to successfully integrate the porous ULK SiCOH thin films.

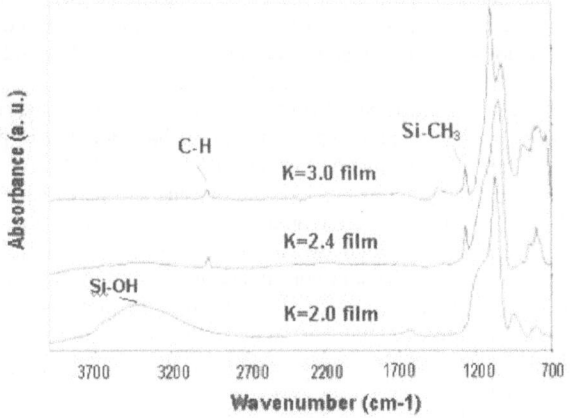

Figure 1: FTIR of PECVD SiCOH films after oxygen plasma treatment.

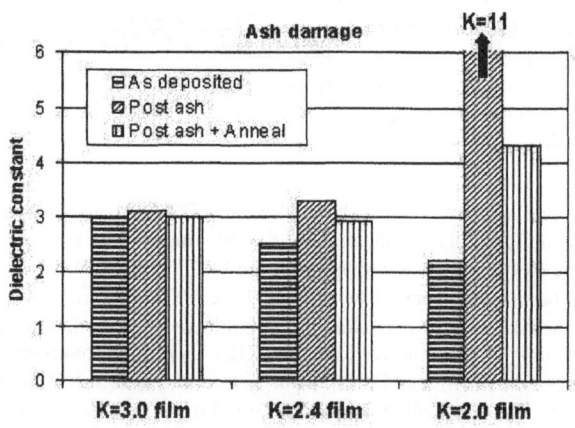

Figure 2: Dielectric constant of SiCOH films before and after oxygen plasma treatment.

Impact of overhead films on κ_{eff}

In addition to plasma damage, a variety of process and reliability challenges have emerged as the ULK ILD materials become more porous [5-7]. The use of overhead dielectric films with the porous ULK ILD helps to mitigate some key process and reliability issues. Figure 3 is a schematic cross sectional view of the interconnect system showing the overhead films that can be used with an ULK ILD. Such a interconnect system is referred to as a "fully-loaded" system in this paper. For example, a cap layer on top of the porous ULK ILD is used to protect the mechanically weak porous ILD from being damaged during CMP and other downstream processes. The deposition of a thin dielectric liner material after the dual-damascene (DD) etch helps to seal the pores on the sidewalls of porous ULK ILD and significantly alleviate the challenges of metallic barrier deposition process. For porous ILD, having smooth trench bottoms without "grass" formation and maintaining consistent trench depths also become challenges. The use of a middle etchstop effectively addresses these two issues.

These overhead films, however, usually have dielectric constants higher than that of the ILD and their use tends to increase the overall effective dielectric constant (κ_{eff}) of the BEOL interconnect. System-level simulations have been done to evaluate the contributions of these overhead films on the total interconnect capacitance and critical path delay at different technology nodes using the physical geometries and film parameters shown in table II. Simulation experimental details are available in previous publications and will not be given here [10-14].

The simulation results in Fig. 4 show that the overhead films have a significant impact on the interconnect capacitance. By the 45 nm technology node, roughly 69% of the total delay is due to the BEOL interconnect, of which 23% of delay is due to the overhead films [10]. To achieve a low κ_{eff}, the excessive use of overhead films with high κ values should be avoided. For the necessary overhead films, the use of films with lower κ values helps to alleviate the κ value requirement pressure of the ILD materials and thus ease the ULK integration challenges while still meeting the κ_{eff} targets of future technology nodes.

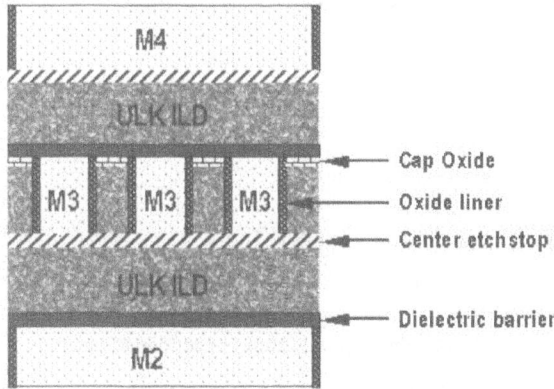

Figure 3: Schematic cross sectional view of a fully loaded interconnect architecture

ULK silicon carbide films

PECVD hydrogenated silicon carbide (H:SiC) or hydrogenated silicon carbon nitride (H:SiCN) thin films with κ values of 4.8-5.0 have been used as the dielectric barriers between Cu and ILD [14,15]. These films act as a Cu diffusion barrier between the ILD and Cu, serve as a bottom etchstop layer during the DD via etch, and provide a good adhesion to Cu for good interconnect reliability. In this work, a PECVD ULK H:SiC film with a κ value of ~ 3.0 has been developed and studied for etch-stop applications. This ULK H:SiC film was also compared to the PECVD H:SiC and H:SiCN films with κ values of 4.8-5.0.

Table II: Interconnect dielectric film properties for simulations [10]

Technology	90 nm	65 nm	45 nm
Etch Stop thickness [nm]	50	50	40
Etch Stop k	4.5	4.5	3.5
Cap Oxide thickness [nm]	50	50	40
Cap Oxide k	4.2	4.2	3.5
Liner thickness [nm]	5	5	5
Liner k	4	4	4
Barrier thickness [nm]	65	50	40
Barrier k	5.5	4.5	3.5
IMD Bulk k thickness [nm]	91	56	28
Bulk k	2.4	2.1	1.9

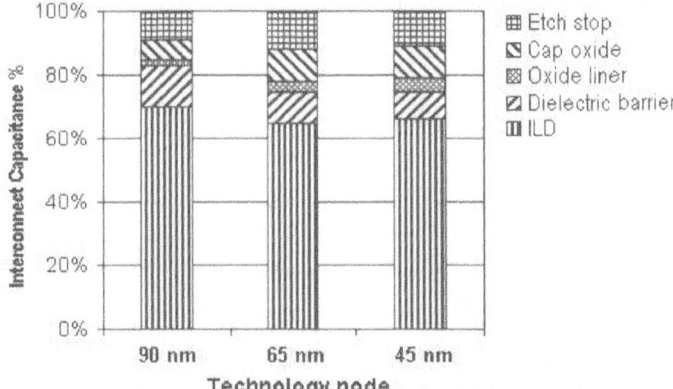

Figure 4: The impact of overhead films on the total interconnect capacitance

The atomic composition of the ULK H:SiC films determined using RBS and HFS is summarized in table III. Also shown for comparison are the chemical compositions of the H:SiCN and H:SiC film with higher κ values. The C/Si atomic ratio of ULK H:SiC film is much higher than the H:SiC film with κ of 4.8. It is proposed that more C atoms exist in the ULK H:SiC film in the form of CH_3 groups single-bonded to Si, resulting in more microporosity in the dense Si-C film lattice. The absence of N in the film helps to reduce the chance of resist poisoning during lithography processes.

Table III: Chemical composition of the ULK H:SiC, H:SiC and H:SiCN films

Films	Si (%)	C (%)	H (%)	N (%)
ULK H:SiC (k=3.0)	16	28	56	0
H:SiC (k=4.8)	25.3	24.7	50	0
H:SiCN (k=5.0)	24.5	17.5	46	12

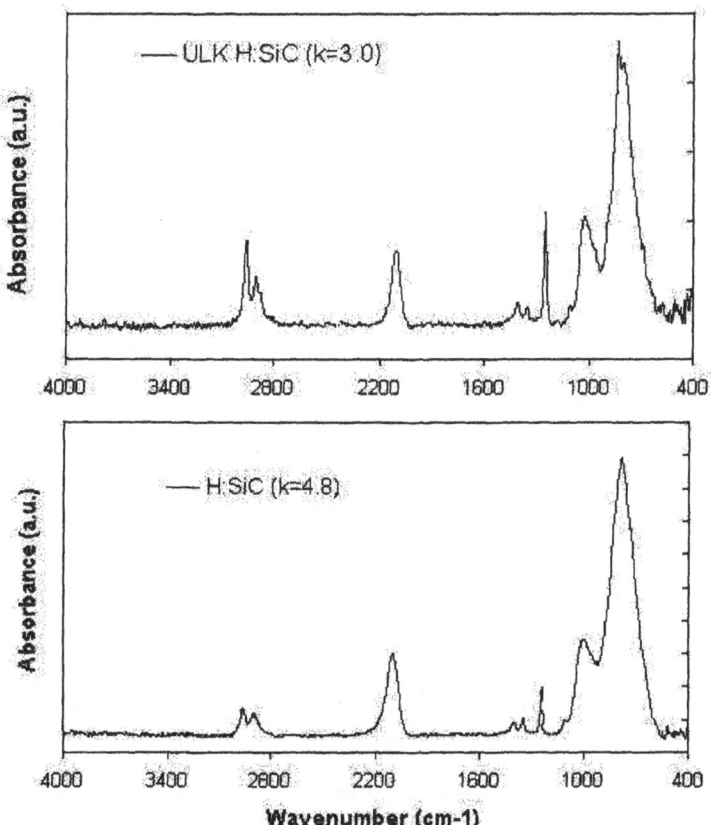

Figure 5: FTIR of the ULK H:SiC film and H:SiC film with a κ of 4.8.

Figure 5 shows the FTIR absorbance spectrum of the ULK H:SiC film and the H:SiC film with a κ of 4.8. These two films have very similar spectrum patterns. They both show a dominant absorbance peak at 780-790 cm^{-1} associated with the Si-C stretching or Si-CH$_3$ rocking. Other significant absorbance peaks are Si-CH$_2$-Si bending at ~1000 cm^{-1}, Si-CH$_3$ stretching at ~1250 cm^{-1}, Si-H stretching at ~2100 cm^{-1}, and C-H stretching at 2890-2950 cm^{-1}. Compared to the film with a κ of 4.8, a major difference in the FTIR spectrum of the ULK H:SiC film is the much higher peak intensities of the Si-CH$_3$ stretching and C-H stretching normalized to the main peak at 780-790 cm^{-1}. This suggests that more methyl-related groups (CH$_3$, CH$_2$) are attached to the silicon atoms, resulting in less formation of the dense Si-C network in the ULK H:SiC film. The incorporation of more methyl-related groups make the H:SiC film less dense and cause the reduction in κ. This is also confirmed by the higher C/Si atomic ratio of the ULK H:SiC film.

The κ value at 1 MHz and RI at 633 nm of the ULK H:SiC film were measured to be 3.0 and 1.6 respectively. The κ and RI comparisons between the ULK H:SiC film and the H:SiC and H:SiCN films with higher κ values are summarized in table IV, showing an approximately 40% reduction in the κ value for the ULK H:SiC film.

Table IV: Dielectric constant and refractive index of the ULK H:SiC, H:SiC and H:SiCN films

Films	Dielectric constant	Refactive index
ULK H:SiC	3.0	1.60
H:SiC	4.8	2.03
H:SiCN	5.0	1.94

The pore characteristics of this ULK H:SiC film have been studied using PALS. This film was found to have only closed micro-pores with an average pore size less than 1.2 nm without the presence of meso-pores. The mechanism of forming the H:SiC film with such a low κ value, yet free of meso-pores is not well understood at this time.

Low leakage current is an important requirement for the etchstop films in order to have good dielectric isolation between interconnect wires. The J-V characteristics of this ULK H:SiC film at room temperature have been studied using MIS capacitors with Cu gate electrodes. Figure 6 showed the J-V characteristics of different PECVD silicon carbide thin films. The leakage current of the ULK H:SiC film at 1.0 MV/cm was measured at 6.2×10^{-10} A/cm^2, which is about 5 times lower than the SiC and SiCN films with higher κ values.

The bottom etchstop film protects the underlying Cu during the DD via etch while the middle etchstop films help to achieve consistent and controlled trench depth. A good etch selectivity to the ILD film is a key requirement for the etchstop films to fulfill their functions. The κ value of the PECVD H:SiC films can be reduced by introducing oxygen sources during the deposition, producing oxygen-doped H:SiC films. However, the etchstop film to ILD etch selectivity usually deteriorates as the oxygen content increases. The etch selectivity between a PECVD SiCOH ILD film with a κ value of 3.0 and the ULK H:SiC films was studied and compared to the H:SiCN film with a κ value of 5. In this work, the etch selectivity was evaluated in a magnetically enhanced RIE system using a C$_4$F$_8$/Ar/N$_2$ ILD etch chemistry. The H:SiC film removals inside 0.2 um vias (dense and isolated patterns) after a 30-s etch were measured using high definition cross section SEM analysis and are summarized in table V. Using a SiCOH ILD film with a κ of 3.0 and a H:SiCN film with a κ of 5.0, such a ILD etch chemistry gives a SiCOH-to-H:SiCN etch selectivity of approximately 10:1. As shown in table V, the removal of ULK H:SiC film after the ILD etch is negligible compared to the H:SiCN film, indicating an excellent etch selectivity to the ILD material. Such an extremely high etch selectivity makes the ULK H:SiC film a very good etch-stop layer for SiCOH ILD films.

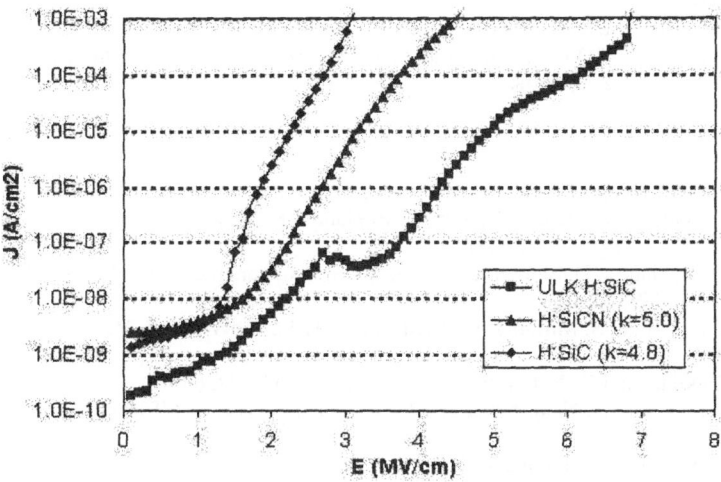

Figure 6: J-V characteristics of the ULK H:SiC, H:SiC and H:SiCN films at room temperature.

Table V: ULK H:SiC and H:SiCN film removal using a ILD etch chemistry.

Films	Film removal (nm)			
	Wafer center		Wafer edge	
	Dense	Iso	Dense	Iso
ULK H:SiC(k=3.0)	Not detectable	Not detectable	Not detectable	Not detectable
H:SICN (k=5.0)	32	28	43	40

SUMMARY

Porous ILD materials are most likely to be used in order to meet the performance requirements of future generation ULSI interconnects. Pore characteristics and oxygen plasma damage of porous ULK materials have been discussed. Simulations showed that the overhead dielectric films in the interconnect system have a significant impact on the κ_{eff} of the interconnects. The basic properties of the PECVD H:SiC films with a κ of 3.0 have been studied.

Compared to the H:SiC and H:SiCN films with κ values of ~5, the ULK H:SiC film offers a 40% reduction in κ value and has a leakage current that is ~ 5 times less at 1 MV/cm. With a negligible etch rate using a PECVD SiCOH ILD etch process, the SiCOH-to-H:SiC etch selectivity of the ULK H:SiC film is also superior compared to the H:SiCN films with a κ value of 5.0. The high etch selectivity of ULK H:SiC films make them excellent etchstop films.

ACKNOWLEDGEMENT

The authors thank Dr. M. Baklanov at IMEC for the EP analysis. The authors are also grateful to the staff of APMD group and failure analysis lab at LSI Logic for the valuable discussions and processing supports.

REFERENCE

1. International Technology Roadmap for Semiconductor-Interconnects: 1999, 2001 and 2003.
2. A. Grill, V. Patel, K. Rodbell, E. Huang, M. Baklanov, K. Mogilnikov, M. Toney, and K. Kim, J. Appl. Phys., **94**, 3427 (2003)
3. K. Maex, M. Baklanov, D. Shamiryan, F. Iacopi, S. Brongersma, and Z. Yanovitskaya, J. Appl. Phys., **93**, 8793 (2003)
4. H. Lu, H. Cui, I. Bhat, S. Murarka, W. Lanford, W. Hsia, and W. Li, J. Vac. Sci. Technol. B, **20**, 828 (2002)
5. N. Aoi, T. Fukuda, and H. Yanazawa, Proceedings of IITC, 72, (2002)
6. E. Kondoh, T. Asano, A. Nakashima, and M. Komatsu, J. Vac. Sci. Technol. B **18**, 1276 (2000)
7. W. Besling, A. Satta, J. Schuhmacher, T. Abell, V. Sutcliffe, a. Hoyas, G. Beyer, D. Gravesteijn, and K. Maex, Proceedings of IITC, 288 (2002)
8. D. Gidley, W. Frieze, T. Dull, A. Yee, E. Ryan, and H. Ho, Phys. Rev. B **60**, R5157 (1999)
9. M. Baklaonv, K. Mogilnikov, V. Polovinkin, and F. Dultsev, J. Vac. Sci. Technol. B **18**, 1385 (2000)
10. P. Zarkesh-Ha, P. Burke, K. Doniger, W. Loh, V. Sukharev, M. Lu, P. Bendix, W. Catabay, W. Hisa, C-H. Chang, Proceedings of VMIC, 17 (2003)
11. J. Davis, V. De, and J. Meindl, IEEE Transaction on Electron Devices, 580 (1998)
12. J. Davis, V. De, and J. Meindl, IEEE Transaction on Electron Devices, 590 (1998)
13. C. Enz, F. Kerummenacher, and E. Vittoz, Analog Integrated Circuits and Signal Processing, (83) 1995
14. M. Loboda, Microelectronic Engineering, **50**, 15 (1997)
15. S. G. Lee, Y. Kim, S. P. Lee, H. Oh, S. J. Lee, M. Kim, I. Kim, J. Kim, H. Shin, J. Hong, H. Lee, and H. Kang, Jap. J. Appl. Phys., **40**, 2663 (2001)

Mat. Res. Soc. Symp. Proc. Vol. 812 © 2004 Materials Research Society   F5.5

CROSS-SECTION NANO-INDENTATION FOR RAPID ADHESION EVALUATION

S. H. Brongersma,[1] Dominiek Degryse,[1] Jerome Souiller,[1] Bart Vandevelde,[1] and K. Maex[1,2]

[1] IMEC, Kapeldreef 75, B-3001 Leuven, Belgium,
[2] K.U.Leuven, ESAT, Kasteelpark Arenberg 10, B-3001 Leuven, Belgium,

ABSTRACT

Cross-section nano-indentation is a technique that consists of cleaving a wafer, indenting behind the stack on the side surface causing delamination, and measuring the delaminated area in an optical microscope. It is shown to be a reliable rapid technique for adhesion evaluation and process optimization of the SiO_2/barrier/copper stack.

INTRODUCTION

With the introduction of copper and low-k materials in back-end-of-line processing, adhesion issues have regained importance. Especially the properties of copper diffusion barriers are of critical importance, as they need to prevent diffusion into the dielectric as well as ensure copper adhesion to the underlying stack. The rapid succession of new low-k and barrier materials necessitates a strong effort in terms of characterizing integratability. Thus the optimization of barrier layers, e.g. TaN, TiN, and WCN, depends critically on rapid feedback from adhesion measurements.

Over the last few years the 4-point bend technique[1] has become the generally accepted choice for reliable quantitative adhesion determination. However, sample

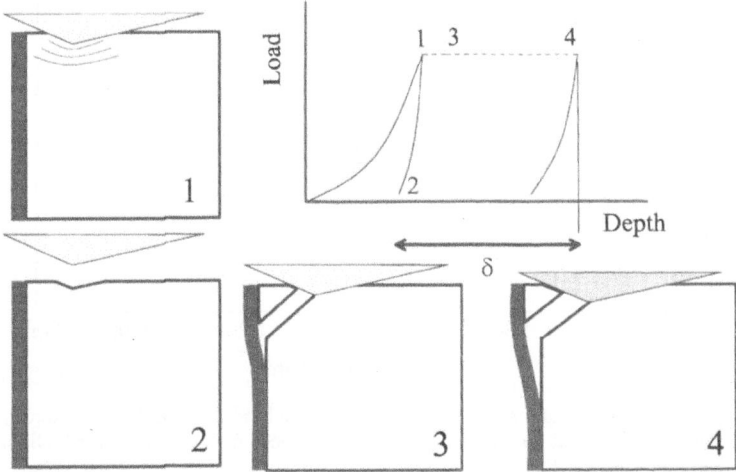

Figure 1. Four stages of cross-section nano-indentation. The initial indent is determined by the silicon properties only. The unloading curve shows how much of this indent is elastic. Above a critical depth cracks running from the indent to the interface are induced and then deflected into the weakest interface.

preparation is cumbersome and time consuming. As an alternative, cross-section nano-indentation provides a rapid alternative as a process optimization tool that was first introduced by Sánchez et al.[2] Here the sample preparation is limited to cleaving the wafer and mounting it on its side in a nano-indentation set-up. Then, a series of indents is performed over a range of distances behind the stack of interest. In contrast to indentation for modulus determination vertical thermal drift of the indent column is not critical and measurements can start within minutes after loading the samples. Finally, the resulting delamination is measured in an optical microscope with 100x objective. Typically, reliable data can be obtained within the hour.

Quantification of the technique depends on a combination of 4-point bend measurements and finite element simulations.

EXPERIMENTS AND RESULTS

Samples are made by blanket layer depositions on Si(001) substrates. For the measurements described here these layers are 50 nm Si_3N_4, 500 nm SiO_2, barrier layer, and 600 or 1000 nm copper. The barrier is either a WCN or Ta/Tan layer.

For indentation a Nano-indenter[XP] from MTS with precision x,y-stage (0.1 µm) is used. Indents are made up to 10 µm's behind the interface of interest.

During indentation several stages can be identified, as is demonstrated in Fig. 1. First the silicon is indented and the resulting loading curve does not depend on how far behind the interface the indent is made (up to point 1). This is illustrated in Fig. 2 where the loading curves for a range of distances are shown. The curves all overlap until the point where cracking out to the interface of interest is initiated. Then

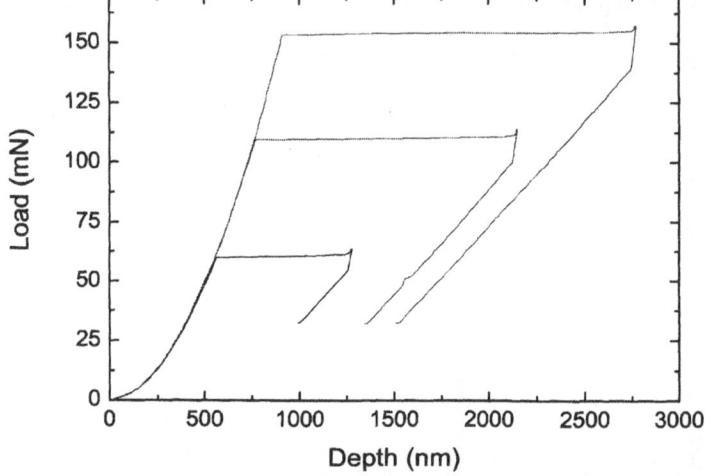

Figure 2. Loading curves obtained for 3 depths behind the interface of interest. Obviously the initial loading curves overlap as these depend only on the silicon substrate properties.

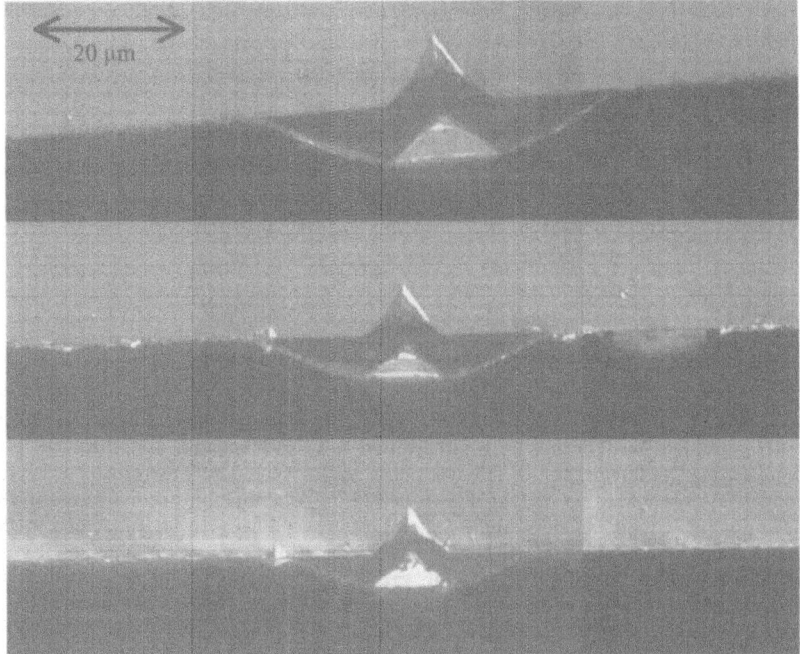

Figure 3. Three optical images of the delamination due to cross-section nano-indentation. Both the delamination length and the chipped-out piece of the silicon substrate are clearly visible. The images correspond to the loading curves in Fig. 2 from top to bottom respectively.

delamination starts (point 3) as the tip penetrates to the point where the applied load is regained (point 4). When unloading from point 1 we see that only part of the energy is released. This part of the energy can help delaminate the layer of interest while another part of the energy has resulted in a permanent deformation of the silicon substrate. Thus the difference in depth δ between points 2 and 4 should be correlated with the length of delamination as can be measured from optical images as shown in Fig. 3.

Since the point just before delamination is practically difficult to predict, a calibration has been performed from a series of loading/unloading curves in bulk silicon. As can be seen in Fig 4. there is a simple linear correlation between points 2 and 3 with a slope of 0.605 that can be used in the extraction of δ from experimental data.

Quantification of the results depends on finite element modeling (FEM) and crosschecks with 4-point bend experiments (which are in process). Analytic linear elastic models do not suffice to describe the system due to the plastic behavior of copper during delamination. In fact, it is this behavior that renders the test feasible because it is the plastic behavior of the copper layer that keeps the test structure together and controls the dynamics of the delamination process. When a brittle material would be used instead of copper the layer would simply break instead of delaminating. As the barrier layer has little influence on the deformation, even though it determines the adhesion, simulations are performed for the simple system of copper

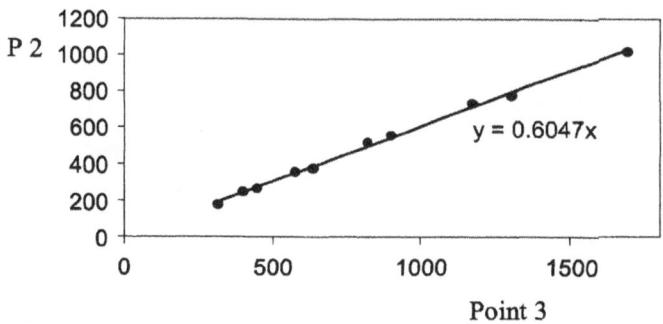

$$y = 0.6047x$$

Point 3

Figure 4. Linear correlation between points 2 and 3 with a slope of 0.605.

on substrate. For FEM we use the MSC MARC simulation software where an initial grid is defined that consists of a simple two-layer system with a triangle cut out, representing the left portion of the symmetric test system. Then, the loose end of the copper layer is forced up and, as shown in Fig. 5, delamination occurs. Stress in the delaminating copper layer is highest at the point of delamination and close to the chipped-out silicon piece where bending is strongest. By performing a series of such simulations with varying degrees of predefined adhesion, and monitoring the delaminated length as a function of displacement, calibration curves can be obtained.

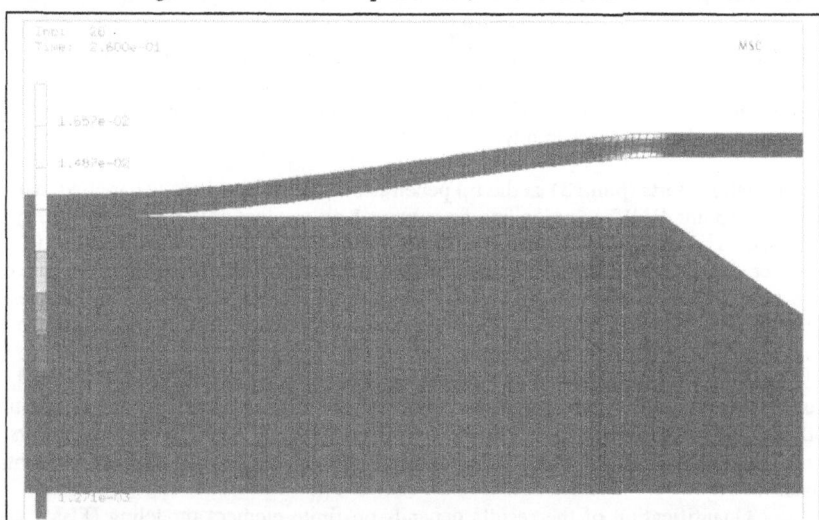

Figure 5. Finite element simulation of a cross-section nano-indentation (the system is symmetric around the right side of the mesh). For a predefined adhesion the top right portion is forced up while the length of delamination is monitored. Stress in the copper layer is highest close to the chipped-out silicon piece and close to the starting point of the delamination.

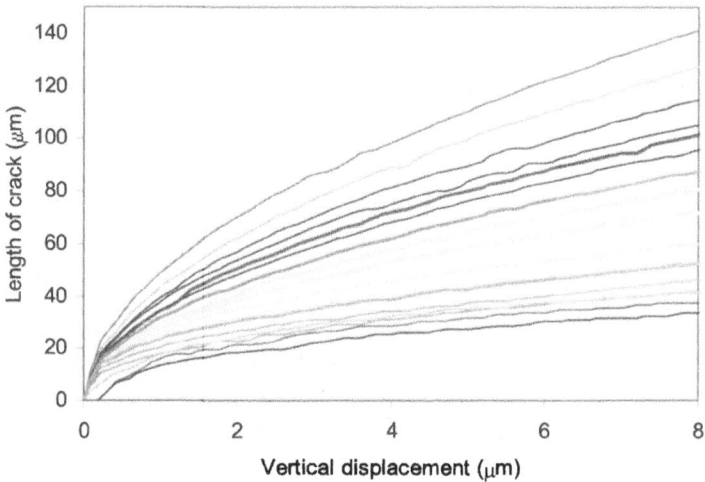

Figure 6. Length of delamination as a function of vertical displacement of the chipped-out piece of silicon for a predefined adhesion energy that varies linearly from 2 J/m^2 (top curve) to 9 J/m^2 (bottom curve).

The results of such simulations with a copper layer of 1µm thick are depicted in Fig. 6 for the adhesion strength ranging between 2 and 9 J/m^2. From the distance between the curves it can be concluded that the length of delamination is a slightly non-linear function of the adhesion force.

The simulations also indicate a complication for the application of this technique. When adhesion is sufficiently strong, plastic deformation of the copper layer tears the copper layer close to where the initial crack reaches the interface. In this case a thicker copper layer is necessary in order to induce delamination. Thus, for each adhesion strength there is a lower limit for the copper thickness. This has also been seen in the experimental data. This is illustrated in Fig 7. where a snap-shot from a simulation for 7 J/m^2 is shown together with a SEM image of the result from a test on a well adhering TaN barrier and 500 nm copper seed.

Figure 7. On the left a simulated picture where the copper layer is too thin to accommodate the vertical displacement. Locally massive plastic deformation occurs. On the right the corresponding real case where the crack induced from a cross-section nano-indentation test does indeed not deflect into the interface.

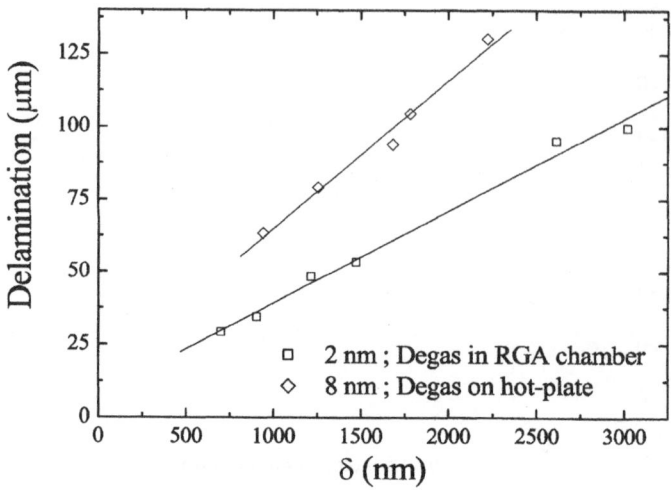

Figure 8. Delamination length vs. parameter δ for two SiO₂/WCN/Cu stacks where both the thickness and degas conditions during barrier deposition were changed.

When performing this test on two WCN barrier layers (2 and 8 nm thick) with 600 nm copper one can see how a few indents are sufficient to characterize the difference in adhesion (Fig. 8). Clearly, the technique is sufficiently sensitive to evaluate the adhesion over a fair range (the details of the optimization for this material's deposition are to be discussed elsewhere).

CONCLUSIONS

Cross-section nano-indentation provides a rapid quantitative method for evaluating adhesion strength. Especially where optimization is still needed the technique is particularly sensitive. For much lower adhesion the layer tends to already delaminate upon cleaving of the wafer and testing is not necessary. For good adhesion, where no integration issues are observed, sensitivity decreases unless much thicker copper layers are deposited (many microns).

REFERENCES

[1] Q. Ma, H. Fujimoto, P. Flinn, V. Jain, F. Adibi-Rizi, F. Moghadam, and R. H. Dauskardt. Mater. Res. Soc. Symp. Proc. **391**, 91 (1995).
[2] J. M. Sanchez, S. El-Mansy, B. Sun, T. Sherban, N. Fang, D. Pantuso, W. Ford, M. R. Elizalde, J. M. Martinez-Esnaola, A. Martin-Meizoso, J. Gil-Sevillano, M. Fuentes, and J. Maiz, Acta Mater. **47**, 4405 (1999).

Mat. Res. Soc. Symp. Proc. Vol. 812 © 2004 Materials Research Society   F5.6

Adhesion Strength Evaluation of Low-k Interconnect Structures Using a Nanoscratch Method

Jiping YE, Kenichi Ueoka, Nobuo Kojima, Junichi Shimanuki, Miyoko Shimada[1] and Shinichi Ogawa[1]
Research Department, NISSAN ARC, LTD., 1 Natsushima-cho, Yokosuka 237-0061, Japan
[1]Semiconductor Leading Edge Technologies, Inc., 16-1 Onogawa, Tsukuba 305-8569, Japan

ABSTRACT

A convenient nanoscratch method was combined with atomic force microscope (AFM) and transmission electron microscope (TEM) observations to conduct the first-ever evaluation of the adhesion strength of a complicated microstructure Cu/Ta/TaN/pSiO$_2$/low-k/SiC/pSiO$_2$/Si-substrate with the aim of correlating the fracture strength with the results of chemical mechanical polishing (CMP) tests. Concretely, this evaluation focused on the fact that specimens having a low-k layer pretreated with rare-gas plasma prior to the deposition of the SiO$_2$ layer exhibited low delaminated densities in the Cu CMP process. It was found that a specimen with the rare-gas plasma pretreatment exhibited a higher friction coefficient, a higher critical load and brittle adhesive failure resulting from delamination at the interface between the low-k and SiC layers. A specimen without the rare-gas plasma pretreatment displayed a lower friction coefficient, a lower critical load, and ductile cohesive failure in the low-k layer. Because less plastic deformation was observed in the low-k layer subjected to the rare-gas plasma pretreatment, it is assumed that the pretreatment reinforced the mechanical properties of the low-k layer, making it more resistant to ductile cohesive failure. These results agreed with the CMP test data and indicated that the nanoscratch method makes it possible to predict the ability of complicated Cu/low-k interconnect structures to withstand the CMP process.

INTRODUCTION

Low-k dielectric materials have attracted a great deal of attention for application to multilevel interconnects and packaging structures [1, 2]. Integration of microelectronic components results in various interfaces between the low-k material and inorganic dielectric layers, such as SiO$_2$ and SiC. Thus, two parts with obviously different thermal, mechanical and chemical properties are combined at the interface. Abrupt changes in these properties are likely to cause delamination at the interface, whereas the low mechanical properties exhibited by low-k dielectric materials are apt to cause debonding of the low-k layer during a chemical mechanical polishing (CMP) process. In either event, poor interfacial adhesion will lower the reliability of the interconnect structures. It is important to evaluate the adhesion strength of low-k interconnect structures for screening out unacceptable dielectric materials and optimizing the CMP process conditions.

Usually, four-point bending tests and stud pull tests are widely used to estimate thin-film fracture strength [3-5]. The former test involves sandwiching and epoxy-bonding the test layer side of a specimen to a Si backing and then making a pre-notch to facilitate crack propagation from the top surface to the interface of interest, while in the latter test a stud is epoxy-bonded perpendicular to the test layer. Both methods induce normal tensile stress by mechanically bending the sandwiched specimen or by pulling the stud to cause debonding within the multi-layer structure. Although these methods have been reported to be suitable for estimating the fracture strength of Cu/low-k interconnect structures, their application requires considerable care to ensure that the specimens can withstand the mechanical and chemical damage that can occur in the sampling process.

Recently, we have purposed a novel and convenient nanoscratch method that is combined with atomic force microscope (AFM) and transmission electron microscope (TEM) observations [6]. This method only involves the application of a diamond stylus to scratch the film on the substrate and thus no additional specimen preparation is needed and also no sampling damage occurs. Since this method is combined with AFM and TEM observations, not only the fracture strength but also the fracture type can be determined. In the case of a SiC/low-k/Si-sub stacked layer [6], a specimen having a low-k layer with a lower modulus and hardness displayed lower fracture strength and suffered ductile adhesive failure resulting from delamination at the interface between the low-k layer and Si substrate. In contrast, a specimen having a low-k layer with a higher modulus and higher hardness exhibited higher fracture strength and suffered brittle cohesive failure that occurred in the low-k layer. The measurements corresponded with the results of the CMP process, indicating that the process characteristics can be predicted with this nanoscratch method.

This paper presents a study concerning a further application of this convenient nanoscratch method to evaluate the adhesion strength of a complicated microstructure $Cu/Ta/TaN/pSiO_2/low-k/SiC/pSiO_2/Si$-substrate. The effect of applying a rare-gas plasma pretreatment to the low-k film on its fracture strength and critical failure sites was investigated. The correlation between the nanoscratch evaluation and the results of chemical mechanical polishing (CMP) tests was examined.

EXPERIMENTAL PROCEDURE

Specimens with the microstructure $Cu/Ta/TaN/pSiO_2/low-k/SiC/pSiO_2/Si$-substrate were examined. The low-k layer of one specimen was pretreated with rare-gas plasma before the SiO_2 barrier layer was grown whereas that of another specimen was not pretreated. The low-k film was fabricated by spin-coating a methylsilsesquioxane (MSQ) porous low-k layer on the SiC layer. From the surface, the thicknesses of the layers were 300 nm for Cu, 15 nm for Ta, 15 nm for TaN, 50 nm for $pSiO_2$, 250 nm for the low-k material, 50 nm for SiC and finally 500 nm for $pSiO_2$ on the Si substrate, respectively. The investigation focused on the fact that the pretreated specimen (denoted as L) exhibited lower delaminated densities in the CMP polishing test than the non-pretreated specimen (denoted as H).

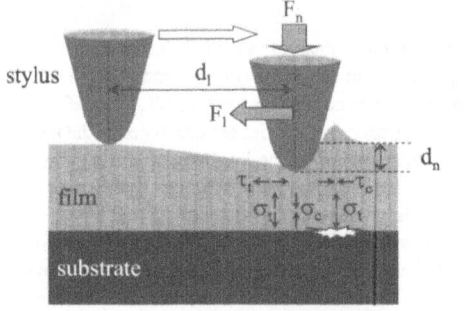

F_n : normal force
F_l : lateral force
τ_t : tensile lateral stress
τ_c : compressive lateral stress
σ_t : tensile normal stress
σ_c : compressive normal stress
d_n : normal displacement
d_l : lateral displacement

Figure 1 A schematic of the nanoscratch method for estimating the normal critical load and showing the estimated stress distribution induced in the film and the interface by nanoscratching.

Nanoscratch measurements were made using a TriboScope (Hysitron, Inc., Minneapolis, MN). Fracture strength was characterized on the basis of the critical normal load at the first abrupt decrease in the friction coefficient (so-called critical point). As shown in figure 1, this nanoscratching process applied a normal force F_n by pressing a diamond stylus on the sample surface and then a lateral displacement d_l by displacing the stylus to scratch the surface. By detecting the lateral force F_l and the normal displacement d_n, the friction coefficient μ for a given normal load or normal displacement was determined by the ratio of F_l to the normal force F_n. A conical diamond tip was used as the stylus. The cone angle on the tip was 60 deg and the radius was about 1 μm. The normal force F_n was applied in a range from zero to 15 mN at a loading rate of 0.125 mN/s and the lateral displacement was applied in a range from zero to 10 μm at a speed of 0.083 μm/s. To facilitate observation of the fracture morphology, AFM surface images were observed under a Nanoscope-IIIa/D3100 microscope (Digital Instruments, Ltd., Santa Barbara, CA). A tapping mode using a Si strap cantilever at a resonance frequency of around 280 kHz was applied. This cantilever possessed a spring constant of approximate 50 N/m, a tip radius of around 10 nm, and a half cone angle of less than 18 deg. TEM cross-sectional images were observed under a H-9000UHR microscope (Hitachi, Ltd.) operated at 200 kV. Specimens for TEM observation were prepared by using a SMI9200 focused ion beam system (Seiko, Ltd.).

RESULTS

Friction coefficient and critical load

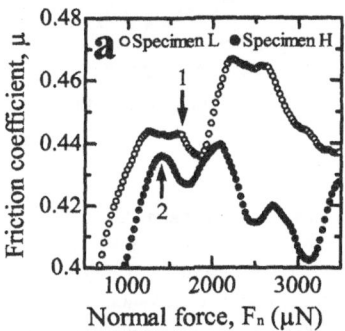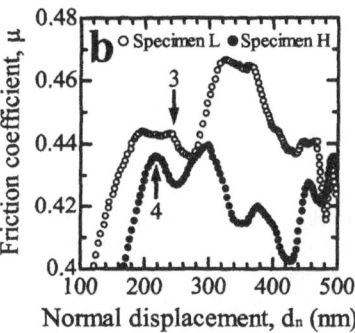

Figure 2 Dependence of friction coefficient μ on (a) normal force F_n and on (b) normal displacement d_n for specimens L and H around the areas at the critical points indicated by arrows 1~4.

The friction coefficients of specimens L and H are shown in figure 2 as a function of the normal force and normal displacement. Specimen L exhibited a larger critical load of 1650 μN at a normal displacement of 245 nm than specimen H, which showed a critical load of 1400 μN at a normal displacement of 215 nm. Thus, the $Cu/Ta/TaN/pSiO_2/low$-$k/SiC/pSiO_2/Si$-substrate stacked structure with the pretreated low-k layer was found to possess higher fracture strength than the structure with the non-pretreated low-k layer. These results agreed with the observations of the delaminated densities in the CMP test. It was also observed that specimen L possessed a higher friction coefficient than specimen H. This result indicated that some mechanical properties in the low-k layer were changed by the rare-gas plasma pretreatment.

Fracture morphology and site

AFM images of the nanoscratch surfaces of specimens L and H are shown in figure 3. For specimen L, higher pile-ups were observed on the sides of the nanoscratch rather than ahead of it. In contrast, for specimen H, a residual peel was observed ahead of the nanoscratch and lower pile-ups were observed on the sides.

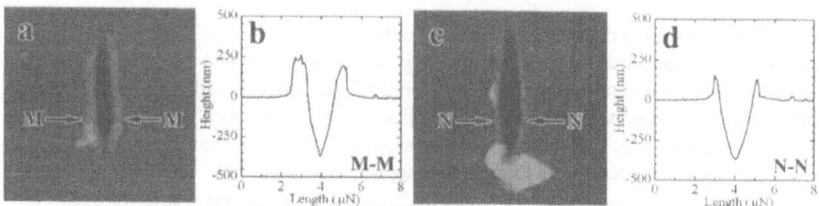

Figure 3 AFM images of the nanoscratch surfaces of (a) specimen L and (c) specimen H, and corresponding cross-sectional profiles indicated by (b) M-M and (d) N-N.

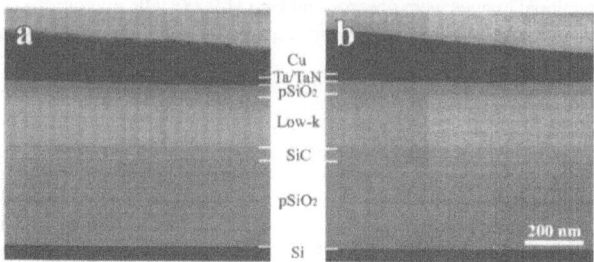

Figure 4 TEM cross-sectional views of the area at the critical point along and in the middle of the nanoscratch for (a) specimen L and (b) specimen H.

Observation was first made of TEM cross-sectional views along and in the middle of the nanoscratch. The nanoscratch areas round the critical point for specimens L and H are shown in figure 4. Both specimens exhibited no crack or failure within the low-k layer and no delamination at the interfaces in the Cu/Ta/TaN/pSiO₂/low-k/SiC/pSiO₂/Si-substrate stacked structure. Further observation was made of TEM cross-sectional views perpendicular to the nanoscratch direction. As shown in figure 5, both specimens exhibited a pair of cracks slightly far from the center of the nanoscratch, but the cracks differed in their failure morphology and location. For specimen L, a crack originated from the interface between the low-k and SiC layers and then propagated along the same interface. Less plastic deformation was observed in the low-k layer, but large plastic deformation in the Cu layer caused large Cu pile-ups on the sides of the nanoscratch. In contrast, specimen H exhibited a crack that originated from the low-k layer and propagated close to the interface between the low-k and SiC layers. The low-k layer in specimen H exhibited more plastic deformation than that in specimen L. The Cu layer in specimen H peeled off from the surface and the remaining Cu layer displayed less plastic deformation than that in specimen L, which is thought to be the reason for the small Cu pile-ups on the nanoscratch sides. These results corresponded with the AFM observations.

As a result, it was found that specimen L experienced brittle adhesive failure resulting from delamination at the interface between the low-k and SiC layers, while specimen H suffered ductile cohesive failure in the low-k layer due to plastic deformation of the layer.

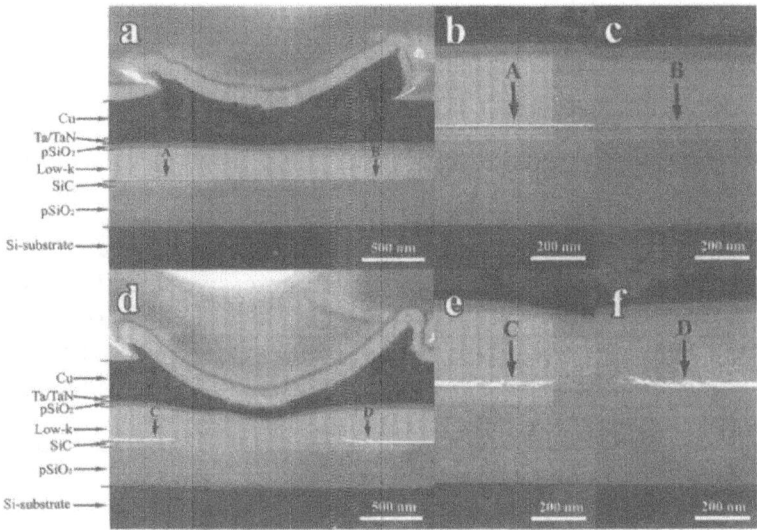

Figure 5 TEM cross-sectional views perpendicular to the nanoscratch direction for (a) specimen L and enlarged failure areas indicated by (b) arrow A and (c) arrow B and for (d) specimen H and enlarged failure areas indicated by (e) arrow C and (f) arrow D.

DISCUSSION

The nanoscratch measurements showed that specimen L with the rare-gas plasma pretreatment displayed a higher friction coefficient than specimen H without the pretreatment. TEM and AFM observations revealed that the low-k layer with the rare-gas plasma pretreatment exhibited less plastic deformation around the $pSiO_2$/low-k layer interface as compared with that without the pretreatment. These results indicated that the rare-gas plasma pretreatment reinforced the mechanical properties of the low-k layer, making it more resistant to ductile cohesive failure. It is thought that the rare-gas plasma pretreatment may reduce the organic concentration or produce some new crosslinks in the MSQ material, thereby increasing the mechanical strength of the low-k layer. To investigate the reasons further, it will be necessary to elucidate the mechanical properties such as hardness and modulus and the chemical state of the low-k layer subjected to the rare-gas plasma pretreatment.

The stress distribution induced by nanoscratching in a cross-section perpendicular to the nanoscratch direction is estimated as shown in figure 6. Compressive normal stress and tensile lateral stress are distributed in the center area just under the stylus while tensile normal stress and compressive lateral stress are distributed on both sides of the center area. For both specimens, TEM observations revealed neither failure in the low-k layer nor delamination at the interfaces in the center area just under the nanoscratch but a pair of cracks located on both sides of the center area. The failure location indicates that the cracks were caused by pure tensile normal stress without any influence from the lateral stress. The fracture types for both specimens were identified to be the pure tensile mode.

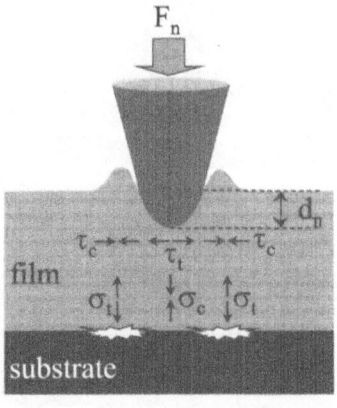

F_n : normal force
τ_t : tensile lateral stress
τ_c : compressive lateral stress
σ_t : tensile normal stress
σ_c : compressive normal stress
d_n : normal displacement

Figure 6 A schematic of the stress distribution induced by nanoscratching as seen in a cross-sectional view perpendicular to the nanoscratch direction.

CONCLUSION

A convenient nanoscratch method combined with AFM and TEM observations was applied to evaluate the adhesion strength of a complicated microstructure Cu/Ta/TaN/pSiO$_2$/low-k/SiC/pSiO$_2$/Si-substrate. The investigation focused on the fact that a specimen (L) having a low-k layer pretreated with rare-gas plasma exhibited lower delaminated densities in a CMP polishing test than a specimen (H) having a low-k layer without the rare-gas plasma pretreatment. Specimen L exhibited a higher friction coefficient, a higher critical load and brittle adhesive failure resulting from delamination at the interface between the low-k and SiC layers. Specimen H displayed a lower friction coefficient, a lower critical load, and ductile cohesive failure in the low-k layer. These results indicated that the rare-gas plasma pretreatment reinforced the mechanical properties of the upper low-k layer, making it more resistant to ductile cohesive failure. These results corresponded well with the CMP process, suggesting that this nanoscratch method can be used to predict the characteristics of complicated Cu/low-k microstructures in the CMP process.

REFERENCES

[1] M. Bohr, Int. Electron Device Meet. Technical Dig., p. 241 (IEEE, New York, 1995).
[2] The National Technology Roadmap for Semiconductors, Technology Needs, 1997 Edition, Semiconductor Industry Association.
[3] Q. Ma, H. Fujimoto, P. Flinn, V. Jain, F. Adibi-Rizi, F. Moghadam and R. H. Dauskardt, MRS Symp. Proc., Materials Reliability in Microelectronics V, Vol. 391, p. 91 (1996).
[4] Q. Ma, C. Pan, H. Fujimoto, B. Triplett, P. Coon and C. Chiang, MRS Symp. Proc., Stresses and Mechanical Properties VI, Vol. 436, p. 379 (1996).
[5] Q. Ma, J. Bumgarner, H. Fujimoto, M. Lane, and R. H. Dauskardt, MRS Symp. Proc., Materials Reliability in Microelectronics VII, Vol. 473, p. 3 (1997).
[6] J. Ye, N. Kojima, K. Ueoka, J. Shimanuki, T. Nasuno and S. Ogawa, J. Appl. Phys., Vol. 95, No. 8 (in press)

Effect of Aqueous Solution Chemistry on the Accelerated Cracking of Lithographically Patterned Arrays of Copper and Nanoporous Thin-Films

E. P. Guyer and R. H. Dauskardt
Department of Materials Science and Engineering, Stanford University
416 Escondido Mall, Stanford, CA 94305

ABSTRACT

The effect of moisture and aqueous solution chemistry on the rate of crack growth in lithographically patterned arrays of copper and low dielectric constant (LKD) materials is reported. Crack growth in the direction orthogonal to the features is demonstrated to fail at significantly increased loads compared to parallel cracking. Decreasing feature width is also shown to increase the structures resistance to fracture, particularly at reduced loads. The chemical interactions and mechanisms of energy dissipation are discussed.

INTRODUCTION

Considerable research is being directed at integrating nanoporous LKD layers into the interconnect structures of high-density integrated circuits. The reliable fabrication of devices containing these extremely fragile materials is, however, a significant technological challenge due to their high propensity for mechanical failure and susceptibility to stress corrosion cracking in moist environments. Survival through chemical mechanical planarization (CMP) and subsequent device packaging is of particular concern as the device structures are subjected to additional loads in the presence of highly reactive solution chemistries [1]. Although significant efforts have been directed at formulating CMP slurries that optimize polishing rates and minimize dishing, little consideration is given to the affect these harsh solutions have on accelerating the rate of stress corrosion cracking [2, 3]. The synergistic effect of mechanical loads and reactive aqueous solutions was recently demonstrated to have a significant effect on the rate of crack growth in blanket thin-films of nanoporous methylsilsesquioxane (MSSQ) [4, 5]. Interconnect devices, however, consist of a number of dissimilar materials that are arranged in lithographically patterned arrays typically containing copper, LKDs, and various diffusion barriers (e.g. TiN, SiC, SiCN, SiO_2). Although the interfacial adhesion of patterned films has been investigated [6], virtually nothing is known about the effect aqueous solution chemistry has on accelerating crack growth in these structures. An advanced understanding of this fundamental mode of failure is necessary for MSSQ and other LKDs to be considered viable candidates for next generation interconnect devices.

EXPERIMENTAL TECHNIQUES

The patterned structures selected for study are shown schematically in Figure 1 (a). The MSSQ and copper lines were approximately equal in width and were 300 nm in height. Features widths of 300 nm and 500 nm were examined. Blanket thin-films of MSSQ capped with a technologically relevant SiC barrier layer were also investigated. The MSSQ (JSR 5109) was ~20% porous, with pores of ~1.5 nm in diameter, and a dielectric constant of 2.2. Double cantilever beam (DCB) sample geometries were fabricated for crack growth testing by sandwiching the thin-films between two elastic silicon substrates using previously reported techniques [7]. Testing was conducted in an environmental chamber with temperature $\pm$ 0.5°C and humidity control $\pm$1% relative humidity (RH). Aqueous environments investigated included water (pH 6), buffered solutions of pH 3 and 11, and a 3 wt%. H_2O_2 solution (~pH 5). Crack growth rates, da/dt, were characterized as a function of the applied strain energy release rate, G (J/m^2), over the range of 10^{-4} m/s to ~10^{-10} m/s using load relaxation fracture mechanics techniques [7]. This involved loading the specimen at a constant displacement rate to a predetermined load then fixing the displacement. The ensuing time-dependent load relaxation resulting from crack growth increases the specimen compliance from which the crack length, a, da/dt and G may be determined. After testing, the fractured surfaces were carefully examined using scanning electron microscopy (SEM), focused ion beam (FIB), and atomic force microscopy (AFM) for imaging and X-ray photoelectron spectroscopy (XPS) for determining the debond path.

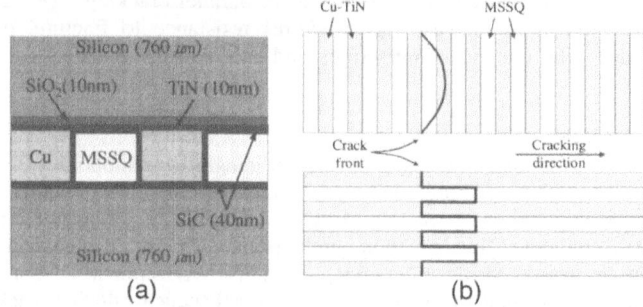

Figure 1. (a) Cross sectional schematic illustration of the lithographically patterned structures sandwiched between two elastic silicon substrates (copper bond layer not shown). (b) Top view schematic showing the likely crack front configuration in both orientations.

RESULTS AND DISCUSSION
Parallel orientation
        Crack growth rate as a function of G observed in water and 30% RH is shown in Figure 2 for 500 nm width lines where cracking was parallel to the features. In this orientation, crack growth behavior is similar to that observed in blanket thin-films of MSSQ, also presented in Figure 2 (a). The primary rate limiting steps of stress corrosion cracking are the transport of environmental species to the crack tip and a concerted chemical reaction resulting in bond rupture [3, 8]. The slope of the reaction dominated regime (apparent in Figure 2 (a) for da/dt > 10^{-7} m/s) is related to the activation volume for the reaction. At high growth rates, the slope was ~35 (~ 9.4 Å^3/molecule), much greater than that observed for the blanket thin-films ~18 (~ 4.8 Å^3/molecule) [3]. In addition, a crack growth threshold, G_{th}, representing values of G below which crack growth is presumed dormant, was evident at low growth rates with a very steep slope of ~ 90.

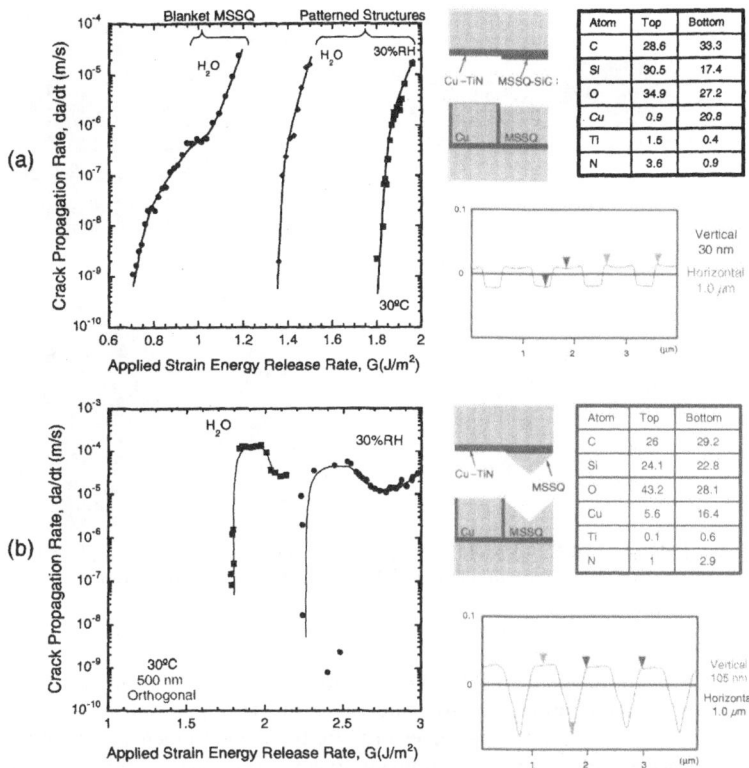

Figure 2. (a) Accelerated crack growth for the 500nm width lines in the parallel orientation as well as the blanket MSSQ. (b) Accelerated cracking orthogonal to 500nm width lines. Schematic representations of fracture path and AFM cross sectional average of fracture surfaces with the relevant heights. Compositional information (XPS) is shown for each fracture surface of the patterned films. Cohesive failure of the blanket MSSQ was observed.

While XPS has been used successfully for determining the debond path of thin-film structures, in the present study the x-ray spot size is greater than the relevant feature dimensions and makes interpretation of the results more challenging. Nevertheless, this quantitative data can still provide useful information about the debond path selection and is given in Figure 2. The morphology of the fractured surfaces was also characterized using FIB, SEM and AFM (Figure 3 (a)). The high-resolution images, compositional information, prior knowledge of the film thicknesses and pattern design specifications allowed us to determine that debonding occurred at the TiN – Cu interface and the MSSQ – SiC interface, represented schematically in Figure 2 (a). The difference in height between the two interfaces was ~30 nm (determined using AFM) and is consistent with the design specifications.

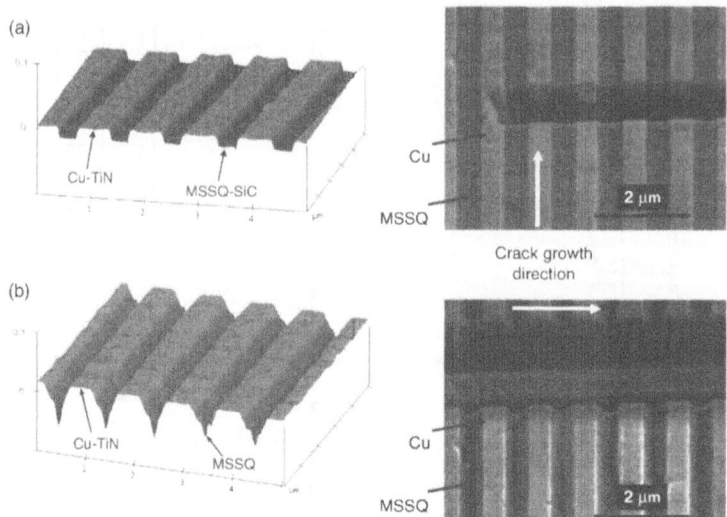

Figure 3. AFM and FIB-SEM reveal the marked changes in fracture morphology based on feature orientation (a) crack growth parallel and (b) orthogonal to features. Images were taken for samples exposed to 30%RH.

Orthogonal orientation

Cracking orthogonal to the 500 nm width features differs markedly from that observed in the parallel orientation for water and 30%RH (Figure 2 (b)). A high velocity plateau is clearly apparent at large values of G for both environments. As the value of G decreases, however, growth rates drop precipitously. The slope of this region was ~80, indicative of a crack growth threshold. There was also a significant difference in the value of G_{th} depending on orientation. We note that in water, G_{th} decreases from 1.8 J/m^2 (orthogonal) to 1.35 J/m^2 (parallel) and results in an increase of over six orders of magnitude in growth rates. This unique cracking behavior is likely due to the crack front bowing (Figure 1 (b), resulting in the fracture of multiple interfaces simultaneously.

The morphology of the fractured surfaces also differs significantly from that observed in the parallel configuration, Figure 3 (b). Here debonding of the TiN–Cu interface was again observed, along with failure of the MSSQ. This resulted in a regular array of plateaus (TiN–Cu) and valleys (MSSQ). Cracking in this orientation results in a much more tortuous crack path and may contribute to the increased resistance to accelerated debonding due to excess energy dissipated by frictional sliding of asperities behind the debond tip [9].

Effect of line width

Figure 4 demonstrates the significant effect varying the line width has on accelerated crack growth orthogonal to the features in a 1% RH environment. The increase may be attributed to the excess energy dissipated by frictional sliding of the crack flank asperities. This dissipative process scales with the contact area of the vertical asperities and thus with decreasing feature width the energy dissipated by friction increases significantly resulting in an increased resistance to fracture [6]. While the effect of the excess energy dissipated is apparent over the entire range of G, the most significant effect is at low growth rates. Here, G_{th} increases from 2.9 J/m^2 (500 nm) to 3.4 J/m^2 (300 nm), resulting in a 4 order of magnitude decrease in the crack growth rate.

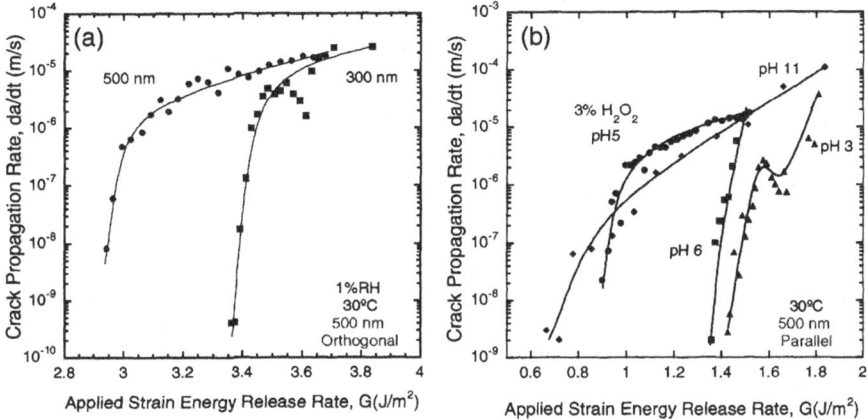

Figure 4. (a) Effect of line width on accelerated cracking in a 1% RH environment for the orthogonal orientation. (b) Effect of solution chemistry on accelerated cracking of 500nm width lines and cracking parallel to the features.

Effect of aqueous solution chemistry

Solution pH has been shown to greatly effect the rate of cracking in blanket LKD thin-films [4, 5]. In general, basic solutions accelerate cracking while acidic solutions inhibit cracking. This is attributed to the enhanced reaction kinetics between the strained Si-O crack tip bonds and the highly nucleophilic hydroxide ions compared to the less nucleophilic H_2O molecules. This same trend was observed in patterned structures for cracking parallel to the features subjected to water and buffered solutions of pH 3 and 11 (Figure 4 (b)). These observations indicate that the same mechanisms of cracking observed in blanket films are also active in patterned structures. An increase in the slope of the reaction dominated regime is also typically observed with decreasing pH for blanket thin-films. Here we find the slope for the pH 11 buffered solution is ~8 (~2.4 Å^3/molecule) and ~25 (~6.7 Å^3/molecule) for the pH 3 buffered solution, however, the most important effect of the solution pH is again the decrease in G_{th}. This value is deceased from 1.4 J/m^2 in pH 3 to 0.7 J/m^2 in pH 11 solutions.

A weakly acidic hydrogen peroxide solution was also employed to more closely simulate CMP solution chemistries. As is clearly apparent in Figure 4 (b), this solution exhibits anomalously high growth rates with respect to solution pH owing to the more vigorous reaction between Si-O bonds and hydrogen peroxide molecules [4]. The threshold observed for this solution is ~ 0.9 J/m^2 and may be due to steric hindrance of the hydrogen peroxide [4]. This has significant implications for the life of devices processed in aqueous solutions, as crack growth could simply be driven by residual stresses trapped in the structure during processing or by the loads applied during CMP.

CONCLUSIONS

This study clearly reveals the substantial effect environmental conditions can have on the fracture of lithographically patterned thin-film structures containing copper and nanoporous MSSQ glass. Crack-growth rates were accelerated for cracking parallel to patterned features and inhibited for growth orthogonal to the features. The resistance to crack growth was increased with decreasing feature width. Basic solutions were demonstrated to accelerate crack growth rates, whereas acidic solutions inhibited cracking. Anomalously high growth rates were observed for cracking in hydrogen peroxide solutions. Cracking behavior in all environments was observed to be analogous to that observed for blanket thin films containing MSSQ.

ACKNOWLEDGEMENTS

This work was supported in part by the Director, Office of Energy Research, Office of Basic Energy Sciences, Materials Sciences Division of the U.S. Department of Energy, under Contract No. D3-FG03-95ER45543. Materials were supplied by JSR and LETI and the authors gratefully acknowledge the assistance of Drs. M. Patz (JSR) and S. Maitrejean (LETI).

REFERENCES

1.   Iacopi, F., M. Patz, I. Vos, Z. Tokei, B. Sijmus, Q. Le, E. Sleeckx, B. Eyckens, H. Struyf, A. Das, and K. Maex, *Impact of LKD5109 low-k to cap/liner interfaces in single damascene process and performance.* Microelectronic Engineering, 2003. **70**(2/4): p. 293-301.

2.   Wiederhorn, S.M. and H. Johnson, *Effect of electrolyte pH on crack propagation in glass.* Journal of the American Ceramic Society, 1973. **56**(4): p. 192-7.

3.   Wiederhorn, S.M., *Influence of water vapor on crack propagation in soda-lime glass.* Journal of American Ceramic Society, 1967. **50**(8): p. 407-14.

4.   Guyer, E.P. and R.H. Dauskardt, *Fracture of nanoporous thin-film glasses.* Nature Materials, 2004. **3**(1).

5.   Lin, Y., J. Vlassak, T. Tsui, and A. McKerrow. *Environmental effects on subcritical delamination of dielectric and metal films from organosilicate glass (OSG) thin films.* 2003: Warrendale, PA, USA : Mater. Res. Soc, 2003.

6.   Litteken, C.S., *The effects of multi-dimensional constraint on the adhesion of thin-films and patterned structures*, Ph.D. Dissertation in Materials Science & Engineering. 2004, Stanford University.

7.   Dauskardt, R.H., M. Lane, Q. Ma, and N. Krishna, *Adhesion and debonding of multi-layer thin film structures.* Engineering Fracture Mechanics, 1998. **61**(1): p. 141-62.

8.   Michalske, T.A. and S.W. Freiman, *A molecular interpretation of stress corrosion in silica.* Nature, 1982. **295**(5849): p. 511-12.

9.   Lane, M., *Interface fracture.* Annual Review of Materials Research, 2003. **33**: p. 29-54.

Mat. Res. Soc. Symp. Proc. Vol. 812 © 2004 Materials Research Society   F6.10

Fundamental Limits for 3D Wafer-to-Wafer Alignment Accuracy

M. Wimplinger, J.-Q. Lu*, J. Yu*, Y. Kwon*, T. Matthias**, T.S. Cale*, and R.J. Gutmann*

EV Group Inc., 3701 E. University Dr., Phoenix, AZ 85034, m.wimplinger@EVGroup.com
*Rensselaer Polytechnic Institute, 110 8th Street, Troy, NY 12180, luj@rpi.edu
** EV Group, E. Thallner Str. 1, 4780 Schaerding, Austria

ABSTRACT

Wafer-level three-dimensional (3D) integration as an emerging architecture for future chips offers high interconnect performance by reducing delays of global interconnects and high functionality with heterogeneous integration of materials, devices, and signals. Various 3D technology platforms have been investigated, with different combinations of alternative alignment, bonding, thinning and inter-wafer interconnection technologies. Precise alignment on the wafer level is one of the key challenges affecting the performance of the 3D interconnects. After a brief overview of the wafer-level 3D technology platforms, this paper focuses on wafer-to-wafer alignment fundamentals. Various alignment methods are reviewed. A higher emphasis lies on the analysis of the alignment accuracy. In addition to the alignment accuracy achieved prior to bonding, the impacts of wafer bonding and subsequent wafer thinning will be discussed.

INTRODUCTION

It is an acknowledged fact in the IC industry, that one of the major challenges for further enhancing Moore's will be Interconnect Technology [1-3]. 3D Interconnect using wafer-to-wafer chip stacking offers an approach to vertically connect two integrated circuits for the purpose of shortening the wiring distance between them. Wafers are aligned, bonded, and interconnected face-to-face, then thinned-back prior to dicing.

The most critical challenge in interconnect technology are the so-called "global" interconnects, that are routed across the whole chip and distribute time critical signals. Prominent examples for that are "clock distribution" trees and the communication lines with e.g. an on–chip memory. [2-4] Stacking of functional modules such as logic and memory or logic and sensing element reduces the length of these critical connections from millimeters or centimeters to a couple of microns. That reduction of the interconnect length by orders of magnitude reduces the interconnect related delays in these lines by the same factor. Besides this tremendous advantage, 3 D integration also offers additional advantages: Today's high performance microprocessors include a huge on-chip memory. In many cases, this memory is as big as the logic part of the processor itself. 3 D integration will allow for manufacture of the memory on a separate wafer with optimised processing. (E.g., the wafer only needs to be processed through the mask levels that are required to manufacture the memory) This will allow for a more cost effective manufacturing process at higher yields. This integration of modules produced with optimised processes can be even further expanded to the integration of dissimilar substrate materials. This is a true advantage as soon as optical or RF functions should be integrated on a chip. These possibilities allow for the creation of unimaginable devices that may penetrate our daily life. New devices that incorporate complex sensing elements, signal processing, logic functions, a memory with a program and reference values and outputs will be possible. The high bandwidth that is offered by 3D integration will allow for parallel processing of data (e.g., of an image sensor). Standardization of building blocks such as processing unit, memories, I/Os, sensing arrays, etc., will enable a broad variety of devices with a fast time to market at a competitive cost.

WAFER TO WAFER ALIGNMENT FUNDEMENTALS

There are two different approaches for aligned wafer bonding. The first one is to perform alignment and bonding in the same tool, whereas the second approach is to split these two techniques and use a bond chuck for transfer between two separate tools. This article will focus on the second approach due to technical and economical reasons. The technical reason is that the bonding methods for 3D interconnect applications, especially thermo compression bonding, require a large bonding pressure. The combination of a sub-micron high precision alignment stage and of high bonding pressure does not seem feasible. The economic reason is that the process time for aligning is considerably lower than the process time for bonding, whereas the costs for an aligner are in the same order or larger than for the bonder. It is therefore reasonable to operate several bonders with only one aligner. Therefore the approach with split tools is the only path for high volume production. The exception is silicon-direct-bonding (SDB) where the pre-bonding occurs in the aligner and the bonding is performed in an oven at elevated temperatures.

ALIGNMENT METHODS

Over the last decades a number of techniques have been developed to achieve wafer-to-wafer alignment for bonding applications. It can be distinguished between direct alignment and indirect alignment methods.

The direct alignment methods allow visual alignment of the two wafers and offer a simultaneous control loop for the alignment result. These methods require that at least one of the wafers is transparent for visible or infrared (IR) light. For some MEMS applications this can be achieved by drilling holes in one wafer and aligning these holes to alignment keys on the second wafer. Silicon gets transparent for wavelengths above 1050nm, but the transparency decreases with the concentration of dopants. The second restriction is that metal layers cannot be used in the area of the alignment keys, which limits the use of IR alignment for CMOS applications.

Indirect alignment describes methods where the two wafers are aligned by means of an external reference positioning system. For bottom side (or backside) alignment the first wafer needs alignment keys on the front side and the second wafer needs them on the backside. The alignment marks of the first wafer are imaged, the position is stored and the backside the second wafer is then aligned to the front side of the first wafer. This method, now the industry standard for MEMS wafer bonding applications, is not applicable for CMOS applications as the wafers are strictly single-side processed.

SMARTVIEW® ALIGNMENT

The SmartView® alignment is a method for face-to-face alignment, where two wafers with the alignment keys within the bond interface are aligned. Compared to IR alignment it does not require IR transparency of the wafers, and compared to bottom side alignment it is fully compatible with single-side processed wafers.

A schematic of the SmartView® working principle is shown in Figure 1. Instead of using a single microscope in between the wafers, as is used in other substrate alignment platforms, the

SmartView® system employs two microscopes for alignment. One microscope is placed above and the other below the wafer stack. The dual microscopes focus on a common focal plane calibrated for each alignment. Each microscope objective observes one alignment key on the surface of the wafer. The detailed process flow of SmartView alignment is presented below (figure 4).

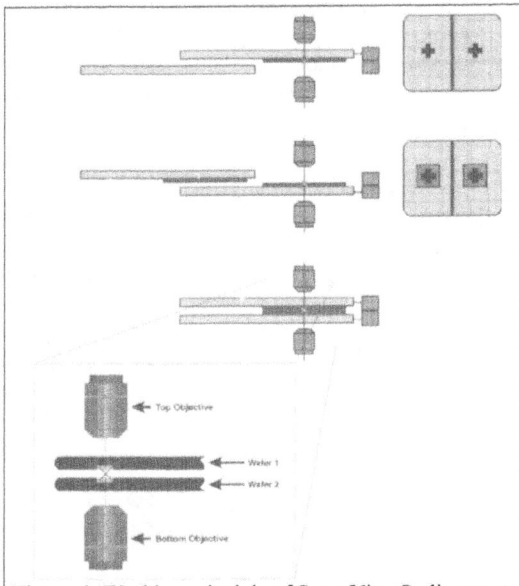

Step 1: First the top wafer is loaded face up and is then rotated into face down position. Next it is observed by the bottom optics. The optics is adjusted to bring the alignment keys into view, and then the image is digitized and stored electronically.

Step 2: The top wafer is retracted, allowing the bottom wafer to be brought into position beneath the top optics and then aligned to the existing digitized

Figure 1. Working principle of SmartView® alignment

image of the top wafer. Once complete the two wafers are automatically moved into alignment by calculating the relative X and Y locations of the alignment keys on each wafer and moving the wafers into the final alignment position.

Step 3: Once alignment is complete the wafers are brought into contact and secured for bonding.

Wafer alignment is accomplished using encoded stage motors allowing X and Y movements in increments of 0.1 µm steps and minimized Z-axis travel controlled by three software controlled spindle motors to preserve planarization between the top and bottom wafers.

ANALYSIS OF ALIGNMENT ACCURACY

High precision alignment is an enabling technology for 3D interconnects and therefore a critical parameter. The higher the alignment accuracy, the lower the required area for the interconnects enabling highest functional density. The alignment accuracy is investigated by means of optical- or IR microscopy (Figure 2). The wafers should have multiple alignment marks with a first print of highest accuracy. Multiple point analysis (e.g. the center point plus 1-3 circles with 4-8 points each) enables to distinguish between shift, rotational error and run-out error (Figure 4).

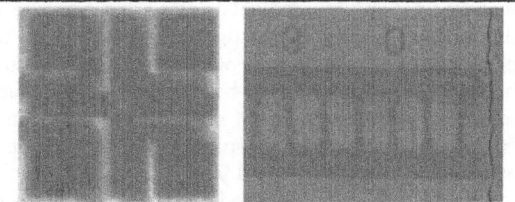

Figure 2. IR microscopy for alignment accuracy measurements; left side: analyzing alignment keys, right side: using Vernier structures.

Whereas shift, the translational misalignment in x or y, is a constant error for all the dies of the wafer, rotation and run-out correlate with the distance from the center and are therefore critical for 300mm applications. The rotational error is correlated to the position of the alignment marks: the closer to the wafer edge, the smaller the rotational error (Table 1).

Table 1. Influence of the alignment mark position on the rotational error; the maximal alignment error in y depends on the alignment capability of the equipment whereas the resulting rotational error also depends on the alignment mark position

Rotational error assuming a y-error of +1µm on the left side and -1µm on the right side			
distance between alignment marks (symmetric) [mm]	100	200	300
rotational error [mrad]	0,020	0,010	0,007

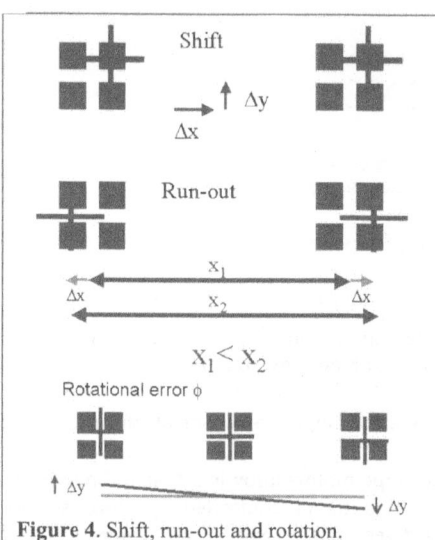

Figure 4. Shift, run-out and rotation.

20 alignments of Si wafers to glass wafers (200mm) have been analyzed in order to determine the distribution of shift, rotation and run-out on the pre-bond alignment accuracy (Figure 3).

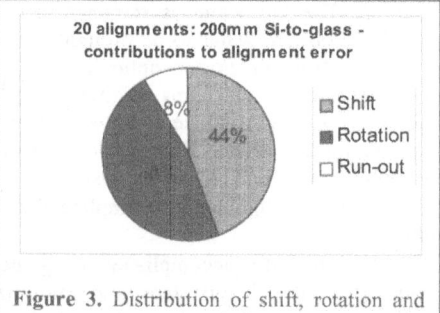

Figure 3. Distribution of shift, rotation and run-out on the pre-bond alignment error.

CONTRIBUTIONG FACTORS TO ALIGNMENT ACCURACY

Run-out describes the misalignment due to distance deviations between the corresponding alignment keys of top and bottom wafer. This may be a result from the front-end processing or due to temperature gradients during the bonding process. The total alignment error can be divided into the pre-bond alignment error, which reflects the accuracy of the aligner, and the post-bond alignment error, which includes the impact of the bonding process. Comparison between pre-bond and post-bond alignment accuracy allows determining the origin of run-out error. Usually the alignment algorithms handle run-out mismatch according to the least-square method. The alignment is typically performed by using two pairs of alignment keys. Therefore asymmetric run-out in x and y cannot be taken into account by the alignment algorithm and has to be adjusted by offsets based on statistical process control.

The alignment keys properties are very important especially for automatic alignment processes. Brightness, contrast and the accuracy of the pattern shape are the key parameters. The desired alignment accuracy defines the needed magnification of the microscopes and therefore the size of the patterns.

20 pre-bond and post-bond alignments of Si-to-Si wafers (200mm) have been analyzed in order to determine the distribution of shift, rotation and run-out on the post-bond alignment accuracy (Figure 5).

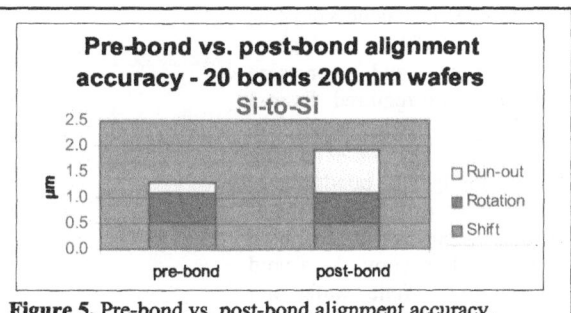

Figure 5. Pre-bond vs. post-bond alignment accuracy.

IMPACT OF WAFER BONDING

There are three potential impacts of the bonding process on the alignment accuracy. The first is a shift or a rotation induced by applying the pressure, the second is related to the use of separation flags and the third one is run-out due to temperature gradients in the wafer stack.

Modern wafer bonder apply the pressure solely perpendicular to the wafer surface, thereby preventing shearing forces which might result in shift or rotation. However, bowed wafers are flattened at the beginning of the bonding process, which might result in a small shift proportional to the bow.

The use of separation flags typically decreases the post-bond alignment accuracy. The separation flags, typically 3, can be pulled out simultaneously or one after another. However, the frictional forces between wafers and flags induce an additional error on the alignment accuracy resulting in an increased variance (Figure 6).

A high temperature uniformity of top and bottom side heater is a must in order to avoid run-out induced by temperature gradients.

The research and development during the last decades resulted in a variety of well-established bonding methods [5]. However, the specific requirements of CMOS processing limit the potential bonding methods. Only low temperature methods can be used in order that the active devices are not destroyed. Table 2 shows a comparison of the various low temperature bonding techniques. The presence of Pb (glass fritt bonding), Na (Anodic bonding) or Au (eutectic or thermal compression bonding) is a major concern for CMOS applications. The most interesting bonding methods for 3D interconnections application are metal-to-meal diffusion bonding and adhesive bonding.

Diffusion bonding is generally applicable to systems in which the diffusion coefficient is high at relatively low temperature. This occurs for some fcc metals like Cu. Thus it is possible to create Cu-Cu bonds.

Adhesive bonding with low-k dielectrics like Benzocyclobutene (BCB) gains more and more attention since they allow for the creation of electrical interconnects between different functional modules [6-8].

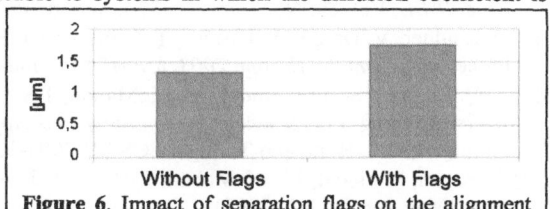

Figure 6. Impact of separation flags on the alignment accuracy; experiment: 20 SmartView® alignments with 200mm Si wafers; measurement of alignment accuracy before and after pulling the flags out.

In order to enable void-free bond interfaces, highest temperature and pressure uniformity are required. Even a small void can result in breakage of the wafer during back thinning. The bond quality can be investigated by means of IR microscopy and by scanning acoustic microscopy. As a void impacts the whole wafer not only the dies covered by the void, void-free bonding is an enabling technology for 3D integration.

Table 2. Overview of low-temperature bonding methods.

General B. Method	Anodic						Intermediate layer		
Specific Bonding Methode	Anodic	SOG Anodic	Dry Activated Direct	Metal to Metal Diffusion	Spin On dielectric	Thermal cure Epoxy	UV cure Epoxy	Glass Fritt	Eutectic Si-Au
CMOS Compatible			•	•	•	•	•	-	-
Temperature Range:									
<440°C	•	•	•	•	•	•	•	•	•
<200°C	-	-				•	•	•	•
Room Temperature			-					•	
Hermetic Seals (<1mbar)	•	•	•	•	•	-	-	•	•
High Vacuum Package (<1E-2mbar)	•	•	•	•					•
Vertical electric interconnects	•	•	•	•	•				•

• fully complies
– not compliant
- to some degree with certain limitations or boundry conditions

CONCLUSION

Aligned wafer bonding offers a viable solution for 3D integration. The analysis of the contributing factors to the alignment accuracy shows that micron-level and sub-micron alignment can be achieved. Impact of the actual bonding step on alignment has to be further investigated.

ACKNOWLEDGMENTS

The work at Rensselaer is supported in part by DARPA, MARCO, and NYSTAR through the Interconnect Focus Center (IFC).

REFERENCES

[1] International Technology Roadmap for Semiconductors (public.itrs.net).

[2] J.-Q. Lu, Y. Kwon, J.J. McMahon, A. Jindal, B. Altemus, D. Cheng, E. Eisenbraun, T.S. Cale, and R.J. Gutmann, in 20th International VLSI Multilevel interconnection Conference (VMIC 2003), 227 (2003).

[3] R.J. Gutmann, J.-Q. Lu, S. Pozder, Y. Kwon, A. Jindal, M. Celik, J.J. McMahon, K. Yu, and T.S. Cale, in Advanced Metallization Conference in 2003 (AMC 2003), 19 (2003).

[4] J. A. Davis et al, "Interconnect Limits on Gigascale Integration (GSI) in the 21st Century," Proc. IEEE, Vol. 89, No. 3, pp.305-324, March 2001.

[5] P. Lindner, V. Dragoi, S. Farrens, T. Glinsner and P. Hangweier, "Advanced Wafer Bonding Processes", Semicon Europe 2003, MEMS/MST Industry Forum, March 2003.

[6] J.-Q. Lu, Y. Kwon, A. Jindal, J.J. McMahon, T.S. Cale, and R.J. Gutmann, in Semiconductor Wafer Bonding VII: Science, Technology, and Applications, edited by F.S. Bengtsson, H. Baumgart, C.E. Hunt, and T. Suga, ECS PV 2003-19, 2003, pp. 76-86.

[7] Y. Kwon, A. Jindal, J.J. McMahon, J.-Q. Lu, R.J. Gutmann, and T.S. Cale, in MRS Symp. Proc., Vol. 766, E5.8.1 (2003).

[8] J.-Q. Lu, A. Jindal, Y. Kwon, J.J. McMahon, M. Rasco, R. Augur, T.S. Cale, and R.J. Gutmann, in 2003 IEEE International Interconnect Technology Conference (IITC), IEEE, 2003, pp. 74-76.

Mat. Res. Soc. Symp. Proc. Vol. 812 © 2004 Materials Research Society

Optical Interconnect Components for Wafer Level Heterogeneous Hyper-Integration

P. D. Persans, M. Ojha, R. J. Gutmann, J.-Q. Lu, A. Filin, J. Plawsky
Focus Research Center: Interconnections for Gigascale Integration
Rensselaer Polytechnic Institute, Troy, New York 12180, U.S.A

ABSTRACT
Optical waveguides for three-dimensionally stacked chip fabrication technologies, in which optical connection between layers plays a central role, are described. CMOS-compatible approaches to optical via and waveguide fabrication are addressed. Detailed modeling is used for design optimization and also addresses how manufacturing variations from ideal design may affect device performance.

INTRODUCTION
Three-dimensional integration in which wafers are stacked is an attractive technology because it decreases RC delays by allowing shorter electrical wire lengths [1, 2]. This is particularly important as feature size reduction results in interconnect delays that limit advances in high-speed digital ICs at future technology nodes. Stacking wafers also enables hyper-integration in which devices of different functionality and structure can be included in CMOS-compatible ways [3].

Future computer chips may also require novel components that may include quantum spin state, quantum dot, magnetic, and/or magneto-optic components, some of which may be based on molecular-scale structures. It is likely that these components will be integrated with CMOS-like drivers and with significant portions of the chip based on CMOS manufacturing technologies. Connection to, or reading of, some of these components will be facilitated by direct optical interaction. Integrating optics onto the chip will also facilitate high-speed communication off the chip.

In this paper we describe example structures where optical beams are coupled vertically from a horizontal waveguide layer through stacked CMOS wafer layers to an active device layer as shown schematically in Fig. 1. We have previously described the fabrication and characterization of forty-five degree self-aligned aluminum mirrors on 2-micron polymer waveguides [4, 5]. The remaining problem is to couple this beam into a device in the lowest level of the structure. In the first case, the light is coupled vertically into a waveguide via. In the second, the vertical beam passes directly though the multilayer stack to a device in the lowest CMOS layer. We have modeled the optical behavior of these structures using BeamProp (TM), and FullWave (TM) software, as well as basic plane-wave calculations.

Many issues must be addressed in the design such structures. Optical considerations include: What is the transmission of the structure? What operating wavelengths are available or recommended? Is the transmission of the structure sensitive to wavelength? What are the limitations on beam size and angular range? Fabrication and processing considerations include: How does the available choice of materials affect optical functionality (i.e.- absorption and index)? Are thermal and chemical properties compatible? Are processing properties (e.g.- etching) compatible? Is the processing sequence difficult?

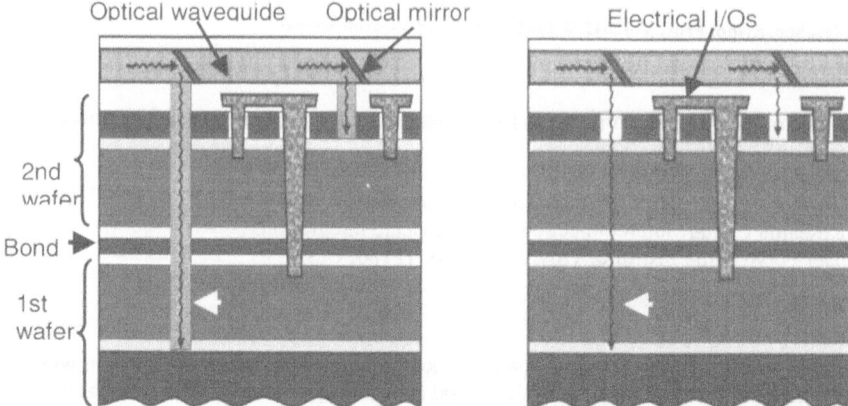

Fig. 1 - Schematic stacked- wafer structures. In both figures, the top layer is an optical waveguide which includes mirrors for coupling the optical beam from the horizontal waveguide into a vertical direction. In the left-hand example, the vertical beam is confined in a waveguide. In the right hand example, the light passes though the multilayer stack without confinement.

The fabrication of a clad-waveguide optical via (as shown in the left side of Fig. 1) provides an example of a complex fabrication process. After the wafers are bonded and aligned, a high aspect ratio via must be etched through several layers of different materials including silicon, interconnect layer dielectric, and bonding polymer. (The anticipated dimensions of the via are, 2 micron x 2 micron width with a length of 6 to 20 microns.) The via could then be backfilled with cladding dielectric and the waveguide core would have to be etched through the center of the via. The top waveguide layer cladding and core would then be deposited and the aligned mirror would be fabricated. Some aspects of this process have been demonstrated, including high aspect via drilling and bonding of wafers [1, 2, 3]. Other aspects, such as backfilling with polymer and subsequent high aspect etching, have not.

The beam via on the right in Fig. 1 presents a more straightforward fabrication sequence, much of which has already been demonstrated [1-5]. Electrical vias involve high aspect etching, but back-filling involves only metal deposition [1-3]. The addition of the top waveguide layer and mirrors has also been demonstrated [4, 5].

Optical Modelling Results

In the following, we describe optical modelling results for four cases. We consider the waveguide via (left side of Fig. 1) and the beam via (right side of Fig. 1) for the two cases where the upper active Si layer is either a thinned wafer (~ 5 microns thick) or silicon-on-insulator (~ 70 nm thick). As the basis for calculations, we assume a stack starting at the bottom in Fig. 1 as follows: Si wafer with active device layer 1, ILD layer (~6 microns thick), BCB bonding layer (thickness 1 micron), ILD layer (thickness 6 microns), active Si device layer 2 (thickness as described above), oxide cladding (0.5 micron), polymer waveguide (2 microns), and top cladding. Calculations were carried out in the wavelength regions near 830 nm and 1550 nm, corresponding to commonly available semiconductor laser.

For both wavelengths, the loss in a 2-micron core waveguide via is very small. The worst-case loss is less than 5%, even when the lateral positions of the sidewalls of the via are allowed to vary due to realistic differences in RIE etch rates. The major losses for the waveguide via case occur when light is coupled into and out of the waveguide via. At the top mirror there are scattering and coupling losses as the beam is reflected by 45 degrees, resulting in about 15% loss. At the interface between the waveguide and the bottom Si device layer, the reflection loss is significant. In the absence of an antireflection coating, it can exceed 30%. <u>Thus, total losses for the waveguide via exceed 50%.</u>

Transmission in the absence of a fabricated via is more interesting. In Fig. 2 we show an example of the transmission near 1550 nm for the thinned wafer case where the upper Si layer is about 5 microns thick. The transmission exhibits interference (etalon) fringes due to reflections at the Si/oxide interfaces. Optimization of the Si thickness at ~4900 nm yields 99% coupling between the top of the stack and the bottom Si device layer. We note that the transmission in this case is clearly sensitive the details of the structure. As an example, in Fig 3 we show how the transmission at 1550 nm depends on the thickness of the upper Si layer. A variation of only 100 nm (2%) causes the transmission to drop below 50%. The transmission is similarly dependent on the dielectric (ILD/BCB) thickness.

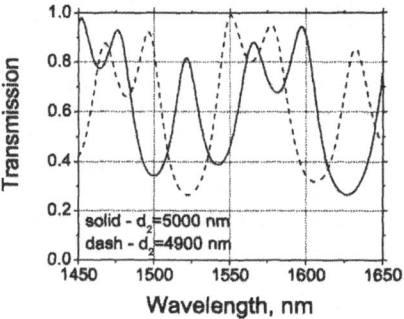 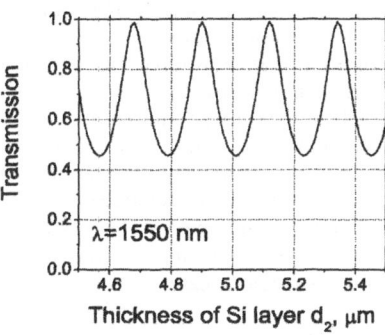

Fig. 2 (left above) - Plane-wave transmission through a multilayer stack as described in the text. The upper Si layer thickness is 4900 or 5000 nm, consistent with a thinned wafer in 3D integration.

Fig. 3 (right above) - Plane-wave transmission at wavelength=1550 nm where the thickness of the upper Si layer is varied.

A similar calculation is shown in Fig. 4 for λ=1550 nm with a thin upper Si layer, consistent with an SOI wafer. Optimum coupling is found when the Si thickness is ~ 30 nm, rather thinner than SOI currently available. When the ILD/BCB thickness is varied, it is possible to achieve over 80% transmission at λ=1550 nm for Si thickness of 70 nm.

Transmission at λ=830 nm is decreased by absorption in the Si. For the thicker (~ 5 micron) Si layer, the maximum transmission is only ~ 55%, even when Si and ILD thicknesses are varied. The situation for SOI at 830 nm is more promising. In Fig. 5, the Si layer thickness has been fixed at 70 nm and the ILD/BCB thickness was varied. Maximum transmission of over 80% is found.

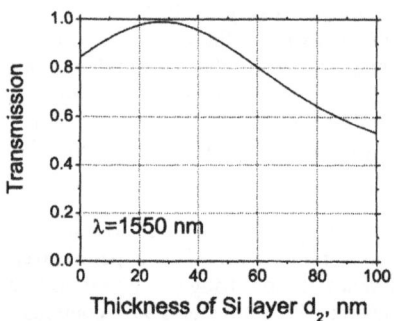 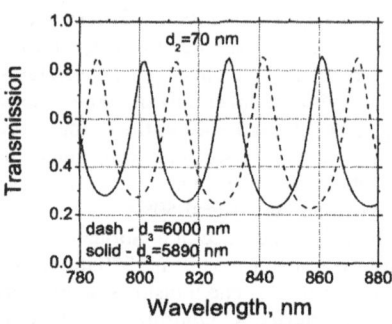

Fig. 4 (left above)- Plane-wave transmission at 1550 nm as a function of Si layer thickness for an SOI-type upper Si layer.
Fig. 5 (right above) - Plane wave transmission near 830 nm for upper Si layer thickness of 70 nm. Dashed and solid lines correspond to different dielectric thicknesses as described in the legend.

The second important consideration for waveguide and beam vias is the divergence and width of the vertical beam. In the waveguide via case, the beam can be well confined and has an effective width that is close to the width of the waveguide core - two microns in this case. In the unconfined beam via case, the beam divergence depends on the width of the mirror. Divergence can be easily approximated by assuming that the mirror acts as an aperture and that the beam diffracts through the aperture. The beam width at the lower Si layer depends ~ linearly on the propagation distance (about 10-20 microns) and inversely on the mirror width. An example in which the beam profile was computed using BeamProp is given in reference [3] and several examples are given in reference [6]. For a propagation distance of 20 microns, the narrowest beam width for λ=830 nm at the substrate is ~ 5 microns. This is significantly larger than the width for the waveguide case.

CONCLUSIONS
We have reviewed several considerations for optical beam propagation in 3D wafer stacks for hyper-integration. Unconfined optical beam transmission through the wafer stack yields excellent transmission at λ=1550 nm for both SOI and thinned wafer stacks. Multilayer stack interference effects lead enhance transmission, but produce sensitivity to details of the layer thicknesses. For waveguide vias transmission is independent of details of the structure and scattering and waveguide propagation losses are small. Unfortunately, coupling between waveguide modes and the Si device layer yields large reflection losses. For λ=830 nm, transmission for SOI upper layer is excellent, but absorption losses decrease transmission for thinned wafers.

ACKNOWLEDGEMENTS
We gratefully acknowledge preliminary work by Shom Ponoth and Navnit Agarwal.
This work was funded by MARCO, DARPA, SIA, and Semi/Sematech through the Focus Research Center for Interconnections for Gigascale Integration, and by New York State through the Focus Center - New York: Interconnections for Gigascale Integration.

REFERENCES
1. J.-Q. Lu, Y. Kwon, J.J. McMahon, A. Jindal, B. Altemus, D. Cheng, E. Eisenbraun, T.S. Cale, and R.J. Gutmann, in Proceedings of 20th International VLSI Multilevel interconnection Conference (VMIC 2003), edited by T. Wade, IMIC, 2003, pp. 227-236.
2. R.J. Gutmann, J.-Q. Lu, S. Pozder, Y. Kwon, A. Jindal, M. Celik, J.J. McMahon, K. Yu and T.S. Cale, in Advanced Metallization Conference in 2003 (AMC 2003), Eds. G.W. Ray, T. Smy, T. Ohta and M. Tsujimura, MRS, 2004, pp. 19-26.
3. J.-Q. Lu, A. Jindal, P.D. Persans T.S. Cale, and R.J. Gutmann, in 2003 International Conference on Compound Semiconductor Manufacturing Technology, GaAs MANTECH 2003 Scottsdale, pp. 91-94.
4. S. Ponoth, N. T. Agarwal, P. D. Persans, and J. L. Plawsky, J. Vac. Soc. Tech. B **21**, 1240, (2003).
5. S. Ponoth, N. T. Agarwal, P. D. Persans, and J. L. Plawsky, Mat. Res. Soc. Symp. Proc. **744** (2003).6. S. Ponoth, Ph.D. Thesis, Rensselaer Polytechnic Institute (2003)

Mat. Res. Soc. Symp. Proc. Vol. 812 © 2004 Materials Research Society   F6.16

Evaluation of Thin Dielectric-Glue Wafer-Bonding for
Three Dimensional Integrated Circuit-Applications

Y. Kwon, J. Yu, J.J. McMahon, J.-Q. Lu, T.S. Cale, and R.J. Gutmann
Focus Center - New York, Rensselaer: Interconnections for Hyperintegration
Rensselaer Polytechnic Institute, Troy, New York 12180-3590, luj@rpi.edu

ABSTRACT

The critical adhesion energy of benzocyclobutene (BCB)-bonded wafers is quantitatively investigated with focus on BCB thickness, material stack and thermal cycling. The critical adhesion energy depends linearly on BCB thickness, increasing from 19 J/m^2 to 31 J/m^2 as the BCB thickness increases from 0.4 μm to 2.6 μm, when bonding silicon wafers coated with plasma enhanced chemical vapor deposited (PECVD) silicon dioxide (SiO_2). In thermal cycling performed with 350 and 400 °C peak temperatures, the significant increase in critical adhesion energy at the interface between BCB and PECVD SiO_2 during the first thermal cycle is attributed to relaxation of residual stress in the PECVD SiO_2 layer. On the other hand, the critical adhesion energy at the interface between BCB and PECVD silicon nitride (SiN_x) decreases due to the increase of residual stress in the PECVD SiN_x layer during the first thermal cycle.

INTRODUCTION

Monolithic wafer level three-dimensional (3D) integration is an emerging technology to increase interconnect performance and functionality of integrated circuits (ICs) [1,2]. In our approach, fully processed wafers are aligned and bonded with a dielectric adhesive glue, followed by top-wafer thinning and inter-wafer interconnection [1,2]. We have developed a baseline wafer bonding process using benzocyclobutene (BCB) glue [1-3]. In this paper, emphasis is placed on thin BCB wafer bonding, which is desirable for reducing via aspect ratio of inter-wafer interconnect for 3D ICs. The critical adhesion energies of wafers bonded with various BCB thicknesses are determined by four-point bending. The effects of high temperature thermal cycling on critical adhesion energy at the interfaces between BCB and plasma enhanced chemical vapor deposition (PECVD) silicon dioxide (SiO_2), and between BCB and PECVD silicon nitride (SiN_x) are evaluated, since both SiO_2 and SiN_x are commonly used on top of interconnects.

EXPERIMENTAL PROCEDURES

Material preparation

Five sets of 200 mm wafers were used in this work: (a) silicon wafer, (b) glass wafer whose coefficient of thermal expansion (CTE, 3.78 ppm/°C) is matched with that of silicon (CTE, 2.6 ppm/°C), (c) silicon wafer with 2 μm thermal SiO_2, (d) silicon wafer with 1 μm PECVD SiO_2, and (e) silicon wafer with 170 nm PECVD SiN_x. Silicon wafers, glass wafers, and thermally oxidized silicon wafers are obtained from commercial vendors. PECVD SiO_2 layers are deposited using SiH_4 and N_2O at 300 °C in a Plasmatherm 73 (Plasma-Therm Inc., Voorhees, NJ) with a chamber pressure of 0.9 Torr and rf power of 25 W. PECVD SiN_x layers are deposited using $SiH_4/N_2/He$ gases at 300 °C in the Plasmatherm 73 with a chamber pressure of

0.89 Torr and power of 30 W. The SiO_2 and SiN_x thicknesses are measured using ellipsometry. The standard deviation of the measured thickness across the wafer is less than 5%.

BCB thin film is deposited using a FlexiFab™ spin coater track. Prior to deposition of BCB on a substrate, the wafers are dipped into a sulfuric acid (H_2SO_4) and hydrogen peroxide (H_2O_2) solution of a ratio of 2:1 for 20 min. at room temperature (piranha clean) and are rinsed with DI water and spin-dried in a nitrogen ambient. An adhesion promoter (Dow Chemical AP3000) is spun on, followed by BCB spin coating. Several BCB resins are used for these experiments [4]. Two of them are Dow Chemical Cyclotene 3022-35 and 3022-46 and others are BCB resins diluted by mixing with solvent. After BCB spin coating, the wafers are baked to drive off volatile solvents. Ellipsometry is used to measure the BCB thickness, showing a uniform distribution across 200 mm wafers (< 2 % standard deviation). Wafer bonding is conducted using a baseline process described elsewhere [3] with an EVG EV501 wafer bonder (EVGroup Inc., Austria).

Material characterization

The critical adhesion energy is determined using four-point bending (see Ref. [3] for detailed information). For the results reported here, at least four samples were tested with the standard deviation of the critical adhesion energy less than 10% of the average value. Residual stress in each silicon wafer coated with PECVD SiO_2, PECVD SiN_x and BCB is estimated by measuring wafer curvature (see Ref. [5] for detailed information). Thermal cycling tests are performed with peak temperatures of 350 and 400 °C. One and five thermal cycles were chosen because preliminary experiments indicated that both residual stress and critical adhesion energy reached a saturation point in less than five cycles. A nitrogen ambient is used during the thermal cycling to prevent oxidation of BCB during thermal cycling. A ramp rate (both ramp up and cool down) of 25 °C/min was used, with a dwell time of 20 minutes at the peak temperature. Chemical properties of the PECVD SiO_2 and SiN_x layers before and after thermal cycling are examined using Nicolet MAGNA 560 Fourier-transform infrared spectroscopy (FTIR) (Nicolet, Madison, WI) [5].

RESULTS AND DISCUSSION

After bonding between a silicon and a glass wafer using a thin layer of BCB (<1 µm), the bonding interface is visually inspected; defect-free (visible) bonding is routinely obtained. Based on this robust bonding process, the four point bending technique was used to measure the critical adhesion energy after processing at different conditions.

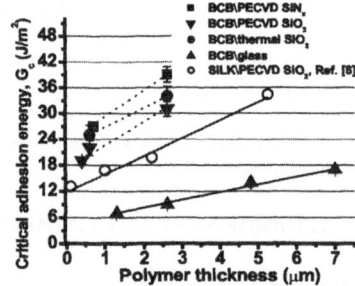

Critical adhesion energy dependence on BCB thickness and material stack

Beam specimens with BCB thicknesses of 0.4 to 7 µm and four sets of stack materials are prepared for four point bending tests. Figure 1 shows the thickness dependence of the critical adhesion energy for these four sets of samples: (a) Si/ PECVD SiN_x (170 nm)/ BCB/ SiN_x (170 nm)/ Si, (b) Si/ thermal SiO_2 (2 µm)/ BCB/ thermal SiO_2 (2 µm)/

Figure 1. Critical adhesion energy for different BCB thicknesses and material stacks (this work) and at the interface between SiLK and PECVD SiO_2 (Ref. [8]).

Si, (c) Si/ PECVD SiO_2 (1 μm)/ BCB/ PECVD SiO_2 (1 μm)/ Si, and (d) Si/ BCB/ glass. As can be seen from Fig. 1, critical adhesion energy, G_c, increases with increasing BCB thickness.

The critical adhesion energy can be primarily attributed to the following three contributions: the energy used to break atomic bonds, G_0, the plastic deformation energy, G_p, and the deformation energy due to the residual stress, $G_{residual}$ [6]. Without high temperature post-curing thermal cycling, $G_{residual}$ was negligible. For example, in a structure consisting of a 2.6 μm thick BCB layer sandwiched between two silicon wafers, the calculated $G_{residual}$ value is less than 0.5 J/m^2, while the measured G_c is approximately 30 J/m^2. G_c is therefore expressed as the sum of G_0 and G_p in the following manner:

$$G_c = G_0 + G_p = G_0 (1 + G_p / G_0) \qquad (1)$$

G_c versus BCB thickness, t, can be expressed as shown in Fig. 1:

$$G_c \approx G_0(1+0.3t) \qquad 0.4 \text{ μm} \le t \le 7 \text{ μm} \qquad (2)$$

While a liner relationship can only be assumed for sample sets with only two data points, a similar fitting approach is used as that for sample set (d). This linear relationship between G_c and BCB thickness is similar to the results previously published [7,8] (a linear relationship between G_c and SiLK thickness presented in Ref. [8] is shown in Fig. 1) and is expected if plastic zone height, R_p, formed in the BCB is thicker than the BCB thickness (i.e., large scale yielding) [7]. Varias et al. [9] predicted that R_p depends on material properties of a ductile thin film and elastic material stack, with the assumptions that: 1) the mismatch of elastic properties between the two layers is large, and 2) yield stress of the ductile thin film does not vary with the film thickness. The theory in Ref. [9] is applicable since the mismatch of Young's modulus between BCB and the adjacent substrate is more than a factor of twenty (Young's modulus is 131 GPa in silicon, 62.75 GPa in glass, and 2.9 GPa in BCB) and the yield stress of BCB (~48 MPa) is independent of the BCB thickness [10]. The R_p of BCB is 7 to 86 μm and exceeds the BCB thickness so that large scale yielding can be applied. Decrease in BCB thickness therefore induces a linear decrease in R_p, and linear decreases in G_p and G_c.

Effect of thermal cycling on critical adhesion energy

A bonded wafer stack usually undergoes multiple high temperature (over 350 °C) processes during 3D IC integration. One concern is degradation of the bonding during these processes. Thus, it is critical to understand the effects of the high temperature thermal cycling on critical adhesion energy of the wafer stacks.

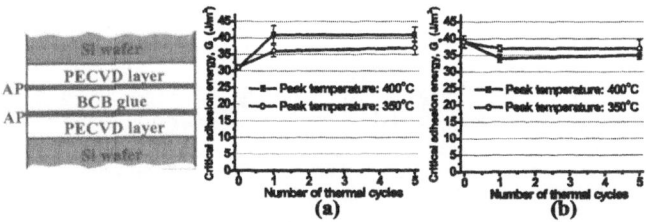

Figure 2. Beam specimen configurations and critical adhesion energy for 1 and 5 thermal cycles at peak cycle temperature of 350 and 400 °C: (a) BCB layer is sandwiched by PECVD SiO_2 layer and (b) BCB layer is sandwiched by PECVD SiN_x layer.

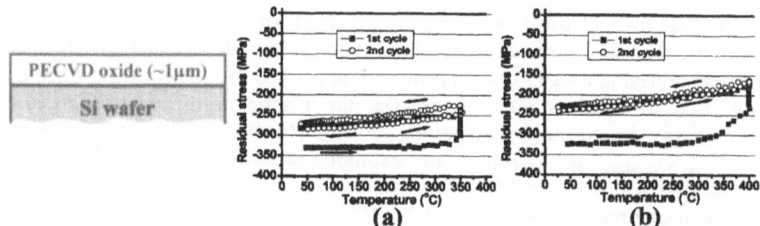

Figure 3. Geometry of PECVD SiO$_2$ on a silicon wafer and temperature dependence of residual stress in a PECVD SiO$_2$ film during thermal cycling: (a) peak temperature of 350 °C and (b) peak temperature of 400 °C.

A BCB layer (1.3 µm on each wafer, for 2.6 µm total thickness) is sandwiched between two silicon wafers that are coated with either 1 µm PECVD SiO$_2$ or with 170 nm PECVD SiN$_x$ (Fig. 2). The effect of thermal cycling on critical adhesion energy at both bonding interfaces is also shown in Fig. 2. In the interface between BCB and PECVD SiO$_2$ (Fig. 2a), a substantial increase (at least 16%) in critical adhesion energy was observed after the first thermal cycle with peak temperature of 350 and 400 °C, and remains almost unchanged after further thermal cycling. In the bonding interface between BCB and PECVD SiN$_x$ (Fig. 2b), the critical adhesion energy decreased after thermal cycling for both peak temperatures. The first cycle effect observed at the interface between BCB and PECVD SiO$_2$ is attributed to the decrease in G$_{residual}$ [6]. To evaluate the effect of thermal cycling on G$_c$ and G$_{residual}$, the dependences of residual stress on the thermal cycle process for each film (PECVD SiO$_2$, PECVD SiN$_x$ and BCB) are examined individually.

The temperature dependence of residual stress of a PECVD SiO$_2$ layer on a silicon wafer during thermal cycling is shown in Fig. 3. The magnitude of released compressive stress in the SiO$_2$ layer is large during the first cycle due to the change in chemical composition by condensation reaction within the SiO$_2$ layer [5]. The condensation reaction was analyzed by FTIR. Figures 4a and 4b present enlarged FTIR traces showing changes in peaks of siloxane and silanol bonds, respectively. After thermal cycling, the Si-O-Si peaks (Si-O-Si rocking at ~450 cm^{-1}, the Si-O-Si bending at ~805 cm^{-1}, and the Si-O-Si stretching at ~1070 cm^{-1}) increase and the silanol peak (3670~3680 cm^{-1}) decreases, indicating that the condensation reaction occurs during thermal cycling [11]. This large decrease in the residual stress of PECVD SiO$_2$ layer after the first thermal cycle leads to stress relaxation at the interface between BCB and the SiO$_2$ layers. This decrease of G$_{residual}$ results in an increase in critical adhesion energy.

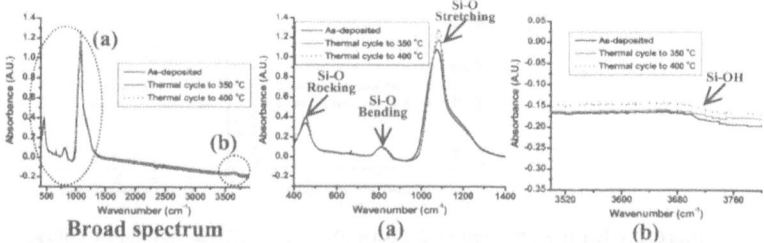

Figure 4. FTIR absorbance spectra of a PECVD SiO$_2$ layer before and after cycling: broad spectrum, (a) FTIR traces showing changes in peaks of siloxane bonds, and (b) FTIR traces showing changes in peaks of silanol bonds.

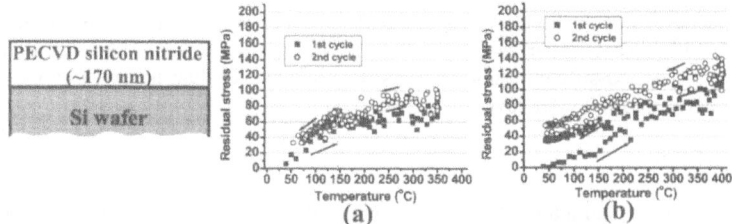

Figure 5. Geometry of PECVD SiN$_x$ on silicon wafer and temperature dependence of residual stress in the PECVD SiN$_x$ film during temperature cycling: (a) peak temperature of 350 °C and (b) peak temperature of 400 °C.

The temperature dependence of residual stress of a SiNx layer on a silicon wafer during thermal cycling is shown in Fig. 5. After deposition, a tensile stress in the layer was observed and the residual stress increased as thermal cycling proceeds. The increased residual stress is mainly caused by a decrease in the Si/N ratio induced by chemical reactions within the SiN$_x$ layer. When a SiH$_4$ + N$_2$ plasma is used for deposition of SiN$_x$, the main dissociation products in the plasma are SiH$_3$, Si$_2$H$_6$ and N [12]. During thermal cycling, desorption of a weak Si-H group is expected because SiH$_3$ radical and Si$_2$H$_6$ molecules may decompose and desorb as SiH$_4$ and H$_2$ [12]. On the other hand, since NH or NH$_2$ groups are not formed in the film during deposition (no peak in the FTIR spectra, as shown in Fig. 6), desorption of volatile NH$_3$ is not expected to take place and the N composition does not change during thermal cycling [12]. The chemical reaction of the SiN$_x$ layer was tracked by FTIR as seen in Fig. 6. After thermal cycling, the Si-H peak (2100~2150 cm^{-1}) decreases, indicating that desoprtion of Si-H occurs during thermal cycling. According to Toivola et $al.$ [13], as the Si/N ratio in a SiN$_x$ film decreases, tension in the film increases because bonds of the film are stretched relative to their equilibrium lengths. This increase in the residual tensile stress in PECVD SiN$_x$ layers after thermal cycling causes an increase in stress at the interface between BCB and the SiN$_x$ layer. The increase in G$_{residual}$ leads to a decrease in critical adhesion energy.

Changes in the residual stress in the BCB-on-silicon configuration are small during the thermal cycling because the BCB has mostly stabilized after curing. Post-cure thermal cycling does not significantly affect the residual stress of BCB, while further curing of the BCB may slightly enhance the critical adhesion energy after more thermal cycles (see Fig. 2 and Ref. [5]).

In summary, experiments on single films on silicon wafers lead us to conclude that the increase of the critical adhesion energy for BCB-bonded silicon wafers, which are coated with PECVD SiO$_2$ films, is attributed to the dominant stress relaxation in the SiO$_2$ layer during the first thermal cycle. The decrease of the critical adhesion energy for BCB-bonded silicon wafers, which are coated with PECVD

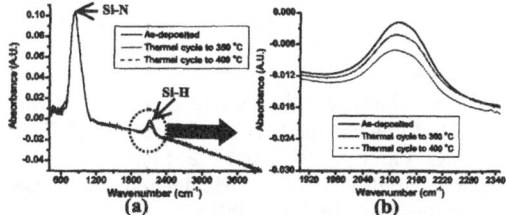

Figure 6. FTIR absorbance spectra for PECVD SiN$_x$ layer before and after cycling: (a) FTIR traces over the entire wavelength and (b) FTIR trace showing change in peak of Si-H group.

SiN$_x$ films, is attributed to the increases in tensile stress of the SiN$_x$ layer during the first thermal cycle. Even with a BCB layer as thin as 0.4 µm, the critical adhesion energy of BCB-bonded wafers (except glass-to-silicon bonded wafers) is much higher than that of a copper interconnect structure with ultra low-k inter-level dielectric (6 J/m^2) [3].

CONCLUSIONS

Our quantitative characterizations have provided a more precise evaluation of the thermal and mechanical stability of BCB-bonded wafer stacks. The first thermal cycle causes a relatively large change in critical adhesion energy (at least ~16% for wafers coated with PECVD SiO$_2$ and peak temperature of 400 °C), while limited change occurs with further thermal cycling. The FTIR and thin film residual stress analyses suggest that the residual stress changes within the PECVD SiO$_2$ and SiN$_x$ layers due to the chemical reactions are the main factor impacting critical adhesion energy at both BCB interfaces during thermal cycling. The BCB bonding glue is further cured without degradation during post-bonding thermal processing, thus enhancing the critical adhesion energy slightly. The critical adhesion energy (even after thermal cycling) is sufficiently high compared to that of copper / ultra low-k interconnect structure that failures during the four-point bending test occurs within the interconnect structure as demonstrated previously.

ACKNOWLEDGMENTS
This work was supported by DARPA, MARCO, and NYSTAR through the Interconnect Focus Center.

REFERENCES
1. J.-Q. Lu, Y. Kwon, J.J. McMahon, A. Jindal, B. Altemus, D. Cheng, E. Eisenbraun, T.S. Cale, and R.J. Gutmann, in *20th International VLSI Multilevel interconnection Conference (VMIC 2003)*, 227 (2003).
2. R.J. Gutmann, J.-Q. Lu, S. Pozder, Y. Kwon, A. Jindal, M. Celik, J.J. McMahon, K. Yu, and T.S. Cale, in *Advanced Metallization Conference in 2003 (AMC 2003)*, 19 (2003).
3. Y. Kwon, A. Jindal, J.J. McMahon, J.-Q. Lu, R.J. Gutmann, and T.S. Cale, in *MRS Symp. Proc.*, **Vol. 766**, E5.8.1 (2003).
4. Processing Procedures for Dry-Etch Cyclotene Advanced Electronics Resins, Dow Chemical Company, Midland, MI, 1997.
5. Y. Kwon, J.-Q. Lu, T.S. Cale, and R.J. Gutmann, in *International Conference on Microelectronics and Interfaces (ICMI'04)*, 40 (2004).
6. Y. Sha, C.Y. Hui, E.J. Kramer, S.F. Hahn, and C.A. Berglund, *Macromolecules*, **29**, 4728 (1996).
7. A.A. Volinsky, N.R. Moody, and W.W. Gerberich, *Acta Met.*, **50**, 441 (2002).
8. C.S. Litteken, and R.H. Dauskardt, *Int. J. Fract.*, **119/120**, 475 (2003).
9. A.G. Varias, Z. Suo, and C.F. Shih, *J. Mech. Phys. Solids*, **39**, 963 (1991).
10. R.J. Hohlfelder, D.A. Maidenberg, R.H. Dauskardt, Y.G. Wei, and J.W. Hutchinson, *J. Mater. Res.*, **16**, 243 (2001).
11. F. Chen, B.Z. Li, T.D. Sullivan, C.L. Gonzalez, C.D. Muzzy, H.K. Lee, M.W. Dashiell, J. Kolodzey, and M.D. Levy, *J. Vac. Sci. Technol. B*, **18**, 2826 (2000).
12. T. Karabacak, Y.-P. Zhao, G.-C. Wang, and T.-M. Lu, *Phys. Rev. B*, **66**, 075329 (2002).
13. Y. Toivola, J. Thurn, R.F. Cook, G. Cibuzar, and K. Roberts, *J. Appl. Phys.*, **94**, 6915 (2003).

Mat. Res. Soc. Symp. Proc. Vol. 812 © 2004 Materials Research Society   F6.25

Three Dimensional Interconnect Stress Modeling for Back End Process

Xiaopeng Xu and Victor Moroz
TCAD R&D, Synopsys, Inc.
700 E. Middlefield Rd, Mountain View, CA 94043, USA

ABSTRACT

A process oriented approach is demonstrated for modeling the stress evolution during the entire back end process flow. No *ad hoc* assumptions regarding stress states are made for layer deposition and etching. Intrinsic stresses from material formation, thermal mismatch stresses from temperature ramps, stress relaxation due to viscous deformation, and stress profile redistribution upon deposition and etching are all considered at each process. Parametric studies are carried out to examine the effects of viscous flow, material selection and layout variations. It is found that the viscous flow of the passivation and dielectric materials have large impact on stress evolution. A TCAD assisted design approach is suggested for lowering stress levels of critical stress components. The implications of the stress modeling results on reliability issues like stress-triggered void formation are discussed.

INTRODUCTION

In modern integrated circuits, interconnects consist of multilevel metal lines that are embedded within several layers of different dielectric materials above the silicon transistors. The materials in interconnects have very different mechanical properties. These differences can lead to large thermal mismatch stresses during the fabrication process. The residual stresses in interconnects are known to cause yield and reliability issues. Predicting stress and deformation fields in interconnects for reliability assessments and yield improvements has been a major effort in the semiconductor industry and in the research community [1,2].

Historically, interconnect stresses are modeled with various assumptions. For example, 2D plane strain or 2D axisymmetric simplifications are made for a 3D structure; viscous relaxation in low-k dielectric insulators are ignored; and *ad hoc* numerical treatments are introduced to activate and deactivate elements in the mesh instead of layer deposition and etching [3,4]. Stress simulations with such assumptions become increasingly inadequate to predict stress distributions inside interconnects that are fabricated with complex process flows.

In this study we use our three-dimensional process simulator to simulate the stress evolution during the entire interconnect fabrication process which includes material deposition and etching steps at room temperature and elevated temperatures, and temperature ramps in between. Stresses are generated during the temperature ramps due to thermal expansion mismatches, relaxed in the meantime because of viscous deformations, and redistributed during deposition and etching due to the rebalance of forces. The thermal stress generation, relaxation and redistribution are considered at each process step by solving stress equations for realistic evolving geometry during deposition and etching steps.

MODELS

A typical interconnect structure consists of materials with strikingly different mechanical behavior. The metal interconnects and vias are encapsulated inside dielectric or passivation materials such as oxide and nitride. While crystalline materials like silicon can be modeled with linear elasticity, it has been shown that amorphous materials like oxide and nitride should be treated as non-linear visco-elastic solids [5]. According to Maxwell model for visco-elastic solids, the stress σ is related to strain ε by:

$$\frac{\partial \sigma}{\partial t} = 2G \frac{\partial \varepsilon}{\partial t} - \frac{\sigma}{\tau}, \qquad\qquad \tau = \frac{\mu}{G} \qquad\qquad (1)$$

Here G is the shear modulus, τ is the stress relaxation time and μ is the viscosity. This relation states that the change in stress is proportional to the change in strain, minus a relaxation term that is proportional to the stress itself.

The viscosity μ is also stress dependent and follows the Eyring's model given as:

$$\mu = \mu_0 \frac{(\sigma_{max}/\sigma_0)}{\sinh(\sigma_{max}/\sigma_0)}, \qquad\qquad \sigma_0 = \frac{2kT}{V_0} \qquad\qquad (2)$$

Where μ_0 is the low stress viscosity, σ_{max} is the maximum effective stress, σ_0 is the linear elastic limit of the effective stress, V_0 is the unit activation volume in which stress-induced local atomic displacements occur, k is the Boltzmann's constant, and T is the temperature.

Figure 1 shows the stress evolution in a thin planar oxide layer during a thermal cycle. The oxide layer is deposited on top of the silicon substrate. The temperature is ramped up to 900°C, kept at the elevated temperature for 10 minutes, and then ramped down to the room temperature. The material properties used are listed in Table 1. For comparison purposes, the oxide material was also simulated using an elastic model that lacks viscous flow in addition to using its accurate representation as a visco-elastic solid that exhibits viscous flow and stress relaxation at high temperature.

For the visco-elastic film, the resulting residual stress after 10 minutes relaxation is only 49% of it original value, and the residual stress at the end of the temperature ramp cycle changes its state from tensile to compressive and has nearly 56% of its peak tensile value. This comparison shows that the stress relaxation effect for a visco-elastic material like oxide can be crucial for accurate modeling of the backend fabrication process.

CASE STUDIES

The materials used in our case studies include silicon, silicon dioxide, silicon nitride, low-k dielectric, and copper. Some of the mechanical and thermal properties are listed in Table 1. The low stress viscosity values at room temperature are taken as 76.36 (MPa·s) for oxide and low-k material, and as 1.89 (GPa·s) for nitride. The activation volume values are taken as $3 \cdot 10^{-28}$ (m^3) for all visco-elastic materials. During the simulation, the silicon and copper regions are treated as elastic materials, while silicon dioxide, nitride and low-k dielectric regions are treated as visco-elastic materials.

Table 1. Material properties.

	Si	SiO2	Si3N4	Low-k	Cu
Young's modulus (GPa)	187	66	192	2.45	129.8
Poisson's ratio	0.28	0.2	0.3	0.34	0.343
Thermal coefficient (10^{-6}/K)	3.1	1.0	3.0	26.0	16.5
Activation volume (m^3)	-	$3 \cdot 10^{-28}$	$3 \cdot 10^{-28}$	$3 \cdot 10^{-28}$	-
Viscosity (MPa·s)	-	76.36	1890	76.36	-

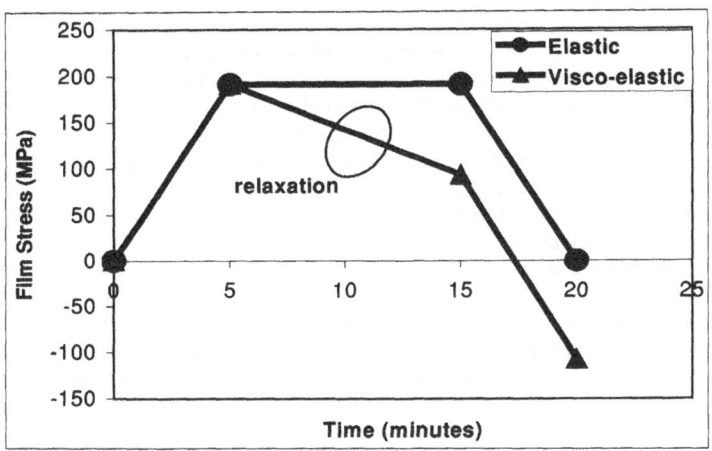

Figure 1. Comparison of stress evolution for an elastic layer vs. a visco-elastic oxide layer during a temperature cycle.

The simulated back end process steps are listed in Table 2. A silicon dioxide layer with intrinsic stress is first deposited on top of the silicon wafer. The structure is heated up to 400°C to deposit the first layer of low-k dielectric. It is then cooled down to 20°C to etch off part of the low-k dielectric. The first metal line (copper, M1) is electroplated into the space that was created by the previous etching step. The structure is again heated up to 400°C to deposit a thin layer of dielectric (silicon nitride), and the second layer of low-k dielectric. The thin layer of silicon nitride is used as a diffusion barrier for the copper layer.

Then the wafer is cooled down to room temperature and dry etching process is performed to etch off part of the nitride and the low-k dielectric. Copper is again electroplated to form the second metal line (M2) and the via that connects the first metal line and the second metal line. After the formation of the second metal line, the interconnect structure is heated up again to 400 degrees to deposit another thin diffusion barrier and then is cooled down to room temperature. The subsequent process steps involve heating up to 400°C to deposit passivation silicon dioxide and silicon nitride, and cooling down to room temperature. One of the typical simulated interconnect structures is shown in Figure 2a.

Table 2. Simulated back-end process steps.

	Process condition		Process condition
Step 1	Oxide deposition	Step 10	Cool down to 20°C
Step 2	Heat up to 400°C	Step 11	Dry etch low-k and nitride
Step 3	Low-k dielectric deposition	Step 12	Copper deposition
Step 4	Cool down to 20°C	Step 13	Heat up to 400°C
Step 5	Dry etch low-k dielectric	Step 14	Nitride deposition
Step 6	Copper deposition	Step 15	Cool down to 20°C
Step 7	Heat up to 400°C	Step 16	Heat up to 400°C
Step 8	Nitride deposition	Step 17	Oxide and nitride deposition
Step 9	Low-k dielectric deposition	Step 18	Cool down to 20°C

The stress evolution is simulated during the entire back end process flow. Figure 2b shows the resulting dilatation stress σ_d field for the structure shown in Figure 2a at the end of step 18. For better clarity, the stress field is only shown inside copper. It is interesting to note that the maximum dilatational stress occurs around the bottom of the via. This is the most troublesome area for void formation – see for example TEM images in [2].

Since interconnects are fabricated on top of the silicon wafer, the overall thermal deformations in interconnects are controlled by that of the thick silicon wafer. The local stress states inside copper lines result from thermal expansion mismatches and constrains from surrounding materials that are experiencing visco-elastic deformations. Thus the final stress states depend on temperature ramping history, the material properties, and the geometry of the etched and deposited layers.

The impact of material properties of the low-k dielectric on the overall stresses is investigated with the following two cases. Case one uses regular silicon dioxide as the low-k dielectric while case two uses a soft low-k dielectric which has much lower Young's modulus but higher thermal expansion coefficient than the regular silicon dioxide, as listed in Table 1.

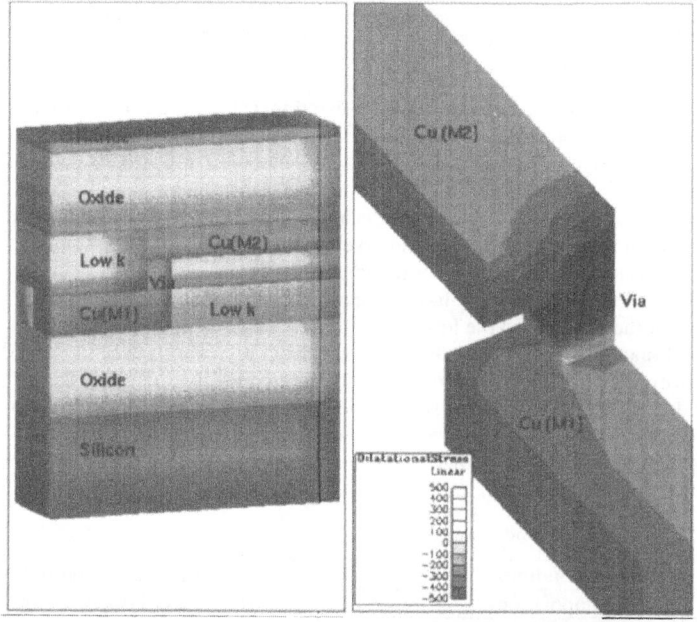

Figure 2. a) The simulated interconnect structure. b) The dilatational stress field in copper M1, via and copper M2 regions at the end of step 18.

Figure 3 shows evolution of the dilatational stresses at the center of the via during the back end process steps for the two cases. Even though the surrounding dielectric material in case two undergoes much larger thermal mismatch strain but the average dilatational stress in the via is much lower than in case one due to its low Young's modulus. Also due to the visco-elastic effect as demonstrated in Figure 1, the dilatation stress state at the end of the process becomes compressive in case two, instead of tensile in case one.

From a mechanical viewpoint, void nucleation and growth are promoted by high tensile dilatational stress. The stress state change from tensile to compressive and the reduction in overall stress level will likely help to reduce the danger of void formation.

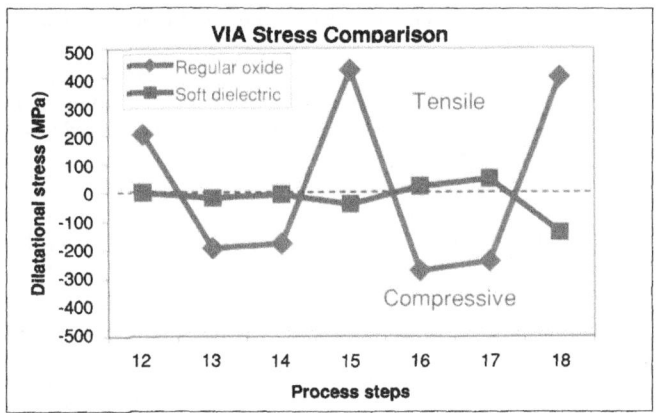

Figure 3. Via stress comparison for two different surrounding dielectrics.

To investigate stress sensitivity to the layout, a "T" shaped interconnect structure is simulated. The fabrication process follows the same steps outlined in Table 2. The dilatational stress σ_d field for the structure is shown in Figure 4 at the end of step 18.

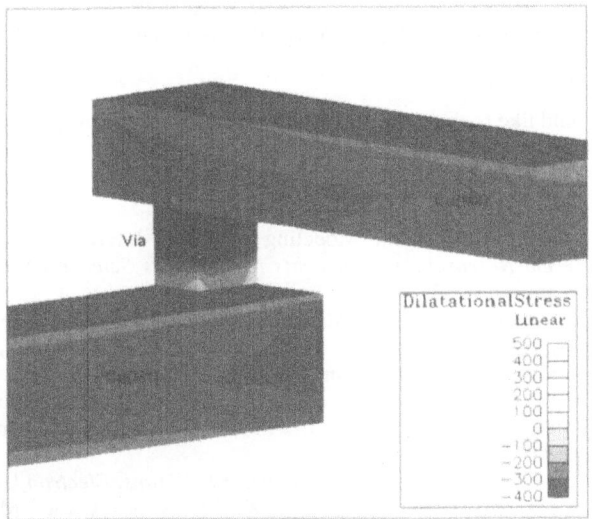

Figure 4. The dilatational stress distribution in copper M1, via and copper M2 regions at the end of step 18 for the "T" shaped interconnect structure.

For better clarity, the non-metal regions are removed and the stress field is only shown in part of the via and metal lines. Comparing to the case shown in Figure 2, the overall stress levels in the "T" shaped interconnect are lower. The maximum tensile and compressive dilatation stress values reduce by about 10%. This may be the result of the structural symmetry. The contributions from the bending moments that are often present in the non-symmetric structures do not exist in a symmetric one.

The effect of typical etching mask misalignment can be considerable. For the "T" shaped interconnect structure, if the etching mask for the via is shifted by 50 nm in the z direction towards the symmetry center of the cross, the maximum tensile stress is reduced by about 25%, likely due to the effect of higher symmetry.

The simulation of the back end process flow takes less than 10 minutes of CPU time on a 2 GHz Linux machine and the final structure contains about 3k nodes with fine mesh inside the copper regions and coarse mesh elsewhere for the structures reported in this paper.

CONCLUDING REMARKS

In this study we use our three dimensional process simulator Taurus-Process [6] to simulate the stress evolution for the entire interconnect fabrication process flow. No *ad hoc* assumptions regarding stress states are required for region additions and removals during deposition and etching steps. Intrinsic stresses from material formation, thermal mismatch stresses from temperature ramps, stress relaxation from viscous deformation, and stress redistribution from deposition and etching are all considered at each process step by solving stress equilibrium equations with evolving boundary conditions. Parametric studies are carried out to examine the effects of viscous flow and material selection, as well as layout variations. The results suggest that the viscous flow of dielectric materials has to be considered for accurate interconnect stress modeling. It has demonstrated that a TCAD assisted design approach can provide valuable insights into stress sensitivity to the layout, material selection, and thermal history.

ACKONWLEDGMENTS

The authors would like to thank Dr. Gregory Khrenov from Synopsys and Dr. Charlie Zhai from AMD for fruitful discussions.

REFERENCES

1. Y.-L. Shen, "Thermo-Mechanical Modeling of Metal Interconnects in Microelectronic Devices", in *Recent Research Developments in Materials Science IV, Chapter 6*, pp. 125-155, Research Signpost, 2003.
2. M. Hierlemann *et al*, "Impact of Mechanical Stress on BEOL Reliability, Enhanced Understanding by TCAD Simulation", ChiPPs meeting, Oct. 13-17, 2002, Prague.
3. J. Lee and A. S. Mack, *IEEE Trans. Semi. Manufacture*, **11**, 458 (1998).
4. K. Yang, J. J. Waeterloos, J. Im, and M. E. Mills, "Sequential Process Modeling for Determining Process-Induced Thermal Stress in Advanced Cu/Low-k Interconnects", MRS Spring meeting, 2003, San Francisco.
5. V. Senez, D. Collard, P. Ferreira, and B. Baccus, *IEEE Trans. Electron Dev.*, **43**, 720 (1996).
6. Taurus Process Reference Manual, version 2003.12, Synopsys, Inc., Mountain View, CA, USA.

Mat. Res. Soc. Symp. Proc. Vol. 812 © 2004 Materials Research Society   F6.26

Thermomechanical Stresses in Copper Interconnect/Low-*k* Dielectric Systems

Y.-L. Shen and E. S. Ege
Department of Mechanical Engineering, University of New Mexico
Albuquerque, NM 87131, U.S.A.

ABSTRACT

Numerical simulations of thermal stresses in copper interconnect and low-*k* dielectric systems are carried out. The analyses include two- and three-dimensional finite element modeling of the interconnect structure. Various combinations of metal, oxide and polymer-based low-*k* dielectric schemes are considered in the simulation. The evolution of stresses and deformation pattern in copper, barrier layers, and the dielectrics are critically assessed.

INTRODUCTION

This study concerns thermal stress modeling and its implications in the reliability features of copper (Cu) and/or low-*k* dielectric-based systems. Recently two-dimensional (2D) finite element analyses have been reported on idealized Cu interconnect/low-*k* structures [1,2]. A comprehensive 2D study, taking into account the thin-film constitutive properties and a wide variety of geometrical/material features, has also been undertaken [3]. In this paper we first present baseline results from the 2D modeling, and then extend the analyses to a three-dimensional (3D) case where two levels of Cu lines are connected by a via. Attention is devoted to the thermal stresses generated in the Cu lines and via, in the thin barrier layers surrounding the metal structure and part of the dielectric, and in the interlevel dielectric materials. The significance of such an investigation can be understood through the following considerations.

- The propensity of void formation and interface damage in Cu depends strongly on the stress/strain state and its spatial distribution.
- In general polymer-based and porous low-*k* materials are mechanically weak. Stresses carried by these materials and the possible consequences need to be carefully assessed.
- With the weak dielectric materials in place, a legitimate concern is that the thin barrier (including etch stop) layers may bear the brunt of internal mechanical actions so the reliability of the barrier layers themselves becomes an issue.

APPROACH

The 2D model features a series of infinite, parallel, single-level Cu lines embedded within the dielectric on top of the silicon (Si) substrate, Fig. 1(a). The metal lines are perpendicular to the paper. Because of the periodicity and symmetry in the lateral (across-the-line, y) direction, only one half of a unit segment is required in the calculation. The model is based on an embedded scheme that low-*k* dielectric is used only in areas adjacent to the side walls of the metal line while the rest of the dielectric is still silicon oxide (SiO_x). Thin tantalum nitride (TaN) and silicon nitride (SiN_x) layers, all taken to be 20 nm thick in the structure, serve as diffusion barriers (liners) and/or etch stop. Unless otherwise stated, the Cu line width w and pitch p are taken to be 0.24 and 0.56 μm, respectively, in the 2D model as well as the 3D model described below. The choice of these dimensions is somewhat arbitrary. However, as can be seen in Fig. 2 below and in our preliminary calculations, using different dimensions over a relatively wide range will only moderately affect the simulation result. The fundamental features remain unchanged.

Figure 1(b) shows the 3D model. The direction of Cu lines is parallel to the x-axis. The entire model has spans of 3.0 μm, 0.28 μm and 2.0 μm along the x, y and z directions, respectively. The computational domain represents one half of the unit structure, with the xz-plane exposed in Fig. 1(b) being the mirror symmetry plane showing the middle cut of the metal line/via structure. The model is bound by a top and a bottom layer silicon oxide (SiO_x), each having a thickness of 0.38 μm. The thin red layers directly on top of both levels of Cu lines are 0.02 μm-thick SiN_x and the thin orange layers are 0.02 μm-thick TaN. The height of both levels of Cu lines is 0.4 μm. The Cu via, located at the center in the x-direction, is assumed to be a rectangular block with the same width (in the y-direction) as the Cu lines. Its dimensions in the x- and z-directions, not including the barrier layer, are 0.3 μm and 0.38 μm, respectively. The copper material in the via and in the upper line is directly connected. The bottom boundary of the via, however, is a TaN layer. Two types of dielectric are considered: silicon oxide and low-k material. Three different regions of the dielectric, having the same vertical positions as the lower line, via, and upper line, are defined. In one case all three regions of dielectric are filled with oxide. This model is henceforth referred to as "all-oxide." In the second case all three regions are filled with the low-k dielectric, and is referred to as the "all-lowk" model. The third model is based on an embedded scheme that the low-k dielectric is used only in regions at the same levels of the long metal line (for reducing the dominant line-to-line capacitance), while at the via level silicon oxide is still used. This model is referred to as "lowk/oxide." Note that in all cases the topmost and bottommost layers are both silicon oxide.

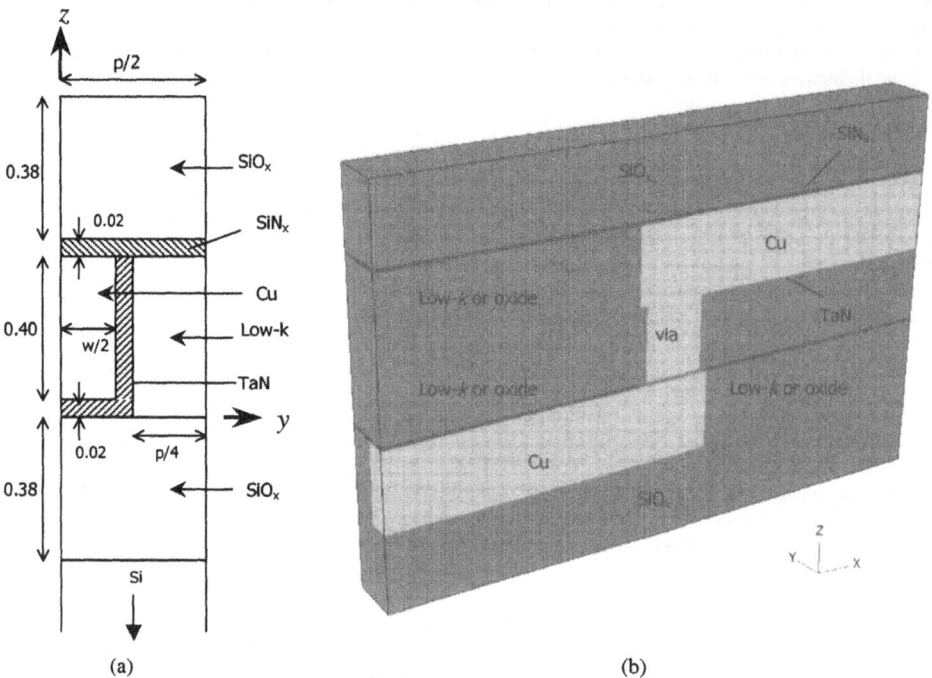

Fig. 1   Model geometries used in the (a) 2D and (b) 3D finite element modeling. (Unit: μm)

The evolution of thermal stresses is modeled by imposing a spatially uniform temperature change from the initial stress-free temperature of 350°C to 20°C. In the modeling all lateral boundaries are taken to be a mirror symmetry plane. The top surface is not constrained. All the interfaces between dissimilar materials are treated as perfectly bonded (i.e., the displacement field is continuous across the interface). The finite element program ABAQUS [4] is employed for all calculations. All material properties used in the modeling are listed in Table I. The properties of the low-k dielectric follow those of divinyl-siloxane-bis-benzo-cyclobutene (BCB) reported in Ref. [1]. Although this is a specific dielectric material, its thermoelastic properties are very similar to other polymer based low-k systems such as SiLK™ [5]. Therefore the modeling results presented in this work are considered as having a high degree of generality. The detailed discussion of material model, especially that of the elastic-plastic response of Cu, can be found in reference [3].

334

Table I   Material properties used in the numerical analysis. E: Young's modulus, ν: Poisson's ratio, α: coefficient of thermal expansion (CTE), σ_y: yield strength. Unless otherwise stated, a linear variation of properties with temperature between the indicated temperatures is assumed.

		Cu	SiO$_x$	Si	low-k	TaN	SiN$_x$
E (GPa) at	20°C	110	71.4	130.0	2.5	200	221
	350°C	104	71.4	130.0	0.3[a]	200	221
ν at	20°C	0.30	0.16	0.28	0.34	0.30	0.27
	350°C	0.30	0.16	0.28	0.34	0.30	0.27
α (10^{-6}/K) at	20°C	17.0	0.52	3.1	63.6	4.7	3.2
	350°C	19.3	0.68	4.4	63.6	4.7	3.2
σ_y (MPa) at	20°C	223[b]	--	--	--	--	--
	350°C	131[c]	--	--	--	--	--

[a]   Linear variation only between 20 and 180°C; constant (0.3 GPa) between 180 and 350°C.
[b,c] The kinematic hardening model adopted here also includes a linear hardening rate of 77 GPa. See reference [3] for details.

RESULTS ANS DISCUSSION

For the purpose of establishing a baseline understanding, we first focus on the stresses in long Cu lines resulting from the 2D simulation. Figure 2 shows the normal stress components in Cu, averaged over the cross section of the line, as a function of the line aspect ratio. Here, various Cu widths (w) are considered and the line height is kept constant at 0.40 μm. The pitch-to-width ratio is fixed ($p/w = 2.33$) in all cases. The hydrostatic stress is defined as $(1/3)$ $(\sigma_{xx} + \sigma_{yy} + \sigma_{zz})$. It can be seen that σ_{xx} and σ_{zz} increase monotonically with increasing line aspect ratio while σ_{yy} displays an opposite trend. The combination of stress variation results in an increasing average hydrostatic stress with increasing aspect ratio, which is qualitatively similar to the case of densely arrayed Al interconnects [6-8]. The overall increase in stress magnitude throughout the range of aspect ratio considered, however, is significantly smaller in the case of Cu. Overall the hydrostatic stress values are also smaller than those experienced by the traditional Al interconnect lines. It is evident from Fig. 2 that the magnitudes of σ_{zz} are considerably smaller than the other components. This is primarily due to the fact that, along the z-direction, the Cu line is essentially free of the influence of the substrate material but is strongly affected by the low-k dielectric, which tends to contract more than Cu itself upon cooling.

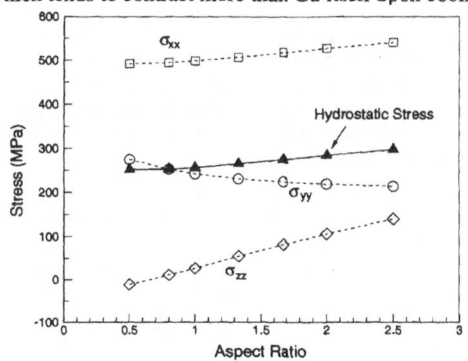

Fig. 2   Average normal stresses (σ_{xx}, σ_{yy} and σ_{zz}) and hydrostatic stress as a function of aspect ratio of the Cu line, upon cooling from 350 to 20°C.

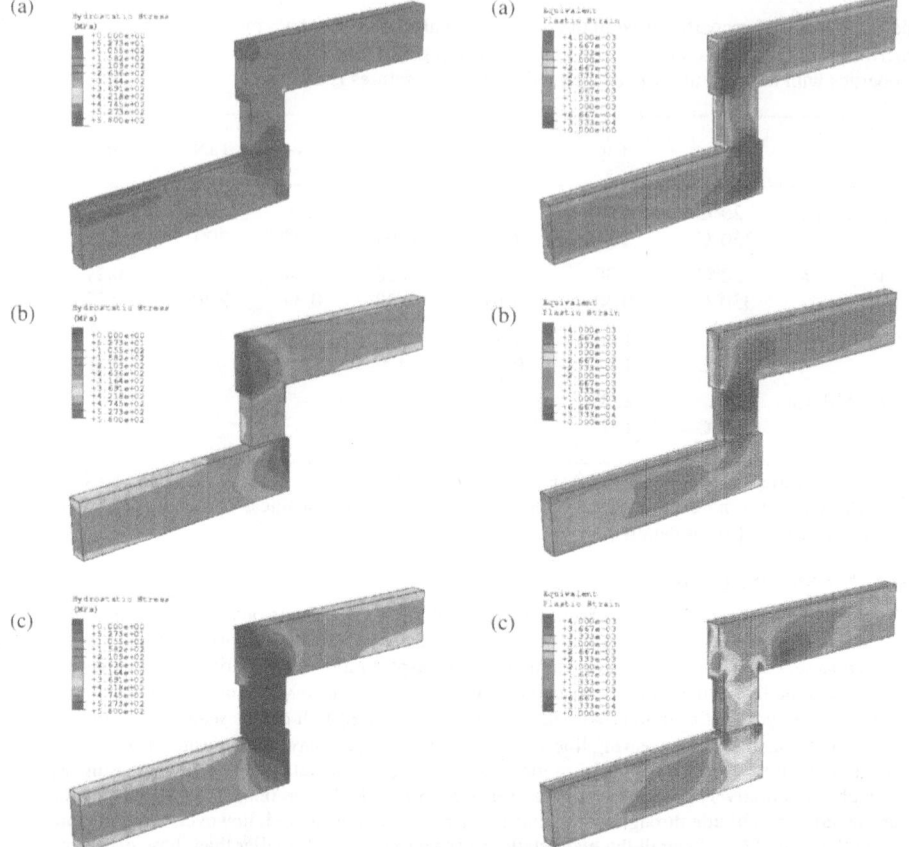

Fig. 3 Contours of hydrostatic stress in the copper line/via assembly for the models of (a) "all-oxide," (b) "lowk/oxide," and (c) "all-lowk."

Fig. 4 Contours of equivalent plastic strain in the copper line/via assembly for the models of (a) "all-oxide," (b) "lowk/oxide," and (c) "all-lowk."

Attention is now turned to the 3D modeling. Figures 3(a)-(c) show the contour plots of hydrostatic stress developed in the line/via structure upon cooling for the models "all-oxide," "lowk/oxide," and "all-lowk," respectively. For easy visualization all the other materials (including the barrier layers) are removed from the figures so the bare Cu is shown. It can be clearly seen that Cu in the "all-oxide" model is under very high tension, due primarily to the low CTE value of SiO_x surrounding the metal structure. The hydrostatic stress in Cu is above 500 MPa in almost the entire Cu structure. Particularly high stresses exist near the bottom of via and some corner and interface regions of the lines.

With the incorporation of low-k dielectric, the magnitude of tensile hydrostatic stress decreases significantly in both the line segments and the via (Figs. 3(b) and 3(c)). In the "all-lowk" model, the via is actually under compression. This arises from the fact that the stress component σ_{zz} is compressive because the surrounding low-k material has a much greater CTE than Cu and is relatively free to contract along the z-direction upon cooling, thus exerting compression on the via. While the via is still largely in tension along x and y (not specifically shown here), net hydrostatic pressure results. The existence of thin barrier layers tends to counter the effect of low-k dielectric, but only slightly. Note that the combination

336

of tensile and compressive stress components leads to a large deviatoric stress field, which enhances plastic deformation as shown below. Figure 3(b) shows that, with an embedded scheme in the "lowk/oxide" model, the via is laterally surrounded by the oxide and therefore a triaxial tensile state still exist, although in this case the hydrostatic tension is much smaller than that in the "all-oxide" model.

Figures 4(a) to 4(c) show the contour plots of equivalent plastic strain in Cu for the models of "all-oxide," "lowk/oxide" and "all-lowk," respectively. It can be seen that strong plastic deformation tends to occur near the interface and corner regions. Compared to the other two models, the "all-lowk" case shows significantly higher plastic strains in the via and its vicinity. This is a natural consequence of the high deviatoric stress field mentioned above. When the embedded scheme is adopted ("lowk/oxide," Fig. 4(b)), the overall plastic strains in the via are even smaller than in the case of "all-oxide" (Fig. 4(a)). In the long line segments, relatively stronger plasticity is seen to exist near the top and bottom interfaces, compared to the interior mid section of the line. This is consistent with the previous 2D analysis [3]. Experiments have shown that atomic diffusion during electromigration takes place primarily along the interfaces between Cu and the adjacent barrier layers, particularly the top interface [9,10]. The present modeling result, together with the fact that plastic deformation generates crystal defects, suggest that the deformation field resulting from the thermal stress effect can enhance such damage, or in fact even be a part of its causes. Additionally, the top interface is subject to contamination resulting from the chemical-mechanical polishing processes. Voiding and local debonding may readily occur.

The numerical analysis presented here can be used to explain some other experimental observations of structural damages in Cu interconnects. When the via is surrounded by the low-k dielectric, the combination of compressive σ_{zz} and a high propensity of plastic deformation brings about the possibility of severe distortion (permanent deformation) of the via. This new failure mode has indeed been reported [11,12], where discrete shearing of the upper part of the via relative to the lower part creates highly discernible shear steps at the interface or shear cracks through the via. The mechanically weak low-k material is unable to prevent this from occurring.

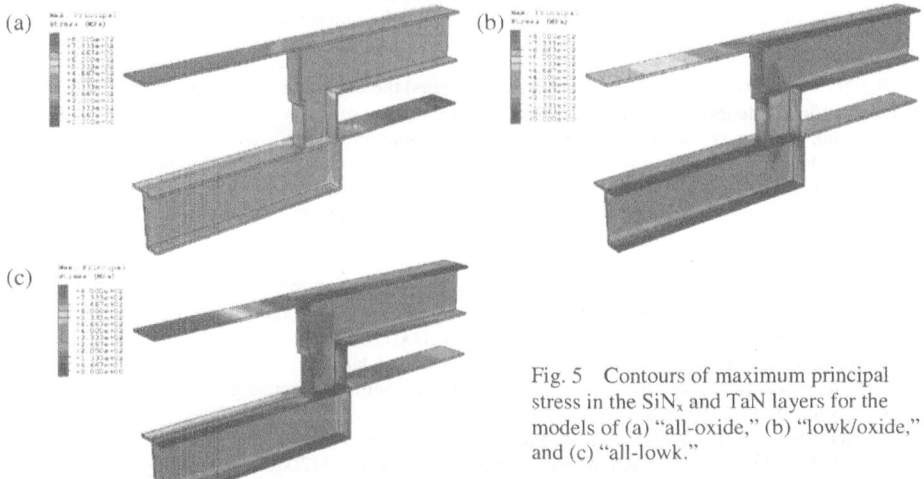

Fig. 5   Contours of maximum principal stress in the SiN$_x$ and TaN layers for the models of (a) "all-oxide," (b) "lowk/oxide," and (c) "all-lowk."

Figures 5(a) to 5(c) show the contours of maximum principal stress in the SiN$_x$ and TaN layers for the "all-oxide," "lowk/oxide," and "all-lowk" models, respectively. (Due to the brittle nature and thin dimensions, one appropriate stress parameter relating to potential failure is the maximum principal stress (if it is tensile).) For clarity all the oxide, dielectric and Cu materials are removed from the presentation. Note that the contours shown in Fig. 5 do not provide information on the principal directions of the stress tensor, but a majority portion of the barrier layers is seen to subject to tensile stresses. The local tensile stresses can reach high magnitudes (well above 800 MPa).

In Fig. 5(a) where "all-oxide" is considered, the TaN layer at the via bottom and the lower SiN_x layer adjacent to the via (partially obstructed in the figure) are under relatively large stresses. In the "lowk/oxide" model (Fig. 5(b)), stresses carried by TaN are relieved to a great extent. However, the upper SiN_x shows a significant increase in stress. This is primarily due to the enhanced contraction in the vertical (z) direction of the upper-level low-k material upon cooling, causing downward deflection of the SiN_x beam near the left boundary. (The right half of upper SiN_x can be thought of as clamped because it is supported by a stiffer structure.) This beam-deflection effect becomes more prominent when the "all-lowk" model is adopted, as in Fig. 5 (c). In Fig. 5(c) the lower SiN_x also displays a narrow region of high stresses, which is due largely to the very high tensile stresses developed in the lateral (y) direction locally. It can be realized, from Figs. 5(a) to 5(c), that as the oxide-based dielectric is replaced by the polymer-based low-k material, the barrier layers are forced to experience increasingly large stresses in the structure. As a consequence, greater attention should be paid to the reliability of these thin layers, not just the metal and dielectric materials.

In our simulations we have also examined other stress components in the barrier structure as well as in the dielectric materials. We have found that the hydrostatic stress in the low-k material can attain 150 MPa. While this level of stress may be tolerated by typical metallic and ceramic materials used in microelectronic devices, great uncertainty exists for polymer-based systems. If the dielectric is porous, local stress concentration in the solid part of the structure will be even considerably higher. It is therefore essential to devote attention to the mechanical strength of low-k dielectric during material development. Note also that the present paper focuses only on polymer-based low-k dielectrics. We have undertaken parametric studies by systematically varying the dielectric modulus and CTE so PECVD-based dielectrics such as SiOCH low-k materials may be represented. The qualitative features such as severe via deformation were also obtained (although to a lesser extent in comparison with the polymer-based low-k systems). Due to space restriction the detailed analyses will be presented elsewhere.

CONCLUSIONS

The 2D and 3D finite element analyses have illustrated the evolution of thermal stresses in the Cu interconnect/low-k dielectric structures. With the incorporation of low-k dielectric, the hydrostatic tension in the Cu lines is reduced. However, plastic deformation in the via and its vicinity is strongly enhanced if the entire metal assembly is surrounded by the low-k material. The thin barrier/etch stop layers become severely stressed. The low-k material can experience a hydrostatic stress well in excess of 100 MPa. Implications of these findings to device reliability are also briefly discussed.

REFERENCES

1. J.-H. Zhao, W.-J. Qi, and P. S. Ho, *Microelectron. Reliab.* **42**, 27 (2002).
2. Y.-L. Shen and U. Ramamurty, *J. Vac. Sci. Technol. B* **21**, 1258 (2003).
3. E. S. Ege and Y.-L. Shen, *J. Electronic Mater.* **32**, 1000 (2003).
4. ABAQUS, Version 6.4, Abaqus, Inc., Pawtucket, Rhode Island.
5. Web site: www.dow.com/silk/lit/index.htm, The Dow Chemical Company.
6. Y.-L. Shen, *J. Appl. Phys.* **82**, 1578 (1997).
7. A. Gouldstone, Y.-L. Shen, S. Suresh, and C. V. Thompson, *J. Mater. Res.*, **13**, 1956 (1998).
8. M. S. Kilijanski and Y.-L. Shen, *Microelectron. Reliab.* **42**, 259 (2002).
9. C. K. Hu, R. Rosenberg, and K. Y. Lee, *Appl. Phys. Lett.* **74**, 2945 (1999).
10. C. S. Hau-Riege, *Microelectron. Reliab.* **44**, 195 (2004).
11. M. Fayolle, G. Passemard, M. Assous, D. Louis, A. Beverina, Y. Gobil, J. Cluzel, L. Arnaud, *Microelectronic Engng.* **60**, 119 (2002).
12. T. M. Shaw, X.-H. Liu, C. Murray, M. Y. Wisniewski, G. Fiorenza, M. Lane, S. Chiras, R. R. Rosenberg, R. Filippi, J. Mcgrath, H. Rathore and V. Mcgahay, presentation U9.1, 2003 MRS Fall Meeting, Boston, MA.

Mat. Res. Soc. Symp. Proc. Vol. 812 © 2004 Materials Research Society

Mortality Dependence of Cu Dual Damascene Interconnects on Adjacent Segment

C.W. Chang[1], C.L. Gan[1,2], C.V. Thompson[1,3], K.L. Pey[1,4], W.K. Choi[1,5], N. Hwang[6]
[1] Advanced Materials for Micro- and Nano-Systems, Singapore-MIT Alliance, 4, Engineering Drive 3, Singapore 117576
[2] School of Materials Engineering, Nanyang Technological University, Nanyang Avenue, Singapore 639798
[3] Department of Materials Science and Engineering, Massachusetts Institute of Technology, Cambridge, Massachusetts 02139
[4] School of Electrical & Electronics Engineering, Nanyang Technological University, Nanyang Avenue, Singapore 639798
[5] Department of Electrical and Computer Engineering, National University of Singapore, 4, Engineering Drive 3, Singapore 117576
[6] Microsystems, Modules & Components Laboratory, Institute of Microelectronics Singapore, 11, Science Park Road, Singapore Science Park II, Singapore 117685

ABSTRACT

Three terminal 'dotted-I' interconnect structures, with vias at both ends and an additional via in the middle, were tested under a variety of test conditions. Failures (mortalities) were observed even when segments were tested under conditions that would not have led to failure in two-terminal structures. Mortalities were found in right segments with jL values as low as 1250 A/cm, which is lower when compared to the immortality condition $(jL)_{cr}$ of 3700 A/cm reported for similar via-terminated structures. Moreover, we found that the mortality of a dotted-I segment is dependent on the direction and magnitude of the current in the adjacent segment. These results suggest that there is not a definite value of the jL product that defines true immortality in individual segments that are part of an interconnect tree. More importantly, the critical jL value for a single segment of Cu interconnects may be reduced or increased by an adjacent segment. Therefore independently determined $(jL)_{cr}$ values cannot be directly applied to interconnects with branched segments, but rather the magnitude as well as the direction of the current flow in the adjoining segments must be taken into consideration in evaluating the immortality of interconnect segments in an interconnect network.

INTRODUCTION

With increasing requirements for high clocking speeds and complex functions for integrated circuits (IC), more metal layers and longer interconnects are incorporated into each single chip. In addition, continued shrinking of interconnect widths leads to higher current densities with each new technology node. The 2003 International Technology Roadmap for Semiconductors [1] predicts that in the year 2005, a total length of 907 m of interconnects in 11 layers will be required in a 1 cm^2 chip. At the same time the tolerable interconnect failure rate must be reduced to 5 FITs (5 failures per 10^9 device-hours) to protect the overall functionality of the chip. This sets a great challenge to the development of a reliable interconnect technology, especially with respect to electromigration resistance.

Electromigration occurs through a net atomic flux as a result of the electron wind force that is countered by a force caused by the resulting mechanical stress gradient

$$J_a = -\frac{D_a C_a}{kT}\left(Z^* e\rho j - \Omega\frac{\partial\sigma}{\partial x}\right) \quad ,$$ (1)

where D_a is the atomic diffusivity, C_a is the atomic concentration, kT is the thermal energy, Z^* is the effective charge number, e is the elementary charge, ρ is the resistivity, j is the current density, Ω is the atomic volume, and σ is the hydrostatic stress as a function of position along the length of the interconnect. For short lines bound by zero-flux boundaries that are electrically stressed at low current densities, the mechanical stress along the line will evolve to a uniform stress gradient, at which the back force due to the stress gradient balances the electron wind force [2]. If the hydrostatic stresses stay low enough that void nucleation or metal extrusion does not occur, the metal line will be immune to electromigration-induced failure (immortal). For zero net atomic flux, equation 1 can be rearranged into the following form:

$$(jL)_{cr} = \frac{\Omega\Delta\sigma}{Z^* e\rho} \quad ,$$ (2)

where $(jL)_{cr}$ is the critical product of current density, j, and the line length, L, and $\Delta\sigma$ is the difference in the hydrostatic stress between the anode and the cathode.

The criterion for immortality, $(jL)_{cr}$, is an important parameter in IC layout design. Immortality has been demonstrated for Al-based metallization systems [2-3] and has also been recently reported for Cu interconnects [4-6], though with a wide range of values from 1500 A/cm to 3700 A/cm, depending on the structure of the interconnects. In addition, Cu-based interconnects are sometimes said to have a 'probabilistic' immortality that is dependent on the location of the void formation [7]. Moreover, the reliability of a segment in a Cu-based interconnect tree has been reported to depend on the stress conditions of the adjoining segments [8]. In this paper, we further investigate the reliability of segments in interconnect trees to test the applicability of the immortality criteria, $(jL)_{cr}$, determined in single-segment trees, for interconnect trees segments.

EXPERIMENTAL DETAILS

Package-level electromigration tests were carried out on straight via-terminated lines with an additional via in the middle that creates two segments of 25 μm length each (dotted-I structures). The test structures were in metal 2 (M2) and connected to metal 1 (M1) leads in via-below configurations, as shown in Figure 1. The test structures were fabricated using a dual damascene process with SiO_2 as the inter-level dielectric, Ta as the diffusion barrier and silicon nitride as the capping dielectric. The current density in the right segment was kept constant at 0.5 MA/cm^2 with the electrons flowing from the right via towards the middle via, while the current density in the left segment was varied, with values of 0, 0.5, 1.5 or 2.5 MA/cm^2, and with electrons flowing either into or out of the central via. The current density in the left segment (j_L) was defined to be positive if the electron flow direction was the same as the right segment (see Figure 1(a)). A sample population of 16 was used for each of the test conditions. All the samples were stressed for 780 hours and the failure criterion was set as a 30% resistance increase.

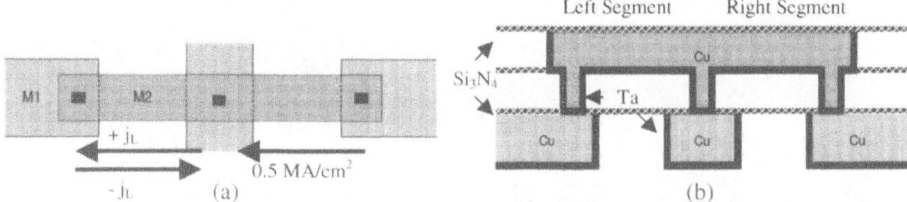

Figure 1. (a) Top and (b) cross-sectional view of the dotted-I test structure. The arrows indicate the directions of the electron flow.

RESULTS

Due to the small number of failures in some of the test configurations, rather than comparing the median-time-to-failure, we present the percentage of the population that had failed when testing was terminated at 780 hrs. The percentages of the population that failed in the left segments, right segments, and overall structure (either the left or right segment) are tabulated in Table 1. Also given in Table 1 are the percentages of the total population that failed first in the left or right segment, or that failed in both the left and right segments simultaneously. The sum of the percentages for these failure sequences gives the overall failure percentage. The percentage of the right and left segments that failed first versus j_L is plotted in Figure 2(a) and the total failure percentage of each type of segment is shown Figure 2(b). Failure

Table 1. Failed populations for individual segments and overall structure.

Configuration (MA/cm²)	Total Failed (% of population)			Failure Sequence (% of population)		
	Right	Left	Overall	Right First	Left First	Right = Left
2.5<-- <--0.5	68.75	68.75	81.25	37.50	31.25	12.50
1.5 <-- <--0.5	62.50	43.75	62.50	50.00	6.25	6.25
0.5<-- <--0.5	37.50	12.50	37.50	37.50	0.00	0.00
0 <--0.5	18.52	0.00	18.52	18.52	0.00	0.00
0.5 --> <--0.5	14.00	14.00	20.00	8.00	8.00	4.00
2.5 --> <--0.5	43.75	56.25	62.50	12.50	50.00	0.00

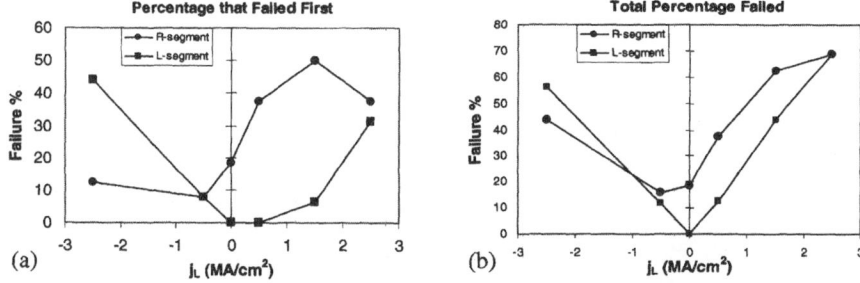

Figure 2. (a) Plot of the percentage of the failures that occurred first in either the left or the right segment, as a function of j_L. (b) Plot of the total percentage of the left and right segments that failed (in some cases, left and right segments failed simultaneously).

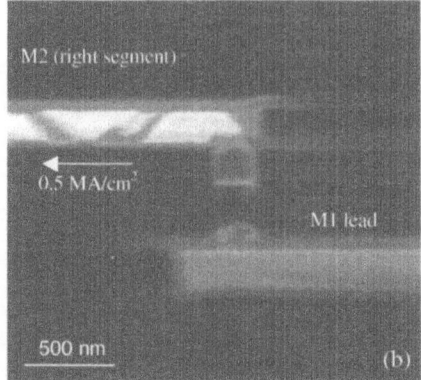 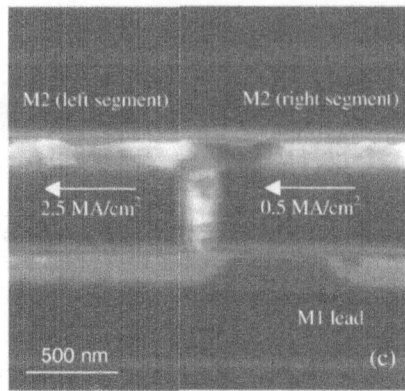

Figure 3. FIB cross-sectional images of (a) the cathodes of the right segments with j_L = 1.5 MA/cm^2, (b) the cathodes of the right segments with j_L = 2.5 MA/cm^2 and (c) the central via area of the dotted-I with j_L = 2.5 MA/cm^2.

analysis on the failed samples using focused-ion beam (FIB) showed that the right segments failed due to voiding either near the cathode or at a region close to the central via, as shown in Figure 3.

DISCUSSION

The jL value for the right segment in our experiments was only 1250 A/cm, compared to the value below which failure is reported not to occur, $(jL)_{cr}$, of 3700 A/cm for similar via-below structures [5], or 1500A/cm for via-above structures [6]. Despite this, immortality was not found in our experiments. Moreover, we found that the mortality of the right segment (total failure percentage) in the dotted-I structure was dependent on the direction and magnitude of the current in the adjacent left segment. The failed population of the right segment increases with increasing current density in the left segment for electron flow in both directions, with the lowest percentage failed corresponding to $j_L = -0.5$ MA/cm^2, the case in which the current into the central via is balanced from the two opposite directions (refer to Figure 2(b), right segment). For the cases in which $j_L = 1.5$ MA/cm^2 and 2.5 MA/cm^2, over 40% of the samples' right segments failed earlier than the left segments, even though the current densities in the left segments were higher. As the right segment acts as an active atomic source for the left segment in these cases (positive j_L), voids can form earlier in the right segment than the left segment either near the terminal via of the right segment or in a region close to the middle via [8], as shown in Figure 3.

The left segment can be treated as an active atom sink for the right segment if the electron flow is in the same direction as the right segment (positive j_L), and as an active atom source if electron flow is in the opposite direction (negative j_L). The left segment is a passive sink when there is no current in it. The mortality of the right segment increases with increasing current density in the left segment when the electrons flow in the same direction. This happens because the left segment is draining more material from the central via region. This leads to a more rapid development of tensile stresses near the central via, which reduces back-stresses to oppose the atomic flux from the right segments.

The closest approach to immortality of the right segment does not correspond to zero current flow in the left segment, but to an equal current flow of 0.5 MA/cm^2 in the opposite direction. This is because when the left-side segment is inactive, it serves as an atomic sink that reduces the build up of the compressive stress at the central via which would otherwise balance the electron wind force in the right segment. When j_L is less than -0.5 MA/cm^2, the failure population of the right segment increases even though the compressive stress generated at the central via is expected to increase when the left segment is active. This is due to the fact that more of the left segments fail before the right segments when the current is higher and in the opposite direction in the left segments (refer to Figure 2(a)). Once failure occurs in the left segment it no longer acts as an active source for atoms for the right limb, and no longer leads to the build up of compressive stress that opposes void growth in the right limb. In fact, once the left segment fails, the void in the left segment serves as an inactive sink for atoms from the right segment and thus promotes void growth (and failure) in the right segment.

The results described above show the limited applicability of the $(jL)_{cr}$ determined from single-segment via-terminated structures, when considering the mortality of individual segments of an interconnect tree. As the mortality of a segment in an interconnect tree is highly dependent on the stress conditions of the adjacent segments, the application of the conventional $(jL)_{cr}$ in filtering the immortal segments of an interconnect tree may be too optimistic. Instead, the magnitude as well as the direction of the current flow in the adjoining segments must be taken into consideration in evaluating the immortality of interconnect segments [8].

CONCLUSIONS

No total immortality was found in Cu dual damascene interconnect segments with constant current flow and very low jL values. Moreover, the percentage of interconnect segments that failed was found to be dependent on the stressing conditions in adjacent segments. These results demonstrate that an immortality criterion can not be defined for an individual segment within an interconnect tree, since the critical jL value for a single segment of Cu interconnect may be reduced or increased depending on the stress conditions in an adjoining segment. Therefore, independently determined $(jL)_{cr}$ values in two-terminal via-terminated lines cannot be directly applied to interconnects with branched segments, but rather the magnitude as well as the direction of the current flow in the adjoining segments must be taken into consideration in evaluating the immortality of interconnect segments.

ACKNOWLEDGEMENTS

The research was supported by the Singapore-MIT Alliance (SMA) and the Semiconductor Research Corporation. C.W. Chang also holds a SMA research scholarship. The authors would like to thank the Institute of Microelectronic Singapore for fabricating the samples as well as providing facilities for testing and failure analysis.

REFERENCES

1. Semiconductor Industry Association Technology Roadmap, 2003 (http://public.itrs.net).
2. I.A. Blech and C. Herring, *Appl. Phys. Lett.* **29**, 131 (1976).
3. R.G. Filippi, R. A. Wachnick, H. Aochi, J. R. Lloyd, and M. A. Korhonen, *Appl. Phys. Lett.* **69**, 2350 (1996).
4. P.-C. Wang and R.G. Filippi, *Appl. Phys. Lett.* **78**, 3598 (2001).
5. Ki-Don Lee, Ennis T. Ogawa, Hideki Matsuhashi, Patrick R. Justison, Kil-Soo Ko and Paul S. Ho, *Appl. Phys. Lett.* **79**, 3236 (2001).
6. C.S. Hau-Riege, A.P. Marathe, V. Pham, *Proc. of the Advanced Metallization Conf.*, 169 (2002)
7. S.P. Hau-Riege, *J. Appl. Phys.* **91**, 2014 (2002).
8. C.L. Gan, C.V. Thompson, K.L. Pey, W.K. Choi, F. Wei, B. Yu and S.P. Hau-Riege, *Mater. Res. Soc. Proc.* **716**, 431 (2002).

Mat. Res. Soc. Symp. Proc. Vol. 812 © 2004 Materials Research Society

Unexpected Mode of Plastic Deformation in Cu Damascene Lines Undergoing Electromigration

Arief S. Budiman[1], N. Tamura[2], B. C. Valek[2], K. Gadre[3], J. Maiz[3], R. Spolenak[4], W. A. Caldwell[2], W. D. Nix[1] and J. R. Patel[1,2]

[1]Department of Materials Science & Engineering, Stanford University, Stanford, California 94305;
[2]Advanced Light Source (ALS), Lawrence Berkeley National Laboratory (LBNL), 1 Cyclotron Rd., Berkeley, California 94720;
[3]Intel Corporation, Hillsboro, Oregon 97124;
[4]Max-Planck-Institut für Metallforschung, Heisenbergstrasse 3, D-70569 Stuttgart, Germany.

ABSTRACT

An unexpected mode of plastic deformation was observed in damascene Cu interconnect test structure during an in-situ electromigration experiment and before the onset of visible microstructural damages (void, hillock formation). We show here, using a synchrotron technique of white beam X-ray microdiffraction, that the extent of this electromigration-induced plasticity is dependent on the line width. The grain texture of the line might also play an important role. In wide lines, plastic deformation manifests itself as grain bending and the formation of subgrain structures, while only grain rotation is observed in the narrower lines. This early stage behavior can have a direct bearing on the final failure stage of electromigration.

INTRODUCTION

Copper is a much better conductor than Aluminum but the difficulty in processing Cu to obtain patterned interconnect lines has long prevented its use in the semiconductor industry. With the advent of the damascene technique, Cu is now quickly replacing Al as the major metallization layer materials. Electromigration, the diffusion of the constitutive atoms of the metal lines under the influence of a high current density, still constitutes a major reliability challenge for the microelectronics industry.[1] Several solutions such as the use of "shunt" layers, were devised to alleviate or completely stop this phenomenon to take place in lines with the current micron and submicron dimensions. However, with the ever decreasing size of interconnect widths, current densities will rise to new high levels and the technologies of today will not be sufficient to prevent the occurrence of electromigration damage. A thorough understanding of the mechanisms leading to electromigration damage is therefore needed.[2-4] Recently, a very early stage of plastic deformation and microstructural evolution during an electromigration test was detected in Al(Cu) interconnect lines, long before any macroscopic damages become visible, by using a synchrotron technique involving white beam X-ray microdiffraction.[5] In particular it was observed that during *in-situ* electromigration a gradient of plastic deformation evolves along the line which results in bending and in polygonization of the largest grains between the cathode and the anode end. Smaller grains do not readily deform but do rotate as electromigration proceeds. Plastic deformation is initiated at the cathode end and gradually progresses toward the anode end while electromigration is occurring until a steady state is reached. The present paper presents preliminary findings on similar experiment performed on Cu damascene interconnect test structures.

The synchrotron technique of scanning white beam X-ray microdiffraction has been described elsewhere.[6] It consists of scanning the sample under a micron size X-ray beam and capturing a Laue diffraction pattern at each step with a CCD detector. Using a small beam allows us to consider each grain of the interconnect sample as a single crystal. The indexing of the Laue pattern gives the orientation of the grain while the shape of the Laue peaks yields information regarding plastic deformation of the individual grain.

EXPERIMENTALS

The test interconnect line here is an electroplated Cu damascene structure. The test line has dimensions of 70 μm in length and approximately 1 μm in thickness, with two variations of width of 1.6 μm and 0.6 μm. The schematic diagram of the test structure is shown in Fig. 1. The lines are passivated with 4 μm of nitride and polymer. Several vias at either end of the line connect to a lower metallization level, which in turn connects to unpassivated bond pads for electrical connection to be made.

The white beam X-ray microdiffraction experiment was performed on beamline 7.3.3. at the Advanced Light Source, Berkeley, CA. The electromigration test was conducted first at 300°C. Current and voltage across the sample were monitored at 10s increments. The sample (width = 1.6 μm) was scanned in 0.5 μm steps, 10 steps across the width of the line and 160 steps along the length of the line, for a total of 1600 CCD frames collected. A complete set of CCD frames takes about 6 to 7 hours to collect. The exposure time was 4 s plus about 10 s of electronic readout time for each frame. In this manner the Laue pattern and information regarding plastic deformation for each grain in the sample was collected for each time step during the experiment. The current was ramped up to 25 mA (j = 1.9 MA/cm^2) over the course of 48 hours, then set at that value for the rest of the test.

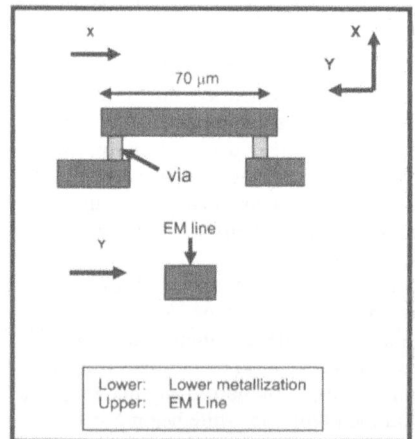
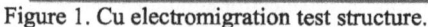

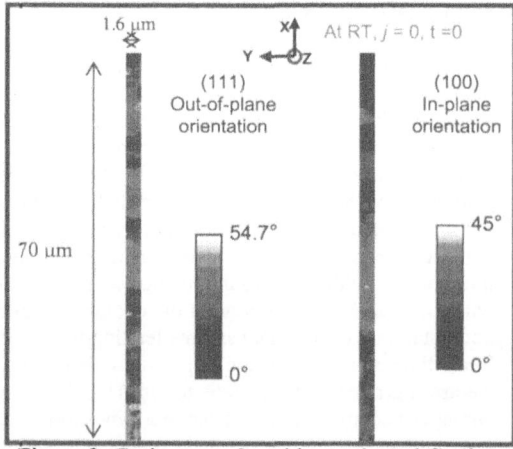

Figure 1. Cu electromigration test structure.   Figure 2. Grain map of a wide passivated Cu line.

346

The second group of tests were conducted at a higher temperature, 360°C, for reasons that will be discussed later in the article. The narrow sample (0.6 μm) was scanned in the same manner as above, except that the current ramp up was up to 15 mA ($j = 2.8$ MA/cm^2) over the course of 48 hours, then set at that value for the rest of the test.

RESULTS & DISCUSSIONS

We first describe the in-situ electromigration studies on the wide (1.6 μm) damascene Cu test structures. A cross-section diagram of these structures with line dimensions is shown in Fig.1. The grain structure in these lines as determined by white beam x-ray microdiffraction through the passivation layer is shown in Fig. 2. The line shows all grains along the interconnect line with the [111] direction of individual grains varying from normal to the line to 54.7° away from the normal of the line.

If we examine the individual diffraction spots after electromigration in some detail we find that in certain grains the spots broaden not in any random direction but always in the y direction across the line. In Fig. 3 we show a Laue spot shape at a position near the anode end of the line under the electromigration conditions indicated. Broadening of the peak is observed, in this particular example, in a large single grain spanning across the width of the line. In that particular grain where broadening is observed we also show the peak profile through the broadening direction for reflection (111). Besides the broadening we note that the peaks are split, an indication of sub-grain formation. We can use the broadening and the peak splitting observed to obtain information on the dislocation structure in the grain induced by electromigration. From the streak length as measured in the CCD camera and the sample to detector distance we obtain the curvature angle of the grain of 0.75°. Since the grain map in Fig. 2 indicates a near bamboo structure the grain width is thus 1.6 μm, from which we get the curvature radius of the grain R =126 μm. The geometrically necessary dislocation density to account for the curvature observed can be calculated from the Cahn-Nye relationship $\rho = 1/Rb$ where b is the Burgers vector. The geometrically necessary dislocation density is then $\rho = 3 \times 10^9$ /cm^2. The total number of dislocations introduced is only 49. Furthermore, the observation of the peak profile at the end of the EM test indicates a splitting of the peak into two subpeaks. This shows that a low angle boundary has formed dividing the grain into two subgrains.

We now describe *in-situ* electromigration studies on the narrow (0.6 μm) damascene Cu test structures. The higher test temperature of this group of experiment was designed to give more pronounced streaking effect of the Laue peaks as the grains undergo electromigration. A previous similar electromigration study on Al(Cu) system has shown an extensive broadening of peaks in the Al(Cu) system.[10] By increasing the test temperature, the homologous temperature (T/T$_M$) of our present experiment is now almost comparable to that of the previous electromigration study on Al(Cu) system.

However, our observation of the peaks of the grains along the narrow Cu line did not show any broadening of the peaks in the line during electromigration, as can be seen in Fig. 4. This is true despite the higher temperature used in the present experiment. The homologous test temperature (T/T$_M$) is now ~ 0.48, which is higher than that of our first group of experiment (~ 0.4). The homologous test temperature of the current study is almost similar to that of the previous study on Al(Cu) system, which was ~ 0.51.

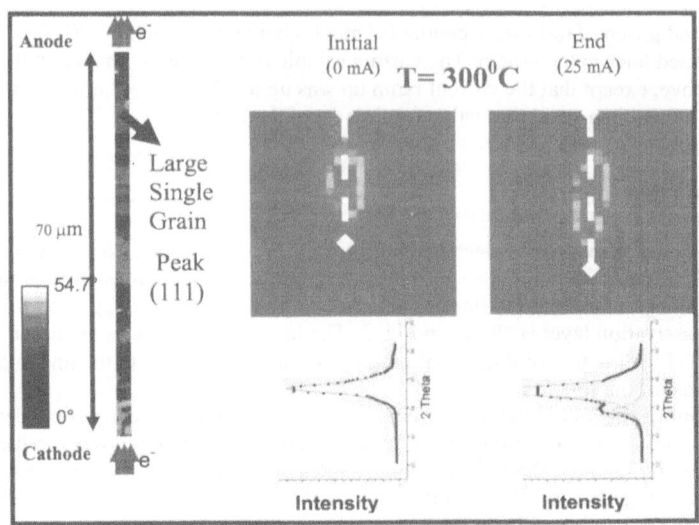

Figure 3. Broadening of a Laue peak of a large single grain near the anode end of the Cu line.

Apparent grain rotations are however observed both in the cathode-end, as shown in Fig. 4, in the middle of the line (at the highly-strained grains), as well as in the anode-end of the line. Grain rotations can be quantitatively detected by the shifts of the Laue peak as indicated in Fig. 4. From the Laue pattern, we have also a precise measure of the orientation matrix of a given grain. By comparing the orientation matrix before and after electromigration, we could deduce the angle of rotation as well as the axis of rotation. For the particular case of Fig. 4, we found that the grain rotated by 6.3 degrees around a (100) direction. The narrow Cu line seems to have higher electromigration resistance to plastic deformation inducing only grain rotation but no streaking of the peaks (compared to our previous results with wider Cu line). This finding and the reasons why grains rotate without the streaking of the Laue peaks in the narrow Cu line are still under investigation.

In terms of texture, this particular Cu narrow line consists of grains of three main groups. Fig. 5 shows the number of grains versus the out-of-plane orientation of the grains along the Cu line in this electroplated damascene test structure. Not many (111) oriented grains in the present sample are observed. This observation is in contrast to the texture known from sputtered Cu films. However, it is in agreement with observations made on annealed electroplated films.[7] The (115) twin texture had been found to be predominant. This is verified in the present experimental result. Almost half of the grains have their [111] directions 38.9° off the normal of the line – the theoretical angle between a (111) plane and a (115) plane. Additionally, there are also (111) components and indications for a sidewall texture (grains having their [111] directions between 15° and 25° off the normal of the line), as shown in Fig. 5. This has been known as a feature of electroplated Cu damascene structures and has been reported previously.[8]

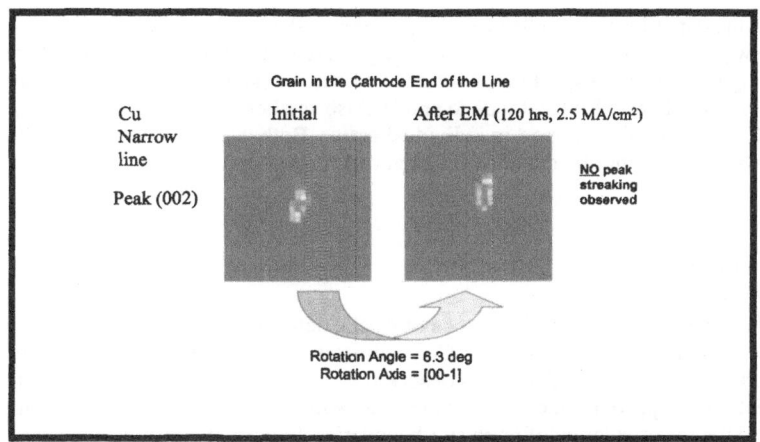

Figure 4. Grain rotation with no peak broadening observed in the EM of Cu narrow line.

Any model proposed must account for the bending of individual Cu grains transverse to the electron flow direction. The general outlines are similar to those proposed for the transverse bending of Al(Cu) grains where vacancies generated at the interface diffuse into grains. If the bottom half of the grain is rigidly constrained by the substrate the grain deforms by bending. To accommodate the bending dislocations of predominantly one sign are introduced with Burgers vector components in the direction of current flow. If we assume that for Cu lines it is the interface at the top of the line, between Cu and SiN_3 where defect generation and motion predominates, then at the anode end where matter accumulates we have a build up of stress in the top half of the grain. To relieve this stress grains deform in bending thereby introducing dislocations predominantly of one sign. The details of the process by which excess matter accumulates in the top half of a Cu grain are not yet completely understood. Since the bending is across the grain width, dislocations generated have components predominantly in the direction of current flow. Thus in addition to the flow of point defects at interfaces, dislocations provide an easy path for point defect transport in the bulk crystal.

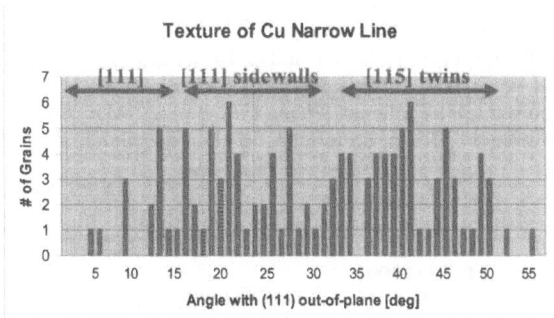

Figure 5. Texture of electroplated Cu narrow line.

From the experimental results available and the model that is proposed we can venture to make a few predictions. Since deformation of a grain requires some compliance of the dielectric surrounding it, we would expect that a stiffer dielectric will suppress deformation, thus retarding the flow of point defects from cathode to anode. Also smaller grains which have a higher yield stress will suppress bulk transport in individual grains. Both these predictions should extend electromigration lifetimes by delaying the final catastrophic events where voids and hillocks are formed.

CONCLUSIONS

We have described a pre-failure mode of plastic deformation during in-situ electromigration of Cu line. Broadening of Laue spots in Cu occurs transverse to the electron flow direction, and long before any macroscopic damage is observed (early stage). From the streak length and its subsequent break-up, we have calculated the dislocation density in an individual grain and determined the misorientation of grain sub-boundaries. In term of early stage plastic deformation, the narrow line seems to have higher electromigration resistance. A mechanism and model of plastic deformation during electromigration involving generation of point defects and their subsequent diffusion into the grains is proposed.

ACKNOWLEDGEMENTS

The authors would like to thank Advanced Micro Devices (AMD) for valuable discussions. One of the authors (ASB) was supported through the SRC Grant: Task ID. 945.001. The Advanced Light Source (ALS) is supported by the Director, Office of Science, Office of Basic Energy Sciences, Materials Sciences Division, of the U.S. Department of Energy under Contract No. DE-AC03-76SF00098 at Lawrence Berkeley National Laboratory.

REFERENCES

1. J. F. Lloyd, J. Phys. D **32**, R109 (1999).
2. I. A. Blech, J. Appl. Phys. **47**, 1203 (1976).
3. M. A. Korhonen, P. Borgesen, K. N. Tu, and C. Li, J. Appl. Phys. **73**, 3790 (1993).
4. R. J. Gleixner and W. D. Nix, J. Appl. Phys. **83**, 3595 (1998).
5. B. C. Valek, N.Tamura, R. Spolenak, W. A. Caldwell, A. MacDowell, R. S. Celestre, H. A. Padmore, J. C. Bravman, B. W. Batterman, W. D. Nix, and J. R. Patel, J. Appl. Phys. **94**, 3757 (2003).
6. N. Tamura, R. S. Celestre, A. A. MacDowell, H. A. Padmore, R. Spolenak, B. C. Valek, N. Meier Chang, A. Manceau, and J. R. Patel, Rev. Sci. Instrum. **73**, 1369 (2002).
7. R. Spolenak, C. A. Volkert, K. M. Takahashi, S. A. Fiorillo, J. F. Miner, and W. L. Brown, Mat. Res. Soc. Proc. **524**, 55 (1999).
8. C. Lingk, M. E. Gross, and W. L. Brown, Applied Physics Letters **74** (5), 682-684 (1999).

Mat. Res. Soc. Symp. Proc. Vol. 812 © 2004 Materials Research Society

Coupling Between Precipitation and Plastic Deformation During Electromigration in a Passivated Al (0.5wt%Cu) Interconnect

R.I. Barabash[1*], G.E. Ice[1], N. Tamura[2], B.C. Valek[3], R. Spolenak[4], J.C. Bravman[3] and J.R. Patel[2]

[1] Metals & Ceramics Divisions, Oak Ridge National Laboratory, Oak Ridge TN 37831
[2] Lawrence Berkeley National Laboratory, 1 Cyclotron Road, Berkeley CA 94720
[3] Dept. Materials Science & Engineering, Stanford University, Stanford CA 94305
[4] Max Planck Institut für Metallforschung, Heisenbergstrasse 3, D-7056 Stuttgart, Germany

ABSTRACT

In the present paper the evolution of the dislocation structure during electromigration in different regions along the Al(Cu) interconnect line is considered. It is shown that plastic deformation increases in the regions close to cathode end of the interconnect line. A coupling between the dissolution, growth and re-precipitation of Al_2Cu precipitates and the electromigration-induced plastic deformation of grains in interconnects is observed. Possible mechanism of the Cu doping effect on the improved electromigration resistance of the Al(Cu) interconnects is discussed.

INTRODUCTION

The scaling of device dimensions with a simultaneous increase in functional density imposes a challenge to materials technology and reliability of interconnects[1,2]. Although the general mechanism of electromigration is understood[3-5], the effect of the atomic flow on the local metallic line microstructure is largely unknown. Recently white beam X-ray microdiffraction[6-17] was used to probe microstructure in interconnects. The first quantitative analysis of the dislocation structure in individual grains of a polycrystalline interconnect line was performed in[15-17] and it was shown that dislocations with their lines almost parallel to the current flow direction are formed first. Electromigration in Al(Cu) interconnects and the effects of Cu were studied extensively since 70's[3,4]. Recent measurements of precipitate evolution during electromigration in Al (Cu) interconnects[17-21], in agreement with earlier results of Rosenberg[3] and Hu *et al*[4] indicates that Cu is preferentially depleted from the cathode end of the line and accumulates at the anode. However it is still not completely understood why doping with small amount of Cu greatly improves the electromigration resistance of Al-based interconnect lines. In the present paper we consider a model describing the possible correlation between Cu drift, precipitation and formation of dislocations in an Al(Cu) interconnect line.

EXPERIMENTAL

Data collection has been carried out at the X-ray microdiffraction end-station on beamline 7.3.3 at the Advanced Light Source. The sample is a patterned Al (0.5% wt.

---

* Corresponding author Dr. Rosa Barabash <barabashr@ornl.gov>

Cu) line (length:30 μm, width: 4.1 μm, thickness: 0.75 μm) sputter deposited on a Si wafer and buried under a glass passivation layer (0.7 μm thick). Electrical connections to the line are made through unpassivated Al (Cu) pads connected to the sample by W vias. Details on the experimental setting and data collection can be found elsewhere[4-6]. A qualitative description and semi-quantitative interpretation of the entire data set collected for the present sample can be found in recent articles[14, 15]. The dislocation structure was determined by the analysis of the intensity distribution of the reflections observed in the experimental Laue images[16, 17, 23].

RESULTS AND DISCUSSION

In agreement with the previous study[13 - 17] white beam analysis of plastic deformation in the Al(Cu) interconnect line demonstrates that the amount of plastic slip increases when the probing location approaches the cathode. The slip systems with dislocation lines almost parallel to the direction of current flow are activated first[15 - 17]. Near the ends of the line plastic activity is coupled with the depletion of Cu from the cathode end of the line. There is practically no plastic activity in the "near anode" end of the line[17].

Such peculiarities in the plastic deformation behavior at the "near anode/cathode" regions can be understood taking into account that Cu is preferentially depleted from the cathode region of the inteconnect[3, 4, 18 - 20]. Al-Cu alloys are known to demonstrate strong tendency to short range order with strong preference to unlike neighbors. After quenching from the solid solution range across the solvus, decomposition occurs via a sequence of metastable phases prior to the formation of the equilibrium precipitate Al_2Cu[21 - 25]. Cu rich clusters are known to form during aging at or near room temperature throughout the alloy. These clusters are usually referred to as "GP zones". These are the mixture of single and multiple {100} Cu rich planes. There is no apparent barrier to GP zone nucleation[25]. It is commonly accepted that in such alloys the sequence of phase transformations is: GP1$\Rightarrow$GP2 (or θ'') $\Rightarrow\theta'$ $\Rightarrow\theta$. GP1 are assumed to consist of a single layer of pure Cu on (100) plane. Due to smaller size of the Cu atoms, the surrounding Al planes collapse towards the Cu layer. GP2 consist of multilayered Cu-rich zones[26 - 28]. The strains around the zones oscillate with distance form them, and vanish near the fourth or fifth {100}Al plane[22]. The tetragonal θ phase is the equilibrium precipitate Al_2Cu. The size of clusters changes from several atomic layers to 1-5 nm. Such small size coherent precipitates in Al rich Al-Cu alloys were observed for example, by x-ray or neutron diffuse scattering measurements, high resolution electron microscopy or field ion microscopy[21-28].

The diffusion mechanism in the presence of such precipitates may differ from the diffusion in the homogeneous solid solution. This agrees with the viewpoint[18] which strongly suggests that some transport mechanism other than lattice diffusion controls steady state transport in the interconnect lines. Interestingly, during the time of the electromigration test, there is no visible dislocation activity close to the anode end of the line, while the "near cathode" grain is quickly plastically deformed. The dissolution kinetics of Al_2Cu precipitates near the cathode end and its migration in the interconnect line is coupled with the plastic deformation activity. To interpret such a behavior we propose a generalization of the model of electromigration-induced Cu motion and precipitation in Al(Cu) interconnects, described in[3, 4, 18 - 20], and a mechanism which takes into account the coupling between plastic deformation and precipitation formation, dissolution and migration in different regions of the line.

Dissolution of large size precipitates at the cathode end was described in the number of papers[3, 4, 18 - 20, 29]. With the decrease of precipitate size they finally reach some critical "nanometer" size (cluster) and start to migrate in the matrix relative to the lattice due to directional diffusion of atoms under applied electric field. Such migration results in the gradual departure (dissolution) of atomic layers of the Al matrix on one side of the precipitate and their simultaneous growth on the opposite side of the precipitate. As a result the precipitate migrates relatively to the matrix in the applied electric field. As is shown below, the velocity of migration, v, is inversely proportional to the precipitate size, $v \propto 1/R$ (see Eq.10). This means that the migration rate is negligible for large precipitates and they just dissolve (in agreement with a number of experimental observations[4, 18 - 20]). However when the precipitate diameter reduces to the size of about 5-10 atoms the migration rate becomes essential. For such small precipitate size the diffusivity related to precipitate migration is about 10^{-3} times smaller than the diffusivity of individual Cu atoms (such ratio between the bulk diffusivity and experimentally observed during electromigration was found in [18]).

Diffusivity for Cu and Al atoms at the matrix/precipitate interface is different. This is why an applied electric field causes more rapid motion of one type of atoms along the surface while a different atomic species would move more slowly. This creates an inhomogeneous distribution of interface concentration within the matrix at the precipitate boundary, which in turn causes additional diffusion fluxes. They compensate the difference in diffusivity of Cu and Al atoms. After a relaxation time ε, the process stabilizes and steady state is reached. The relaxation time can be estimated by:

$$\varepsilon \sim R^2 \big/ D_S \tag{1}$$

Here R is the average precipitate size and D_S the interface diffusivity. After that time, the interface concentration distribution becomes stationary and the precipitate/matrix interface migrates with constant velocity in the matrix. We note that for large precipitates this time is huge and such precipitates dissolve rather than migrate (as observed experimentally[3, 4, 18 - 20]). To quantitatively describe this process, we define the unit vector perpendicular to the precipitate/matrix interface, n_s. The product $I_s n_s$ between n_s and the surface atomic flux I_s (across the precipitate surface) gives the number of atoms passing in the near surface region through the unit length tangential to the vector n_s per unit of time. This flux depends on the coordinate of the position at the surface of the precipitate. This way the number of atoms coming in through some surface element and leaving the precipitate is different. Under the applied electric field such diffusive fluxes result in the motion of atoms from the front to the back side of the precipitate. As a result small coherent precipitates are migrating in the interconnect line (Fig. 1).

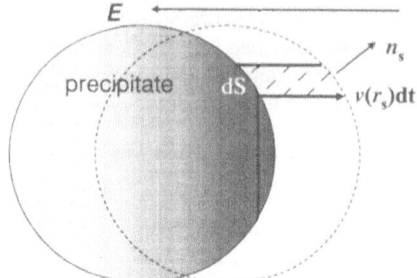

Figure 1. Volume change due to the migration of the surface element of precipitate dS on the distance $v(r_s)dt$. n_s is a surface normal unit vector at the position r_s at the surface of the precipitate.

This mechanism is especially important at relatively low temperatures when the ratio of surface to volume diffusivity is high, and for small size of precipitates when the ratio of their surface atoms to their volume atoms is not too small. For example for a typical size of Al_2Cu precipitates in the interconnect line of the order of 5 - 10 atoms in diameter the number of surface atoms is approximately equal to their number in the volume of precipitate.

The volume diffusion of Cu and Al atoms under the applied electric field in the interconnect can be written as[30]

$$I_i = N_0 c_i \frac{D_i^*}{f_i kT} e Z_i E \tag{2}$$

Here D_i^* is the volume self-diffusion coefficient for atom i (i=1, 2, for Cu and Al respectively), Z_i is an effective charge of ions caused both by the direct force applied to the ion in the electric field $\mathbf{E}$ and by the force of the electron wind, N_0 is a number of atoms per unit volume, c_i concentration of Cu/Al atoms in the alloy, f_i is a correlation factor for ions in the solid solution; e is the electron charge.

The interface diffusion flux $\vec{I}_{Si}$ for Cu and Al atoms of the Al (0.5 wt% Cu) alloy in the electric field $\mathbf{E_s}$ can be written

$$\vec{I}_{Si} = N_0 c_{Si} \frac{D_{Si}^* a}{f_{Si} kT} e Z_{Si} \vec{E}_S \tag{3}$$

Here D_{Si}^* is interface self-diffusion coefficient for atom i, Z_{Si} is effective charge, f_{Si} is a correlation factor; a is the thickness of one atomic surface layer, e is the electron charge. In both equations for volume and surface diffusion all parameters are semi-phenomenological as they are obtained by averaging over possible configurations and depend strongly on the temperature T and the concentration c_i and c_{si}.

In the coordinate system connected to the interconnect crystal lattice the number of lattice sites is constant. This means that the total flux of Cu and Al atoms and vacancies is equal zero:

$$(\mathbf{I}_{Cu} + \mathbf{I}_{Al} + \mathbf{I}_{v}) = 0 \tag{4}$$

At the same time the material of the interconnect line (Cu and Al atoms) is moving in the direction opposite to the direction of vacancies flux. Average velocity of atomic movement $\mathbf{v}_a(\mathbf{r})$ is equal to the product of the resulting atomic flux $(\mathbf{I}_{Cu} + \mathbf{I}_{Al})$ and the atomic volume:

$$\mathbf{v}_a(\mathbf{r}) = \omega(\mathbf{I}_{Cu} + \mathbf{I}_{Al}) = -\omega\mathbf{I}_v \tag{5}$$

As the grain boundary is moving together with the materials of the grain the last equation determines the rate of the boundary's displacement in the crystal lattice coordinate system. Vacancy flux corresponds to the particular point of the boundary. The continuous appearance and formation of vacancies in the volume of the crystal grain or at its boundary during diffusion, results in the formation of new and disappearance of existing crystal planes. This means that continuous diffusion deformation of the lattice is taking place.

It is convenient to relate the coordinate system to the boundary of the interconnect line. Continuous reconstruction of the lattice during electromigration results in the motion of each point of the lattice with a velocity $\omega(\mathbf{r})$. If diffusion fluxes are not homogeneous in space, the lattice will not only be moving but deforming as well. Relative volume change of a given volume $\delta V(\mathbf{r})$ during such a deformation is equal to

$$\frac{1}{\delta V}\frac{d\delta V}{dt} = div\omega(\mathbf{r}) \tag{6}$$

We consider relatively early stages of electromingration when there are no voids or hillocks in the line and number of lattice sites does not change. Under such conditions volume change can take place only if during diffusion some excess number of vacancies gets into the volume δV and simultaneously the same number of atoms leaves this volume. As a result its volume reduces. In the opposite situation when some part of vacancies is leaving the element δV and in exchange new atoms come into this element of the lattice. Such volume will increase. Then the relative volume change of the certain volume $\delta V(\mathbf{r})$ during this process is equal to

$$\frac{1}{\delta V}\frac{d\delta V}{dt} = \omega\frac{\partial N_v}{\partial t} = \omega div\mathbf{I}_v \tag{7}$$

From the comparison between the last two equations, we obtain the differential equation for the velocity of the lattice displacement, $\omega(\mathbf{r})$:

$$divo(\mathbf{r}) = \omega div\mathbf{I}_v \tag{8}$$

The problem can be considered as one dimensional so that,

$$\omega_x(x) = \omega[I_{vx}(x) - I_{vx}(x_0)] \tag{9}$$

Here $I_{vx}(x)$ and $I_{vx}(x_0)$ are vacancy fluxes in the position x and at the fixed boundary x_0. In the stationary stage of precipitate migration, the velocities of different points at the surface of the precipitate are equals, $v(r_s) = v$. Using Darken relations for surface diffusion coefficients[31] we derive an approximate equation for the velocity of precipitate as follows:

$$v = -2(1 + \chi)\frac{a}{R}\frac{c_{SCu}Z_{SCu} + c_{SAl}Z_{SAl}}{f_s kT}D_{Seff}eE \tag{10}$$

Here the effective surface diffusivity is:

$$D_{Seff} = \frac{D^*_{SCu}D^*_{SAl}}{c_{Al}D^*_{SCu} + c_{Cu}D^*_{SAl}} \tag{11}$$

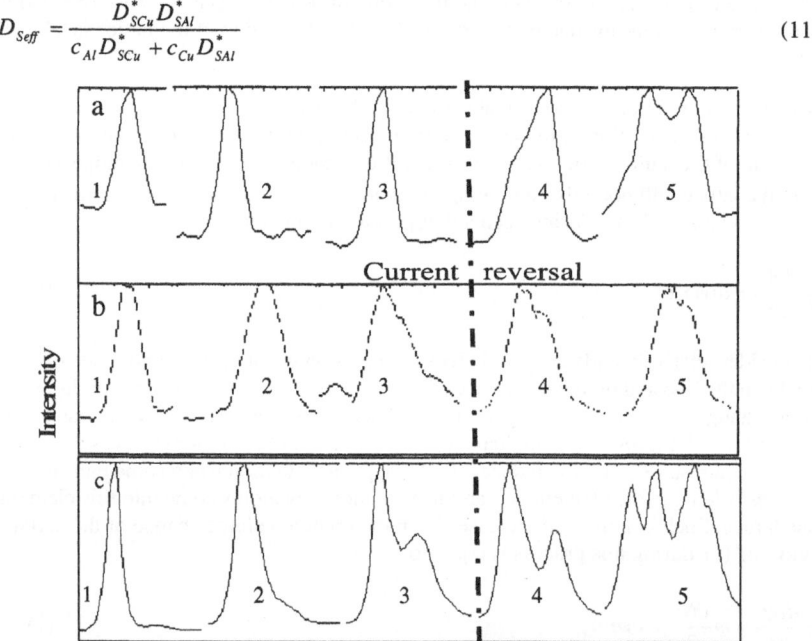

Figure 2. Intensity distribution I (in arbitrary units) along the streak direction (in pixel) for grains located near the anode (a) and cathode (b) ends of the interconnect line before the current reversal and grain (c) in the middle of the interconnect line with time of current flow: 1)initial state before the current flow; 2)5 hours; 3)16 hours;4)9 hours reversed current; 5)19 hours reversed current.

From the last two equations, we see that a small addition of Cu atoms with smaller diffusivity relative to the Al matrix atoms may essentially decrease the velocity of electromigration. Such strong decrease of D_{Seff} and of v takes place if D_{SCu} is smaller than D_{Sal} so that even the value $c_{Cu}(D_{SAl}/D_{SCu}-1)$ becomes of the order of unity (or more) at small concentration (0.5wt%). In this region of concentrations when

$1 \gg c_{Cu} \gg \dfrac{D_{SCu}}{D_{SAl}}$ the velocity of precipitate movement is inversely proportional $\sim c_{Cu}$.

This explains why the addition of small (0.5wt%) amount of Cu strongly improves the reliability of interconnects.

The presence of small precipitates is known to strengthen the Al-based alloys and increase their yield stress. Cu depletion in the near cathode region causes diffusion and partial dissolution of precipitates. Decrease of precipitates size first causes symmetric broadening of the Laue spot, as observed here (Fig.2b, curve2). After precipitates disappear in the near cathode region the critical shear stress decreases and plastic deformation is activated. This is accompanied by streaking and further splitting of Laue spots (Fig2b, curve 3). With the reversal of current flow, Cu concentration in this grain starts to increase and precipitates form again. Critical shear stress is high and plastic activity is suppressed. Reversal of current flow is accompanied with dissolution of precipitates. Critical shear stress decreases and plastic deformation occurs (Fig. 2a, curves 4, 5). This supports the idea[18] that Al (0.5% wt. Cu) interconnects are most reliable when Cu depletion from the cathode end is the slowest.

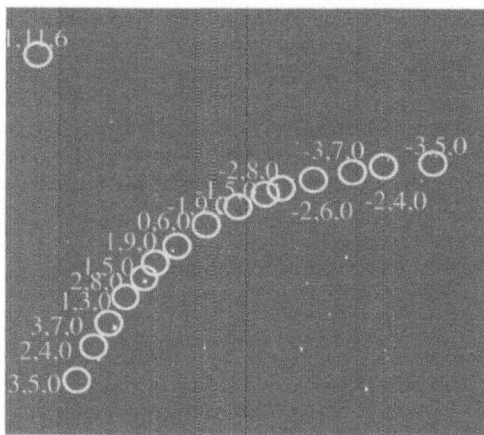

Figure 3. Laue pattern with reflections of Al Cu particle in the metallization.

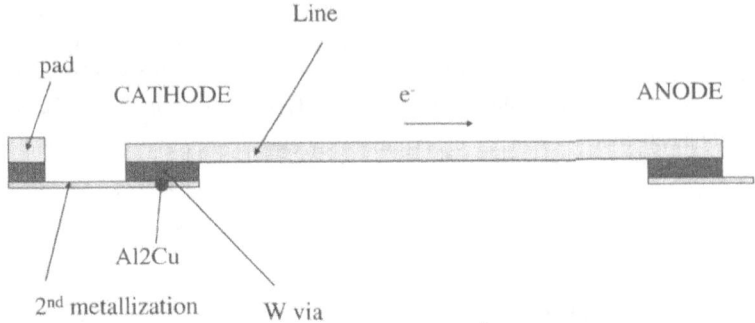

Figure 4. Location of the Al$_2$Cu precipitate found below the via which connect to the

Precipitates inside the line were too small to be observed by the white beam X-ray microdiffraction technique. However a huge (about 2 micron) Al$_2$Cu particle was clearly visible in fluorescence and diffraction in the metallization (which has a different aspect ratio than the line) below the via which connect to the pad (Figs. 3, 4). This precipitate appeared after 25 hrs under a current of 30 mA. Although not in the line, this precipitate is still electromigration induced and slowly dissolves after current is turned off. This finding is in agreement with the results of Witt et al[18] who also observed precipitates in the contact pad. The proposed mechanism for precipitates growth, dissolution and migration of precipitates is consistent with the results[3, 4, 18 - 21] which showed the growth of precipitates at the anode. It explains why the addition of small amounts of Cu to Al conductor lines effectively retards damage of the interconnect line.

Analysis of the orientation of the activated dislocation slip systems shows that the slip systems with dislocation lines almost parallel to the direction of current flow are activated first. Dislocations and dislocation walls cause additional scattering of the electrons by deformation potential related to them. The major part of the interaction potential between electrons and dislocations comes from the deformation field in the matrix around the dislocation, and not from the dislocation core. Around an edge dislocation the deformation field decreases very slowly, as $1/r_\perp$ ($r_\perp$ is the distance to the dislocation line), leading to a large scattering matrix element. However, scattering only occurs in the direction perpendicular to the dislocation line, because of translation invariance along the dislocation line. The scattering of the electrons can be described as[32]

$$(\mathbf{k}_0 - \mathbf{k})\tau = 2\pi n / d_\tau \_, \quad n=0,+(-)1.... \tag{12}$$

Here d_τ is a lattice parameter along the dislocation line, τ, and k_0, k are impulses of incident and scattered electrons. There exists a strong anisotropy of scattering. The probability of scattering depends on the difference (k_0-k) as well as on the direction of initial electron momentum k_0. When almost all unpaired dislocations with the density $n+$ are parallel (as in the case of Al-based interconnects) the anisotropy of scattering becomes essential and electrical properties of the interconnect depend on the direction of the electric current relative to the orientation of dislocation network. Relaxation time of the transverse component of electron's momentum in the presence of a parallel set of edge dislocations with the density $n+$ in the first approximation can be estimated as follows[33]

$$\frac{1}{t(k)} = \frac{1}{8} n^{+} b^{2} \Omega^{2} \frac{m}{\hbar^{3} k_{\perp}^{2}}$$ (13)

Here t is the time for a single large angle scattering of the electron, Ω is an effective parameter of deformation potential of the edge dislocation, $k_{\perp}$ is a transverse component of electron's momentum, m is the electron mass. The largest scattering occurs in the plane transverse to dislocation line. In the described experiments dislocations are formed in thin Al-based of interconnect under the applied electric field during electromigration. Applied electric field creates additional constraints for dislocations. If a dislocation is formed with the line perpendicular to the applied electric field, scattering of electrons leads to an increased heat dissipation, and such dislocation will be annealed[16]. On the other hand, a dislocation network with the lines close to the direction of the electric current creates the smallest resistance for the electric current, and is therefore more likely to persist. When the above slip planes saturates the next possible slip system is activated which has the next smallest inclination angle to the current direction.

CONCLUSIONS

The existence of a coupling between the density of GNDs and the Cu depletion from the cathode is demonstrated. There is practically no plastic activity in the "near anode" end of the line. Our model is consistent with the dissolution, migration and re-growth of Cu rich precipitates, on opposite ends of the interconnect line during electromigration.

ACKNOWLEDGEMENT

Research is supported by the Director, Office of Science, Office of Basic Energy Sciences, U.S. Department of Energy, under Contract DE-AC05-00OR22725 with UT-Battelle, LLC and with the Advanced Light Source, Materials Science Division, under the Contract No. DE-AC03-76SF00098 at Lawrence Berkeley National Laboratory.

REFERENCES

[1] I.A. Blech, J. Appl. Phys., **47**, 1203 (1976).
[2] C.V. Thompson and J.R. Lloyd, Mater. Res. Soc., Bull. **18**, 19 (1993).
[3] R. Rosenberg, JVST, **9**, 1, 263 (1972)
[4] C.-K. Hu, M.B. Small, P.S. Ho, J. Appl. Phys. **74**,(2), 969 (1993)
[5] M.A. Korhonen, P. Borgesen, K.N. Tu, and C.-Y. Li, J. Appl. Phys. **73**, 3790 (1993).
[6] A.A.MacDowell, R.S.Celestre, N.Tamura, R.Spolenak, B.C. Valek, W.L.Brown, J.C.Bravman, H.A.Padmore, B.W.Batterman and J.R.Patel, Nuclear Instruments and Methods in Physics Research A **467-468,** 936 (2001).
[7] G.E. Ice and B. C. Larson, Advanced Engineering Materials, **2**, 10, 643 (2002).
[8] B.C.Larson, Wenge Yang, G.E.Ice, J.D.Budai and J.Z.Tischler, Nature, **415**, 887 (2002).
[9] P.-C. Wang, I. C. Noyan, S. K. Kaldor, J. L. Jordan-Sweet, E. G. Liniger, and C.-K. Hu, Appl. Phys. Lett., **78**, 2712 (2001).
[10] P. C. Wang, G. S. Cargill III, I. C. Noyan,C. K. Hu, Appl. Phys. Lett.,**72**, 1296 (1998).

[11] N. Tamura, A.A. MacDowell, R.S. Celestre, H.A. Padmore, B.C. Valek, J.C. Bravman, R. Spolenak, W.L. Brown, T. Marieb, H. Fujimoto, B.W. Batterman and J.R. Patel, Appl. Phys. Lett. **80,** 3724 (2002).

[12] N. Tamura; R. Spolenak, B.C. Valek; A. Manceau; M. Meier Chang; R.S. Celestre; A.A. MacDowell; H.A. Padmore and J.R. Patel; Review of Scientific Instruments **73,** 1369 (2002) .

[13] B.C. Valek, N. Tamura, R. Spolenak, J.C. Bravman, A.A. MacDowell, R.S. Celestre, H.A. Padmore, W.L. Brown, B.W. Batterman and J.R. Patel, Appl. Phys. Lett. **81,** 4168 (2002).

[14] B.C. Valek, N. Tamura, R. Spolenak; W.W. Caldwell, A.A. MacDowell; R.S. Celestre; H.A. Padmore; J.C. Bravman; B.W. Batterman; W.D. Nix, J.R. Patel, J. Appl. Physics, **94,** 6, 3757 (2003).

[15] R.I. Barabash, G.E. Ice, N. Tamura, B.C. Valek, J. C. Bravman, R. Spolenak and J.R. Patel, Mater.Res.Soc.Symp.Proc.,**738,** (2003)

[16] R.I. Barabash, G.E. Ice, N. Tamura, B.C. Valek, J. C. Bravman, R. Spolenak and J.R. Patel, J. Appl. Physics, **93,** 5701 (2003).

[17] R.I. Barabash, G.E. Ice, N. Tamura, B.C. Valek, J. C. Bravman, R. Spolenak and J.R. Patel, Mater.Res.Soc.Symp.Proc.,**766,** 107 (2003).

[18] C. Witt, C. Volkert, E. Arzt, Acta Materialia, **51,** 49 (2003) .

[19] R. Spolenak, O. Kraft, and E. Arzt, AIP. Con. Proc., **491,** 126 (1999).

[20] R. Spolenak, PhD Thesis, Stuttgart, (1999)

[21] E. Matsubara and J.B. Cohen, Acta Metall. **31,**12,2129-2135 (1983)

[22] E. Matsubara and J.B. Cohen, Acta Metall. **33,**11,1945-1955,1957-1969 (1985)

[23] V. Gerold, Scripta metallurgica,**22,** 927-932, (1988)

[24] P.P.Muller, B. Schonfeld, G. Kostorz and W. Buhrer, Acta metall, **37,** 8, 2125-2132 (1989)

[25] D. Haeffner, A. Winholtz, Jr. And J.B. Cohen, Scripta Metallurgica, **22,** 1821-1822, (1988)

[26] X. Auvray, P. Georgopolus and J.B. Cohen, Acta metal. **29,**1669 (1983)

[27] K. Osamura, Y. Murakami, T. Sato, T. Takahashi, T. Abe and K. Hirano, Acta metal. **31,** 1669 (1983)

[28] C. Wolverton, Acta mater.**49,** 3129-3142, (2001)

[29] J.P. Dekker, C.A. Volkert, E. Arzt, and P. Gumbsch, PRL,**87(3),** 035901 (2001)

[30] I. Geguzin, M. Krivoglaz, *Electromigration of macriscopic inclusions in solids*, Naukova Dumka, Kiev, (1972)

[31] Darken L.S. Trans AIME **175,**184 (1948).

[32] J. Ziman, *principles of the theory of solids*, Cambridge University Press, (1972)

[33] V. Gantmaher, I. Levinson, *Scattering of electrons in metals and semiconductors*, Nauka, Moscow (1984)

Mat. Res. Soc. Symp. Proc. Vol. 812 © 2004 Materials Research Society   F7.5

Effect of mass transport along interfaces and grain boundaries on copper interconnect degradation

Ehrenfried Zschech, Moritz A. Meyer, Eckhard Langer
AMD Saxony LLC & Co. KG, Materials Analysis Department,
P. O. Box 11 01 10, D-01330 Dresden, Germany

ABSTRACT

In-situ SEM electromigration studies were performed at fully embedded via/line interconnect structures to visualize the time-dependent void evolution in inlaid copper interconnects. Void formation, growth and movement, and consequently interconnect degradation, depend on both interface bonding and copper microstructure. Two phases are distinguished for the electromigration-induced interconnect degradation process: In the first phase, agglomerations of vacancies and voids are formed at interfaces and grain boundaries, and voids move along weak interfaces. In the second phase of the degradation process, they merge into a larger void which subsequently grows into the via and eventually causes the interconnect failure. Void movement along the copper line and void growth in the via are discontinuous processes, whereas their step-like behavior is caused by the copper microstructure. Directed mass transport along inner surfaces depends strongly on the crystallographic orientation of the copper grains. Electromigration lifetime can be drastically increased by changing the copper/capping layer interface. Both an additional CoWP coating and a local copper alloying with aluminum increase the bonding strength of the top interface of the copper interconnect line, and consequently, electromigration-induced mass transport and degradation processes are reduced significantly.

INTRODUCTION

For leading edge microprocessors that provide a performance benefit compared to previous generations, the continued shrink of on-chip interconnect structures is requiring both advanced process technologies and new combinations of materials [1-3]. Recently implemented thin film materials are causing new reliability challenges to on-chip interconnects: other types of interfaces, different metal microstructure and new degradation phenomena. Electromigration, stress-induced migration and - in case of low-k materials - thermo-mechanical weakness are reliability concerns for inlaid copper interconnects [4].

Currently, delay time reduction in on-chip interconnect structures is focused on reducing the effective dielectric constant of the isolating material between the copper interconnects. The substitution of SiO_2 by so-called low-k dielectric materials can be accompanied by the introduction of a new etch stop layer. The resulting changed bonding strength of the interface between the chemical-mechanically polished (CMP) copper line and the capping etch stop layer may alter the reliability of the interconnect. The introduction of an additional thin layer on top of the polished copper line [5-8] and interface strengthening (i. e. increase of the bonding strength of the copper/capping layer interface) by copper alloying [9] are possible approaches to reduce the mass transport along the copper/capping layer interface.

Although a lot of theoretical and experimental work has been done on electromigration, it is still not fully understood how the interconnect degradation takes place [10,11]. To ensure the required reliability of microelectronic products, electromigration test structures are stressed in lifetime experiments (at high temperature and high current density). The resistance of particular test structures is measured as a function of time, and a failure distribution curve is determined considering a certain failure criterion. Lifetime and activation energy are calculated from the experimental data. In addition to standard reliability tests, both a careful process control based on a large number of data to reach statistically relevant conclusions and the study of solid-state physical degradation mechanisms at representative samples are needed to understand weaknesses in the interconnect technology and to exclude reliability-related failures in copper interconnects. One approach to control the stability of the interconnect manufacturing process is to monitor microstructure parameters like grain size, texture and stress for copper interconnects on a routine basis [12]. Complementary, an experimental setup was designed to study reliability-limiting degradation mechanisms in interconnects in such a way that mass transport and interconnect degradation are visualized *in-situ* at fully embedded copper via/line test structures using scanning electron microscopy (SEM) and X-ray microscopy (XRM) [13,14].

The focus of this paper is to describe void formation, void growth and void movement along interfaces and grain boundaries and to discuss the effect of interface bonding strength and copper microstructure on electromigration-induced interconnect degradation mechanisms.

THEORY – MASS TRANSPORT IN COPPER INTERCONNECTS

Electromigration-induced degradation processes of inlaid copper interconnects are connected with directed mass transport. The gradient of the electrical potential which is proportional to interconnect resistivity and current density gives atoms that migrate along the interconnects a preferred direction to the anode. In addition, local temperature peaks caused by increased electrical current densities lead to temperature gradients during processor operation, and therefore, thermomigration is closely connected with electromigration. Stress gradients are expected for via/line structures, depending on the used material combinations and process parameters [15]. The atomic transport can be described by Fick's first law considering the transport terms for electromigration, thermomigration and stress-induced migration. Interdiffusion or mass flux along the copper interconnects due to a gradient of the chemical potential is not expected. Degradation occurs due to local divergences in the mass flux, which are described by Fick's second law.

For the directed atomic transport in a polycrystalline material, so-called effective diffusion coefficients have to be analyzed. The effective diffusion coefficient for an inlaid copper line with a width w and a height h is given in equation (1) as the sum of several contributions [16,17]:

$$D_{eff} = \left[v_l D_l + \frac{\delta_{gb}}{d} D_{gb} + \delta_{Cu/l} D_{Cu/l} \left(\frac{2}{w} + \frac{1}{h} \right) + \delta_{Cu/c} D_{Cu/c} \frac{1}{h} + \delta_s D_s l \right] \tag{1}$$

D_l, D_{gb}, $D_{Cu/l}$ and $D_{Cu/c}$ and D_s are the diffusion coefficients within the grains (lattice diffusion), along grain boundaries (grain boundary diffusion), along Cu/liner and Cu/capping layer interfaces, respectively (interface diffusion), and along inner surfaces of voids (surface diffusion). v_l is the volume fraction of single crystalline regions (v_l is nearly 1), d the average grain size. l is a measure for the length of diffusion paths along inner surfaces. δ_{gb}, $\delta_{Cu/l}$, $\delta_{Cu/c}$ and δ_s are the effective thickness of the grain boundaries, the interface layers and the surface layer.

For $T < 0.5\ T_{melt}$, which is realistic for all process and operating temperatures, the diffusion along defects is at least one order of magnitude larger than the diffusion within single crystalline grains, i. e. D_l can be neglected. That means, D_{eff} depends strongly on the diffusion along interfaces, grain boundaries and inner surfaces. As will be shown below, the activation energy for surface diffusion is significantly smaller than for all other diffusion contributions. The effect of interface bonding and microstructure on the individual contributions to the directed mass transport has to be discussed since the fastest pathway for mass transport dominates the interconnect degradation.

IN-SITU SEM DEGRADATION EXPERIMENT AND GENERAL OBSERVATIONS

Void formation, growth and movement in a copper interconnect during an electromigration test can be continuously monitored in an *in-situ* SEM degradation experiment [18]. This is one approach to study the mass transport along interconnects. Liniger et al. performed top-down experiments at unpassivated dual-inlaid copper lines, which allow to investigate the void growth kinetics and to measure an average growth rate for voids of a copper line [19]. The aim of that study was to correlate void growth behavior and line resistance change.

In the *in-situ* SEM degradation experiment applied in this study, cross-sections with fully embedded dual-inlaid copper via/line test structures at the cathode-end region were investigated. This kind of experiment enables to study mass transport and degradation along both Cu/capping layer and Cu/liner interfaces. The experimental setup was described in [13]. Test structures are stressed and continuously imaged in a SEM Gemini 1550 (Leo Elektronenoptik Oberkochen, Germany) equipped with a custom-made heating stage. The investigated structures consist of copper lines in two layers connected by one copper via at each end. The vias (diameter about 0.3 μm) and the upper line with a length of several hundred microns are the structure under test. The lower metal lines (width several microns) provide the link to bond pads to feed the current into the upper line. In contrast to most of the *in-situ* electromigration experiments described in literature, the samples were prepared in such a way that the via/line structure of interest in the cathode-end region is fully embedded in dielectric material. Focused ion beam (FIB) cross-sections were cut very precisely to make sure that a thin oxide layer with a thickness of about 50 to 100 nm is left over for interconnect passivation. Secondary electron imaging provides sufficient resolution and reasonable image quality even for imaging through a passivation layer with a thickness up to about 100 nm.

All test structures were stressed at elevated temperatures and current densities. In the experiments described here, the dc current direction was from the lower to the upper level, the current density applied to the test structures was between 10 and 20 MA/cm^2. The samples were tested at temperatures ranging from 200 to 350 °C.

During the copper interconnect degradation process caused by electromigration, two phases can be distinguished: In the first phase, agglomerations of vacancies and voids are formed at interfaces and grain boundaries [12,20,21]. Initial shallow voids at interfaces remain at their position until a certain (critical) size is reached. Depending on the interface bonding energy, voids move along weak interfaces, eventually towards the cathode end of the line. In the second phase of the degradation process, they merge into a larger void which subsequently grows into the via [12,20,21]. During both phases of the degradation process, void movement and void growth are discontinuous processes which depend on the microstructure of the copper interconnect [22].

The results demonstrate clearly that void formation, growth and movement, and consequently degradation, depend on both the interface bonding and the copper microstructure of inlaid interconnect structures. The individual processes during the two phases of electromigration degradation as defined above are described in the following two chapters.

COPPER INTERCONNECT DEGRADATION: EFFECT OF INTERFACE BONDING

For inlaid copper interconnects, interfaces dominate usually the electromigration-induced mass transport, and therefore, degradation of copper interconnects is a function of the bonding strength of the weakest interface [23,24]. Numerous papers have been published on electromigration of copper inlaid lines that were embedded in metal liners (typically Ta or Ta-based compounds or Ta/TaN layer stacks) at the bottom and at the sides, and their surface was covered by a dielectric material (typically SiN_x or $a-SiC_xH_y$ etch stop layers) after the CMP process. Most of the previous electromigration studies have shown that the void growth is dependent on mass transport along the copper/dielectric interface but not dependent on mass transport along the copper/metal interface (the metal liner provides a good adhesion, i. e., high bonding strength). That means, the top copper/dielectric interface has been found to be the weakest interface in most cases [11,12,21,23,25-29]. The reason that copper/dielectric interfaces are rapid pathways for mass transport is the weak bonding and the resulting poor adhesion between copper and the adjacent material from which the interface is composed.

Lane et al. established relationships between atomic transport along interfaces, adhesion and electromigration-induced drift velocity for backend-of-line (BEoL) structures [1,2]. Since both activation energy of atomic transport processes and work of adhesion for the Cu/(dielectric) capping layer and/or Cu/(metal) liner interfaces are related to the bond strength, interface strengthening - particularly for the top Cu/capping layer interface - seems to be the consequent way to prolong copper electromigration lifetime. Recently, some approaches have been published to improve the electromigration lifetime by increasing the bonding strength of the top interface:

- Selective electroless CoWP, CoSnP or Pd coatings (10...20 nm thick) on top of the polished copper line [5-7]
- Ta/TaN layer stack deposition on top of the polished copper line [8]
- Local alloying of copper interconnects, e. g. using CuAl [9].

In all cases, the mass transport along the Cu/capping layer interface ("conventionally" Cu/SiN_x or $Cu/a-SiC_xH_y$) is reduced and the and void growth is slowed down.

Figures 1 and 2 show schematically typical sequences of interconnect degradation during the *in-situ* SEM experiment for a standard Cu/SiN_x capping layer interface, and for a strengthened top interface (introduction of an additional CoWP coating or copper lines locally alloyed with aluminum), respectively. In all cases, tantalum was used as liner material. Besides the general observations described in the previous chapter, significant differences were observed for these test structures.

In case of the standard interface, the voids were formed initially at the Cu/SiN_x interface (often away from the cathode end), then evolved to deplete the cathode end and the via bottom interface [13]. Consequently, the Cu/capping layer interface was the weakest interface, i. e., the pathway with the highest transport rate of copper atoms. The individual processes were described in [12]: Cu/SiN_x interface diffusion and diffusion along the inner surface of a void, grain boundary diffusion and void agglomeration, as well as void growth and Cu redeposition.

The CoWP coated copper lines and the copper interconnects locally alloyed with aluminum show void formation at both the Cu/capping layer and at the Cu/liner interfaces, and subsequent movement of the voids. The copper microstructure and the position of grain boundaries seem to play a more important role than for the standard structure. Both kinds of Cu/capping layer interface strengthening slowed down the mass transport along the top interface of the interconnect line, which indicates a stronger interface bonding. In some cases, the mass transport along this interface was reduced to an extend that the Cu/liner interface became the dominant diffusion path.

The experimental result for the sample with the CoWP coating confirms those in [1,2], where a significantly increased interface debond energy for Cu/CoWP compared to Cu/SiN_x and a reduced void growth rate for the CoWP coated copper lines were observed. The reason for the increased bond strength of the Cu/CoWP interface compared to the Cu/Si_3N_4 interface is the high strength of the metallic Cu-Co bonding. Even if details have not been fully understood, metallic bonding seems to be the preferred solution since it is stronger than metal-oxygen and metal-nitrogen bonding. Since the electromigration-induced mass transport is shifted from the weak Cu/SiN_x interface to the strong Cu/CoWP interface, the activation energy for directed mass transport along polycrystalline copper lines is lower for the lines with the CoWP coating. From standard electromigration test, an activation energy of about 1.2 eV was reported for the lines with CoWP coating where a $Cu/CoWP/SiN_x$ layer stack is formed compared to 0.8 eV for standard structures with Cu/SiN_x interface [1,2]. Since the activation energy for Cu/CoWP interface diffusion is comparable to the activation energy for copper grain boundary diffusion [30-32], mass transport along grain boundaries has to be taken into consideration too for interconnects with polycrystalline copper microstructure.

The locally alloying of the copper interconnect with aluminum increased the electromigration lifetime significantly too. Obviously, the interface diffusion is significantly reduced by the alloying element since the solute segregation increases the bonding strength. The local alloying seems to provide a compromise between interconnect material resistivity and electromigration behavior.

As a result of the two approaches of interface strengthening described here, the mass transport along the changed Cu/capping layer interface is not faster than those along the Cu/liner interface anymore. This observation implicates that the bonding strength of the Cu/capping layer interface is at least as high as of the Cu/liner interfaces.

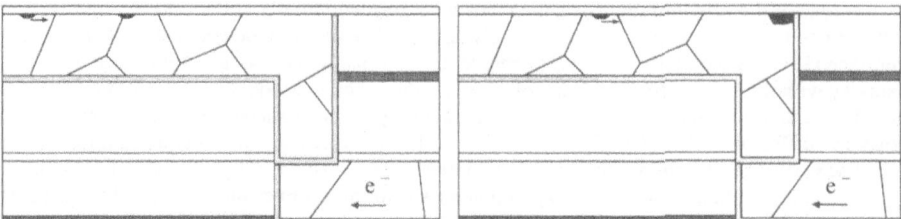

Interface diffusion: Initial void formation at the copper/capping layer interface

Void movement and agglomeration at the line end, inner surface diffusion

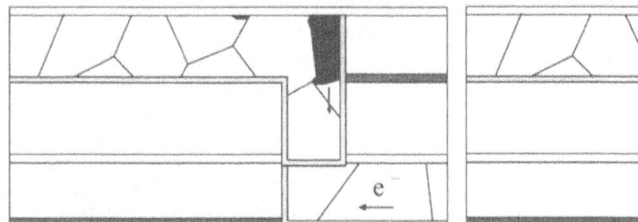

Void growth into the via from top to bottom, discontinuous process

Material redeposition in the upper region of the via.
Failure occurs when the remaining cross-section is reduced to a critical size

Figure 1. Degradation sequence in copper via/line structures with Cu/SiN top interface

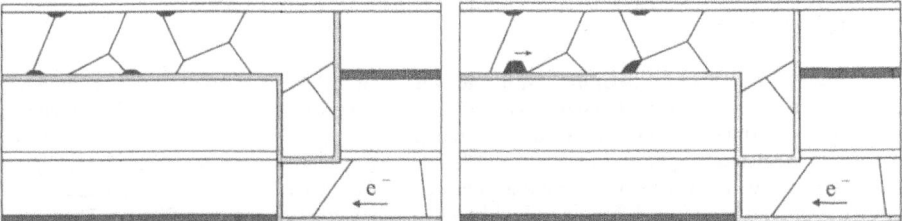

Void formation at the Cu/liner and the
Cu/capping layer interfaces, interface diffusion

Void movement and agglomeration,
void growth, interface diffusion

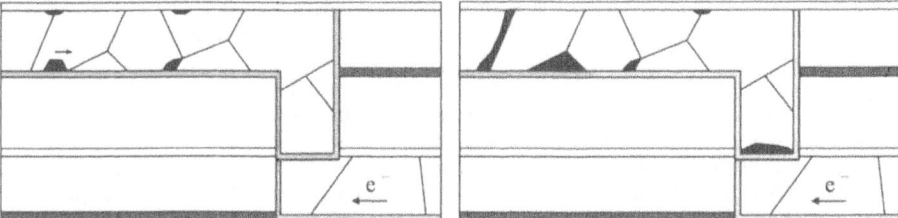

Void shape modification, pinning at grain
boundaries and triple points, void growth,
interface and grain boundary diffusion

Failure occurs if a pinned void grows across
the entire cross-section or if a void reaches
the via, thereby reducing the cross-section
to a critical size.

Figure 2. Degradation sequence in copper via/line structures with strengthened Cu/SiN$_x$ top
interface: CoWP coating or local alloying with Al

COPPER INTERCONNECT DEGRADATION: EFFECT OF COPPER MICROSTRUCTURE

Since at the temperatures of interest the activation energy for copper lattice diffusion and for copper diffusion along twin boundaries is much higher than for copper grain boundary diffusion [33], these contributions to the directed mass transport have usually not been considered in the discussion of electromigration-induced degradation of copper interconnect structures [3,17].

As discussed above, electromigration-induced mass transport is usually dominated by interface and inner surface diffusion rather than by copper grain boundary diffusion. This observation confirms experimental results published elsewhere [19,23,24]. That means, the electromigration behavior is relatively insensitive to microstructure [1]. More precisely, the microstructure effect is reduced for the first part of the degradation process as long as the activation energy for electromigration-induced mass transport along interfaces is significantly lower than along grain boundaries. If the Cu/capping layer interface is strengthened by an additional CoWP film or by local alloying of the Cu/capping layer interface region with aluminum, the changed top interface is as strong as the Cu grain boundaries [1,2], i. e., mass transport along grain boundaries has to be considered too, and the copper microstructure becomes important.

The electromigration-induced transport of copper atoms along interfaces and grain boundaries results in mass transport, depletion and voids. Independently of the strength of the interfaces, triple points which are created by copper grain boundaries and interfaces act as heterogeneous nucleation sites for void formation in the first phase [19,21], i. e., the copper microstructure is influencing this phase of the degradation process. Subsequently, a void movement along the Cu/capping layer and/or Cu/liner interfaces is observed. Depending on the used material combinations and the interface bonding, the mass transport dominates along one of these interfaces. The discontinuous, step-like void movement - in opposite direction to the electron flow and copper migration - depends on the position of grain boundaries. Particularly, voids are fixed for a relatively long time at points where grain boundaries hit interfaces. This observation gives evidence for theoretical models postulating that voids grow at sites of flux divergences like grain-boundary triple points [34]. In case of a different crystallographic orientation of neighbored copper grains, the activation energies for diffusion along copper surfaces differ (see below), and consequently, mass flux divergences occur [35].

In [12] it was shown that an electromigration-induced mass transport in dual-inlaid copper structures results in void formation, movement, agglomeration and growth, and eventually in partial via depletion and interconnect failure. The image sequence shown in figure 3 (and more clearly the respective video sequence) reveals that both the void movement along the copper line in the first phase and the grain-consuming void growth in the via (caused by grain thinning or edge displacement) in the second phase of degradation are step-like processes [13,20,22]. To make this time-dependent void evolution in the second phase of the degradation experiment more visible, the projected void area was determined from the SEM image for two different regions of the via/line interconnect structure. Figure 4 shows that the growth of the void within the upper metal line is non-linear until the void reaches a maximum size (black curve). The size of the "top void" remains constant for some time. Subsequently, this void starts to grow into the via (red curve). The red curve shows sections with different slopes, interrupted by certain steps. Simultaneously, the size of the "top void" is reduced, indicating a coalescence of voids and a redeposition of material into the upper part.

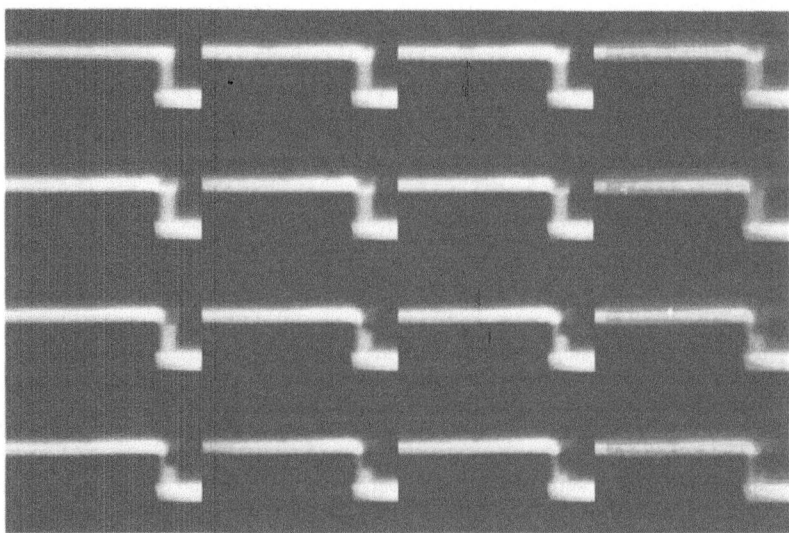

Figure 3. Secondary electron image sequence that shows the development of voids at the cathode of a test structure (The time difference between two images is 2 hours. For a video sequence, images were taken in 2-min intervals) [13]

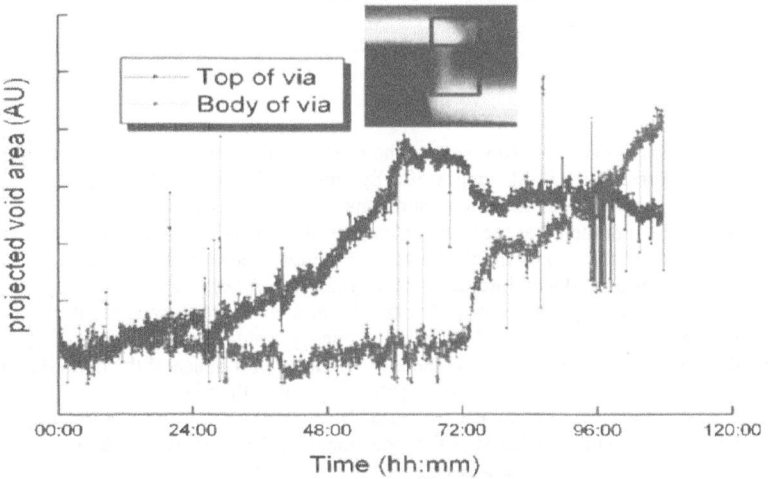

Figure 4. Void growth during an in-situ electromigration experiment for two different regions of a via/line interconnect structure [22]

Voids in interconnects are surrounded by crystalline copper grains, and consequently, the electromigration-induced transport of copper atoms is directed relative to the crystallographic orientation of these grains. Since the diffusion rate of copper atoms along copper surfaces depends significantly on their crystallographic orientation, grains with different surface orientation will be disintegrated at different speeds. This effect results in a discontinuous void growth. That means, the step-like behavior of void growth in the via (particularly in the "Body of via" region) during the second part of the degradation process is likely caused by the copper microstructure in the via. Once a void, and consequently an inner surface, has been formed, the void growth process seems to be dominated by diffusion along inner surfaces in the direction of the electron wind.

This experimental observation and the subsequent numerical analysis of the void growth during the second phase of the degradation experiment show varying slopes in the void growth curves. The significantly different slopes (factor of 10) are indications for different surface diffusion rates for the inner copper surfaces of the degradation-dominating void possibly caused by different crystallographic orientation. This explanation is supported by literature data: Experimentally determined activation energies for the self-diffusion of copper atoms are 0.04 eV only along a Cu(111) surface [36], and between 0.28 and 0.40 eV along a Cu(100) surface [37]. For self-diffusion along a Cu(110) surface, the energy barrier for the diffusion of copper atoms is about 0.35 eV along the (1-10) direction and 0.84 eV along the (001) direction [38]. Theoretical data based on several types of calculations vary significantly [37], however, nevertheless they confirm the experiments at least semi-quantitatively.

Since the activation energies for directed mass transport along interfaces, copper grain boundaries and inner surface of a void depend on interface bonding and interconnect microstructure, these material-specific data are generally different for each degradation process, and consequently, multi-mode failure distribution with strong and weak mode failures may occur [39,40].

CONCLUSIONS

In-situ SEM studies were performed to visualize the time-dependent void evolution in inlaid copper interconnect structures. From SEM images of fully embedded via/line interconnect structures, void formation, growth and movement, and consequently interconnect degradation were observed. It could be concluded that these electromigration-induced phenomena depend on both the interface bonding and the copper microstructure of the interconnect structures.

During the electromigration-induced interconnect degradation process, two phases can be distinguished: In the first phase, agglomerations of vacancies and voids are formed at interfaces and grain boundaries. Depending on the interface bonding energy, voids move along weak interfaces, eventually towards the cathode end of the line. In the second phase of the degradation process, they merge into a larger void which subsequently grows into the via.

Electromigration lifetime can be drastically increased by changing the copper/capping layer interface. In both cases described here - the use of an additional CoWP coating and the local copper alloying with aluminum - the bonding strength of the top interface of the copper interconnect line was increased, and consequently, electromigration-induced mass transport and all the related degradation processes like void formation, growth and movement were reduced significantly.

The *in-situ* SEM experiments show that void movement along the copper line and void growth in the via are discontinuous step-like processes. The root cause for this behavior is likely the copper microstructure. In the copper lines, the directed mass transport depends on the position of grain boundaries. Particularly, voids are fixed for a relatively long time at points where grain boundaries hit interfaces. This observation gives evidence for theoretical models postulating that voids grow at sites of flux divergences like grain-boundary triple points. As soon as voids have been formed in the via or near the via, the mass transport along inner surfaces is dominating the interconnect degradation process. The copper surface diffusion depends strongly on the crystallographic orientation of the surfaces of the grains that surround the void, and consequently, grains with different surface orientation are disintegrated at different speeds.

ACKNOWLEDGEMENTS

One of the authors (Ehrenfried Zschech) would like to thank Paul S. Ho, University of Texas, Austin/TX, C. K. Hu, Michael Lane and Eric Liniger, IBM Research Center, Yorktown Heights/NY, and Klaus Kern, Max Planck Institute for Solid State Research, Stuttgart, Germany, for encouraging discussions.

REFERENCES

[1] M. W. Lane, E. G. Liniger, J. R. Lloyd; *Appl. Phys. Lett.* **93**, 1417 (2003)
[2] M. Lane, R. Rosenberg, *Mater. Res. Soc. Proc.* **766**, (2003) (in press)
[3] R. Spolenak, E. Zschech, „Interconnects for Microelectronics" in „*Metal Based Thin Films for Electronics*", ed. K. Wetzig, C. M. Schneider (Wiley-VCH, Berlin, 2003) pp. 7 - 24
[4] P. S. Ho, K. D. Lee, E. T. Ogawa, S. Yoon, X. Lu, in *Characterization and Metrology for ULSI Technology*, ed. D. G. Seiler et al, AIP Conf. Proc. (AIP: Melville, NY, 2003), pp. 533 - 539
[5] Y. Schacham Diamand, Y. Sverdlov, N. Petrov, Z. Li, N. Croitoru, A. Inberg, E. Gileadi, A. Kohn, M. Eizenberg, *Proc. Electrochem. Soc.* **99-34**, 102 (2000)
[6] C. K. Hu, L. Gignac, R. Rosenberg, E. Liniger, J. Rubino, C. Sambucetti, A. Domenicucci, X. Chen, A. K. Stamper, *Appl. Phys. Lett.* **81**, 1782 (2002)
[7] C. K. Hu, L. Gignac, R. Rosenberg, E. Liniger, J. Rubino, C. Sambucetti, A. K. Stamper, A. Domenicucci, X. Chen, *Microelectronic Engineering* **70**, 406 (2003)
[8] C. K. Hu, D. Canaperi, S. T. Chen, L. M. Gignac, B. Herbst, S. Kaldor, E. Liniger, D. L. Rath, D. Restaino, R. Rosenberg, J. Rubino, A. Simon, S. Smith, W. T. Tseng, submitted to *Appl. Phys. Lett.*
[9] M. A. Meyer, M. Grafe, H. J. Engelmann, E. Zschech, *Materials for Advanced Metallization Conference (MAM)*, Brussels (2004)
[10] C. K. Hu, S .G. Malhotra, L. Gignac, *Electrochem. Soc. Proc.* **31**, 206 (1999)
[11] E. T. Ogawa, K. D. Lee, V. A. Blaschke, P. S. Ho, *IEEE Transaction on Reliability* **51**, 403 (2002)
[12] E. Zschech, H. Geisler, I. Zienert, H. Prinz, E. Langer, M. A. Meyer, G. Schneider, *Proc. of the Advanced Metallization Conference (AMC)*, San Diego, 305 (2002)

[13] M. A. Meyer, M. Herrmann, E. Langer, E. Zschech, *Microelectronics Engineering* **64**, 375 (2002)

[14] G. Schneider, G. Denbeaux, E. H. Anderson, B. Bates, A. Pearson, M. A. Meyer, E. Zschech, D. Hambach, E. A. Stach, *Appl. Phys. Lett.* **81**, 2535 (2002)

[15] V. Sukharev, R. Choudhury, C. W. Park, *Proc. ECD Meeting*, Paris (2003) (in press)

[16] R. E. Reed-Hill, *Physical Metallurgy Principles*, (D. Van Nostrand Co., New York, 1973), p. 626

[17] E. Zschech, W. Blum, I. Zienert, P. R. Besser, *Z. Metallkd.* **92**, 803 (2001)

[18] M. A. Meyer, E. Zschech, E. Langer, *US patent application* 10/677,911 (2003)

[19] E. Liniger, L. Gignac, C. K. Hu, S. Kaldor, *J. Appl. Phys.* **92**, 1803 (2002)

[20] E. Langer, M. A. Meyer, E. Zschech, M. Herrmann, in *Proc. Int. Symp. for Testing and Failure Analysis (ISTFA)*, (2002) (in press)

[21] A. V. Vairagar, A. Krishnamoorthy, K. N. Tu, S. G. Mhaisalkar, A. M. Gusak, M. A. Meyer, E. Zschech, submitted to *J. Appl. Phys.*

[22] E. Zschech, E. Langer, M. A. Meyer, *Proc. IPFA*, (2003) (in press)

[23] C. K. Hu, R. Rosenberg, K. L. Lee, *Appl. Phys. Lett.* **74**, 2945 (1999)

[24] D. Edelstein, C. Uzoh, C. Cabral Jr., P. Dehaven, P. Buchwalter, A. Simon, E. Cooney, S. Malhotra, D. Klaus, H. Rathore, B. Agarwala, D. Nguyen, *Proc. of the International Interconnect Technology Conference (IITC)*, (Piscataway, NJ, 2001), p. 9

[25] C. S. Hau-Riege, C. V. Thompson, *Appl. Phys. Lett.* **78**, 3451 (2001)

[26] K. D. Lee, E. T. Ogawa, H. Matsuhashi, P. R. Justison, K. S. Ko, P. S. Ho, V. A. Blaschke, *Appl. Phys. Lett.* **79**, 3236 (2001)

[27] C. K. Hu, L. Gignac, S. G. Malhotra, R. Rosenberg, Appl. Phys. Lett. 904 (2001)

[28] S. Yokogawa, N. Okada, Y. Kakuhara, H. Takizawa, *Microelectronics Reliability* 1409 (2001)

[29] Q. Guo, A. Krishnamoorthy, N. Y. Huang, P. D. Foo, in *Proc. of the Advanced Metallization Conference (AMC)*, San Diego, 191 (2002)

[30] *Handbook of Grain and Interphase Boundary Diffusion Data*, ed. I. Kaur and W. Gust (Ziegler Press, Stuttgart, 1989)

[31] D. Gupta, C. K. Hu, K. L. Lee, *Defect Diffusion Forum* **143**, 1397 (1997)

[32] N. I. Peterson, *J. Nucl. Mater.* **69&70**, 3 (1970)

[33] D. Gupta, in *Diffusion Phenomena in Thin Films and Microelectronics Materials*, ed. D. Gupta, P. S. Ho (Noyes, Park Ridge, NJ, 1988), Chapter 1

[34] R. Spolenak, *Alloying Effects in Electromigration*, PhD Thesis, University Stuttgart (1999)

[35] H. P. Bonzel, *Surface Physics of Materials II*, ed. By J. M. Blakely (Academic, New York, 1975), Chapter 6

[36] N. Knorr, H. Brune, M. Epple, A. Hirstein, M. A. Schneider, K. Kern, *Phys. Rev. B* **65**, 115420 (2002)

[37] F. Montalenti, R. Ferrando, *Phys. Rev.* B **59**, 5881 (1999)

[38] I. K. Robinson, K. L. Whiteaker, D. A. Walko, *Physica B* **221**, 70 (1996)

[39] A. Krishnamoorthy, G. Qiang, A. V. Vairagar, S. Mhaisalkar, *Mater. Res. Soc. Proc.* **766**, (2003) (in press)

[40] P. S. Ho, private communication

Mat. Res. Soc. Symp. Proc. Vol. 812 © 2004 Materials Research Society

Fatal Void Size Comparisons in Via-Below and Via-Above Cu Dual-Damascene Interconnects

Z. -S. Choi[a] , C. L. Gan[b,c] , F. Wei[a], C. V. Thompson[a,b] , J. H. Lee[d], K. L. Pey[b], and W. K. Choi[b]
a. Department of Materials Science and Engineering, Massachusetts Institute of Technology, Cambridge, Massachusetts 02139, USA
b. Singapore-MIT Alliance, 4 Engineering Drive 3, Singapore 117576
c. School of Materials Engineering, Nanyang Technological University, 50 Nanyang Avenue, Singapore 639798
d. Department of Electrical Engineering and Computer Science, Massachusetts Institute of Technology, Cambridge, Massachusetts 02139, USA

ABSTRACT

The median-times-to-failure (t_{50}'s) for straight dual-damascene via-terminated copper interconnect structures, tested under the same conditions, depend on whether the vias connect down to underlaying leads (metal 2, M2, or via-below structures) or connect up to overlaying leads (metal 1, M1, or via-above structures). Experimental results for a variety of line lengths, widths, and numbers of vias show higher t_{50}'s for M2 structures than for analogous M1 structures. It has been shown that despite this asymmetry in lifetimes, the electromigration drift velocity is the same for these two types of structures, suggesting that fatal void volumes are different in these two cases. A numerical simulation tool based on the Korhonen model has been developed and used to simulate the conditions for void growth and correlate fatal void sizes with lifetimes. These simulations suggest that the average fatal void size for M2 structures is more than twice the size of that of M1 structures. This result supports an earlier suggestion that preferential nucleation at the Cu/Si_3N_4 interface in both M1 and M2 structures leads to different fatal void sizes, because larger voids are required to span the line thickness in M2 structures while smaller voids below the base of vias can cause failures in M1 structures. However, it is also found that the fatal void sizes corresponding to the shortest-times-to-failure (STTF's) are similar for M1 and M2, suggesting that the voids that lead to the shortest lifetimes occur at or in the vias in both cases, where a void need only span the via to cause failure. Correlation of lifetimes and critical void volumes provides a useful tool for distinguishing failure mechanisms.

I. INTRODUCTION

It is well understood that electromigration, atomic diffusion driven by a momentum transfer from conducting electrons, leads to serious reliability concerns for integrated circuits. The failure mechanisms due to electromigration are well characterized in aluminum (Al) interconnect technology [1-3]. As industry has migrated to lower resistance copper (Cu), different failure mechanisms have been discovered, due to different material properties and processes. One of the failure characteristics of Cu interconnects that differs from that of Al is the asymmetry of the median-times-to-failure (t_{50}'s) in M1 and M2 structures, terminating in vias to upper levels (via-above) and vias to lower levels (via-below), respectively. The terminating vias for these two types of dual-damascene copper interconnects, are illustrated in Figure 1. In earlier studies [4], it has been shown that the t_{50} for M2 structures of various lengths, widths, and numbers of vias are higher than those of analogous M1 structures tested under the same current density and temperature conditions (Table I). It has been proposed that the cause of this asymmetry in lifetimes is due to the preferential nucleation of voids at the Cu-overlayer (Si_3N_4) interface [4]. The critical tensile stress for void nucleation at the Cu/Si_3N_4 interface is reported to be about 40MPa [5]. The stress in an interconnect evolves non-uniformly when it is subjected to continuous electromigration stressing. The critical tensile stress for void nucleation will first be reached at the cathode end of the line [6]. For M2 structures, the maximum tensile stress is expected to develop at the base of the Cu-filled via. However, if the critical stress required for void nucleation at the Cu/Si_3N_4 interface is significantly lower than that at the Cu-liner (Ta) interface, a void will nucleate and grow on the Cu/Si_3N_4 interface instead. To cause failure, a void that nucleates at the Cu/Si_3N_4 interface in M2 structures must span the width and thickness of the line (Figure 2b). It is assumed that Joule heating rapidly leads to failure once the current is forced to shunt through the thin Ta liner around the full-spanning void. In M1 structures, voids that nucleate at the Cu/Si_3N_4 interface need only grow to span the base of the via to cause failure [4,5], because the Si_3N_4 overlayers do not provide a shunting path for current as TiN anti-reflection coating (ARC) layers do in Al technology (Figure 2a). In this paper, the critical void sizes at the times-to-failure (TTF's) of M1 and M2 structures are calculated using a numerical simulation tool.

The experimental results reported earlier for failure times for straight Cu dual-damascene lines of type M1 and M2 are listed in Table I [4]. The median-times-to-failure (t_{50}'s) and shortest times to failure (STTF) are listed for each test population. These Cu test structures were fabricated by IME in Singapore and had Ta liners and Si_3N_4 overlayers. The lengths for both M1 and M2 lines were 50μm, 100μm, and 800μm. The thicknesses for M1 and M2 lines were 0.36μm and 0.24μm, respectively. The width of all lines was 0.28μm and the cylindrically shaped vias had diameters of 0.26μm. For the analysis, only data for lines terminated with a single via was used. The structures were stressed in an electromigration test system at various current densities and a temperature of 350°C.

In a separate set of studies [7], we have investigated the electromigration drift velocities as measured in M1 (via-above) and M2 (via-below) structures prior to failure. After voids have nucleated, but before they grow large enough to cause failure, their growth often leads to a

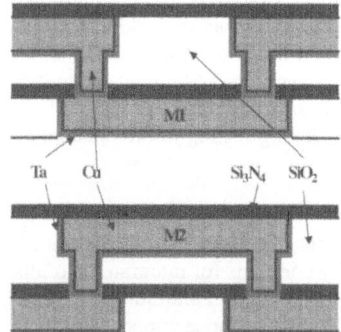

Figure 1. Schematic diagram of M1 and M2 structures.

TABLE I
t_{50}'S AND STTFs

L (μm)	Type/j ($\times 10^6$ A/cm^2)	MTTF / STTF (hours)
50	M1 / 2.3	20.8 / 1.38
50	M2 / 3.6	68.7 / 3.13
100	M1 / 2.3	25.2 / 4.53
100	M2 / 3.6	116 / 5.73
100	M1 / 2.5	20.5 / 7.33
100	M2 / 2.5	122.8 / 10.01
800	M1 / 2.3	14.5 / 7.36
800	M2 / 3.6	48.5 / 11.67
800	M1 / 2.5	28.7 / 4.44
800	M2 / 2.5	107 / 3.98

All lines are single via terminated; the widths of all lines are 0.28μm; the thickness of M1 is 0.34μm and of M2 is 0.24μm. All tests were done at T = 350 °C. Assumed failure when resistance increases more than 30% of the initial resistance.

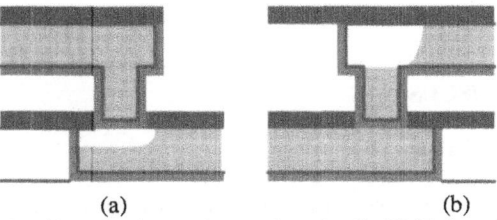

Figure 2. Preferential void nucleation and growth at the Cu/Si₃N₄ interface in (a) M1 and (b) M2 structures.

steady resistance increase. Characterization of this resistance increase can be related to the rate of void growth, which is also often referred to the rate of electromigration drift. The rate of drift of the void edge is related to the drift velocity, which can in turn be related to the effective electromigration induced diffusivity, or more specifically the effective Z^*D product, where Z^* and D are the effective charge and diffusivity, respectively. The values of drift velocity, v_d, obtained using the M1 interconnects are in agreement with the value measured using M2 interconnects, even among the samples fabricated by different organizations. *Therefore, even though the median lifetimes are significantly different in M1 and M2 structures, the drift velocities are not.* This indicates that electromigration in these two types of structures can be modeled using the same effective Z^*D product.

II. NUMERICAL SIMULATION

Korhonen *et al.* have proposed an equation describing the electromigration-evolution of stress in a one-dimensional conductor [8],

$$\frac{\partial \sigma}{\partial t} = \frac{\partial}{\partial x}\left[\frac{D_{\text{eff}}B}{kT}\left(\Omega\frac{\partial\sigma}{\partial x} + Z^*_{\text{eff}}e\rho j\right)\right], \quad (1)$$

where σ is the stress, x is the position along the line length, Ω is the atomic volume, B is an effective modulus (that depends on the conducting material as well as the material surrounding the conducting material), k is the Boltzmann's constant, T is the temperature, e is the fundamental electron charge, ρ is the electrical resistivity of the conducting material, and j is the current density. D_{eff} is expressed as [9]

$$D_{\text{eff}} = D_o \exp\left(-\frac{\Delta H}{kT}\right)\exp\left[\left(\frac{\Omega}{kT} + \frac{1}{B}\right)\sigma\right], \quad (2)$$

where ΔH is the activation energy for diffusion and D_o is a temperature independent constant. Because D_{eff} depends on stress, Eq. (1) is non-linear. Therefore, rather than solving Eq. (1) analytically, we have developed a simulation tool to predict the time evolution of stress as well as the atomic concentration.

Assuming each atom has the atomic volume $\Omega = 1.18 \times 10^{-29}\text{m}^3$, the size of the void is estimated as the total volume of the atoms removed from the void nucleation site. Based on the experiments described earlier, D_{eff} was calculated to be 2.99×10^{-16} m²/s for M1 structures and 4.23×10^{-16} m²/s for M2 structures at T = 350°C and σ = 0. Although the dominant diffusivity for both M1 and M2 structures is along the Cu/Si₃N₄ interface, the difference in thicknesses of M1 (0.34μm) and M2 (0.24μm) causes a slight difference in D_{eff} [10]. The following values were used for the simulation: $\sigma_{crit} = 40MPa$ [5]; $Z_{eff}^* = 1$ [11]; B = 28GPa; $\Omega = 1.18 \times 10^{-29}$ m^3; $\rho = 4.1 \times 10^{-8}$ Ω-m; $\Delta H = 0.8eV$.

III. ANALYSIS AND DISCUSSION

A. ANALYSIS WITH SIMULATION

The number of atoms removed from the void nucleation site for each structure was simulated and the fatal void size at t_{50} and at the shortest-times-to-failure (STTFs) of each structure have been determined and are shown in Table II. From the calculated fatal void sizes, it is clear that the M2 structures in general require larger void sizes for failure than the M1 structures. This supports the earlier conclusion that the reason for the difference in void sizes is due to preferential void nucleation at the Cu/ Si$_3$N$_4$ interface, which causes the void size required for failure in M2 to be larger than that of M1, as shown in Figure 2. The void in M2 must fully span the line in order to cause an open circuit failure, whereas only a partially spanning void directly below the via can cause failure in M1 structures. It is important to note that the thickness of M2 is smaller than that of M1. If the thickness of M2 is increased to that of M1, the t_{50} of M2 should also increase. This will result in an even larger difference in fatal void sizes.

When comparing the void sizes at the STTFs, however, it can be seen that failure of M2 structures does not *necessarily* require a larger void than for failure of M1 structures. We suggest that the reason for the similarity in critical void sizes for failures at the STTF is due to the void forming near or in the via in the M2 structures and at the base of the vias in M1 structures. If the void forms in the via in M2 structures, even a small void size can lead to failure. Moreover, in the case of M1, as shown in Figure 2, if a void nucleates at the end of the line it must grow first to the via, and then across it to cause failure. If the void nucleates "downwind" of the via, it may grow to span the line width and thickness, as in the case of M2 structures, or the void can grow towards the via and cover the bottom of the via to cause an open circuit failure at relatively small volumes. Fatal void volumes depend critically on the site for void nucleation, so that critical void volumes and lifetimes can vary over broad ranges for both M1 and M2 structures. While the Cu/Si$_3$N$_4$ interface is the most probable site for void

TABLE II
NUMBER OF REMOVED ATOMS AND VOID VOLUMES

L (μm)	Type/j ($\times 10^6$ A/cm^2)	N($\times 10^8$) /V(μm^3) (t_{50})	N ($\times 10^8$) /V(μm^3) (STTF)
50	M1 / 2.3	3.05 / 0.0052	0.20 / 0.0003
50	M2 / 3.6	11.1 / 0.0190	0.51 / 0.0009
100	M1 / 2.3	3.69 / 0.0063	0.67 / 0.0011
100	M2 / 3.6	11.9 / 0.0203	0.93 / 0.0016
100	M1 / 2.5	3.27 / 0.0056	1.17 / 0.0020
100	M2 / 2.5	11.4 / 0.0195	1.13 / 0.0019
800	M1 / 2.3	2.13 / 0.0036	1.08 / 0.0018
800	M2 / 3.6	5.47 / 0.0094	1.90 / 0.0032
800	M1 / 2.5	4.58 / 0.0078	0.71 / 0.0012
800	M2 / 2.5	12.1 / 0.0207	0.45 / 0.0007

j = current density, N = number of atoms removed, V = void volume, L = length of the interconnect

nucleation in M2 structures, it is possible to have a defect inside the via, such as a pre-existing void present before the test, that can lead to void growth and the resulting line failure.

B. COMPARISON BETWEEN SIMULATED AND EXPERIMENTAL RESULTS

In order to verify that the simulated void volumes correctly approximate the actual void volumes created in the electromigration experiments, several cross-sectional images of failed interconnects were obtained using focused ion beam/scanning electron microscopy as shown in Figure 3. The void volumes were estimated from the cross-sectional images, assuming that the observed voids spanned across the width of the interconnects. The measured and simulated void volumes are shown in Table III, and compare reasonably well. The discrepancies in void volumes, especially for the cases in Figures 3(b) and 3(c), could be due to the fact that the voids formed away from the cathode. In the simulation, voids form at the end of the cathode because it is the location where the critical stress for void nucleation is first reached. In Figures 3(b) and 3(c), pre-existing voids away from the cathode may have been present before the current was applied for the electromigration test, which led to a growth as current was applied in the interconnect. The growth rate of the voids away from the cathode is expected to be lower than that of voids at the cathode. This is because while an atomic flux is only going out from the void that forms at the end of the cathode (Figure 4(b)), there is an additional atomic flux which goes in to the void when the void is away from the cathode (Figure 4(a)). The imbalance of fluxes going out from and into the void may lead to the drift of the void towards the cathode as well as to growth of the void.

To observe the void volume location for the STTF samples, cross-sectional images of

TABLE III
VOID VOLUME COMPARISONS

Figure	Simulation ($\times 10^{-20}$ m^3)	Experiment ($\times 10^{-20}$ m^3)	V_{ex}/V_{sim}
4(a)	4.12	3.73	~ 90%
4(b)	9.31	6.44	~ 69%
4(c)	3.61	1.92	~ 53%

The error range of measured void volumes is within 0.4×10^{-20} m^3. V_{ex}/V_{Sim} corresponds to ratio of the voids between experiment and simulation.

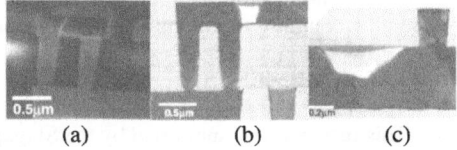

(a)   (b)   (c)

Figure 3. (a) M2 L = 800μm, w = 0.28μm, current terminated after 312 hours. (b) M2 L = 800μm, w = 1.00μm, current terminated after 55 hours. (c) M1 L = 100μm, w = 0.28μm, current terminated after 192 hours.

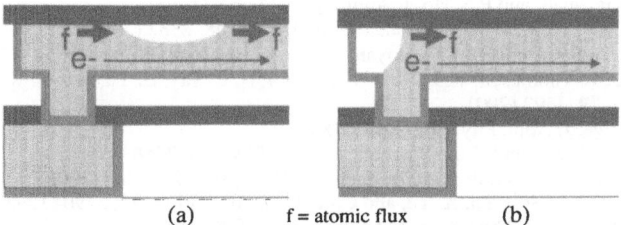

(a)   f = atomic flux   (b)

Figure 4. (a) Void nucleation and growth away from the cathode in M2. (b) Void nucleation and growth at the end of the cathode in M2.

800μm and 100μm long M2 lines stressed at 2.5MA/cm^2 were also sectioned and imaged. In both cases, the location of the failures was at the via, as expected. However, in these experiments the current was set to continue to flow to open circuit failure so that the resulting voids were larger than calculated for the 30% resistance increase failure criterion.

IV. CONCLUSION

We have developed a simulation tool for electromigration modeling. The tool was used to simulate the number of atoms removed from the site of void nucleation, and this number was used to estimate the fatal void size. The simulation results corresponding to observed t_{50}'s for straight lines terminated with either via-below (M2) or via-above (M1) structures confirm that the average void sizes required for failure in M2 structures is larger than that for M1 structures. This is due to the preferential nucleation of voids at the Cu/Si_3N_4 interface. On the other hand, the simulation results for STTFs show that the fatal void sizes for M1 and M2 do not show asymmetric behavior. The similarity in void sizes suggests that the minimum fatal void sizes correspond to nucleation in the via in M2 structures, and at the base of the vias in M1 structures. These results were consistent with experimental observations of void sizes and locations. The simulation suggests that while in most cases M2 structures are more reliable than M1 structures, the minimum lifetimes for the two structures might generally be similar if a few flaws exist in the vias. These results show that variations in the location of void nucleation lead to large variations in the lifetimes of Cu-based interconnects, not only from line-to-line, but from structure-to-structure. These variations are larger than in the case of Al, and complicate the task of accurate reliability projections for large populations of interconnects. These results also demonstrate the usefulness of simulations, in combination with experimental data, in differentiating complex variations in failure mechanisms.

ACKNOWLEDGEMENTS

This research was supported by the Singapore-MIT Alliance (SMA) and the Semiconductor Research Corporation. The authors would like to thank Institute of Microelectronics DSIC Cu BEOL for processing the test samples, and Prof. J. White for assistance in development of the numerical simulation tool.

REFERENCE

[1] C.-K.Hu, M.B Small, and P.S. Ho, J. Appl. Phys. **74**, 969 (1993).
[2] S. P. Hau-Riege, and C. V. Thompson, J. Appl. Phys. **88**, 2382 (2000).
[3] P.-C. Wang, G. S. Cargill III, I. C. Noyan, and C.-K. Hu, Appl. Phys. Lett. **72**, 1296 (1998).
[4] C. L. Gan, C. V. Thompson, K. L. Pey, W. K. Choi, H. L. Tay, B. Yu, and M. K. Radhakrishnan, Appl. Phys. Lett. **79**, 4592 (2001).
[5] S. P. Hau-Riege, J. Appl. Phys. **91**, 2014 (2002).
[6] G. Filipi, G. A. Biery, and R. A. Wachnik, J. Appl. Phys. **78**, 3756 (1995).
[7] F. Wei, C. L. Gan, T. Marieb, J. Maiz, and C.V. Thompson, TechCon (2002).
[8] M. A. Korhonen, P. Borgsen, K. Tu, and C.-Y. Li, J. Appl. Phys. **73**, 3790 (1993).
[9] J. J. Clement and C. V. Thompson, J. Appl. Phys. **78**, 900 (1995).
[10] C. L. Gan, C. V. Thompson, K. L. Pey, and W. K. Choi, J. Appl. Phys. **94**, 1222 (2003).
[11] C.-K. Hu, R. Rosenberg, and K. Y. Lee, Appl. Phys. Lett. **74**, 2945 (1999)

Mat. Res. Soc. Symp. Proc. Vol. 812 © 2004 Materials Research Society   F7.7

Statistical Analysis of Electromigration Lifetimes and Void Evolution for Cu Interconnects

M. Hauschildt*, M. Gall, S. Thrasher, P. Justison, L. Michaelson, R. Hernandez, H. Kawasaki, and P. S. Ho*.
*The University of Texas at Austin, PRC/MER, Mail Code R8650, Austin, TX 78712
Freescale Semiconductor, MD K10, 3501 Ed Bluestein Blvd, Austin, TX 78721

ABSTRACT

Electromigration (EM) failure statistics and the origin of the lognormal deviation (σ) for Cu interconnects have been investigated by analyzing the lifetime statistics and void size distributions at various stages during EM testing. Experiments were performed on 0.18 μm wide Cu interconnects with tests terminated after specific amounts of resistance increases, or after a specified test time. Void size distributions of resistance-based, as well as time-based EM tests were obtained using focused ion beam (FIB) microscopy. The lifetime and void size distributions were found to follow lognormal distribution functions. The σ values of EM lifetime and time-based void size distributions decrease with higher percentages of resistance increase, reaching an asymptotic value of $\sigma \sim 0.14$. In contrast, σ values of resistance-based void size distributions are significantly smaller and do not show an obvious dependence on time. The statistics of resistance-based void size distributions can mainly be accounted for by geometrical variations of the void shape, while the statistics of time-based void size distributions requires consideration of kinetic aspects of the EM process. The σ values of EM lifetime distributions at long times can be simulated based on measured void size distributions, taking into account geometrical and experimental factors of EM. In contrast, for short times the statistics of initial void formation and the kinetics of interfacial mass transport have to be considered.

INTRODUCTION

The introduction of Cu and low-k dielectrics and continuing scaling of on-chip interconnects raise serious reliability concerns on electromigration (EM) and stress-induced voiding (SIV). EM failures in Cu interconnects occur mainly by void growth at the cathode end of the line. With continuing void growth, the line resistance increases, eventually leading to failure of the line. The corresponding failure times usually follow a lognormal distribution with the median lifetime depending on the quality of the interface, which controls the mass transport [1-4]. Since EM experiments are conducted at higher current densities and temperatures compared to operating conditions, extrapolations are needed to assess reliability at operating conditions. The extrapolated lifetime depends linearly on the measured median lifetime and exponentially on the activation energy and the lognormal standard deviation σ. Several recent studies have shown that the median lifetime of Cu interconnects can be significantly improved by strengthening the top interface. Such efforts include the inclusion of a thin layer of CoWB between Cu and the passivation layer [5-7], as well as alloying Cu with small percentages of Aluminum [7,8]. While these studies focus on reducing interfacial mass transport to improve the lifetime, a better understanding of σ is still needed. The objective of this study is to identify intrinsic parameters that control the EM statistics and failure mechanism in Cu interconnects, other than processing variations. To accomplish this task, lifetimes, void evolution and void size distributions have been analyzed in detail as a function of failure criterion. Simulations were used to explain the statistics of void sizes and EM lifetimes.

EXPERIMENT

EM experiments were performed on single damascene, one-link Cu interconnects. The Cu lines were surrounded by a thin Ta-based barrier layer on the bottom and on the sidewalls, and capped by SiN_x on top. The dielectric material was fluorinated SiO_2. EM test structures consisted of two metal levels connected by vias. The current flows from a wide metal M1 line through the via into the metal M2 line, which was the line of interest in this case. The test structure had a line length and width of 250μm and 0.18μm, respectively. EM testing was performed with a current density of 1.5MA/cm^2 at 300°C. EM experiments were either stopped after a defined test time or after a specified resistance increase had been reached. The former case will be referred to as time-based EM tests. In the latter case, for resistance-based EM tests, the current for each line was turned off individually when that particular line reached the specified resistance change, such as 10%, 30% or 100%. The EM lifetimes were analyzed as a function of resistance increase criterion using a lognormal distribution function to fit the lifetime data. In order to examine the correlation between EM lifetimes and void sizes, the void areas of at least 10 samples for each EM condition were obtained using FIB cross-sectioning.

RESULTS

The time distributions needed to obtain a first resistance increase, 1%, 5%, 10%, 20%, 30%, 50%, 70% and 100% resistance increase are plotted in Figure 1. A lognormal distribution clearly provides a good fit to the data. It can be seen that the median time to failure increases with increasing resistance criterion, since it takes more time to reach a larger resistance change. Interestingly, σ decreases with increasing resistance criterion from approximately 0.32 to 0.14, indicating a smaller variation in lifetimes with larger median times to failure.

In order to understand this behavior, the void evolution during EM was analyzed. Four stages were identified, namely void formation, void evolution at the interface, first resistance increase and continuous void growth. FIB images typical of each stage are shown in Figures 2(a)-(d). In Cu damascene structures, void formation normally occurs at the interface between Cu and SiN_x, usually away from the cathode end. The initial void formation site appears to be at junctions of grain boundaries with the interface. Different Cu grain orientations can influence Cu diffusivities and thus induce a flux divergence leading to void formation. Here, the dependence of interfacial mass transport on Cu grain orientation can explain the statistical distribution of the initial voids with respect to the distance from the cathode end. In the second stage, the void grows at the interface towards the cathode end of the line. Due to a continuous flux of Cu atoms along the interface, the original voids refill, leading to void

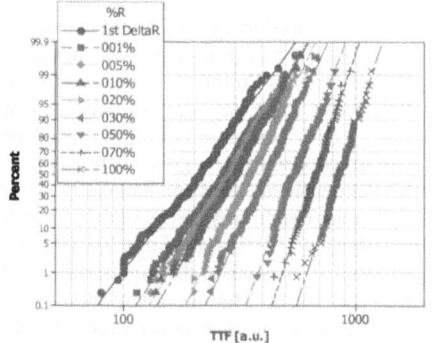

Figure 1. EM lifetime distributions as a function of resistance increase.

evolution toward the cathode end. When the void reaches the cathode end, the final flux divergence site is reached, where the barrier layer between via and line prevents additional Cu to leave the via and fill the void. Subsequently, the void grows at the cathode end until the first resistance increase is reached. The resistance increase occurs when the void spans the whole line and covers most of the via. Figure 2(c) shows that the connection between the via and the Cu portion of the line is very small, forcing the current through a small area. The last stage of void evolution is depicted in Figure 2(d). The void grows past the via and continuously along the line. Since the barrier layer is left intact, the current continues to flow through the structure and a total open is prevented. In this stage, the line resistance increases linearly, proportional to the mass transport rate under EM.

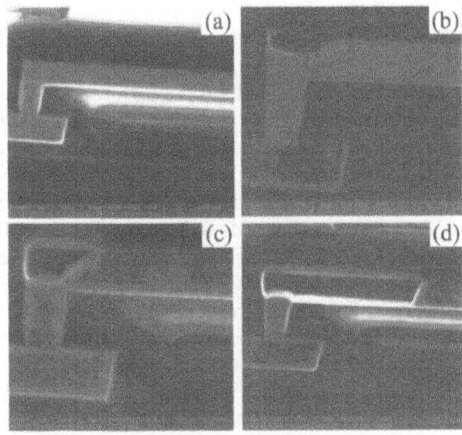

Figure 2. FIB images illustrating four stages of void evolution: (a) Void formation, (b) Void evolution at interface, (c) 1st resistance increase, (d) continuous void growth.

Results for void size distributions from time-based as well as resistance-based EM experiments are shown in Figures 3(a) and 3(b), respectively. All distributions can be well fit by a lognormal distribution function, even though the number of data points does not allow a definite statement about the most appropriate distribution function. In order to examine the σ values of EM lifetime as well as void size distributions, σ is plotted as a function of time in Figure 4. For the EM data as well as the resistance-based void sizes, the time is defined as the median lifetime of the experiment. As mentioned previously, the σ values of EM lifetime distributions decrease as a function of median lifetime. Likewise, the σ values of time-based void size distributions decrease with increasing test time. Both data sets appear to have asymptotic behavior at long times with the minimum σ being approximately 0.14 for the test structure used in this experiment. In contrast to these characteristics, the σ values of resistance-based void size distributions do not show an obvious trend with increasing median lifetime. Furthermore, σ is significantly smaller compared to the values of time-based void sizes.

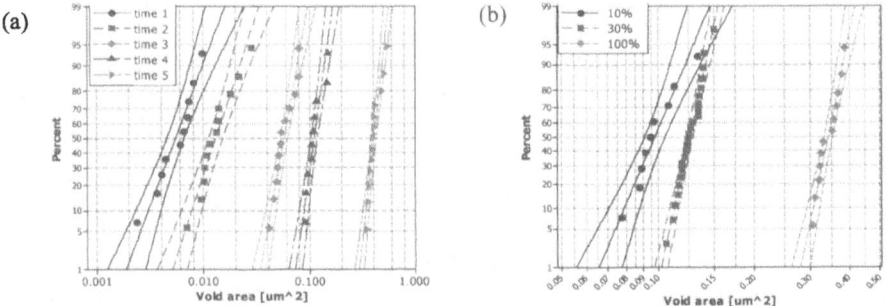

Figure 3. Void area distributions of (a) time-based and (b) resistance-based EM experiments.

DISCUSSION

In order to understand the differences between void size distributions from resistance-based and time-based EM tests, the characteristics of resistance-based void size distributions will be analyzed more closely. To get a specified resistance increase, a certain amount of Cu adjacent to the via needs to be removed, forcing the current through the highly resistive barrier layer. The resulting void areas, as seen in cross sections along the line, have been measured. In the following, geometrical arguments will be used to simulate the void areas as a function of resistance increase. In this model, the total void area is divided into two parts as illustrated in Figure 5(a). The first part is approximated as a rectangle defined by the void length adjacent to the via, which is needed to obtain a certain resistance increase.

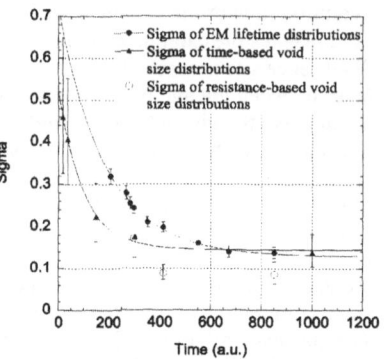

Figure 4. Sigma as a function of time for EM lifetimes, time-based, and resistance-based void size distributions.

The second part is induced by the inclination angle of the void towards the line. The inclination can be considered to represent the difference in mass transport between the interface and the average of the line. The statistics of the inclination arises because the interfacial mass transport may well depend on the local grain orientation. The first part of the void area can be calculated following an analysis of the resistance of the interconnect line. Knowing the resistivities and the cross-sectional areas of barrier layer and Cu, the void length can be calculated. It needs to be mentioned that the cross-sectional areas have a variation due to process variations. Hence, approximately 30 cross-sections have been measured to obtain their statistics. Using the experimental data, random distributions were generated. The void length distributions for various resistance increases can now be calculated and subsequently the first part of the void area can be obtained. For the second part, the statistical details of the void inclination angle distribution have been acquired from measurement. A random distribution of void inclination angles can then be generated, from which the second part of the total void area can be calculated. The total area has been computed by randomly adding parts 1 and 2. Results for this simulation are shown in Figure 5(b) together with experimental data. It can be seen that the simulation

(a)   (b)   (c)

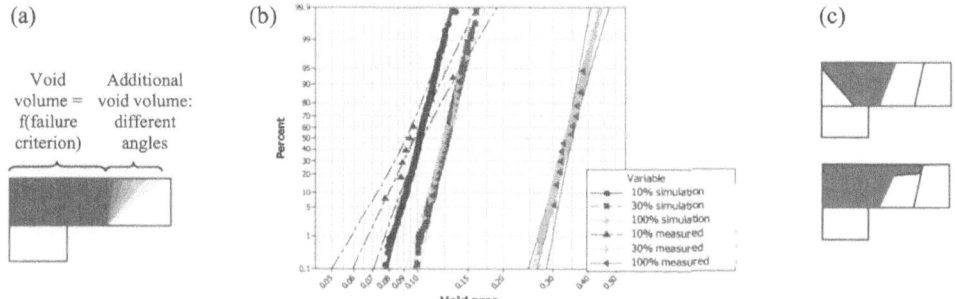

Figure 5. Geometry-based simulation of void size distributions: (a) schematic of void illustrating the simulation, (b) void size distributions obtained from simulation and experiment, (c) schematics of voids with 'artifacts'.

agrees well with experimental measurements for large void sizes, which result from a 30% and a 100% resistance increase. However, for small void sizes corresponding to 10% resistance increase, the agreement is not as good. Possible reasons for this disagreement are 'left-over' Cu above the via at the cathode end or extended interface voids as indicated in Figure 5(c). The effect of these 'artifacts' is much more influential when considering smaller void sizes. In general, however, it can be said that the void size distribution for resistance-based EM tests can mainly be explained by geometrical factors. In contrast, the statistics of void size distributions from time-based experiments depend significantly on kinetic aspects, such as the location of void formation, the time needed for the void to evolve to the cathode end and possibly diffusivity differences from line to line. The influence of kinetics in addition to geometrical factors leads to larger variations in void sizes explaining the larger σ values for time-based void size distributions.

Likewise, σ values from EM lifetime distributions are larger than σ values of resistance-based void size distributions, even though both are resistance-based. The question arises whether it is possible to explain EM lifetime statistics from measured void sizes and additional experimental parameters. Following Hu et al.[1], the void growth rate v_d can be obtained from the following two equations

$$v_d = \frac{l_{void}}{t} \quad (1) \qquad \text{and} \qquad v_d \approx \frac{Z_i^* D_i \delta_i je\rho}{hkT} \quad (2)$$

where l_{void} is the average length of the void, t the lifetime, Z_i^* the effective charge, D_i the diffusivity, δ_i the interface width, j the current density, e the electronic charge, ρ the resistivity, h the line height, k Boltzmann's constant, and T the absolute temperature. Since the major EM diffusion path is the interface between Cu and the SiN_x layer, all other possible diffusion paths, such as grain boundaries, dislocations and Cu/barrier layer interface, are neglected here. Hence, the subscript i denotes interface parameters. From equations 1 and 2 the following formula for the lifetime can be obtained:

$$t = C \frac{A_{void} T A_{Cu}}{D_i^0 \exp(-\frac{Q_i}{kT})} \quad (3)$$

where C is a constant, A_{void} the void area, A_{Cu} the Cu cross sectional area, D_i^0 the prefactor for the diffusivity term and Q_i the activation energy. Thus, the lifetime distribution is a function of the void area distribution, the temperature variation in the EM oven, the Cu cross-sectional area and the variation in diffusivity from line to line. Thermocouple measurements were used to determine the temperature profile in the EM oven. The statistical characteristics of the void area and the Cu cross-sectional area distributions have been discussed above. It is important to mention that the void area distributions, based on resistance increase, need to be used in this calculation because the lifetime values are based on a resistance increase failure criterion as well. The line-to-line variation in diffusivities is assumed to be negligible. Using the experimentally obtained statistics, data sets representing the distributions of void area, temperature and Cu cross-sectional area were randomly created. The results of the simulation as a function of resistance increase together with experimental results are shown in Figure 6. The simulated lifetime distributions are normalized so that their median values agree with the measured median lifetimes of the respective resistance increase value. As can be seen, for large times, this simulation is adequate in describing the lifetime characteristics. However, for 10% and 30% criteria, the experimental σ value is significantly larger compared to the simulated σ value.

Additional variations appear to have a significant influence at short times, however, become negligible at long times. These effects probably include variations in void formation and evolution prior to any resistance increase, i.e. stages 1 and 2 of the void evolution process. For the evolution of the void towards the cathode end, the orientation distribution of Cu grains close to the cathode end might be important, since it could influence the Cu diffusion.

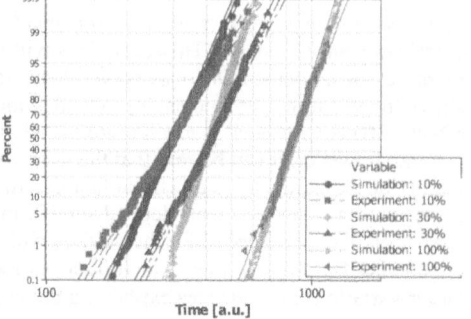

Figure 6. EM lifetime distributions obtained from kinetic simulation and experiment.

CONCLUSION

The statistical details of EM lifetimes and of the corresponding void evolution have been analyzed. σ values of EM lifetime distributions are a function of failure criterion, showing a significant decrease with increasing resistance failure criterion. Likewise, σ values from time-based void size distributions show a decrease with increasing test time. The σ values for both these distributions appear to have asymptotic behavior at large times with a minimum σ of ~0.14. In contrast, the σ values of resistance-based void size distributions are smaller and do not show an obvious dependence on the failure criterion. It has been shown that resistance-based void size distributions depend mainly on geometrical factors, whereas void size distributions from time-based tests depend on geometry as well as kinetics. Furthermore, it has been shown that for large times resistance-based void size distributions and experimental variations, such as temperature and line cross-section distributions, can be used to simulate the σ values of EM lifetimes indicating that they depend mainly on geometric and experimental parameters. However, for small times, σ values of EM lifetimes appear to be influenced significantly by kinetic aspects in the initial stages of void evolution. Further studies are underway to investigate this early stage.

REFERENCES

1. C.-K. Hu, R. Rosenberg, H.S. Rathore, D.B. Nguyen, B. Agarwala, *Proc. IEEE Int. Interconnect Technology Conf.*, 267-269, (1999).
2. C.S. Hau-Riege, C.V. Thompson, *Appl. Phys. Lett.* **78** (22), 3451-3453, (2001).
3. E.T. Ogawa, K.-D. Lee, V.A. Blaschke, P.S. Ho, *IEEE Transactions on Reliability*, **51** (4), 403-419, (2002).
4. E. Zschech, H. Geisler, I. Zienert, H. Prinz, E. Langer, M.A. Meyer, G. Schneider, *Proc. of the Advanced Metallization Conference*, San Diego, 305-311, (2002).
5. C.-K. Hu, L. Gignac, R. Rosenberg, E. Liniger, J. Rubino, C. Sambucetti, A. Domenicucci, X. Chen, A.K. Stamper, *Appl. Phys. Lett.* **81** (10), 1782-1784, (2002).
6. C.-K. Hu, L. Gignac, R. Rosenberg, E. Liniger, J. Rubino, C. Sambucetti, A.K. Stamper, A. Domenicucci, X. Chen, *Microelectronic Engineering* **80**, 406-414, (2003).
7. E. Zschech, M.A. Meyer, E. Langer, to be published in *Mater. Research Soc. Symp. Proc.*, (2004).
8. M.A. Meyer, M. Grafe, H.J. Engelmann, E. Zschech, to be published in *Materials for Advanced Metallization Conference*, Brussels, (2004).

Mat. Res. Soc. Symp. Proc. Vol. 812 © 2004 Materials Research Society   F7.8

Stressmigration studies on dual damascene Cu/oxide and Cu/low k interconnects

Won-Chong Baek[1], Paul. S. Ho[1], Jeong Gun Lee[2], Sung Bo Hwang[2], Kyeong-Keun Choi[2], and Jong Sun Maeng[2]
[1]Microelectronics Research Center, the University of Texas at Austin, Austin, TX 78712, USA,
[2]System IC R&D Center, Hynix Semiconductor Inc., 1 Hyangjeong-dong, Hungduk-gu, Cheongju Si, 361-725, South Korea

ABSTRACT

Stress-induced void formation (SIV) was studied in dual damascene Cu/oxide and Cu/low k interconnects over a temperature range of 140 ~ 350 °C. Two modes of stressmigration were observed depending on the baking temperature and sample geometry. At lower temperatures (T < 290 °C), voids were formed under the periphery of via connecting to narrow lines. This mode of stressmigration showed a typical behavior of stressmigration with peak damage at 240 °C, and an activation energy (Q) of 0.75 eV for Cu/oxide interconnects. At a higher temperature range (T > 290 °C), voids were found in via bottoms which were connected to wide lines. The rate of high temperature stressmigration increased exponentially with temperature up to 350 °C and did not show a peak at a certain temperature. The activation energy was 1.0 eV for Cu/oxide, 0.86 eV for Cu/OSG, and ~1.0 eV for Cu/FSG interconnects. The dependence of stressmigration on linewidth, sample geometry, and ILD material is presented in this paper.

INTRODUCTION

Stressmigration has been a major concern on interconnect reliability. In Al(Cu) interconnects, void formation was known to be driven by a high tensile stress state in narrow lines [1]. Since the driving force, i.e. tensile stress, decreases with increasing temperature while the diffusivity increases with increasing temperature, there is a peak rate of stressmigration at a certain temperature (180 ~ 250 °C). However, in Cu interconnects, it was reported that stressmigration was more significant in wide lines than in narrow lines and it was attributed to vacancy sources due to grain growth [2]. There are several explanations regarding stressmigration in via-to-wide line structures according to their process technologies [3]. In this paper, the stressmigration phenomenon in dual damascene Cu/oxide and Cu/low k interconnect was studied by using high sensitivity *in-situ* resistance measurement technique to investigate its dependence on linewidth, sample geometry, and ILD material in a temperature range of 140 ~ 350 °C. Void formation was verified by FIB micrograph and appropriate kinetics are presented in this paper.

EXPERIMENTAL DETAILS

Stressmigration tests were done using dual damascene Cu/oxide and Cu/low k via-chain structures with via size of 0.16, 0.19, 0.36 μm. To investigate the linewidth dependence, narrow-line and wide-line structures were tested in the same runs (Figure 1). In narrow-line structures, the linewidth was comparable to the via size e.g. 0.36 μm via and 0.44 μm line, respectively. In wide-line structures, linewidth was up to 2.0 μm. The rate of stressmigration was obtained by *in-situ* resistance measurement during high temperature stressing (HTS) for one week. Resistance was measured by a digital multi meter. When the sensitivity was not satisfactory, a small AC sensing current ($I < 0.1$ MA/cm^2) was employed. The sensing current was on only during the measurement and off otherwise to prevent electromigration (EM) effect. The Wheatstone Bridge method was used when it became necessary to enhance the sensitivity further. The measured resistances were then represented relatively to that of 0.36 μm via. 12 ~ 15 samples were used in each set of test and the median resistance increase was used to calculate the activation energy (Q).

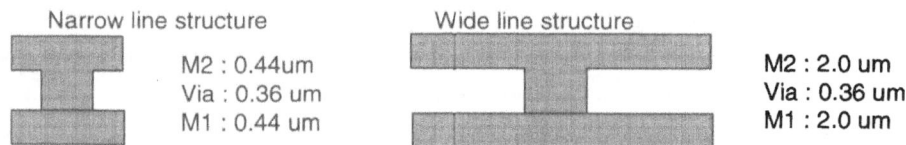

Figure 1. Schematics of test structures with 0.36 μm via

RESULTS AND DISCUSSION

(1) Low temperature stressmigration in Cu/oxide structure

In the low temperature range (T < 290 °C), narrow line structures showed more resistance increase than wide line structures. Figure 2 shows resistance traces of Cu/oxide narrow line structures measured at 210 °C, which was plotted against square root of time. The resistance increase due to stressmigration is attributed to nucleation and subsequent growth of voids. When the voids grow by diffusion mechanism, it is known that resistance increase is proportional to square root of time [4]. The resistance increase versus temperature in figure 3 shows a typical behavior of stressmigration with a peak rate of 1.6 % at 240 °C. However, even the peak rate of 1.6 % was relatively small, amounting to about 2 Ω increase out of 1500 Ω.

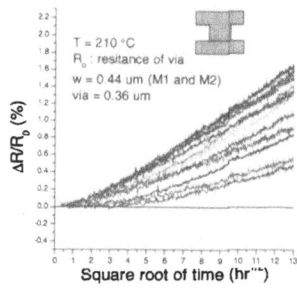

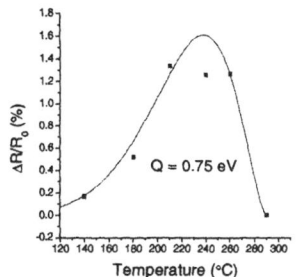

Figure 2. Resistance increases due to low temperature stressmigration in Cu/Oxide narrow line structure at 210 °C

Figure 3. Stressmigration rate vs. temperature

McPherson and Dunn [5] developed a kinetic model for stressmigration based on stress induced creep phenomenon. The kinetics can be represented as equation (1).

$$R = C(T_0 - T)^N \exp(\frac{-Q}{k_B T}) \tag{1}$$

where R – stressmigration rate, N – creep exponent, T_0 – stress free temperature, k_B – the Boltzman constant, and Q – activation energy

By fitting data in figure 3 into equation (1), we obtained $N \cong 1.7$, $Q \cong 0.75$ eV, $T_0 \cong 290$ °C. The activation energy of 0.75 eV is similar to that obtained for electromigration [6]. This suggests that the diffusion path responsible for stressmigration was also along the interface between Cu and capping layer. As this phenomenon became clear in narrow-line structures than in wide-line structures, this suggests that the tensile stress state together with diffusion controls stressmigration in narrow lines, instead of vacancy sources in wide lines. Stress-induced void formation is shown in figure 4. The location of void formation is consistent with previous literature [2, 3, 7]. It is believed that it is less tensile under the periphery of via so that vacancies migrate along the stress gradient and form a void [2].

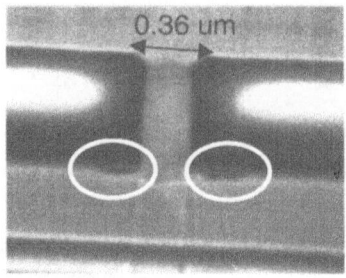

Figure 4. FIB micrograph showing stress-induced void formation under the periphery of via.

Location and shape of voids shown in figure 4 explains the small increase of resistance in low temperature stressmigration. If a void forms under the via, a small void can electrically disconnect via from M1 line causing open failure. However, as observed in figure 4, voids are found under the periphery of via and not right under the via. Resistance increase due to these voids cannot be large because voids are far from open failure. In Cu/low k structures, the resistance increase due to low temperature stressmigration was even smaller.

(2) High temperature stressmigration in Cu/oxide and Cu/low k structure

In a higher temperature range (T > 290 °C), wide-line structures showed more resistance increase than narrow line structures. Figure 5 shows resistance traces of Cu/oxide wide-line structures measured at 350 °C. The behavior of high temperature stressmigration was different from that of the low temperature stressmigration. The stressmigration rate did not show a peak, but increased exponentially with temperature, up to 350 °C (figure 6) with an activation energy of 1.0 eV. Since the rate increased exponentially over temperature, it became significant at high temperatures. At 350 °C, the rate was four times higher than the peak rate of the low temperature stressmigration.

This phenomenon was different from typical stressmigration kinetics. In figure 6, thermally activated kinetics, which is proportional to exponential of temperature, was observed. However, the effect from tensile stress state, which decreases with increasing temperature, was little [3, 7].

In Cu/OSG and Cu/FSG wide-line structures, high temperature stressmigration showed similar trends as in Cu/oxide structures. The rate increased exponentially with temperature with activation energies of 0.86 eV for Cu/OSG and ~ 1.0 eV for Cu/FSG (figure 7). The stressmigration rate was three times higher in Cu/OSG interconnects, but smaller by an order of magnitude for Cu/FSG compared with that of Cu/oxide interconnects. As the material properties of dielectric materials cannot be as different as observed stressmigration rates, the different rates for different ILD materials seem to depend on process technique other than material properties.   Considering the little effect from tensile stress state and the dependency on process technique, the driving force of high temperature stressmigration can be an extrinsic factor e.g. process-induced defects related with ILD material other than stress state.

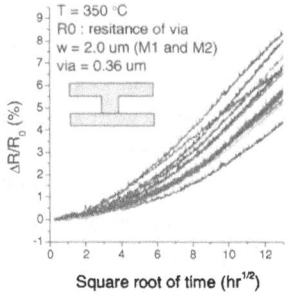

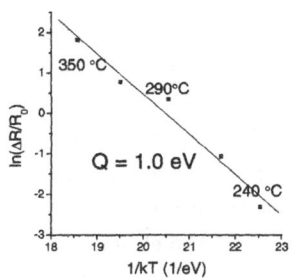

Figure 5.   Resistance increases due to high temperature stressmigration in Cu/oxide wide line structure

Figure 6. Activation energy of high temperature stressmigration of Cu/Oxide

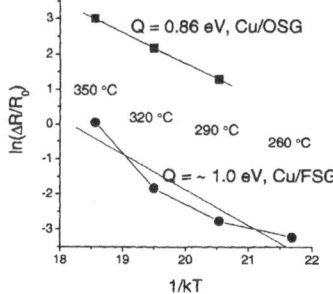

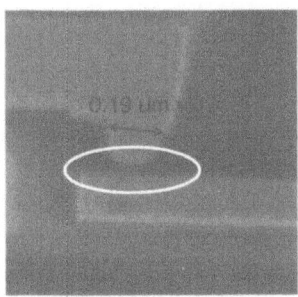

Figure 7.   Activation energy of high temperature stressmigration in Cu/FSG and Cu/OSG

Figure 8. Void formation due to high temperature stress-induced void formation in Cu/OSG

Figure 8 shows void formation observed for high temperature stressmigration in Cu/OSG structure, where void formation was seen in the via bottom. When a void forms in the via bottom, the failure volume for open failure is relatively small [8]. This explains why the rate of high temperature stressmigration was higher than that of low temperature stressmigration.

CONCLUSIONS

Two different modes of stressmigration were investigated as a function of temperature and structure geometry. At the low temperature range (T < 290 °C), a typical behavior of stressmigration was observed in Cu/oxide narrow-line structures with peak rate of resistance increase of 1.6 % at 240 °C. Activation energy was 0.75 eV and voids were seen under the periphery of via. This can be attributed to a kinetic balance between increasing diffusivity and decreasing tensile stress with increasing temperature. At the high temperature range (T > 290 °C), the stressmigration rate increased exponentially with temperature with activation energies of 1.0 eV for Cu/oxide, 0.86 eV for Cu/OSG, and ~1.0 eV for Cu/FSG. The driving force for high temperature stressmigration was attributed to an extrinsic factor other than the stress state. The rate was higher for high temperature stressmigration than the low temperature stressmigration, which was attributed to the location of stress-induced voiding.

ACKNOWGEMENTS

This work has was supported in part by Hynix Semiconductor Inc.

REFERENCES

1. H. Okabayashi et al, *Mat. Sci. & Eng. R11, No 5*, 189 (1993).
2. E. T. Ogawa et al, " Stress-induced voiding under vias connected to wide Cu metal leads", *Proc. IRPS* (2002)
3. S. Matsumoto et al, "Reliability Improvement of 90 nm-node Cu/Low-k interconnects", *Proc. IITC* (2003)
4. C.-K. Hu et al, *IBM J. Res. Devlop. Vol. 39 No. 4*, 465 (1995)
5. J. W. McPherson et al, *J. Vac. Sci & Tech. B5(5)*, 1321 (1987)
6. K. Lee et al, *Appl. Phy. Let., Vol. 79, No. 20*, 3236 (2001)
7. A. von Glascow et al, Proc. "New approaches for the assessment of stress-induced voiding in Cu interconnects", *IITC (2002)*
8. E. T. Ogawa et al, *IEEE Trans. Rel., vol. 51, No. 4*, 403 (2002)

Mat. Res. Soc. Symp. Proc. Vol. 812 © 2004 Materials Research Society   F8.9

Textural Evolution of Cu Damascene Interconnects after Annealing

Jae-Young Cho, Hyo-Jong Lee[1], Hyoungbae Kim and Jerzy A. Szpunar
Department of Mining, Metals and Materials Engineering, McGill University, Montreal, Quebec, H3A 2B2, Canada
[1]School of Materials Science and Engineering and Research Institute of Advanced Materials, Seoul National University, Seoul 151-744, Korea

ABSTRACT

Textural evolution of Cu interconnects having a different line width was investigated after annealing. Texture was measured on the surface of Cu interconnects using EBSD (electron backscattered diffraction) techniques including GBCD (grain boundary character distribution). To analyze a relationship between the stress distribution and textural evolution in the samples investigated, the micro stresses were calculated for the different line width at 200°C using FEM (finite element modeling). In this investigation, it was found that the inhomogeneity of stress distribution in Cu interconnects is an important factor is necessary for understanding textural transformation after annealing. A new interpretation of textural evolution in damascene interconnects lines after annealing is suggested, based on the state of stress and the growth mechanisms of Cu electrodeposits.

INTRODUCTION

As the features of integrated circuitry (IC) chips are scaled down to submicron dimensions, the manufacturer demands new technology to meet performance and reliability requirements for the electronic interconnects. According to these technical demands, the copper damascene process became an important issue in the integrated circuitry (IC) chips industry since it allows a decrease in RC (resistance and capacitance) delay losses, reduces the number of processing operations and increases the lifetime of the interconnect lines [1]. Since the Cu damascene process has been introduced in the IC chips industry, significant research on the relationship between texture and reliability of copper interconnects has been undertaken. It is well known that strong {111} texture increases the resistance of electromigration failure and this failure can be correlated with the frequency of the occurrence of CSL (coincidence site lattice) boundaries or low or high diffusivity boundaries and the strength of {111} texture in aluminum thin films [2]. However, such relationship for the Cu hasn't been firmly established and the driving forces which can affect the textural evolution during annealing were not clearly identified until now [3-5]. In this study, the details of textural evolution will be examined using electron backscattered diffraction (EBSD) techniques and the stress contribution to the evolution of texture and microstructure during annealing will be discussed. A model for the textural transformation as the line width decreases after annealing based on the state of stress and the growth mechanisms of Cu electrodeposits will be proposed.

EXPERIMENTAL PROCEDURE

Two different samples, one as-deposited and another annealed were used for this investigation. Both samples were fabricated using the same conditions and were kept at room temperature for 6 months. TaN, 400A thick, was deposited on the surface of the Si (100) wafer as the barrier layer, and then a copper seedlayer was deposited on the barrier layer. The trenches were filled with copper by electroplating in a sulfuric acid bath using 24 mA/cm^2 current density. Then, the samples were annealed at 200°C for 10 minutes in the vacuum furnace to avoid oxide formation on top of Cu interconnects. Each sample has different line widths from 0.14 to 2 μm and every line has the same trench depth of 0.7 μm. To remove the overburden of Cu damascene interconnects, the samples were electro-polished for 120 seconds in a H$_3$PO$_4$ solution using 17 mA/cm^2 current density. After electropolishing, the top surface area of the trench was analyzed using the orientation imaging microscope (OIM) mounted on a Philips XL30 FEG-SEM to identify the orientation of each grain and type of grain boundaries in the copper interconnects. Misorientation between grains was measured and classified as CSL (The coincidence site lattice) and non-CSL boundaries. The frequency of occurrence of CSL boundaries up to Σ29 was calculated. The stress distribution in interconnects was calculated by FEM (finite element methods) using FEMLAB, commercial software.

RESULTS

Texture and GBCD on the top surface area of Cu damascene interconnects

To obtain quantitative information about texture, three pole figures were measured from the top surface of Cu interconnects using the EBSD technique. The results presented in figure 1, indicate that {111}<110> textures exist in all samples, however it becomes fiber-like textures as the line width increases. Compared to the "as-deposited" sample, the texture of the "annealed" sample becomes stronger. As shown in figure 2a, maximum intensity is the strongest in the narrowest line and it decreases as the line width increases. The difference between the "as-deposited" and "annealed" sample in intensity is highest for the narrow lines and it decreases as the line width increases. In addition, a weak {111} sidewall component was found in the narrow lines, such as 0.14, 0.24 and 0.5 μm line width, as shown in figure 1. To make it clear, the intensity of the sidewall {111} component was plotted, as shown in figure 2b. This figure shows that the strength of the sidewall {111} component changes as the line width increases. In this figure, the strongest intensity was found at the narrowest lines, and it decreases when the line width increases up to 1 μm and then increases to 2 μm line width in both specimens. The intensities of the "as-deposited" sample are always higher than the "annealed" sample. It seems that the annealing process minimizes the sidewall contribution to overall texture in the Cu interconnects.

To analyze GBCD after annealing, the types of grain boundaries on the top area of the Cu interconnect were measured by EBSD technique, as shown in figure 3. The results obtained demonstrate that the fraction of Σ3 boundaries increase as the line width increases. Also, a higher fraction of twin boundaries is observed in 2 μm line. However, after annealing, this fraction decreases, especially in the 0.14 μm line width interconnects sample (figure 3a). On the contrary, figure 3b shows the fraction of the low angle grain boundaries increases after annealing, especially in the 0.14 μm line. From these results, it can be concluded that the annealing process enhanced the grain growth of Cu interconnects consuming the twin boundaries and producing the low angle boundaries.

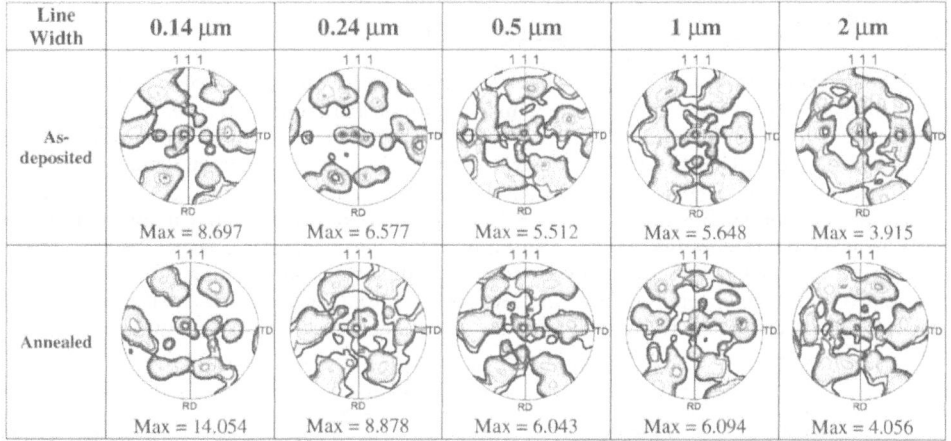

Figure 1. {111} Pole figures of copper interconnects having a different line width: as-deposited samples and after 200°C annealing.

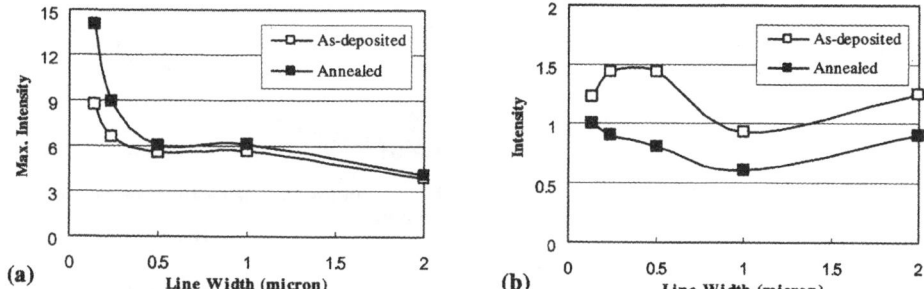

Figure 2. The changes of intensity in as-deposited and annealed sample as a function of line width: (a) the maximum intensity of (111) pole figure, (b) the intensity of the sidewall {111} texture component.

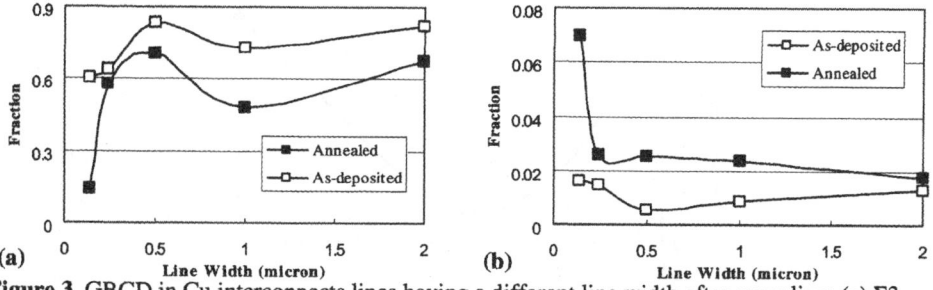

Figure 3. GBCD in Cu interconnects lines having a different line width after annealing: (a) Σ3 CSL boundaries (b) low angle grain boundaries (<15°).

Stress on the top surface area of copper damascene interconnects

   A possible contribution of the stress to the textural and microstructural evolution during annealing was investigated. In order to examine the stress distribution in the trench and overburden area, the FEM (finite element modeling) was used (figure 4a). In this calculation, it is assumed that the finite element mesh can expand freely along the ND (normal direction) when Cu interconnect with overburden is annealed at 200°C. Since the mirror symmetry is applied along the ND (normal direction) and the LD (line direction), only the half of Cu interconnect line was modeled in this investigation. It is also assumed that the copper interconnect has the isotropic mechanical properties and the expansion coefficients of surrounding silicon oxide is compared to that of copper. Therefore, the data obtained from this modeling should only be considered as a qualitative representation of the stress changes at high temperature.

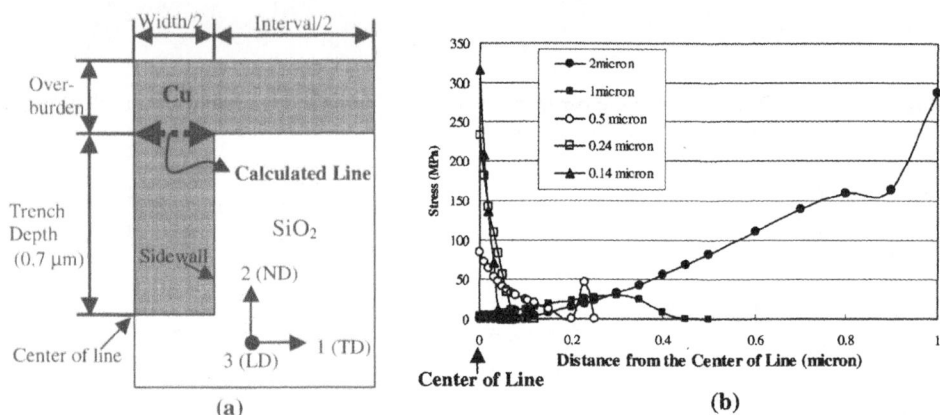

(a)   (b)

Figure 4. The calculation of stress distribution in the Cu interconnects by FEM: (a) the cross-sectional schematic diagram of Cu interconnects (b) S_{11} stress distribution from the center of the Cu line to the sidewall on the top surface area at 200°C.

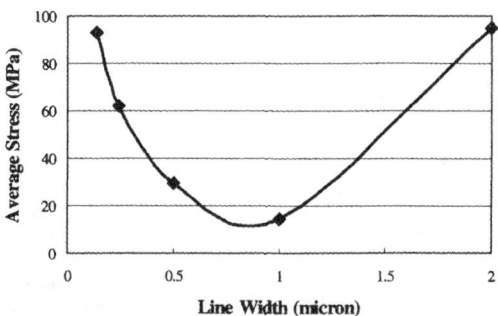

Figure 5. Average S_{11} as a function of line width at 200°C.

Figure 4b shows the stress distribution along width S_{11} from the center of the Cu line to the sidewall on the top surface area at 200°C. From the data in figure 4b, one can see that the thermal stresses change across the line width and the location of the maximum value of S_{11} moves from the corner (sidewall) of the trench to the middle as the line width decreases. To characterize these stress distributions, the average stress is calculated and plotted in figure 5. This figure shows that the absolute magnitudes of the stresses along the trench width are the largest in the 0.14 and 2 μm line width and the smallest in 2 μm line width. From these results, it can be concluded that the stress distribution is important as well as the total amount of stress in each Cu interconnects line. Therefore, the stress state can be a reason for textural and microstructural transformation.

DISCUSSION AND SUMMARY

The importance of stress on the texture evolution in Cu damascene interconnects have been emphasized by several researchers since differences in the thermal expansion coefficients of copper and dielectric (silicon dioxide) generate stress in the interface area between different layers [4-5]. Lee et al. [6] suggested that the strain energy can be minimized when the absolute maximum principal stress direction is parallel to the minimum Young's modulus direction. The minimum Young's modulus direction of copper is the <100> direction. However, the <100> orientations are not on the {111} plane. Therefore, it is most probable that grains having <112> direction, which is on the {111} plane, and is at the smallest angle with the <100> direction, will grow favorably. Therefore, the texture is likely to approach the texture of {111}<112>// trench width direction which is {111}<110>// trench length direction. In addition, the average stress value at 200°C on the top surface of the trench shows the higher value at 0.14 and 2 μm line width. This result can be used to explain the driving force for the textural evolution after annealing as shown in figure 2a and b. In the narrow line, the stress generated from the constraint of sidewalls and the difference in thermal expansion coefficients between layers help the grains to align and grow to {111}<110> orientation and decrease the sidewall {111} component. However, in spite of the high value of stress from the difference of the thermal expansion coefficient between layers, the intensity of {111} texture decreases since it has a lesser constraints in the 2 μm line width and generates a high number of twin type boundaries and a fiber-like texture component develops. In addition, the stress distribution along the trench width influences textural and microstructural evolution. Figure 4b indicates that the highest value of stress was found near the sidewall in the 2 μm line width, however, in the 0.14μm line width, the highest value was found in the middle of the trench width. Therefore, the stress distribution along the trench width is important in the textural evolution since it can produce a different texture inside the trench.

Another important factor to explain the textural evolution in the damascene Cu interconnects after annealing is the growth mechanism of the copper electrodeposits in the trenches. It has been reported in the literature that narrow Cu trenches can be completely filled during electroplating using a super-conformal filling [7]. To eliminate possibilities of void formation in the trench, this electroplating technique assures that the growth from the bottom of the trench is dominant over the growth from the sidewall, as illustrated in figure 6. Several bodies of research [8-9] have reported that the importance of the sidewall {111} component is increased in the narrow lines and the intensity of the overall texture can be minimized by this. However, in super-conformal filling conditions where the bottom growth is much more dominant than the sidewall growth, such an observation is questionable. In our investigation, weak sidewall components were found on the top surface area of the "as-deposited" sample in the narrow line because the relative volume fraction of the sidewall versus the bottom component increases as the line width decreases. This component, however, decreases after the annealing process, as shown in figure 2b. Therefore, it can be concluded that the contribution of the sidewall component is negligible when sample undergoes annealing process which stress influences textural evolution.

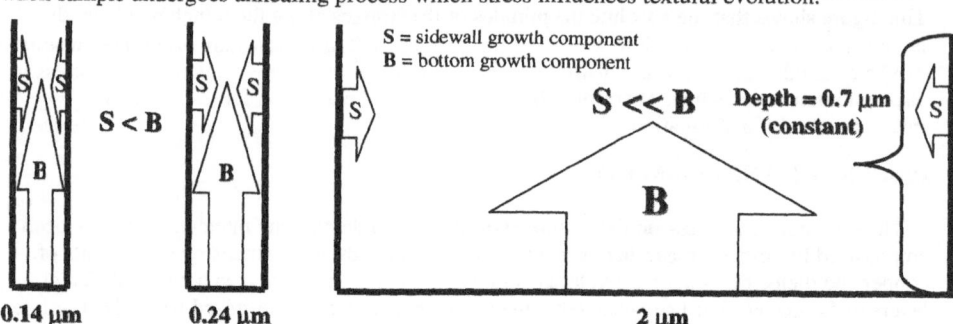

Figure 6. Schematic illustration of texture evolution as the line width decreases in the copper trench according to the condition of the super-conformal filling.

CONCLUSIONS

A relationship between the stress distribution in the trench and textural and microstructural evolution of Cu interconnects after annealing was found: the stress enhances the {111} <110> texture component and decreases the {111} sidewall components. The proposed model of texture evolution after annealing assigns an important role to differences in the stress state, but also demonstrates that the texture of copper electrodeposits can be affected by the growth mechanism in the copper electrodeposits. However, the effect of the sidewall component on the overall texture of Cu interconnects is negligible if stress dominates the textural evolution during annealing. The fraction of the Σ3 boundaries increases as the line width increases, however it decreases after annealing because the grain growth is dominating in the Cu interconnects.

REFERENCES

1. P. Singer, Semicond. Int., **21**, 91(1998).
2. K. T. Lee, J. A. Szpunar, A. Morawiec, D. B. Knorr and K. P. Rodbell, Can. Metall. Quart., **34**, 287 (1995).
3. R. Rosenberg, D.C. Edelstein, C.-K. Hu, and K.P. Rodbell, Annuu. Rev. Mater. Sci., **30**, 229 (2000).
4. C. Lingk, M.E. Gross, W.L. Brown and R. Drese, Solid State Technol., **42**, 47 (1999).
5. S.P. Riege and C.V. Thompson, Scripta Mater., **41**, 403 (1999).
6. D.N. Lee and H.J. Lee, J. Electron. Mater., **32**, 1012 (2003).
7. D. Josell, D. Wheeler, W.H. Huber and T.P. Moffat, Phys. Rev. Lett., **87**, 016102 (2001).
8. C. Lingk, M.E. Gross and W.L. Brown, Appl. Phys. Lett., **74**, 682 (1999).
9. E. Zschech, W. Blum, I. Zienert and P.R. Besser, Z. Metallkd., **92**, 803 (2001).

AUTHOR INDEX

399

SUBJECT INDEX

402